This book comes with access to more content online.

Quiz yourself, track your progress, and improve your grade!

Register your book or ebook at **www.dummies.com/go/getaccess.**

Select your product, and then follow the prompts to validate your purchase.

You'll receive an email with your PIN and instructions.

Calculus

ALL-IN-ONE

by Mark Ryan

Calculus All-in-One For Dummies®

Published by: **John Wiley & Sons, Inc.,** 111 River Street, Hoboken, NJ 07030-5774, www.wiley.com

Published simultaneously in Canada

For general information on our other products and services, please contact our Customer Care Department within the U.S. at 877-762-2974, outside the U.S. at 317-572-3993, or fax 317-572-4002. For technical support, please visit https://hub.wiley.com/community/support/dummies.

Wiley publishes in a variety of print and electronic formats and by print-on-demand. Some material included with standard print versions of this book may not be included in e-books or in print-on-demand. If this book refers to media such as a CD or DVD that is not included in the version you purchased, you may download this material at http://booksupport.wiley.com. For more information about Wiley products, visit www.wiley.com.

Library of Congress Control Number: 2023930245

ISBN 978-1-119-90967-5 (pbk); ISBN 978-1-119-90969-9 (ebk); ISBN 978-1-119-90968-2 (ebk)

SKY10044175_030823

Contents at a Glance

Introduction 1

Unit 1: An Overview of Calculus 5

CHAPTER 1: What Is Calculus? 7

CHAPTER 2: The Two Big Ideas of Calculus: Differentiation and Integration — Plus Infinite Series 13

CHAPTER 3: Why Calculus Works 21

Unit 2: Warming Up with Calculus Prerequisites 25

CHAPTER 4: Pre-Algebra, Algebra, and Geometry Review 27

CHAPTER 5: Funky Functions and Their Groovy Graphs 67

CHAPTER 6: The Trig Tango 95

Unit 3: Limits 117

CHAPTER 7: Limits and Continuity 119

CHAPTER 8: Evaluating Limits 141

Unit 4: Differentiation 181

CHAPTER 9: Differentiation Orientation 183

CHAPTER 10: Differentiation Rules — Yeah, Man, It Rules 215

CHAPTER 11: Differentiation and the Shape of Curves 259

CHAPTER 12: Your Problems Are Solved: Differentiation to the Rescue! 321

CHAPTER 13: More Differentiation Problems: Going Off on a Tangent 367

Unit 5: Integration and Infinite Series 397

CHAPTER 14: Intro to Integration and Approximating Area 399

CHAPTER 15: Integration: It's Backwards Differentiation 439

CHAPTER 16: Integration Techniques for Experts 479

CHAPTER 17: Who Needs Freud? Using the Integral to Solve Your Problems 521

CHAPTER 18: Taming the Infinite with Improper Integrals 557

CHAPTER 19: Infinite Series: Welcome to the Outer Limits 581

Index 623

Table of Contents

INTRODUCTION 1

About This Book 1
Foolish Assumptions 2
Icons Used in This Book 3
Beyond the Book 3
Where to Go from Here 4

UNIT 1: AN OVERVIEW OF CALCULUS 5

CHAPTER 1: **What Is Calculus?** 7
What Calculus Is Not 7
So What Is Calculus, Already? 8
Real-World Examples of Calculus 9

CHAPTER 2: **The Two Big Ideas of Calculus: Differentiation and Integration — Plus Infinite Series** 13
Defining Differentiation 13
The derivative is a slope 14
The derivative is a rate 15
Investigating Integration 15
Sorting Out Infinite Series 17
Divergent series 17
Convergent series 17

CHAPTER 3: **Why Calculus Works** 21
The Limit Concept: A Mathematical Microscope 21
What Happens When You Zoom In 22
Two Caveats; or, Precision, Preschmidgen 24
I may lose my license to practice mathematics 24
What the heck does "infinity" really mean? 24

UNIT 2: WARMING UP WITH CALCULUS PREREQUISITES 25

CHAPTER 4: **Pre-Algebra, Algebra, and Geometry Review** 27
Fine-Tuning Your Fractions 28
Some quick rules 28
Multiplying fractions 28
Dividing fractions 29
Adding fractions 29
Subtracting fractions 30
Canceling in fractions 31
Miscellaneous Algebra 34
Absolute value — absolutely easy 34
Empowering your powers 34

Rooting for roots . . . 35
Logarithms — This Is Not an Event at a Lumberjack Competition . . . 37
Factoring Schmactoring — When Am I Ever Going to Need It? . . . 38
Solving Quadratic Equations . . . 39
Geometry Refresher . . . 43
Handy-dandy geometry formulas . . . 44
Two special right triangles . . . 45
Practice Questions Answers and Explanations . . . 52
Whaddya Know? Chapter 4 Quiz . . . 59
Answers to Chapter 4 Quiz . . . 61

CHAPTER 5: **Funky Functions and Their Groovy Graphs** . . . 67
What Is a Function? . . . 67
The defining characteristic of a function . . . 68
Independent and dependent variables . . . 69
Function notation . . . 69
Composite functions . . . 70
What Does a Function Look Like? . . . 71
Common Functions and Their Graphs . . . 74
Lines in the plane in plain English . . . 74
Parabolic and absolute value functions — even-steven . . . 77
A couple of oddball functions . . . 78
Exponential functions . . . 78
Logarithmic functions . . . 79
Inverse Functions . . . 81
Shifts, Reflections, Stretches, and Shrinks . . . 83
Horizontal transformations . . . 83
Vertical transformations . . . 85
Practice Questions Answers and Explanations . . . 87
Whaddya Know? Chapter 5 Quiz . . . 91
Answers to Chapter 5 Quiz . . . 92

CHAPTER 6: **The Trig Tango** . . . 95
Starting off with SohCahToa . . . 95
Two Important Trig Triangles . . . 96
Circling the Enemy with the Unit Circle . . . 97
Angles in the unit circle . . . 98
Measuring angles with radians . . . 98
Honey, I shrunk the hypotenuse . . . 99
Putting it all together . . . 100
Graphing Sine, Cosine, and Tangent . . . 106
Inverse Trig Functions . . . 107
Identifying with Trig Identities . . . 108
Practice Questions Answers and Explanations . . . 109
Whaddya Know? Chapter 6 Quiz . . . 113
Answers to Chapter 6 Quiz . . . 114

UNIT 3: LIMITS 117

CHAPTER 7: **Limits and Continuity** 119
Take It to the Limit — NOT 119
Using three functions to illustrate the same limit 120
Sidling up to one-sided limits 121
The formal definition of a limit — just what you've been waiting for 122
Limits and vertical asymptotes 123
Limits and horizontal asymptotes 123
Calculating instantaneous speed with limits 126
Linking Limits and Continuity 129
Continuity and limits usually go hand in hand 130
The hole exception tells the whole story 130
Sorting out the mathematical mumbo jumbo of continuity 131
The 33333 Limit Mnemonic 131
Practice Questions Answers and Explanations 135
Whaddya Know? Chapter 7 Quiz 137
Answers to Chapter 7 Quiz 139

CHAPTER 8: **Evaluating Limits** 141
Easy Does It — Easy Limits 141
Limits to memorize 142
Plugging and chugging 142
The "Real Deal" Limit Problems 143
Figuring a limit with your calculator 143
Solving limit problems with algebra 147
Take a break and make yourself a limit sandwich 153
Evaluating Limits at ±Infinity 157
Limits of rational functions at ±infinity 158
Solving limits at ±infinity with a calculator 160
Solving limits at ±infinity with algebra 161
Practice Questions Answers and Explanations 165
Whaddya Know? Chapter 8 Quiz 175
Answers to Chapter 8 Quiz 176

UNIT 4: DIFFERENTIATION 181

CHAPTER 9: **Differentiation Orientation** 183
Differentiating: It's Just Finding the Slope 184
The slope of a line 186
The derivative of a line 188
The Derivative: It's Just a Rate 188
Calculus on the playground 188
Speed — the most familiar rate 190
The rate-slope connection 190
The Derivative of a Curve 191
The Difference Quotient 195
Average Rate and Instantaneous Rate 201

To Be or Not to Be? Three Cases Where the Derivative Does Not Exist 202
Practice Questions Answers and Explanations 205
Whaddya Know? Chapter 9 Quiz 209
Answers to Chapter 9 Quiz 211

CHAPTER 10: **Differentiation Rules — Yeah, Man, It Rules** 215
Basic Differentiation Rules 216
The constant rule 216
The power rule 216
The constant multiple rule 217
The sum rule — hey, that's some rule you got there 218
The difference rule — it makes no difference 218
Differentiating trig functions 220
Differentiating exponential and logarithmic functions 220
Differentiation Rules for Experts — Oh, Yeah, I'm a Calculus Wonk 222
The product rule 222
The quotient rule 222
Linking up with the chain rule 225
Differentiating Implicitly 231
Differentiating Inverse Functions 234
Scaling the Heights of Higher-Order Derivatives 237
Practice Questions Answers and Explanations 240
Whaddya Know? Chapter 10 Quiz 253
Answers to Chapter 10 Quiz 254

CHAPTER 11: **Differentiation and the Shape of Curves** 259
Taking a Calculus Road Trip 259
Climb every mountain, ford every stream: Positive and negative slopes 260
I can't think of a travel metaphor for this section: Concavity and inflection points 260
This vale of tears: A local minimum 261
A scenic overlook: The absolute maximum 261
Car trouble: Teetering on the corner 261
It's all downhill from here 261
Your travel diary 262
Finding Local Extrema — My Ma, She's Like, Totally Extreme 262
Cranking out the critical numbers 263
The first derivative test 264
The second derivative test — no, no, anything but another test! 268
Finding Absolute Extrema on a Closed Interval 273
Finding Absolute Extrema over a Function's Entire Domain 278
Locating Concavity and Inflection Points 281
Looking at Graphs of Derivatives Till They Derive You Crazy 285
The Mean Value Theorem — Go Ahead, Make My Day 289
Practice Questions Answers and Explanations 292
Whaddya Know? Chapter 11 Quiz 310
Answers to Chapter 11 Quiz 311

CHAPTER 12: **Your Problems Are Solved: Differentiation to the Rescue!** 321

Getting the Most (or Least) Out of Life: Optimization Problems 321
The maximum volume of a box 322
The maximum area of a corral — yeehaw! 323
Yo-Yo a Go-Go: Position, Velocity, and Acceleration 326
Velocity, speed, and acceleration 328
Maximum and minimum height 329
Velocity and displacement 330
Speed and distance traveled 331
Burning some rubber with acceleration 333
Tying it all together 333
Related Rates — They Rate, Relatively 336
Blowing up a balloon 336
Filling up a trough 339
Fasten your seat belt: You're approaching a calculus crossroads 341
Try this at your own risk 343
Practice Questions Answers and Explanations 346
Whaddya Know? Chapter 12 Quiz 362
Answers to Chapter 12 Quiz 363

CHAPTER 13: **More Differentiation Problems: Going Off on a Tangent** 367

Tangents and Normals: Joined at the Hip 368
The tangent line problem 368
The normal line problem 370
Straight Shooting with Linear Approximations 374
Business and Economics Problems 378
Managing marginals in economics 378
Practice Questions Answers and Explanations 385
Whaddya Know? Chapter 13 Quiz 393
Answers to Chapter 13 Quiz 394

UNIT 5: INTEGRATION AND INFINITE SERIES 397

CHAPTER 14: **Intro to Integration and Approximating Area** 399

Integration: Just Fancy Addition 400
Finding the Area Under a Curve 401
Approximating Area 403
Approximating area with left sums 404
Approximating area with right sums 406
Approximating area with midpoint sums 408
Getting Fancy with Summation Notation 411
Summing up the basics 411
Writing Riemann sums with sigma notation 414
Finding Exact Area with the Definite Integral 417
Approximating Area with the Trapezoid Rule and Simpson's Rule 421
The trapezoid rule 421
Simpson's rule — that's Thomas (1710–1761), not Homer (1987–) 423

Practice Questions Answers and Explanations . . . 427
Whaddya Know? Chapter 14 Quiz . . . 435
Answers to Chapter 14 Quiz . . . 436

CHAPTER 15: **Integration: It's Backwards Differentiation** . . . 439
Antidifferentiation . . . 439
Vocabulary, Voshmabulary: What Difference Does It Make? . . . 441
The Annoying Area Function . . . 441
The Power and the Glory of the Fundamental Theorem of Calculus . . . 446
The Fundamental Theorem of Calculus: Take Two . . . 449
Why the theorem works: Area functions explained . . . 454
Why the theorem works: The integration-differentiation connection . . . 456
Why the theorem works: A connection to — egad! — statistics . . . 458
Finding Antiderivatives: Three Basic Techniques . . . 460
Reverse rules for antiderivatives . . . 460
Guessing and checking . . . 463
The substitution method . . . 465
Finding Area with Substitution Problems . . . 469
Practice Questions Answers and Explanations . . . 471
Whaddya Know? Chapter 15 Quiz . . . 476
Answers to Chapter 15 Quiz . . . 477

CHAPTER 16: **Integration Techniques for Experts** . . . 479
Integration by Parts: Divide and Conquer . . . 479
Picking your u . . . 482
Integration by parts: Second time, same as the first . . . 484
Tricky Trig Integrals . . . 486
Integrals containing sines and cosines . . . 486
Integrals containing secants and tangents (or cosecants and cotangents) . . . 489
Your Worst Nightmare: Trigonometric Substitution . . . 491
Case 1: Tangents . . . 492
Case 2: Sines . . . 494
Case 3: Secants . . . 496
The *A*'s, *B*'s, and *Cx*'s of Partial Fractions . . . 497
Case 1: The denominator contains only linear factors . . . 497
Case 2: The denominator contains irreducible quadratic factors . . . 498
Bonus: Equating coefficients of like terms . . . 500
Case 3: The denominator contains one or more factors raised to a power greater than 1 . . . 501
Practice Questions Answers and Explanations . . . 503
Whaddya Know? Chapter 16 Quiz . . . 517
Answers to Chapter 16 Quiz . . . 519

CHAPTER 17: **Who Needs Freud? Using the Integral to Solve Your Problems** . . . 521
The Mean Value Theorem for Integrals and Average Value . . . 522
The Area between Two Curves — Double the Fun . . . 525
Volumes of Weird Solids: No, You're Never Going to Need This . . . 529
The meat-slicer method . . . 529
The disk method . . . 532
The washer method . . . 533

Analyzing Arc Length . . . 537
Surfaces of Revolution — Pass the Bottle 'Round . . . 540
Practice Questions Answers and Explanations . . . 544
Whaddya Know? Chapter 17 Quiz . . . 551
Answers to Chapter 17 Quiz . . . 552

CHAPTER 18: **Taming the Infinite with Improper Integrals** . . . 557
L'Hôpital's Rule: Calculus for the Sick . . . 558
  Getting unacceptable forms into shape . . . 559
  Looking at three more unacceptable forms . . . 560
Improper Integrals: Just Look at the Way That Integral Is Holding Its Fork! . . . 563
  Improper integrals with vertical asymptotes . . . 563
  Improper integrals with one or two infinite limits of integration . . . 565
  Blowing Gabriel's horn . . . 568
Practice Questions Answers and Explanations . . . 570
Whaddya Know? Chapter 18 Quiz . . . 575
Answers to Chapter 18 Quiz . . . 576

CHAPTER 19: **Infinite Series: Welcome to the Outer Limits** . . . 581
Sequences and Series: What They're All About . . . 582
  Stringing sequences . . . 582
  Summing series . . . 584
Convergence or Divergence? That Is the Question . . . 586
  A no-brainer divergence test: The nth term test . . . 586
  Three basic series and their convergence/divergence tests . . . 588
  Three comparison tests for convergence/divergence . . . 592
  The two "R" tests: Ratios and roots . . . 598
Alternating Series . . . 602
  Finding absolute versus conditional convergence . . . 602
  The alternating series test . . . 603
Keeping All the Tests Straight . . . 606
Practice Questions Answers and Explanations . . . 607
Whaddya Know? Chapter 19 Quiz . . . 616
Answers to Chapter 19 Quiz . . . 617

INDEX . . . 623

Introduction

The mere thought of having to take a required calculus course is enough to make legions of students break out in a cold sweat. Others who have no intention of ever studying the subject have this notion that calculus is impossibly difficult unless you happen to be a direct descendant of Einstein.

Well, I'm here to tell you that you *can* master calculus. It's not nearly as tough as its mystique would lead you to think. Much of calculus is really just very advanced algebra, geometry, and trig. It builds upon and is a logical extension of those subjects. If you can do algebra, geometry, and trig, you can do calculus.

But why should you bother — apart from being required to take a course? Why climb Mt. Everest? Why listen to Beethoven's Ninth Symphony? Why visit the Louvre to see the *Mona Lisa*? Why watch *South Park?* Like these endeavors, doing calculus can be its own reward. There are many who say that calculus is one of the crowning achievements in all of intellectual history. As such, it's worth the effort. Read this jargon-free book, get a handle on calculus, and join the happy few who can proudly say, "Calculus? Oh, sure, I know calculus. It's no big deal."

About This Book

Calculus All-in-One For Dummies is intended for three groups of readers: students taking their first calculus course, students who need to brush up on their calculus to prepare for other studies, and adults of all ages who'd like a good introduction to the subject either to satisfy their own curiosity or perhaps to help someone else with calculus.

If you're enrolled in a calculus course and you find your textbook less than crystal clear, this is the book for you. It covers the most important topics in the first year of calculus: differentiation, integration, and infinite series.

If you've had elementary calculus, but it's been a couple of years and you want to review the concepts to prepare for, say, some graduate program, *Calculus All-in-One For Dummies* will give you a thorough, no-nonsense refresher course.

Non-student readers will find the book's exposition clear and accessible. *Calculus All-in-One For Dummies* takes calculus out of the ivory tower and brings it down to earth.

This is a user-friendly math book. Whenever possible, I explain the calculus concepts by showing you connections between the calculus ideas and easier ideas from algebra and geometry. I then show you how the calculus concepts work using concrete examples. Only later do I give you the fancy calculus formulas. All explanations are in plain English, not math-speak.

The following conventions keep the text consistent and oh-so-easy to follow:

- Variables are in *italics.*
- Calculus terms are italicized and defined when they first appear in the text.
- In the step-by-step problem-solving methods, the general action you need to take is in bold, followed by the specifics of the particular problem.

It can be a great aid to true understanding of calculus — or any math topic for that matter — to focus on the *why* in addition to the *how-to*. With this in mind, I've put a lot of effort into explaining the underlying logic of many of the ideas in this book. If you want to give your study of calculus a solid foundation, you should read these explanations. But if you're really in a hurry, you can cut to the chase and read only the important introductory stuff, the example problems, the step-by-step solutions, and all the rules and definitions next to the icons. You can read the remaining exposition later only if you feel the need.

I find the sidebars interesting and entertaining. (What do you expect? I wrote them!) But you can skip them without missing any essential calculus. No, you won't be tested on that stuff.

The most important thing is for you to work out the "Your Turn" example problems and the problems in the end-of-chapter quizzes (solutions are provided). You can't learn calculus (or any type of math) without working out dozens or hundreds of problems. If you want even more practice problems, you can find more chapter quizzes online.

Foolish Assumptions

Call me crazy, but I assume

- You know at least the basics of algebra, geometry, and trig.

  If you're rusty, Unit 2 (and the online Cheat Sheet) contains a good review of these pre-calculus topics. Actually, if you're not currently taking a calculus course, and you're reading this book just to satisfy a general curiosity about calculus, you can get a good conceptual picture of the subject without the nitty-gritty details of algebra, geometry, and trig. But you won't, in that case, be able to follow all the problem solutions. In short, without the pre-calculus stuff, you can see the calculus *forest*, but not the *trees*. If you're enrolled in a calculus course, you've got no choice — you've got to know the trees as well as the forest.

- You're willing to do some w_ _ _.

  No, not the dreaded *w*-word! Yes, that's w-o-r-k, *work*. I've tried to make this material as accessible as possible, but it is calculus after all. You can't learn calculus by just listening to a tape in your car or taking a pill — not yet anyway.

Is that too much to ask?

Icons Used in This Book

Keep your eyes on the icons:

MATH RULES

Next to this icon are calculus rules, definitions, and formulas.

REMEMBER

These are things you need to know from algebra, geometry, or trig, or things you should recall from earlier in the book.

TIP

The lightbulb icon appears next to things that will make your life easier. Take note.

WARNING

This icon highlights common calculus mistakes. Take heed.

EXAMPLE

Each example is a calculus problem that illustrates the topic just discussed, followed by a step-by-step solution. Studying these example problems and their solutions will help you solve the "Your Turn" practice problems and the problems in the end-of-chapter quizzes.

YOUR TURN

This icon means it's time to put on your thinking cap. It appears next to practice problems for you to work out. Some of these problems will be quite similar to the example problems from the same section. Others will challenge you by going a bit beyond the garden-variety examples. Solutions are provided.

Beyond the Book

There's some great supplementary calculus material online that you might want to check out.

To view this book's online Cheat Sheet, simply go to www.dummies.com and type **Calculus All in One For Dummies Cheat Sheet** in the Search box. You'll find a nice list of important formulas, theorems, definitions, and so on from algebra, geometry, trigonometry, and calculus. This is a great place to go if you forget a formula.

You'll also have access to additional online quizzes for each chapter of the book, starting with Unit 2. To access the quizzes, follow these steps:

1. **Register your book or ebook at Dummies.com to get your PIN.** Go to www.dummies.com/go/getaccess.
2. **Select your product from the drop-down list on that page.**
3. **Follow the prompts to validate your product, and then check your email for a confirmation message that includes your PIN and instructions for logging in.**

If you do not receive this email within two hours, please check your spam folder before contacting us through our Technical Support website at `http://support.wiley.com` or by phone at 877-762-2974.

Where to Go from Here

Why, Chapter 1, of course, if you want to start at the beginning. If you already have some background in calculus or just need a refresher course in one area or another, then feel free to skip around. Use the table of contents and index to find what you're looking for. If all goes well, in a half a year or so, you'll be able to check calculus off your list:

- [ ] Run a marathon
- [ ] Go skydiving
- [ ] Write a book
- [x] Learn calculus
- [ ] Swim the English Channel
- [ ] Cure cancer
- [ ] Write a symphony
- [ ] Pull an unnatural double cork 1260° at the X Games

For the rest of your list, you're on your own.

1

An Overview of Calculus

In This Unit . . .

CHAPTER 1: **What Is Calculus?**

What Calculus Is Not
So What Is Calculus, Already?
Real-World Examples of Calculus

CHAPTER 2: **The Two Big Ideas of Calculus: Differentiation and Integration — Plus Infinite Series**

Defining Differentiation
Investigating Integration
Sorting Out Infinite Series

CHAPTER 3: **Why Calculus Works**

The Limit Concept: A Mathematical Microscope
What Happens When You Zoom In
Two Caveats; or, Precision, Preschmidgen

IN THIS CHAPTER

» You're only in Chapter 1 and you're already going to get your first calc test

» Calculus — it's just souped-up regular math

» Zooming in is the key

» The world before and after calculus

Chapter 1
What Is Calculus?

"My best day in Calc 101 at Southern Cal was the day I had to cut class to get a root canal."

— MARY JOHNSON

"I keep having this recurring dream where my calculus professor is coming after me with an axe."

— TOM FRANKLIN, COLORADO COLLEGE SOPHOMORE

"Calculus is fun, and it's so easy. I don't get what all the fuss is about."

— SAM EINSTEIN, ALBERT'S GREAT-GRANDSON

In this chapter, I answer the question, "What is calculus?" in plain English, and I give you real-world examples of how calculus is used. After reading this and the following two short chapters, you *will* understand what calculus is all about. But here's a twist: Why don't you start out on the *wrong* foot by briefly checking out what calculus is *not*?

What Calculus Is Not

No sense delaying the inevitable. Ready for your first calculus test? Circle True or False.

True or False: Unless you actually enjoy wearing a pocket protector, you've got no business taking calculus.

True or False: Studying calculus is hazardous to your health.

True or False: Calculus is totally irrelevant.

False, false, false! There's this mystique about calculus that it's this ridiculously difficult, incredibly arcane subject that no one in their right mind would sign up for unless it was a required course.

Don't buy into this misconception. Sure, calculus is difficult — I'm not going to lie to you — but it's manageable, doable. You made it through algebra, geometry, and trigonometry. Well, calculus just picks up where they leave off — it's simply the next step in a logical progression.

REMEMBER

Calculus *is* relevant. Calculus is not a dead language like Latin, spoken only by academics. It's the language of engineers, scientists, and economists. Okay, so it's a couple steps removed from your everyday life and unlikely to come up at a cocktail party. But the work of those engineers, scientists, and economists has a huge impact on your day-to-day life — from your microwave oven, cellphone, TV, and car to the medicines you take, the workings of the economy, and our national defense. At this very moment, something within your reach or within your view has been impacted by calculus.

So What Is Calculus, Already?

Calculus is basically just very advanced algebra and geometry. In one sense, it's not even a new subject — it takes the ordinary rules of algebra and geometry and tweaks them so that they can be used on more complicated problems. (The rub, of course, is that darn *other* sense in which it *is* a new and more difficult subject.)

Look at Figure 1-1. On the left is a man pushing a crate up a straight incline. On the right, the man is pushing the same crate up a curving incline. The problem, in both cases, is to determine the amount of energy required to push the crate to the top. You can do the problem on the left with regular math. For the one on the right, you need calculus (assuming you don't know the physics shortcuts).

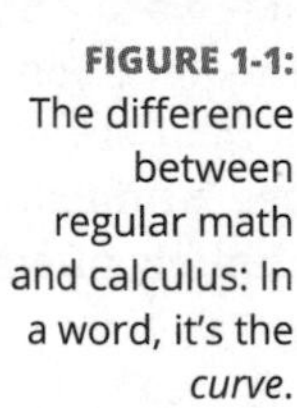

FIGURE 1-1: The difference between regular math and calculus: In a word, it's the *curve*.

For the straight incline, the man pushes with an *unchanging* force, and the crate goes up the incline at an *unchanging* speed. With some simple physics formulas and regular math (including algebra and trig), you can compute how many calories of energy are required to push the crate up the incline. Note that the amount of energy expended each second remains the same.

For the curving incline, on the other hand, things are constantly changing. The steepness of the incline is *changing* — and not just in increments like it's one steepness for the first 3 feet then a different steepness for the next 3 feet. It's *constantly changing*. And the man pushes with a *constantly changing* force — the steeper the incline, the harder the push. As a result, the amount of energy expended is also changing, not every second or every thousandth of a second, but *constantly changing* from one moment to the next. That's what makes it a calculus problem.

REMEMBER

Calculus is the mathematics of change. By this time, it should come as no surprise to you that calculus is described as "the mathematics of change." Calculus takes the regular rules of math and applies them to fluid, evolving problems.

For the curving incline problem, the physics formulas remain the same, and the algebra and trig you use stay the same. The difference is that — in contrast to the straight incline problem, which you can sort of do in a single shot — you've got to break up the curving incline problem into small chunks and do each chunk separately. Figure 1-2 shows a small portion of the curving incline blown up to several times its size.

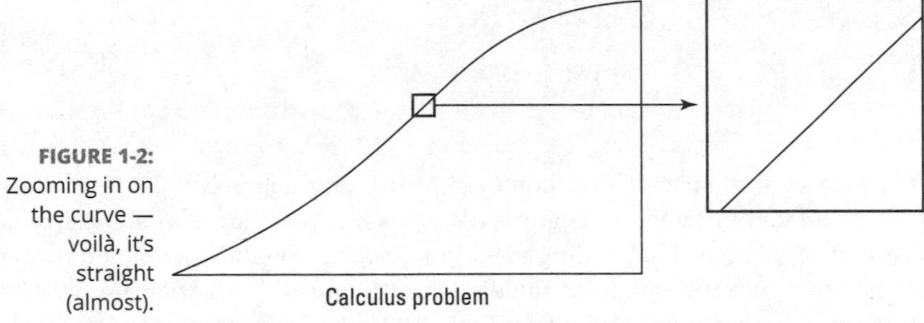

FIGURE 1-2: Zooming in on the curve — voilà, it's straight (almost).

When you zoom in far enough, the small length of the curving incline becomes practically straight. Then, because it's straight, you can solve that small chunk just like the straight incline problem. Each small chunk can be solved the same way, and then you just add up all the chunks.

That's calculus in a nutshell. It takes a problem that can't be done with regular math because things are constantly changing — the changing quantities show up on a graph as curves — it zooms in on the curve till it becomes straight, and then it finishes off the problem with regular math.

What makes the invention of calculus such a fantastic achievement is that it does what seems impossible: It zooms in *infinitely*. As a matter of fact, everything in calculus involves infinity in one way or another, because if something is constantly changing, it's changing infinitely often from each infinitesimal moment to the next.

Real-World Examples of Calculus

So, with regular math you can do the straight incline problem; with calculus you can do the curving incline problem. Here are some more examples.

With regular math you can determine the length of a buried cable that runs diagonally from one corner of a park to the other (remember the Pythagorean Theorem?). With calculus you can determine the length of a cable hung between two towers that has the shape of a *catenary* (which is different, by the way, from a simple circular arc or a parabola). Knowing the exact length is of obvious importance to a power company planning hundreds of miles of new electric cable. See Figure 1-3.

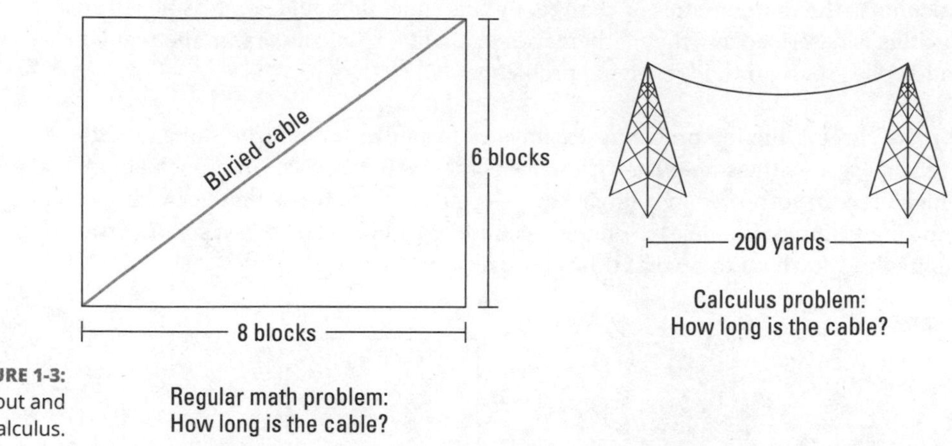

FIGURE 1-3: Without and with calculus.

You can calculate the area of the flat roof of a home with ordinary geometry. With calculus you can compute the area of complicated, nonspherical shapes like the shapes of some sports arena domes. Architects designing such a building need to know the dome's area to determine the cost of materials and to figure the weight of the dome (with and without snow on it). The weight, of course, is needed for planning the strength of the supporting structure. Check out Figure 1-4.

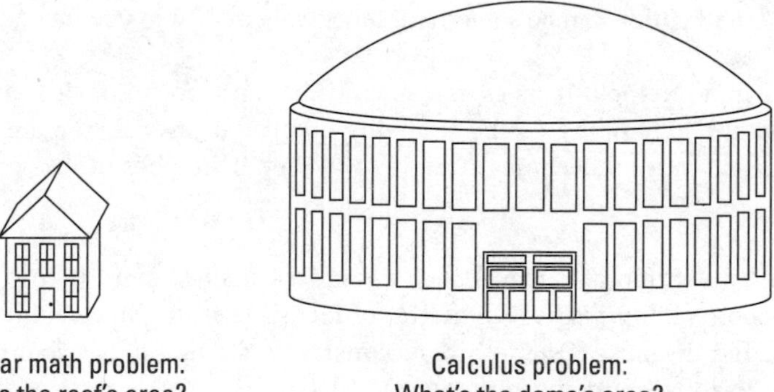

Regular math problem:
What's the roof's area?
Calculus problem:
What's the dome's area?

FIGURE 1-4: Sans and avec calculus.

With regular math and some simple physics, you can calculate how much a quarterback must lead his receiver to complete a pass. (I'm assuming here that the receiver runs in a *straight* line and at a *constant* speed.) But when NASA, in 1975, calculated the necessary "lead" for aiming the Viking I at Mars, it needed calculus because both the Earth and Mars travel on *elliptical* orbits (of different shapes) and the speeds of both are *constantly changing* — not to mention the fact that on its way to Mars, the spacecraft is affected by the different and *constantly changing* gravitational pulls of the Earth, the Moon, Mars, and the Sun. See Figure 1-5.

You see many real-world applications of calculus throughout this book. The differentiation problems in Unit 4 all involve the steepness of a curve — like the steepness of the curving incline in Figure 1-1. In Unit 5, you do integration problems like the cable-length problem shown back in Figure 1-3. These problems involve breaking up something into little sections, calculating each section, and then adding up the sections to get the total. More about that in Chapter 2.

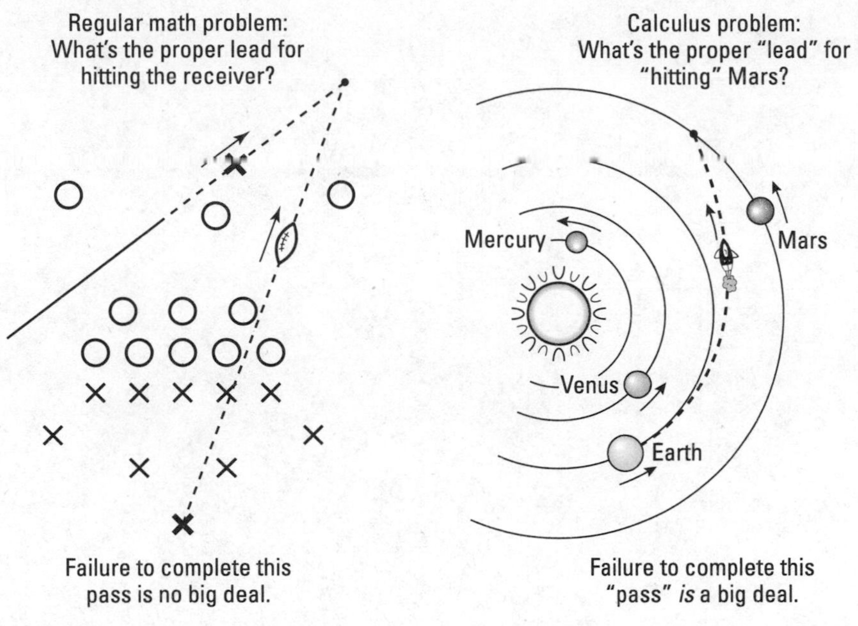
Regular math problem:
What's the proper lead for
hitting the receiver?
Failure to complete this
pass is no big deal.
Calculus problem:
What's the proper "lead" for
"hitting" Mars?
Mercury
Venus
Earth
Mars
Failure to complete this
"pass" *is* a big deal.

FIGURE 1-5: B.C.E. (Before the Calculus Era) and C.E. (the Calculus Era).

IN THIS CHAPTER

» Delving into the derivative: It's a rate and a slope

» Investigating the integral — addition for experts

» Infinite series: Achilles versus the tortoise — place your bets

Chapter 2

The Two Big Ideas of Calculus: Differentiation and Integration — Plus Infinite Series

This book covers the two main topics in calculus — differentiation and integration — as well as a third topic, infinite series. All three topics touch the earth and the heavens because all are built upon the rules of ordinary algebra and geometry, and all involve the idea of infinity.

Defining Differentiation

Differentiation is the process of finding the *derivative* of a curve. And the word "derivative" is just the fancy calculus term for the curve's slope or steepness. And because the slope of a curve is equivalent to a simple rate (like *miles per hour* or *profit per item*), the derivative is a rate as well as a slope.

The derivative is a slope

In algebra, you learned about the slope of a line — it's equal to the ratio of the *rise* to the *run*. In other words, $Slope = \frac{rise}{run}$. See Figure 2-1. Let me guess: A sudden rush of algebra nostalgia is flooding over you.

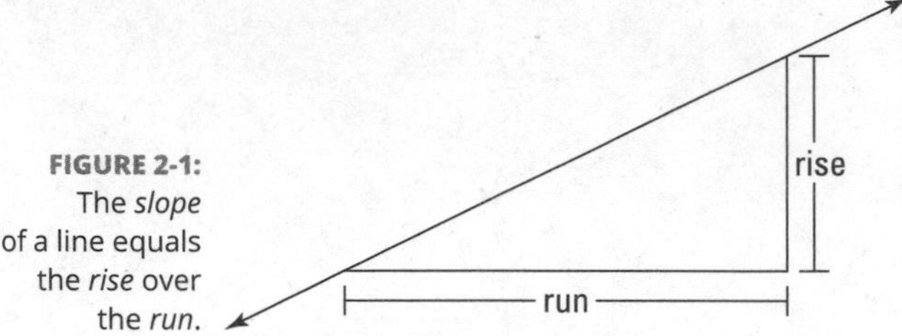

FIGURE 2-1: The *slope* of a line equals the *rise* over the *run*.

In Figure 2-1, the *rise* is half as long as the *run*, so the line has a slope of $\frac{1}{2}$.

On a curve, the slope is constantly *changing*, so you need calculus to determine its slope. See Figure 2-2.

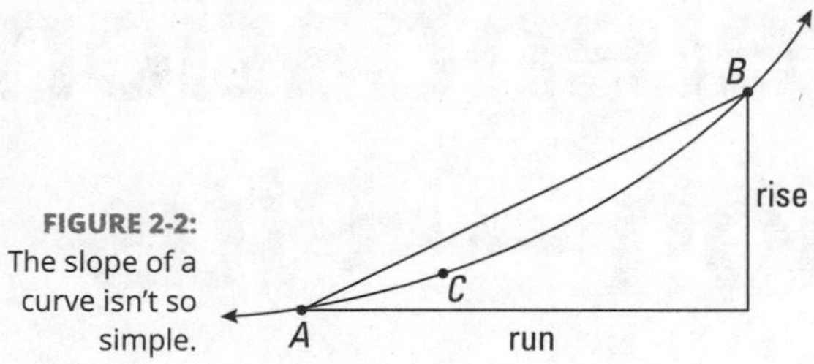

FIGURE 2-2: The slope of a curve isn't so simple.

Just like the line in Figure 2-1, the straight line between *A* and *B* in Figure 2-2 has a slope of $\frac{1}{2}$. And the slope of this line is the same at every point between *A* and *B*. But you can see that, unlike the line, the steepness of the curve is changing between *A* and *B*. At *A*, the curve is less steep than the line, and at *B*, the curve is steeper than the line. What do you do if you want the exact slope at, say, point *C*? Can you guess? Time's up. Answer: You zoom in. See Figure 2-3.

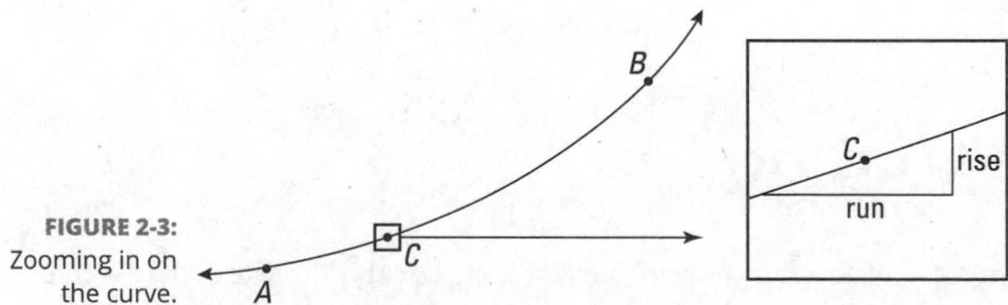

FIGURE 2-3: Zooming in on the curve.

When you zoom in far enough — really far, actually *infinitely* far — the little piece of the curve becomes straight, and you can figure the slope the old-fashioned way. That's how differentiation works.

The derivative is a rate

Because the derivative of a curve is the slope — which equals $\frac{rise}{run}$ or *rise per run* — the derivative is also a rate, a *this per that* like *miles per hour* or *gallons per minute* (the name of the particular rate simply depends on the units used on the x- and y-axes). The two graphs in Figure 2-4 show a relationship between distance and time — they could represent a trip in your car.

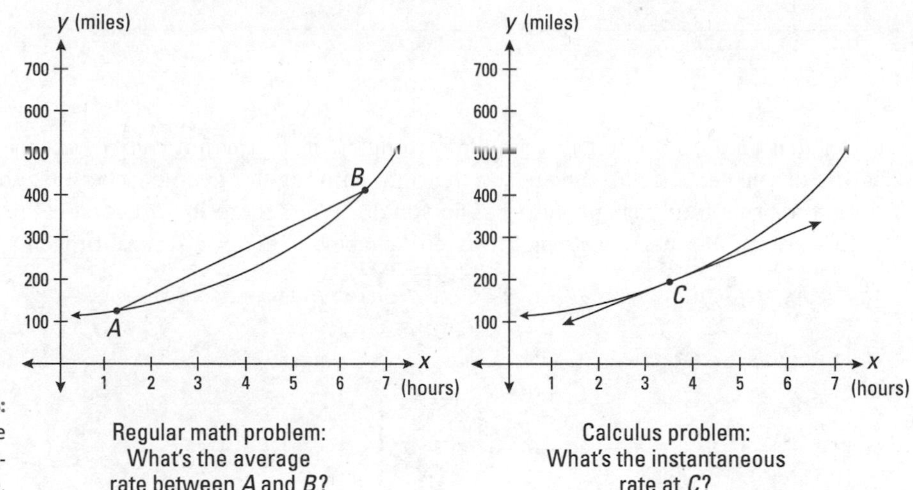

FIGURE 2-4: Average rate and instantaneous rate.

A regular algebra problem is shown on the left in Figure 2-4. If you know the x- and y-coordinates of points A and B, you can use the slope formula $\left(Slope = \frac{rise}{run} = \frac{y_2 - y_1}{x_2 - x_1}\right)$ to calculate the slope between A and B, and, in this problem, that slope gives you the *average* rate in *miles per hour* for the interval from A to B.

For the problem on the right, on the other hand, you need calculus. (You can't use the slope formula because you've only got one point.) Using the derivative of the curve, you can determine the *exact* slope or steepness at point C. Just to the left of C on the curve, the slope is slightly lower, and just to the right of C on the curve, the slope is slightly higher. But precisely at C, for a single infinitesimal moment, you get a slope that's different from the neighboring slopes. The slope for this single infinitesimal point on the curve gives you the *instantaneous* rate in *miles per hour* at point C.

Investigating Integration

Integration is the second big idea in calculus, and it's basically just fancy addition. Integration is the process of cutting up an area into tiny sections, figuring the areas of the small sections, and then adding up the little bits of area to get the whole area. Figure 2-5 shows two area problems — one that you can do with geometry and one where you need calculus.

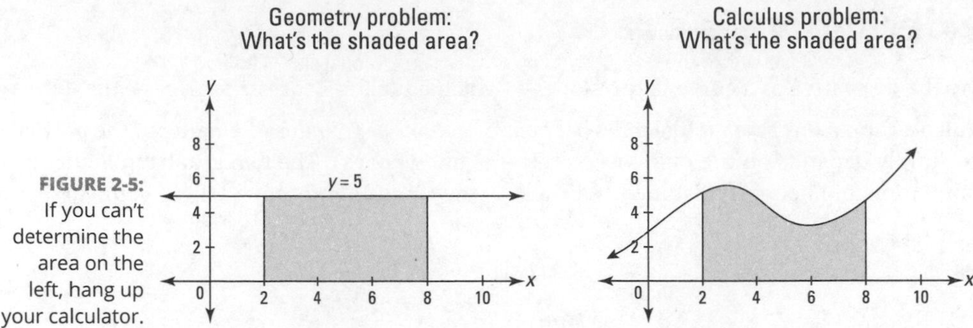

FIGURE 2-5: If you can't determine the area on the left, hang up your calculator.

The shaded area on the left is a simple rectangle, so its area, of course, equals length times width. But you can't figure the area on the right with regular geometry because there's no area formula for this funny shape. So what do you do? Why, zoom in, of course. Figure 2-6 shows the top portion of a narrow strip of the weird shape blown up to several times its size.

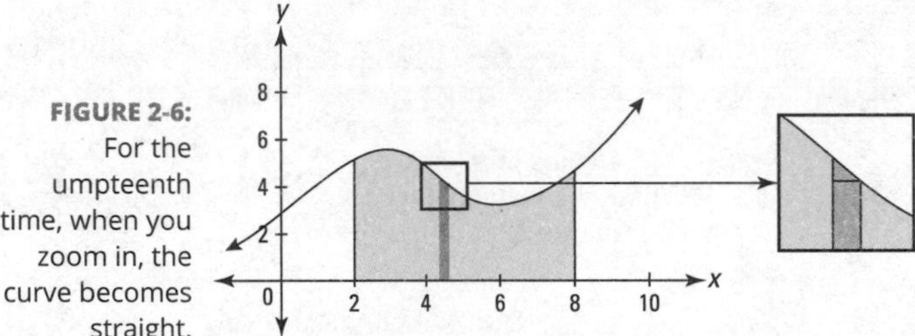

FIGURE 2-6: For the umpteenth time, when you zoom in, the curve becomes straight.

When you zoom in as shown in Figure 2-6, the curve becomes practically straight, and the further you zoom in, the straighter it gets. After zooming in, you get the shape on the right in Figure 2-6, which is practically an ordinary trapezoid (its top is still slightly curved). Well, with the magic of integration, you zoom in *infinitely* close (sort of — you can't really get infinitely close, right?). At that point, the shape is exactly an ordinary trapezoid — or, if you want to get really basic, it's a triangle sitting on top of a rectangle. Because you can compute the areas of rectangles, triangles, and trapezoids with ordinary geometry, you can get the area of this and all the other thin strips and then add up all these areas to get the total area. That's integration.

Figure 2-7 has two graphs of a city's electrical energy consumption on a typical summer day. The horizontal axes show the number of hours after midnight, and the vertical axes show the amount of power (in kilowatts) used by the city at different times during the day.

The crooked line on the left and the curve on the right show how the number of kilowatts of power depends on the time of day. In both cases, the shaded area gives the number of kilowatt-hours of energy consumed during a typical 24-hour period. The shaded area in the oversimplified and unrealistic problem on the left can be calculated with regular geometry. But the true relationship between the amount of power used and the time of day is more complicated than a crooked straight line. In a realistic energy-consumption problem, you'd get something like the graph on the right. Because of its weird curve, you need calculus to determine the shaded area. In the real world, the relationship between different variables is rarely as simple as a straight-line graph. That's what makes calculus so useful.

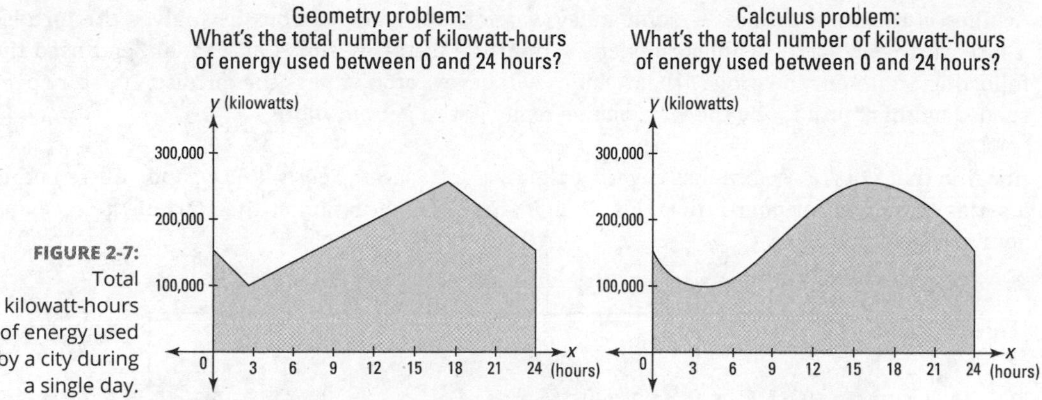

FIGURE 2-7: Total kilowatt-hours of energy used by a city during a single day.

Sorting Out Infinite Series

Infinite series deal with the adding up of an infinite number of numbers. Don't try this on your calculator unless you've got a lot of extra time on your hands. Here's a simple example. The following sequence of numbers is generated by a simple doubling process — each term is twice the one before it:

$$1, 2, 4, 8, 16, 32, 64, 128, \ldots$$

The infinite *series* associated with this *sequence* of numbers is just the sum of the numbers:

$$1 + 2 + 4 + 8 + 16 + 32 + 64 + 128 + \ldots$$

Divergent series

The preceding series of doubling numbers is *divergent* because if you continue the addition indefinitely, the sum will grow bigger and bigger without limit. And if you could add up "all" the numbers in this series — that's all *infinitely many* of them — the sum would be infinity. *Divergent* usually means — there are exceptions — that the series adds up to infinity.

Divergent series are rather uninteresting because they do what you expect. You keep adding more numbers, so the sum keeps growing, and if you continue this forever, the sum grows to infinity. Big surprise.

Convergent series

Convergent series are much more interesting. With a convergent series, you also keep adding more numbers, the sum keeps growing, but even though you add numbers forever and the sum grows forever, the sum of all the infinitely many terms is a *finite* number. This surprising result brings me to Zeno's famous paradox of Achilles and the tortoise. (That's Zeno of Elea, of course, from the 5th century B.C.)

Achilles is racing a tortoise — some gutsy warrior, eh? Our generous hero gives the tortoise a 100-yard head start. Achilles runs at 20 mph; the tortoise "runs" at 2 mph. Zeno used the following argument to "prove" that Achilles will never catch or pass the tortoise. If you're persuaded by this "proof," by the way, you've really got to get out more.

Imagine that you're a journalist covering the race for *Spartan Sports Weekly*, and you're taking a series of photos for your article. Figure 2-8 shows the situation at the start of the race and your first two photos.

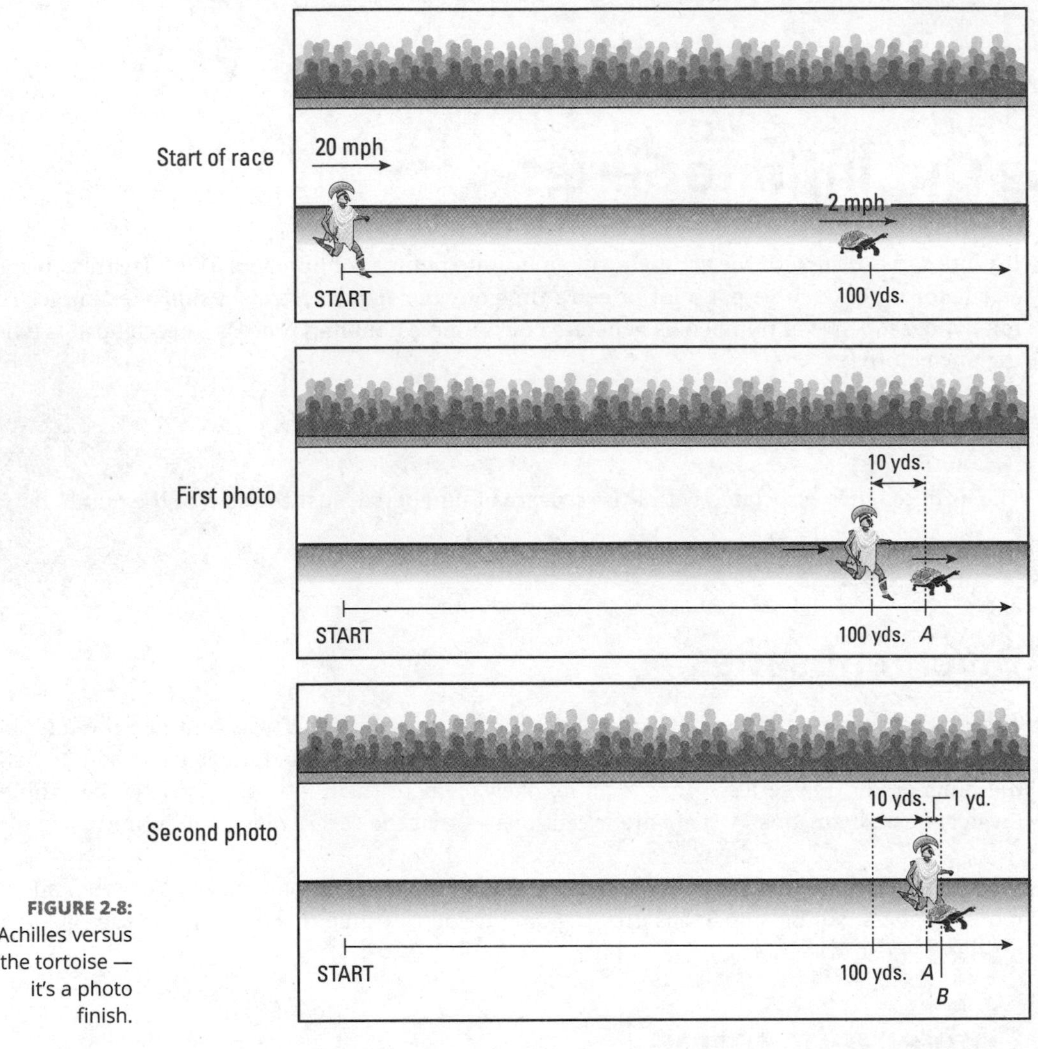

FIGURE 2-8: Achilles versus the tortoise — it's a photo finish.

You take your first photo the instant Achilles reaches the point where the tortoise started. By the time Achilles gets there, the tortoise has "raced" forward and is now 10 yards ahead of Achilles. (The tortoise moves a tenth as fast as Achilles, so in the time it takes Achilles to travel 100 yards, the tortoise covers a tenth as much ground, or 10 yards.) If you do the math, you find that it took Achilles about 10 seconds to run the 100 yards. (For the sake of argument, let's call it exactly 10 seconds.)

You have a cool app that allows you to look at your first photo and note precisely where the tortoise is as Achilles crosses the tortoise's starting point. The tortoise's position is shown as point *A* in the middle image in Figure 2-8. Then you take your second photo when Achilles reaches point *A*, which takes him about one more second. In that second, the tortoise has moved ahead 1 yard to point *B*. You take your third photo (not shown) when Achilles reaches point *B* and the tortoise has moved ahead to point *C*.

Every time Achilles reaches the point where the tortoise was, you take another photo. There is no end to this series of photographs. Assuming you and your camera can work infinitely fast, you will take an infinite number of photos. And *every single time* Achilles reaches the point where the tortoise was, the tortoise has covered more ground — even if only a millimeter or a millionth of a millimeter. This process never ends, right? Thus, the argument goes, because you can never get to the end of your infinite series of photos, Achilles can never catch or pass the tortoise.

Well, as everyone knows, Achilles does in fact reach and pass the tortoise — thus the paradox. The mathematics of infinite series explains how this infinite series of time intervals sums to a *finite* amount of time — the precise time when Achilles passes the tortoise. Here's the sum for those who are curious:

$$\begin{aligned}&10\text{ sec.} + 1\text{ sec.} + 0.1\text{ sec.} + 0.01\text{ sec.} + 0.001\text{ sec.} + \ldots\\&= 11.111\ldots\text{ sec., or } 11\frac{1}{9}\text{ seconds.}\end{aligned}$$

Achilles passes the tortoise after $11\frac{1}{9}$ seconds at the $111\frac{1}{9}$-yard mark.

Infinite series problems are rich with bizarre, counterintuitive paradoxes. You see more of them in Unit 5.

IN THIS CHAPTER

» Using limits to zoom in on curves

» Slope equals rise over run

» Area of a triangle equals one-half base times height

» The Pythagorean Theorem: $a^2 + b^2 = c^2$

Chapter 3
Why Calculus Works

In Chapters 1 and 2, I talk a lot about the process of zooming in on a curve till it looks straight. The mathematics of calculus works because of this basic nature of curves — that they're *locally straight* — in other words, curves are straight at the microscopic level. The earth is round, but to us it looks flat because we're sort of at the microscopic level when compared to the size of the earth. Calculus works because after you zoom in and curves look straight, you can use regular algebra and geometry with them. The zooming-in process is achieved through the mathematics of limits.

The Limit Concept: A Mathematical Microscope

The mathematics of *limits* is the microscope that zooms in on a curve. Here's how a limit works. Say you want the exact slope or steepness of the parabola $y = x^2$ at the point $(1, 1)$. See Figure 3-1.

With the slope formula from algebra, you can figure the slope of the line between $(1, 1)$ and $(2, 4)$. From $(1, 1)$ to $(2, 4)$, you go over 1 and up 3, so the slope is $\frac{3}{1}$, or just 3. But you can see in Figure 3-1 that this line is steeper than the tangent line at $(1, 1)$ that shows the parabola's steepness at that specific point. The limit process sort of lets you slide the point that starts at $(2, 4)$ down toward $(1, 1)$ till it's a thousandth of an inch away, then a millionth, then a billionth, and so on down to the microscopic level. If you do the math, the slopes between $(1, 1)$ and your moving point would look something like 2.8, then 2.6, then 2.4, and so on, and then, once you get to a thousandth of an inch away, 2.001, 2.000001, 2.000000001, and so on. And with the almost magical mathematics of limits, you can conclude that the slope at (1, 1) is precisely 2, even though the sliding point never reaches (1, 1). (If it did, you'd only have one point left and to use the slope formula, you need two separate points.) The mathematics of limits is all based on this zooming-in process, and it works, again, because the further you zoom in, the straighter the curve gets.

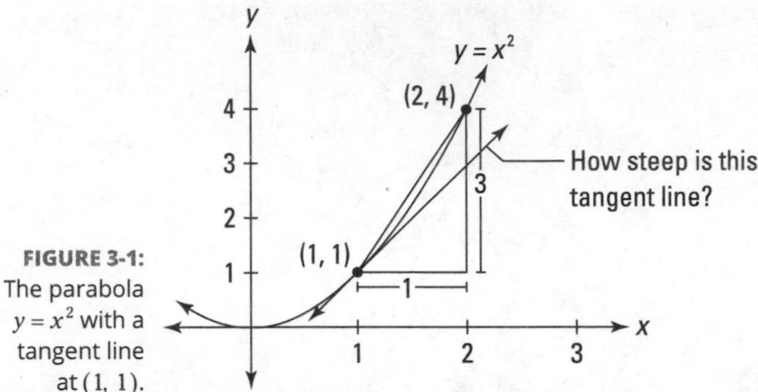

FIGURE 3-1: The parabola $y = x^2$ with a tangent line at (1, 1).

What Happens When You Zoom In

Figure 3-2 shows three diagrams of one curve and three things you might like to know about the curve: 1) the exact slope or steepness at point *C*, 2) the area under the curve between *A* and *B*, and 3) the exact length of the curve from *A* to *B*. You can't answer these questions with regular algebra or geometry formulas because the regular formulas for *slope*, *area*, and *length* work for straight lines (and simple curves like circles), but not for weird curves like this one.

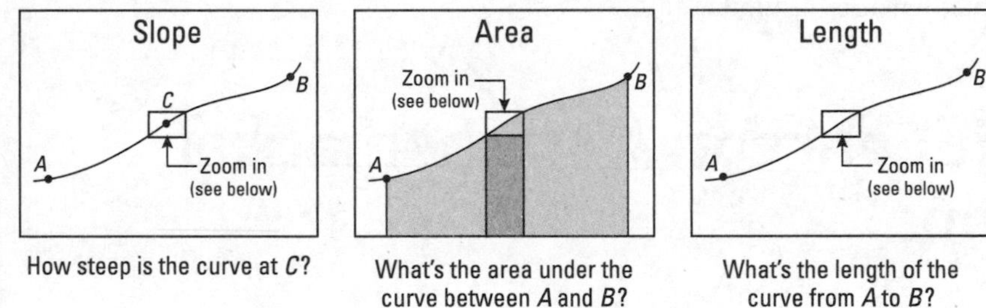

FIGURE 3-2: One curve — three questions.

The first row of Figure 3-3 shows a magnified detail from the three diagrams of the curve in Figure 3-2. The second row shows further magnification, and the third row yet another magnification. For each little window that gets blown up (like from the first to the second row of Figure 3-3), I've drawn in a new dotted diagonal line to help you see how with each magnification, the blown-up pieces of the curves get straighter and straighter. This process is continued indefinitely.

Finally, Figure 3-4 shows the result after an "infinite" number of magnifications — sort of. After zooming in forever, an infinitely small piece of the original curve and the straight diagonal line are now one and the same. You can think of the lengths 3 and 4 in Figure 3-4 (no pun intended) as 3 and 4 millionths of an inch; no, make that 3 and 4 billionths of an inch; no, trillionths; no, gazillionths. . . .

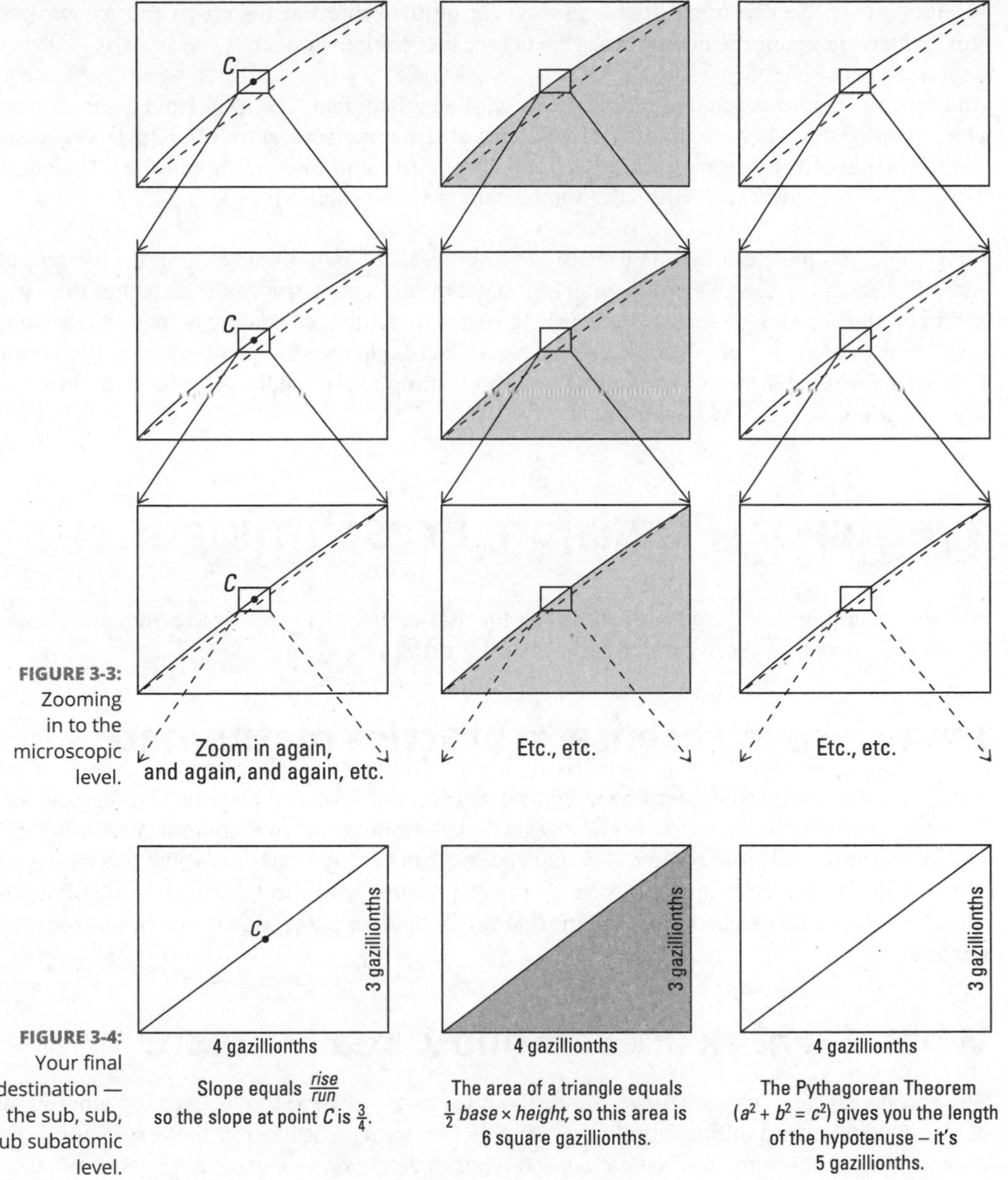
C
C
C
Zoom in again,
and again, and again, etc.
Etc., etc.
Etc., etc.
C
3 gazillionths
4 gazillionths
3 gazillionths
4 gazillionths
3 gazillionths
4 gazillionths
Slope equals $\frac{rise}{run}$
so the slope at point C is $\frac{3}{4}$.
The area of a triangle equals
$\frac{1}{2}$ *base* × *height,* so this area is
6 square gazillionths.
The Pythagorean Theorem
($a^2 + b^2 = c^2$) gives you the length
of the hypotenuse – it's
5 gazillionths.

FIGURE 3-3: Zooming in to the microscopic level.

FIGURE 3-4: Your final destination — the sub, sub, sub subatomic level.

Now that you've zoomed in "forever," the curve is perfectly straight and you can use regular algebra and geometry formulas to answer the three questions about the curve in Figure 3-2.

For the diagram on the left in Figure 3-4, you can now use the regular *slope* formula from algebra to find the slope at point *C*. It's exactly $\frac{3}{4}$ — that's the answer to the first question in Figure 3-2. This is how differentiation works.

For the diagram in the middle of Figure 3-4, the regular triangle formula from geometry gives you an area of 6. Then you can get the shaded area inside the strip shown in Figure 3-2 by adding this 6 to the area of the thin rectangle under the triangle (the dark-shaded rectangle in Figure 3-2). Then you repeat this process for all the other narrow strips (not shown), and finally just add up all the little areas. This is how integration works.

And for the diagram on the right of Figure 3-4, the Pythagorean Theorem from geometry gives you a length of 5. Then, to find the total length of the curve from A to B in Figure 3-2, you do the same thing for the other minute sections of the curve and then add up all the little lengths. This is how you calculate arc length (another integration problem).

Well, there you have it. Calculus uses the limit process to zoom in on a curve till it's straight. After it's straight, the rules of regular-old algebra and geometry apply. Calculus thus gives ordinary algebra and geometry the power to handle complicated problems involving *changing* quantities (which on a graph show up as *curves*). This explains why calculus has so many practical uses, because if there's something you can count on — in addition to death and taxes — it's that things are always changing.

Two Caveats; or, Precision, Preschmidgen

Not everything in this chapter (or this book for that matter) will satisfy the high standards of the Grand Poobah of Precision in Mathematical Writing.

I may lose my license to practice mathematics

With regard to the middle diagrams in Figures 3-2 through 3-4, I'm playing a bit fast and loose with the mathematics. The process of integration — finding the area under a curve — doesn't exactly work the way I explained. My explanation isn't really wrong, it's just a bit sideways. But — I don't care what anybody says — that's my story and I'm stickin' to it. Actually, it's not a bad way to think about how integration works, and, anyhow, this is only an introductory chapter.

What the heck does "infinity" really mean?

The second caveat is that whenever I talk about infinity — like in the last section where I discussed zooming in an infinite number of times — I do something like put the word "infinity" in quotes or say something like "you *sort of* zoom in forever." I do this to cover my butt. Whenever you talk about infinity, you're always on shaky ground. What would it mean to zoom in forever or an infinite number of times? You can't do it; you'd never get there. You can imagine — sort of — what it's like to zoom in forever, but there's something a bit fishy about the idea — and thus the qualifications.

2 Warming Up with Calculus Prerequisites

In This Unit . . .

CHAPTER 4: **Pre-Algebra, Algebra, and Geometry Review**
Fine-Tuning Your Fractions
Miscellaneous Algebra
Geometry Refresher
Practice Questions Answers and Explanations
Whaddya Know? Chapter 4 Quiz
Answers to Chapter 4 Quiz

CHAPTER 5: **Funky Functions and Their Groovy Graphs**
What Is a Function?
What Does a Function Look Like?
Common Functions and Their Graphs
Inverse Functions
Shifts, Reflections, Stretches, and Shrinks
Practice Questions Answers and Explanations
Whaddya Know? Chapter 5 Quiz
Answers to Chapter 5 Quiz

CHAPTER 6: **The Trig Tango**
Starting off with SohCahToa
Two Important Trig Triangles
Circling the Enemy with the Unit Circle
Graphing Sine, Cosine, and Tangent
Inverse Trig Functions
Identifying with Trig Identities
Practice Questions Answers and Explanations
Whaddya Know? Chapter 6 Quiz
Answers to Chapter 6 Quiz

IN THIS CHAPTER

» Winning the fraction battle: Divide and conquer

» Boosting your powers and getting to the root of roots

» Laying down the laws of logarithms and having fun with factoring

» Hanging around the quad solving quadratics

» Getting in shape with two- and three-dimensional shapes

» Getting it right with special right triangles

Chapter 4

Pre-Algebra, Algebra, and Geometry Review

Algebra is the language of calculus. You can't do calculus without knowing algebra any more than you can write Chinese poetry without knowing Chinese. So, if your pre-algebra and algebra are a bit rusty — you know, all those rules for algebraic expressions, equations, fractions, powers, roots, logs, factoring, quadratics, and so on — make sure you review the following algebra basics.

And you've got to review your basic geometry as well. Many calculus problems involve two- and three-dimensional shapes (triangles, rectangles, trapezoids, circles, boxes, cylinders, cones, spheres, and so on) and geometry concepts like length, area, and volume. The two big ideas of calculus are differentiation and integration. Differentiation involves finding the slope of a curve, and slope is a coordinate geometry concept. Integration involves finding the area under a curve. You can't do calculus without a good grasp of basic geometry. Let's get started.

Fine-Tuning Your Fractions

Many, many math students hate fractions. I'm not sure why, because there's nothing especially difficult about them. Perhaps for some students, fraction concepts didn't completely click when they first studied them, and then fractions became a nagging frustration whenever they came up in subsequent math courses. Whatever the cause, if you don't like fractions, try to get over it, because you'll have to deal with fractions in every math course you take.

You can't do calculus without a good grasp of fractions. Open a calculus book to any random page, and you'll likely see a fraction. The very definition of the derivative is based on a fraction called the *difference quotient*. And, on top of that, the symbol for the derivative, $\frac{dy}{dx}$, is a fraction. So, if you're a bit rusty with fractions, get up to speed with the following problems — or else!

Some quick rules

First is a rule that's simple but very important because it comes up time and time again in the study of calculus:

WARNING

You can't divide by zero! The denominator of a fraction can *never* equal zero.

$\frac{0}{5}$ equals zero, but $\frac{5}{0}$ is undefined.

It's easy to see why $\frac{5}{0}$ is undefined when you consider how division works:

$$\frac{8}{2}=4$$

This tells you, of course, that 2 goes into 8 four times; in other words, $2+2+2+2=8$. Well, how many zeros would you need to add up to make 5? You can't do it, and so you can't divide 5 (or any other number) by zero.

Here's another quick rule.

REMEMBER

Definition of *reciprocal*: The reciprocal of a number or expression is its multiplicative inverse — which is a fancy way of saying that the product of something and its reciprocal is 1. To get the reciprocal of a fraction, flip it upside down. Thus, the reciprocal of $\frac{3}{4}$ is $\frac{4}{3}$, the reciprocal of 6, which equals $\frac{6}{1}$, is $\frac{1}{6}$, and the reciprocal of $x-2$ is $\frac{1}{x-2}$.

Multiplying fractions

Adding is usually easier than multiplying, but with fractions, the reverse is true — so I want to deal with multiplication first.

Multiplying fractions is a snap — just multiply straight across the top and straight across the bottom:

$$\frac{2}{5}\cdot\frac{3}{4}=\frac{6}{20}=\frac{3}{10} \text{ and } \frac{a}{b}\cdot\frac{c}{d}=\frac{ac}{bd}$$

Dividing fractions

Dividing fractions has one additional step: You flip the second fraction and then multiply — like this:

$$\begin{aligned}&\frac{3}{10}\div\frac{4}{5}\\&=\frac{3}{10}\cdot\frac{5}{4}=\frac{15}{40} \quad \text{(Now cancel a 5 from the numerator and denominator.)}\\&=\frac{3}{8}\end{aligned}$$

Note that you could have canceled before multiplying. Because 5 goes into 5 one time, and 5 goes into 10 two times, you can cancel a 5:

$$\frac{3}{{}_2\cancel{10}}\cdot\frac{\cancel{5}^1}{4}=\frac{3}{8}$$

Also note that the original problem could have been written as $\frac{\frac{3}{10}}{\frac{4}{5}}$.

Adding fractions

You know that

$$\frac{2}{7}+\frac{3}{7}=\frac{2+3}{7}=\frac{5}{7}$$

You can add up these fractions like this because you already have a common denominator. It works the same with variables:

$$\frac{a}{c}+\frac{b}{c}=\frac{a+b}{c}$$

Notice that wherever you have a 2 in the top equation, an *a* is in the bottom equation; wherever a 3 is in the top equation, a *b* is in the bottom equation; and ditto for 7 and *c*. This illustrates a powerful principle:

TIP

Variables always behave exactly like numbers.

If you're wondering what to do with variables in a problem, ask yourself how you would do the problem if there were numbers in it instead of variables. Then do the problem with the variables the same way, like this:

$$\frac{a}{b}+\frac{c}{d}$$

You can't add these fractions like you did in the previous example because this problem has no common denominator. Now, assuming you're stumped, do the problem with numbers instead of variables. Remember how to add $\frac{2}{5}+\frac{3}{8}$? I'm not going to simplify each line of the solution. You'll see why in a minute.

1. **Find the *least common denominator* (actually, any common denominator will work when adding fractions), and convert the fractions.**

   The least common denominator is 5 times 8, or 40, so convert each fraction into 40ths:

   $$\frac{2}{5}+\frac{3}{8}$$
   $$=\frac{2}{5}\cdot\frac{8}{8}+\frac{3}{8}\cdot\frac{5}{5}$$
   $$=\frac{2\cdot8}{5\cdot8}+\frac{3\cdot5}{5\cdot8}$$ (8 · 5 equals 5 · 8 so you can reverse the order. These fractions are 40ths, but I want to leave the 5 · 8 in the denominators for now.)

2. **Add the numerators and keep the common denominator unchanged:**

   $$=\frac{2\cdot8+3\cdot5}{5\cdot8}\quad\left(\text{You can see this equals }\frac{16+15}{40}\text{, or }\frac{31}{40}.\right)$$

Now you're ready to do the original problem, $\frac{a}{b}+\frac{c}{d}$. In this problem, you have an a instead of a 2, a b instead of a 5, a c instead of a 3, and a d instead of an 8. Just carry out the exact same steps as you do when adding $\frac{2}{5}+\frac{3}{8}$. You can think of each of the numbers in the solution as stamped on one side of a coin with the corresponding variable stamped on the other side. For instance, there's a coin with a 2 on one side and an a on the opposite side; another coin has an 8 on one side and a d on the other side, and so on. Now, take each step of the previous solution, flip each coin over, and voilà, you've got the solution to the original problem. Here's the final answer:

$$\frac{ad+cb}{bd}$$

Subtracting fractions

Subtracting fractions works like adding fractions except instead of adding, you subtract. Insights like this are the reason they pay me the big bucks.

Canceling in fractions

Finishing calculus problems — after you've done all the calculus steps — sometimes requires some pretty messy algebra, including canceling. Make sure you know how to cancel and when you can and can't do it.

In the fraction $\frac{x^5y^2}{x^3z}$, three x's can be canceled from the numerator and denominator, resulting in the simplified fraction, $\frac{x^2y^2}{z}$. If you write out the x's instead of using exponents, you can more clearly see how this works:

$$\frac{x^5y^2}{x^3z} = \frac{x \cdot x \cdot x \cdot x \cdot x \cdot y \cdot y}{x \cdot x \cdot x \cdot z}$$

Now cancel three x's from the numerator and denominator:

$$\frac{\cancel{x} \cdot \cancel{x} \cdot \cancel{x} \cdot x \cdot x \cdot y \cdot y}{\cancel{x} \cdot \cancel{x} \cdot \cancel{x} \cdot z}$$

That leaves you with $\frac{x \cdot x \cdot y \cdot y}{z}$, or $\frac{x^2y^2}{z}$.

Express yourself

An *algebraic expression* or just *expression* is something like xyz or $a^2p^3\sqrt{q} - 6$, basically anything without an equal sign (if it has an equal sign, it's an *equation*). Canceling works the same way with expressions as it does for single variables. By the way, that's a tip not just for canceling, but for all algebra topics.

TIP

Expressions behave exactly like variables.

So, if each x in the preceding problem is replaced with $(xyz - q)$, you've got

$$\frac{(xyz - q)^5 y^2}{(xyz - q)^3 z},$$

and three instances of the expression $(xyz - q)$ cancel from the numerator and denominator, just as the three x's canceled. The simplified result is

$$\frac{(xyz - q)^2 y^2}{z}.$$

The multiplication rule for canceling

Now you know *how* to cancel. You also need to know *when* you can cancel.

REMEMBER

The multiplication rule: You can cancel in a fraction only when it has an *unbroken chain of multiplication* through the entire numerator and the entire denominator.

Canceling is allowed in a fraction like this:

$$\frac{a^2b^3(xy-pq)^5(c+d)}{ab^7z(xy-pq)^3}.$$

Think of multiplication as something that conducts electricity. Electrical current can flow from one end of the numerator to the other, from the a^2 to the $(c+d)$, because all the variables and expressions are connected with multiplication. (Note that an addition or subtraction sign inside parentheses — the "+" in $(c+d)$ for instance — doesn't break the flow of current.) Because the denominator also has an unbroken chain of multiplication, canceling is allowed. You can cancel one a, three b's, and three of the expression $(xy-pq)$. Here's the result:

$$\frac{a(xy-pq)^2(c+d)}{b^4z}$$

WARNING

When you *can't* cancel: Adding an innocuous-looking 1 to the numerator (or denominator) of the original fraction changes everything:

$$\frac{a^2b^3(xy-pq)^5(c+d)+1}{ab^7z(xy-pq)^3}$$

The addition sign in front of the 1 breaks the flow of current, and no canceling is allowed anywhere in the fraction.

I should point out a situation that appears to violate the multiplication rule for canceling. You *can* cancel in a fraction like the following:

$$\frac{15xy+25a-35b}{5p-10q}$$

You can cancel a 5 out of all five terms, resulting in

$$\frac{3xy+5a-7b}{p-2q}$$

This type of canceling is allowed only when you can cancel the same thing out of *every* term in the numerator and denominator. This problem appears to violate the multiplication rule, but it really doesn't, because if you first factor the numerator and denominator like this —

$$\frac{5(3xy+5a-7b)}{5(p-2q)}$$

— you arrive at a fraction that does satisfy the multiplication rule, and you can then cancel the 5's, producing the same answer.

EXAMPLE

Q. Solve: $\frac{a^2b}{d} \cdot \frac{a^3bc}{d} = ?$

A. $\boldsymbol{\frac{a^5b^2c}{d^2}}$. To multiply fractions, you just multiply straight across. You do *not* cross-multiply in a diagonal direction! And you do *not* use a common denominator!

Q. Solve: $\frac{5a}{bc} \div \frac{35a^3}{cd} = ?$

A. $\boldsymbol{\frac{5a}{bc} \div \frac{35a^3}{cd} = \frac{5a}{bc} \cdot \frac{cd}{35a^3} = \frac{5acd}{35a^3bc} = \frac{d}{7a^2b}}$. To divide fractions, you flip the second one, and then multiply.

YOUR TURN

1 Solve: $\frac{12}{0} = ?$

2 Solve: $\frac{0}{10} = ?$

3 Does $\frac{3a+b}{3a+c}$ equal $\frac{a+b}{a+c}$? Why or why not?

4 Does $\frac{3a+b}{3a+c}$ equal $\frac{b}{c}$? Why or why not?

5 Does $\frac{4ab}{4ac}$ equal $\frac{ab}{ac}$? Why or why not?

6 Does $\frac{4ab}{4ac}$ equal $\frac{b}{c}$? Why or why not?

Miscellaneous Algebra

If I could have thought of a nice, mathematical term for the group of topics that follow, I would have used it. But I couldn't. Hence, the oh-so-descriptive *miscellaneous*.

Absolute value — absolutely easy

Absolute value just turns a negative number into a positive and does nothing to a positive number or zero. For example,

$$|-6| = 6, \ |3| = 3, \text{ and } |0| = 0$$

It's a bit trickier when dealing with variables. If x is zero or positive, then the absolute value bars do nothing, and thus,

$$|x| = x$$

But if x is negative, the absolute value of x is positive, and you write

$$|x| = -x$$

For example, if $x = -5$, $|-5| = -(-5) = 5$

REMEMBER

$-x$ can be a positive number. When x is a negative number, $-x$ (read as "negative x," or "the opposite of x") is a *positive*.

Empowering your powers

You are power*less* in calculus if you don't know the power rules:

- $x^0 = 1$

  This is the rule regardless of what x equals — a fraction, a negative, anything — except for zero (zero raised to the zero power is undefined). Let's call it the kitchen sink rule (where the kitchen sink represents zero):

  $(\text{everything but the kitchen sink})^0 = 1$

» $x^{-3} = \frac{1}{x^3}$ and $x^{-a} = \frac{1}{x^a}$

For example, $4^{-2} = \frac{1}{4^2} = \frac{1}{16}$. This is huge! Don't forget it! Note that the power is negative, but the answer of $\frac{1}{16}$ is *not* negative.

» $x^{2/3} = \left(\sqrt[3]{x}\right)^2 = \sqrt[3]{x^2}$ and $x^{a/b} = \left(\sqrt[b]{x}\right)^a = \sqrt[b]{x^a}$

You can use this handy rule backwards to convert a root problem into an easier power problem.

» $x^2 \cdot x^3 = x^5$ and $x^a \cdot x^b = x^{a+b}$

You *add* the powers here. (By the way, you can't do anything to x^2 *plus* x^3. You can't add x^2 to x^3 because they're not *like terms*. You can only add or subtract terms when the variable part of each term is the same, for instance, $3xy^2z + 4xy^2z = 7xy^2z$. This works for exactly the same reason — I'm not kidding — that 3 chairs plus 4 chairs is 7 chairs; and you can't add *unlike* terms, just like you can't add 5 chairs plus 2 cars.)

» $\frac{x^5}{x^3} = x^2$ and $\frac{x^2}{x^6} = x^{-4}$ and $\frac{x^a}{x^b} = x^{a-b}$

Here you *subtract* the powers.

» $\left(x^2\right)^3 = x^6$ and $\left(x^a\right)^b = x^{ab}$

You *multiply* the powers here.

» $(xyz)^3 = x^3y^3z^3$ and $(xyz)^a = x^ay^az^a$

Here you *distribute* the power to each variable.

» $\left(\frac{x}{y}\right)^4 = \frac{x^4}{y^4}$ and $\left(\frac{x}{y}\right)^a = \frac{x^a}{y^a}$

Here you also distribute the power to each variable.

WARNING

» $(x+y)^2 = x^2 + y^2$. **NOT!**

Do *not* distribute the power in this case. Instead, multiply it out the long way: $(x+y)^2 = (x+y)(x+y) = x^2 + xy + xy + y^2 = x^2 + 2xy + y^2$. Watch what happens if you erroneously use the preceding "law" with numbers: $(3+5)^2$ equals 8^2, or 64, *not* $3^2 + 5^2$, which equals $9 + 25$, or 34.

Rooting for roots

Roots, especially square roots, come up all the time in calculus. So, knowing how they work and understanding the fundamental connection between roots and powers is essential. And, of course, that's what I'm about to tell you.

Roots rule — make that, root rules

Any root can be converted into a power, for example, $\sqrt[3]{x} = x^{1/3}$, $\sqrt{x} = x^{1/2}$, and $\sqrt[4]{x^3} = x^{3/4}$. So, if you get a problem with roots in it, you can just convert each root into a power and use the power rules instead to solve the problem (this is a very useful technique). Because you have this option, the following root rules are less important than the power rules, but you really should know them anyway:

- $\sqrt{0} = 0$ and $\sqrt{1} = 1$

  But you knew that, right?

REMEMBER

  No negatives under even roots. You can't have a negative number under a square root or under any other *even* number root — at least not in basic calculus.

- $\sqrt{a} \cdot \sqrt{b} = \sqrt{a \cdot b}$, $\sqrt[3]{a} \cdot \sqrt[3]{b} = \sqrt[3]{ab}$, and $\sqrt[n]{a} \cdot \sqrt[n]{b} = \sqrt[n]{ab}$
- $\dfrac{\sqrt{a}}{\sqrt{b}} = \sqrt{\dfrac{a}{b}}$, $\dfrac{\sqrt[3]{a}}{\sqrt[3]{b}} = \sqrt[3]{\dfrac{a}{b}}$, and $\dfrac{\sqrt[n]{a}}{\sqrt[n]{b}} = \sqrt[n]{\dfrac{a}{b}}$
- $\sqrt[3]{\sqrt[4]{a}} = \sqrt[12]{a}$ and $\sqrt[m]{\sqrt[n]{a}} = \sqrt[mn]{a}$

  You *multiply* the root indexes.

REMEMBER

- $\sqrt{a^2} = |a|$, $\sqrt[4]{a^4} = |a|$, $\sqrt[6]{a^6} = |a|$, and so on.

  If you have an *even* number root, you need the absolute value bars on the answer, because whether *a* is positive or negative, the answer is positive. If it's an odd number root, you don't need the absolute value bars. Thus,

- $\sqrt[3]{a^3} = a$, $\sqrt[5]{a^5} = a$, and so on.

WARNING

- $\sqrt{a^2 + b^2} = a + b$. **NOT!**

  Make this mistake and go directly to jail. Try solving it with numbers: $\sqrt{2^2 + 3^2} = \sqrt{13}$, which does *not* equal $2 + 3$.

Simplifying roots

Here are two last things on roots. First, you need to know the two methods for simplifying roots like $\sqrt{300}$ or $\sqrt{504}$.

The quick method works for $\sqrt{300}$ because it's easy to see a large perfect square, 100, that goes into 300. Because 300 equals 100 times 3, the 100 comes out as its square root, 10, leaving the 3 inside the square root. The answer is thus $10\sqrt{3}$.

For $\sqrt{504}$, it's not as easy to find a large perfect square that goes into 504, so you've got to use the longer method:

1. **Break 504 down into a product of all of its prime factors.**

   $\sqrt{504} = \sqrt{2 \cdot 2 \cdot 2 \cdot 3 \cdot 3 \cdot 7}$

2. **Circle each pair of numbers.**

   $\sqrt{\textcircled{2 \cdot 2} \cdot 2 \cdot \textcircled{3 \cdot 3} \cdot 7}$

3. **For each circled pair, take one number out.**

   $2 \cdot 3\sqrt{2 \cdot 7}$

4. **Simplify.**

$6\sqrt{14}$

The last thing about roots is that, by convention, you don't leave a root in the denominator of a fraction — it's a silly, anachronistic convention, but it's still being taught, so here it is. If your answer is, say, $\frac{2}{\sqrt{3}}$, you multiply it by $\frac{\sqrt{3}}{\sqrt{3}}$:

$$\frac{2}{\sqrt{3}} \cdot \frac{\sqrt{3}}{\sqrt{3}} = \frac{2\sqrt{3}}{3}$$

Logarithms — this is not an event at a lumberjack competition

A *logarithm* is just a different way of expressing an exponential relationship between numbers. For instance,

$2^3 = 8$, so,

$\log_2 8 = 3$ (read as "log base 2 of 8 equals 3").

These two equations say precisely the same thing. You could think of $2^3 = 8$ as the way you write it in English and $\log_2 8 = 3$ as the way you would write it in Latin. And because it's easier to think and do math in English, make sure — when you see something like $\log_3 81 = x$ — that you can instantly "translate" it into $3^x = 81$. The base of a logarithm can be any number greater than zero other than 1, and by convention, if the base is 10, you don't write it. For example, $\log 1000 = 3$ means $\log_{10} 1000 = 3$. Also, log base *e* ($e \approx 2.72$) is written *ln* instead of $\log_e$.

You should know the following logarithm properties:

- $\log_c 1 = 0$
- $\log_c c = 1$
- $\log_c (ab) = \log_c a + \log_c b$
- $\log_c \left(\frac{a}{b}\right) = \log_c a - \log_c b$
- $\log_c a^b = b \log_c a$
- $\log_a b = \frac{\log_c b}{\log_c a}$

  With this property, you can compute something like $\log_3 20$ on a calculator that only has log buttons for base 10 (the "log" button) and base *e* (the "ln" button) by entering $\frac{\log 20}{\log 3}$, using base 10 for *c*. On many newer-model calculators, you can compute $\log_3 20$ directly.
- $\log_a a^b = b$
- $a^{\log_a b} = b$

Factoring schmactoring — when am I ever going to need it?

When are you ever going to need it? For calculus, that's when.

Factoring means "unmultiplying," like rewriting 12 as $2 \cdot 2 \cdot 3$. You won't run across problems like that in calculus, however. For calculus, you need to be able to factor algebraic expressions, like factoring $5xy + 10yz$ as $5y(x + 2z)$. Algebraic factoring always involves rewriting a *sum* (or *difference*) of terms as a *product.* What follows is a quick refresher course.

Pulling out the greatest common factor

The first step in factoring any type of expression is to pull out — in other words, factor out — the greatest thing that all of the terms have in common — that's the *greatest common factor* or GCF. For example, each of the three terms of $8x^3y^4 + 12x^2y^5 + 20x^4y^3z$ contains the factor $4x^2y^3$, so it can be pulled out like this: $4x^2y^3\left(2xy + 3y^2 + 5x^2z\right)$. Make sure you always look for a GCF to pull out before trying other factoring techniques.

Looking for a pattern

After pulling out the GCF if there is one, the next thing to do is to look for one of the following three patterns. The first pattern is *huge;* the next two are much less important.

DIFFERENCE OF SQUARES

Knowing how to factor the *difference of squares* is critical:

$$a^2 - b^2 = (a - b)(a + b)$$

If you can rewrite something like $9x^4 - 25$ so that it looks like $(\text{this})^2 - (\text{that})^2$, then you can use this factoring pattern. Here's how:

$$9x^4 - 25 = \left(3x^2\right)^2 - (5)^2$$

Now, because $(\text{this})^2 - (\text{that})^2 = (\text{this} - \text{that})(\text{this} + \text{that})$, you can factor the problem:

$$\left(3x^2\right)^2 - (5)^2 = \left(3x^2 - 5\right)\left(3x^2 + 5\right)$$

WARNING

A *difference* of squares, $a^2 - b^2$, can be factored, but a *sum* of squares, $a^2 + b^2$, *cannot* be factored. In other words, $a^2 + b^2$, like the numbers 7 and 13, is *prime* — you can't break it up.

SUM AND DIFFERENCE OF CUBES

You might also want to memorize the factor rules for the *sum* and *difference of cubes:*

$$a^3 + b^3 = (a + b)\left(a^2 - ab + b^2\right)$$
$$a^3 - b^3 = (a - b)\left(a^2 + ab + b^2\right)$$

Trying some trinomial factoring

Remember regular old trinomial factoring from your algebra days?

REMEMBER

Several definitions: A *trinomial* is a polynomial with three terms. A *polynomial* is an expression like $4x^5-6x^3+x^2-5x+2$ where, except for the *constant* (the 2 in this example), all the terms have a variable raised to a positive *integral* power. In other words, no fraction powers or negative powers allowed (so, $\frac{1}{x}$ is not a polynomial because it equals x^{-1}). And no radicals, no logs, no sines or cosines, or anything else — just terms with a *coefficient*, like the 4 in $4x^5$, multiplied by a variable raised to a positive integral power. The *degree* of a polynomial is the polynomial's highest power of *x*. The polynomial at the beginning of this paragraph, for instance, has a degree of 5.

It wouldn't be a bad idea to get back up to speed with problems like

$$6x^2+13x-5=(2x+5)(3x-1)$$

where you have to factor the trinomial on the left into the product of the two binomials on the right. A few standard techniques for factoring a trinomial like this are floating around the mathematical ether — you probably learned one or more of them in your algebra class. If you remember one of the techniques, great. You won't have to do a lot of trinomial factoring in calculus, but it does come in handy now and then, so, if your skills are a bit rusty, check out *Algebra II For Dummies*, by Mary Jane Sterling (Wiley).

Solving quadratic equations

A quadratic equation is any *second-degree* polynomial equation — that's when the highest power of *x*, or whatever other variable is used, is 2.

You can solve quadratic equations by one of three basic methods.

Method 1: Factoring

Solve $2x^2-5x=12$.

1. **Bring all terms to one side of the equation, leaving a zero on the other side.**

   $$2x^2-5x-12=0$$

2. **Factor.**

   $$(2x+3)(x-4)=0$$

   You can check that these factors are correct by multiplying them. Does FOIL (First, Outer, Inner, Last) ring a bell?

3. **Set each factor equal to zero and solve (using the *zero product property*).**

$$2x+3=0 \quad \text{or} \quad x-4=0$$
$$2x=-3 \qquad\qquad x=4$$
$$x=-\frac{3}{2}$$

So, this equation has two solutions: $x=-\frac{3}{2}$ and $x=4$.

TIP

The *discriminant* tells you whether a quadratic is factorable. Method 1 works only if the quadratic is factorable. The quick test for that is a snap. A quadratic is factorable if the discriminant, b^2-4ac, is a perfect square number like 0, 1, 4, 9, 16, 25, and so on. (The discriminant is the stuff under the square root symbol in the quadratic formula — see Method 2 in the next section.) In the quadratic equation from Step 1, $2x^2-5x-12=0$, for example, $a=2$, $b=-5$, and $c=-12$; b^2-4ac equals, therefore, $(-5)^2-4(2)(-12)$, which equals 121. Because 121 is a perfect square $\left(11^2\right)$, the quadratic is factorable. Because trinomial factoring is often so quick and easy, you may choose to just dive into the problem and try to factor it without bothering to check the value of the discriminant. But if you get stuck, it's not a bad idea to check the discriminant so you don't waste more time trying to factor an unfactorable quadratic trinomial. (But whether or not the quadratic is factorable, you can always solve it with the quadratic formula.)

Method 2: The quadratic formula

The solution or solutions of a quadratic equation, $ax^2+bx+c=0$, are given by the quadratic formula:

$$x=\frac{-b\pm\sqrt{b^2-4ac}}{2a}$$

Now solve the same equation from Method 1 with the quadratic formula:

1. **Bring all terms to one side of the equation, leaving a zero on the other side.**

   $2x^2-5x-12=0$

2. **Plug the coefficients into the formula.**

   In this example, a equals 2, b is –5, and c is –12, so

$$\begin{aligned}
x&=\frac{-(-5)\pm\sqrt{(-5)^2-4(2)(-12)}}{2\cdot 2}\\
&=\frac{5\pm\sqrt{25-(-96)}}{4}\\
&=\frac{5\pm\sqrt{121}}{4}\\
&=\frac{5\pm 11}{4}\\
&=\frac{16}{4} \text{ or } -\frac{6}{4}
\end{aligned}$$

$$x=4 \text{ or } -\frac{3}{2}$$

This agrees with the solutions obtained previously — the solutions better be the same because you're solving the same equation.

Method 3: Completing the square

The third method of solving quadratic equations is called *completing the square* because it involves creating a perfect square trinomial that you can solve by taking its square root.

Solve $3x^2 = 24x + 27$.

1. **Put the x^2 and the x terms on one side and the constant on the other.**

   $3x^2 - 24x = 27$

2. **Divide both sides by the coefficient of x^2 (unless, of course, it's 1).**

   $x^2 - 8x = 9$

3. **Take half of the coefficient of x, square it, then add that to both sides.**

   Half of –8 is –4 and $(-4)^2$ is 16, so add 16 to both sides:

   $x^2 - 8x + 16 = 9 + 16$

4. **Factor the left side into a binomial squared. Notice that the factor always contains the same number you found in Step 3 (–4 in this example).**

   $(x-4)^2 = 25$

5. **Take the square root of both sides, remembering to put a ± sign on the right side.**

   $$\sqrt{(x-4)^2} = \sqrt{25}$$
   $$x - 4 = \pm 5$$

6. **Solve.**

   $$x = 4 \pm 5$$
   $$x = 9 \text{ or } -1$$

EXAMPLE

Q. Factor $9x^4 - y^6$.

A. $\mathbf{9x^4 - y^6 = (3x^2 - y^3)(3x^2 + y^3)}$. This is an example of the single most important factor pattern: $a^2 - b^2 = (a-b)(a+b)$. Make sure you know it! In this problem, $3x^2$ is your *a* and y^3 is your *b*. That's all there is to it.

Q. Rewrite $x^{2/5}$ without a fraction power.

A. $\mathbf{\sqrt[5]{x^2}}$ **or** $\mathbf{(\sqrt[5]{x})^2}$. Don't forget how fraction powers work!

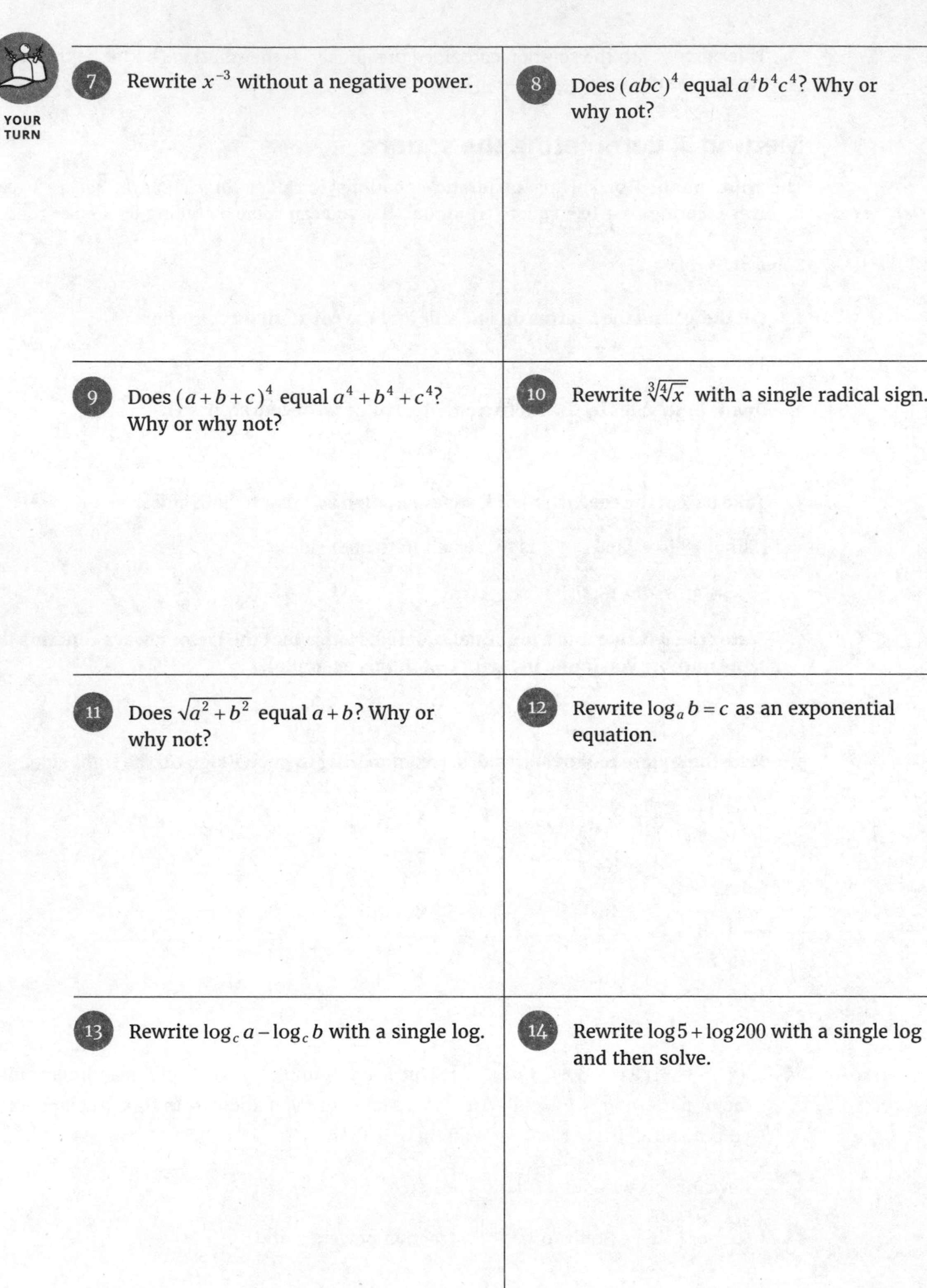

YOUR TURN

7 Rewrite x^{-3} without a negative power.

8 Does $(abc)^4$ equal $a^4b^4c^4$? Why or why not?

9 Does $(a+b+c)^4$ equal $a^4+b^4+c^4$? Why or why not?

10 Rewrite $\sqrt[3]{\sqrt[4]{x}}$ with a single radical sign.

11 Does $\sqrt{a^2+b^2}$ equal $a+b$? Why or why not?

12 Rewrite $\log_a b = c$ as an exponential equation.

13 Rewrite $\log_c a - \log_c b$ with a single log.

14 Rewrite $\log 5 + \log 200$ with a single log and then solve.

15 If $5x^2 = 3x + 8$, solve for x with the quadratic formula.

16 Solve: $|3x + 2| > 14$.

17 Solve: $-3^2 - x^0 + \sqrt{0} - |-1| - 1^0 - 0^1 = ?$

18 Simplify $\sqrt[3]{p^6 q^{15}}$.

19 Simplify $\left(\frac{8}{27}\right)^{-4/3}$.

20 Factor $-x^{10} + 16$ over the set of integers.

Geometry Refresher

You can use calculus to solve many real-world problems that involve two- or three-dimensional shapes and various curves, surfaces, and volumes — such as calculating the rate at which the water level is falling in a cone-shaped tank or determining the dimensions that maximize the volume of a cylindrical soup can. So the geometry formulas for perimeter, area, volume, surface area, and so on will come in handy. You should also know things like the Pythagorean Theorem, proportional shapes, and basic coordinate geometry, like the midpoint and distance formulas. Finally, make sure you know your $45° - 45° - 90°$ and $30° - 60° - 90°$ triangles.

Handy-dandy geometry formulas

The following formulas might come in handy in your calculus studies.

Formulas for two-dimensional shapes

- Pythagorean Theorem for right triangles: $a^2 + b^2 = c^2$
- $Area_{Triangle} = \frac{1}{2} base \cdot height$
- $Area_{Parallelogram} = base \cdot height$ (Works for rhombuses, rectangles, and squares.)
- $Area_{Kite} = \frac{1}{2} diagonal_1 \cdot diagonal_2$ (Works for rhombuses and squares.)
- $Area_{Trapezoid} = \left(\frac{base_1 + base_2}{2} \right) \cdot height$

Formulas for three-dimensional shapes

- $Volume_{Box} = length \cdot width \cdot height$
- $Volume_{Cylinder} = area_{base} \cdot height$
- $Volume_{Cone} = \frac{1}{3} area_{base} \cdot height$
- $Volume_{Pyramid} = \frac{1}{3} area_{base} \cdot height$
- $Volume_{Sphere} = \frac{4}{3} \pi r^3$
- $Surface\ Area_{Box} =$ just add up the areas of all the faces
- $Surface\ Area_{Cylinder} = 2 \cdot area_{base} + circumference \cdot height$
- $Surface\ Area_{Cone} = area_{base} + \frac{1}{2} circumference \cdot slant\ height$
- $Surface\ Area_{Pyramid} =$ just add up the areas of all the faces
- $Surface\ Area_{Sphere} = 4\pi r^2$

Coordinate geometry formulas

- $Slope = \frac{rise}{run} = \frac{y_2 - y_1}{x_2 - x_1}$
- $Midpoint = \left(\frac{x_1 + x_2}{2}, \frac{y_1 + y_2}{2} \right)$ (This is simply the average of the *x*'s and the average of the *y*'s.)
- $Distance = \sqrt{(x_2 - x_1)^2 + (y_2 - y_1)^2}$ (This is mathematically equivalent to the Pythagorean Theorem.)

Two special right triangles

Because so many garden-variety calculus problems involve 30-, 45-, and 60-degree angles, it's a good idea to memorize the two right triangles in Figure 4-1.

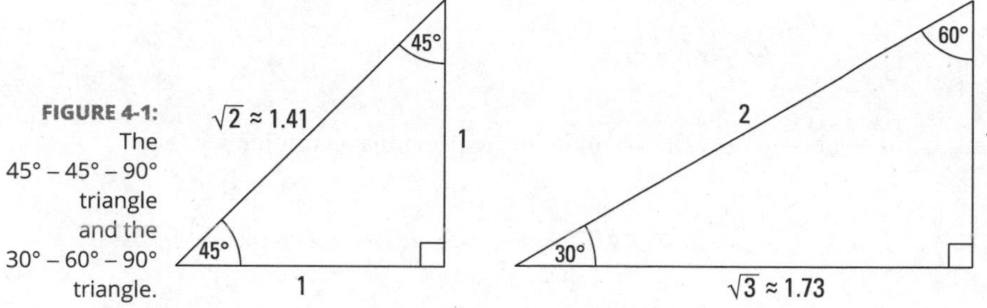

FIGURE 4-1: The 45° – 45° – 90° triangle and the 30° – 60° – 90° triangle.

The 45° – 45° – 90° triangle

Every 45° – 45° – 90° triangle is the shape of a square cut in half along its diagonal. The 45° – 45° – 90° triangle in Figure 4-1 is half of a 1-by-1 square. The Pythagorean Theorem gives you the length of its hypotenuse, $\sqrt{2}$, or about 1.41.

The 30° – 60° – 90° triangle

Every 30° – 60° – 90° triangle is half of an equilateral triangle cut straight down the middle along its altitude.

The 30° – 60° – 90° triangle in Figure 4-1 is half of a 2-by-2-by-2 equilateral triangle. It has legs of lengths 1 and $\sqrt{3}$ (about 1.73), and a 2-unit-long hypotenuse.

WARNING

Don't make the common error of switching the 2 with the $\sqrt{3}$ in a 30° – 60° – 90° triangle. Remember that 2 is more than $\sqrt{3}$ ($\sqrt{4}$ equals 2, so $\sqrt{3}$ be must be less than 2) and that the hypotenuse is always the longest side of a right triangle.

TIP

When you sketch a 30° – 60° – 90° triangle, exaggerate the fact that it's wider than it is tall (or taller than wide if you tip it up). This makes it obvious that the shortest side (length of 1) is opposite the smallest angle (30°).

EXAMPLE

Q. What's the area of the triangle in the following figure?

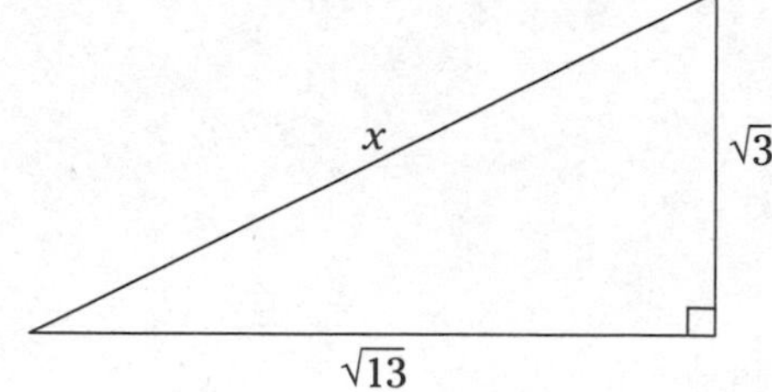

A. $\frac{\sqrt{39}}{2}$

$$\begin{aligned} Area_{Triangle} &= \frac{1}{2} base \cdot height \\ &= \frac{1}{2} \cdot \sqrt{13}\sqrt{3} \\ &= \frac{\sqrt{39}}{2} \end{aligned}$$

Q. How long is the hypotenuse of the triangle in the previous example?

A. $x = 4$.

$$\begin{aligned} a^2 + b^2 &= c^2 \\ x^2 &= a^2 + b^2 \\ x^2 &= \sqrt{13}^2 + \sqrt{3}^2 \\ x^2 &= 13 + 3 \\ x^2 &= 16 \\ x &= 4 \end{aligned}$$

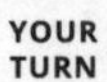

21 Fill in the two missing lengths for the sides of the triangle in the following figure.

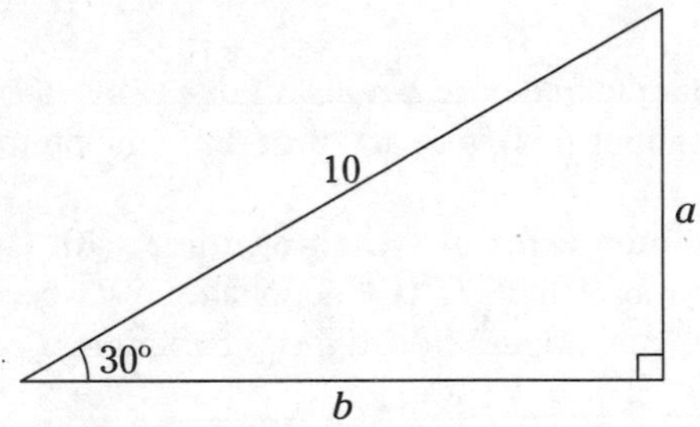

22 What are the lengths of the two missing sides of the triangle in the following figure?

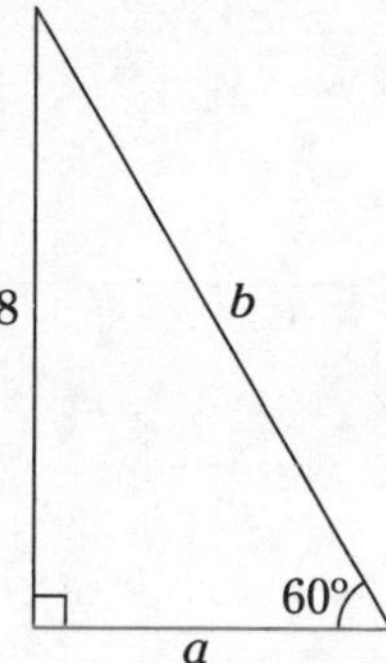

23 Fill in the missing lengths for the sides of the triangle in the following figure.

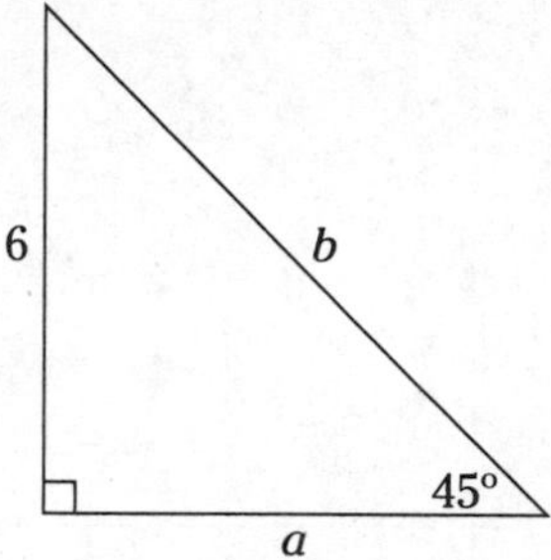

24 **(a)** What's the total area of the pentagon in the following figure (the shape on the left is a square)?

(b) What's the perimeter?

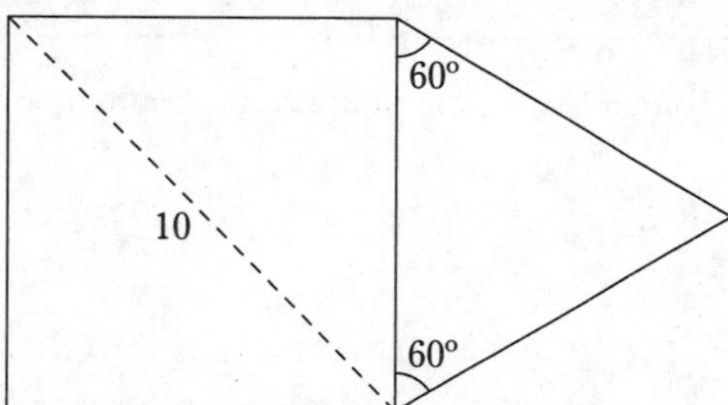

25 Compute the area of the parallelogram in the following figure.

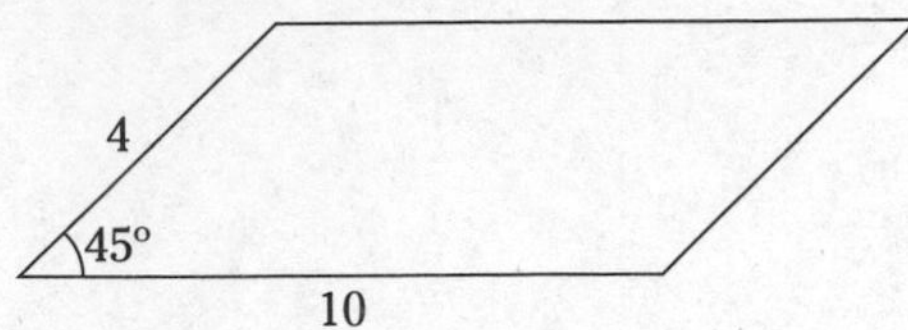

26 What's the slope of $\overline{PQ}$?

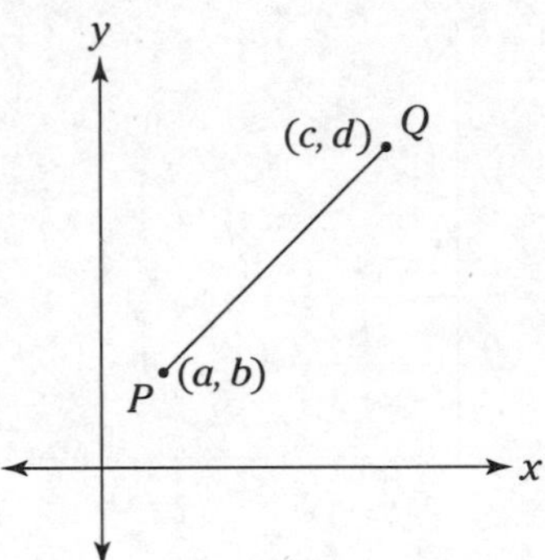

27 How far is it from P to Q in the figure from Problem 26?

28 What are the coordinates of the midpoint of $\overline{PQ}$ in the figure from Problem 26?

29 What's the length of altitude of triangle *ABC* in the following figure?

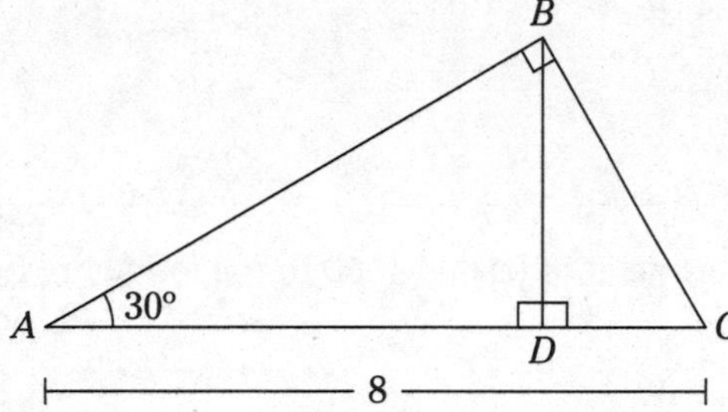

30 What's the perimeter of triangle *ABD* in the figure for Problem 29?

31 What's the area of quadrilateral *PQRS* in the following figure?

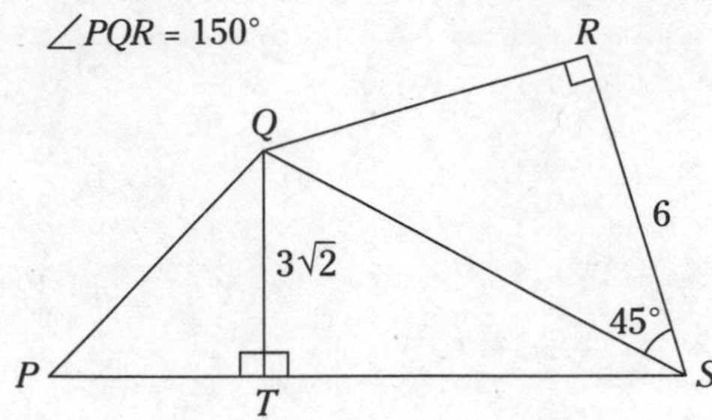

32 What's the perimeter of triangle *BCD* in the following figure?

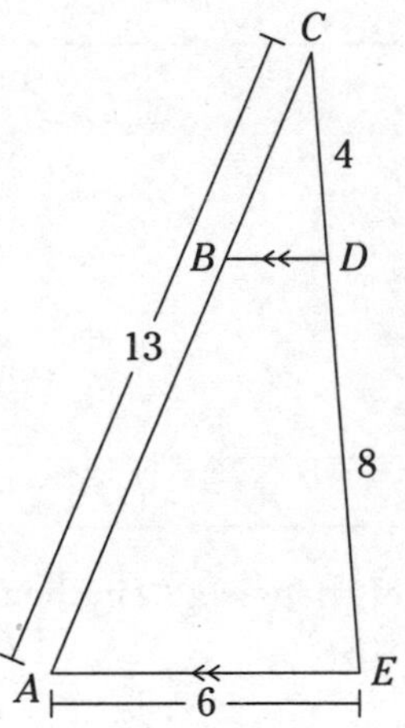

33 What's the ratio of the area of triangle *BCD* to the area of triangle *ACE* in the figure for Problem 32?

34 In the following figure, what's the area of parallelogram *PQRS* in terms of *x* and *y*?

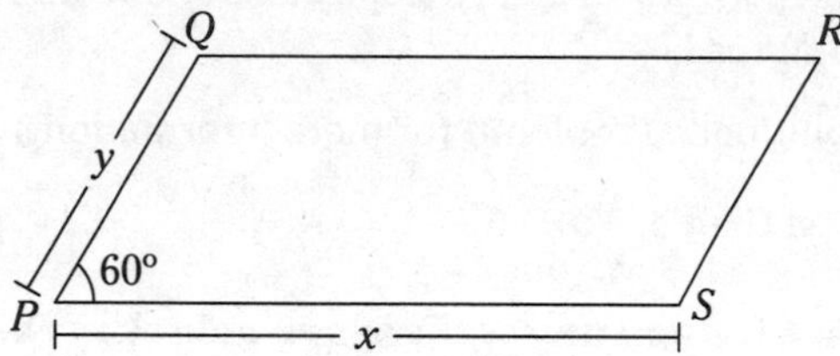

Practice Questions Answers and Explanations

1. $\frac{12}{0}$ **is undefined!** Don't mix this up with something like $\frac{0}{8}$, which equals zero.

   Here's a great way to think about this problem and fractions in general. Consider the following simple division or fraction problem: $\frac{8}{2}=4$. Note the *multiplication* problem implicit here: 2 times 4 is 8. This multiplication idea is a great way to think about how fractions work. So in the current problem, you can consider $\frac{12}{0}=$ ____, and use the multiplication idea: 0 times ______ equals 12. What works in the blank? Nothing, obviously, because 0 times anything is 0. The answer, therefore, is undefined.

   Note that if you think about these two fractions as examples of slope $\left(\frac{rise}{run}\right)$, $\frac{12}{0}$ has a rise of 12 and a run of 0, which gives you a *vertical* line that has sort of an infinite steepness or slope (that's why it's undefined). Or just remember that it's impossible to drive up a vertical road, so it's impossible to come up with a slope for a vertical line. The fraction $\frac{0}{8}$, on the other hand, has a rise of 0 and a run of 8, which gives you a *horizontal* line that has no steepness at all and thus has the perfectly ordinary slope of zero. Of course, it's also perfectly ordinary to drive on a horizontal road.

2. $\frac{0}{10}=0$. (See the solution to Problem 1 for more information.)

3. **No. You can't cancel the 3's.**

WARNING

You can't cancel in a fraction unless there's an unbroken chain of multiplication running across the entire numerator and the entire denominator.

4. **No, they're not equal.** You can't cancel the $3a$'s. (See the warning in Problem 3.) You can also just test this problem with numbers: Does $\frac{3\cdot4+5}{3\cdot4+6}=\frac{5}{6}$? No, they're not equal.

5. **Yes, they're equal.** You can cancel the 4's because the entire numerator and the entire denominator are connected with multiplication.

6. **Yes, they're equal.** You can cancel the $4a$'s.

7. $\frac{1}{x^3}$.

8. **Yes.** Exponents do distribute over multiplication.

9. **No!** Exponents do *not* distribute over addition (or subtraction).

TIP

Use numbers instead of variables! When you're working a problem and can't remember the algebra rule, try the problem with numbers instead of variables. Just replace the variables with simple, round numbers and work out the numerical problem. (Don't use 0, 1, or 2 because they have special properties that can mess up your test.) Whatever works for the numbers will work with variables, and whatever doesn't work with numbers won't work with variables. Watch what happens if you try this problem with numbers:

$$(3+4+6)^4 \stackrel{?}{=} 3^4+4^4+6^4$$
$$13^4 \stackrel{?}{=} 81+256+1296$$
$$28{,}561 \neq 1633$$

10 $\sqrt[12]{x}$.

11 **No!** The explanation is basically the same as for Problem 9. Consider this: If you turn the root into a power, you get $\sqrt{a^2+b^2}=\left(a^2+b^2\right)^{1/2}$. But because you *can't* distribute the power over addition, $\left(a^2+b^2\right)^{1/2} \neq \left(a^2\right)^{1/2}+\left(b^2\right)^{1/2}$, or $a+b$, and thus $\sqrt{a^2+b^2} \neq a+b$.

12 $a^c = b$.

13 $\log_c \frac{a}{b}$.

14 $\log 5 + \log 200 = \log(5 \cdot 200) = \log 1000 = 3$.

When you see "log" without a base number, the base is 10.

REMEMBER

15 $x = \frac{8}{5}$ or -1.

Start by rearranging $5x^2 = 3x+8$ into $5x^2-3x-8=0$ because when solving a quadratic equation, you want just a zero on one side of the equation.

The quadratic formula tells you that $x=\frac{-b\pm\sqrt{b^2-4ac}}{2a}$. Plugging 5 into a, -3 into b, and -8 into c gives you $x=\frac{-(-3)\pm\sqrt{(-3)^2-4(5)(-8)}}{2\cdot 5}=\frac{3\pm\sqrt{9+160}}{10}=\frac{3\pm 13}{10}=\frac{16}{10}$ or $\frac{-10}{10}$, so $x=\frac{8}{5}$ or -1.

16 $x=-\frac{16}{3}$, $x=4$, *or* $x<-\frac{16}{3} \cup x>4$.

1. **Turn the inequality into an equation:**

$$|3x+2|=14$$

2. **Solve the absolute value equation.**

$$3x+2=14 \qquad \text{or} \qquad 3x+2=-14$$
$$3x=12 \qquad\qquad 3x=-16$$
$$x=4 \qquad\qquad x=-\frac{16}{3}$$

3. **Place both solutions on a number line (see the following figure).**

(You use hollow dots for > and <; if the problem had involved ≥ or ≤, you would use solid dots.)

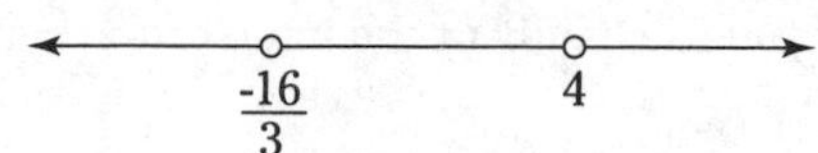

4. **Test a number from each of the three regions on the line (left of the left dot, between the dots, and right of the right dot) in the original inequality.**

For this problem, you can use –10, 0, and 10.

$$|3\cdot(-10)+2| \overset{?}{>} 14$$
$$|-28| \overset{?}{>} 14$$
$$28 \overset{?}{>} 14$$

True, so you shade the left-most region.

$$|3\cdot(0)+2| \overset{?}{>} 14$$
$$2 \overset{?}{>} 14$$

False, so you don't shade the middle region.

$$|3\cdot(10)+2| \overset{?}{>} 14$$
$$|32| \overset{?}{>} 14$$
$$32 \overset{?}{>} 14$$

True, so you shade the region on the right. The following figure shows the result. *x* can be any number where the line is shaded. That's your final answer.

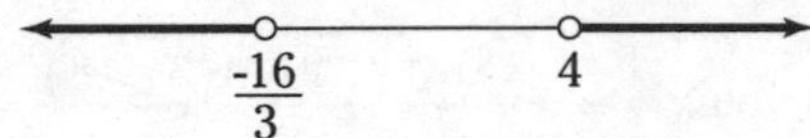

5. **You may also want to express the answer symbolically.**

Because *x* can equal a number in the left region *or* a number in the right region, this is an *or* solution, which means *union* (∪). When you want to include everything from both regions on the number line, you want the union of the two regions. So, the symbolic answer is

$$x < -\frac{16}{3} \cup x > 4$$

(You can write this using the word "or" instead of the union symbol.) If only the middle region were shaded, you'd have an *and* or *intersection* problem (∩). Using the number line points in this example, you would write the middle-region solution like this:

$$x > -\frac{16}{3} \cap x < 4$$

(You can use the word "and" instead of the intersection symbol.) Note that in this solution (whether you use "and" or the intersection symbol), the two inequalities overlap or intersect in the middle region. You can avoid the intersection issue by simply writing the solution as

$$-\frac{16}{3} < x < 4$$

You say "to-*may*-to," I say "to-*mah*-to."

17 **The answer is −12.**

Funny-looking problem, eh? It's just meant to help you review a few basics. Take a look at the six terms:

Don't forget, $-3^2 = -9$. If you want to square a negative number, you have to put it in parentheses: $(-3)^2 = 9$. Next, anything to the zero power (including a variable) equals 1. That takes care of the second and fifth chunks of the problem. The square root of zero is just zero, of course, because zero squared equals zero. And you know that the absolute value of −1 is 1; you just have to be careful not to goof up with all those negative signs and subtraction signs. Finally, zero to any *positive* power equals zero. That does it:

$$\begin{aligned}&-3^2 - x^0 + \sqrt{0} - |-1| - 1^0 - 0^1\\ &= -9 - 1 + 0 - 1 - 1 - 0\\ &= -12\end{aligned}$$

18 **The answer is p^2q^5.**

Most people prefer working with power rules to working with root rules, so that's the way I solve the problem here. First, rewrite the root as a power: $\sqrt[3]{p^6q^{15}} = \left(p^6q^{15}\right)^{1/3}$. Now, just distribute the power to the p^6 and the q^6, and then use the power-to-a-power rule:

$$\begin{aligned}&\left(p^6q^{15}\right)^{1/3}\\ &= \left(p^6\right)^{1/3}\left(q^{15}\right)^{1/3}\\ &= p^{6(1/3)}q^{15(1/3)}\\ &= p^2q^5\end{aligned}$$

19 **The answer is $\frac{81}{16}$.**

I'll give you the longer version of the solution and then show you a shortcut. First, use the definition of a negative exponent to rewrite the problem as $\frac{1}{\left(\frac{8}{27}\right)^{4/3}}$. Next, change the power to a root: $\frac{1}{\sqrt[3]{\frac{8}{27}}^{\,4}}$ (instead, you could first distribute the fraction power to the numerator and denominator). The rest shouldn't be too bad: $\frac{1}{\sqrt[3]{\frac{8}{27}}^{\,4}} = \frac{1}{\left(\frac{\sqrt[3]{8}}{\sqrt[3]{27}}\right)^4} = \frac{1}{\left(\frac{2}{3}\right)^4} = \frac{1}{\left(\frac{16}{81}\right)} = \frac{81}{16}$.

The shortcut is to use the fact that when you have a fraction raised to a negative power, you can flip the fraction and make the power positive, like this $\left(\frac{8}{27}\right)^{-4/3} = \left(\frac{27}{8}\right)^{4/3}$. Then proceed as follows: $\left(\frac{27}{8}\right)^{4/3} = \frac{27^{4/3}}{8^{4/3}} = \frac{\sqrt[3]{27}^{\,4}}{\sqrt[3]{8}^{\,4}} = \frac{3^4}{2^4} = \frac{81}{16}$.

20 $(4-x^5)(4+x^5)$.

To factor $-x^{10}+16$, you use the oh-so-important a^2-b^2 rule. a^2-b^2 factors into $(a-b)(a+b)$. Make sure you know this factoring rule (and the corresponding FOILing rule, which is the factoring rule in reverse). Whenever you see a binomial with a subtraction sign (in the current problem, you have to switch the two terms to see the subtraction sign), ask yourself whether you can rewrite the binomial as $(\quad)^2-(\quad)^2$, in other words, as something squared minus something else squared. If you can, then the first blank is your a, and the second blank is your b.

The binomial in this problem can be rewritten as $(4)^2-(x^5)^2$. Now just plug the 4 into the a and the x^5 into the b in $(a-b)(a+b)$, and you're done.

21 $a=5$ **and** $b=5\sqrt{3}$.

This is a $30°-60°-90°$ triangle.

22 Another $30°-60°-90°$ triangle.

$$a=\frac{8}{\sqrt{3}} \text{ or } \frac{8\sqrt{3}}{3}$$

$$b=\frac{16}{\sqrt{3}} \text{ or } \frac{16\sqrt{3}}{3}$$

23 $a=6$ **and** $b=6\sqrt{2}$.

Make sure you know your $45°-45°-90°$ triangle.

24 a. **The total area of the pentagon is** $50+\frac{25\sqrt{3}}{2}$.

The square is $\frac{10}{\sqrt{2}}$ by $\frac{10}{\sqrt{2}}$ (because half a square is a $45°-45°-90°$ triangle), so the area is $\frac{10}{\sqrt{2}}\cdot\frac{10}{\sqrt{2}}=\frac{100}{2}=50$. The equilateral triangle has a base of $\frac{10}{\sqrt{2}}$, or $5\sqrt{2}$, so its height is $\frac{5\sqrt{6}}{2}$ (because half of an equilateral triangle is a $30°-60°-90°$ triangle). So, the area of the triangle is $\frac{1}{2}(5\sqrt{2})\left(\frac{5\sqrt{6}}{2}\right)=\frac{25\sqrt{12}}{4}=\frac{50\sqrt{3}}{4}=\frac{25\sqrt{3}}{2}$. The total area is thus $50+\frac{25\sqrt{3}}{2}$.

b. **The perimeter is** $25\sqrt{2}$.

The sides of the square are $\frac{10}{\sqrt{2}}$, or $5\sqrt{2}$, as are the sides of the equilateral triangle.

The pentagon has five sides, so the perimeter is $5\cdot 5\sqrt{2}$, or $25\sqrt{2}$.

25 **The answer is** $20\sqrt{2}$.

The height of the parallelogram is $\frac{4}{\sqrt{2}}$, or $2\sqrt{2}$, because its height is one of the legs of a $45°-45°-90°$ triangle. The parallelogram's base is 10. So, because the area of a parallelogram equals base times height, the area is $10\cdot 2\sqrt{2}$, or $20\sqrt{2}$.

26 $\frac{d-b}{c-a}$. Remember that $slope=\frac{rise}{run}=\frac{y_2-y_1}{x_2-x_1}$.

27 $\sqrt{(c-a)^2+(d-b)^2}$. Remember that $distance=\sqrt{(x_2-x_1)^2+(y_2-y_1)^2}$.

28 $\left(\frac{a+c}{2},\frac{b+d}{2}\right)$. The midpoint of a segment is given by the average of the two x-coordinates and the average of the two y-coordinates.

29 **$2\sqrt{3}$.**

There are a few ways to solve this problem, all of which use your knowledge of $30°-60°-90°$ triangles. Here's a quick and easy way. Triangle *ABC* is a $30°-60°-90°$ triangle, and the short leg of a $30°-60°-90°$ triangle is half as long as its hypotenuse, so $\overline{BC}$ is 4. Triangle *BCD* is another $30°-60°-90°$ triangle, so its short leg is half as long as its hypotenuse. That gives $\overline{DC}$ a length of 2. Then, because $\overline{BD}$ is the long leg of $30°-60°-90°$ triangle *BCD*, it's $\sqrt{3}$ times its short leg. That gives you the answer of $2\sqrt{3}$, for altitude $\overline{BD}$.

30 **$6+6\sqrt{3}$.**

Triangle *ABD* is yet another $30°-60°-90°$ triangle, so its hypotenuse is twice as long as its short leg, $\overline{BD}$. That gives you a length of $4\sqrt{3}$ for $\overline{AB}$. Next, $\overline{AD}$ is $8-2$, or 6. The perimeter of triangle *ABD* is therefore $6+2\sqrt{3}+4\sqrt{3}$, or $6+6\sqrt{3}$.

31 **$27+9\sqrt{3}$.**

Piece o' cake. Begin with triangle *QRS*, which you can see is a $45°-45°-90°$ triangle. The legs of a $45°-45°-90°$ triangle are equal, so $\overline{QR}$ is 6, and the hypotenuse of a $45°-45°-90°$ triangle is $\sqrt{2}$ times either leg, so $\overline{QS}$ is $6\sqrt{2}$.

Now you see that the hypotenuse of triangle *TQS* is twice as long as its short leg, $\overline{QT}$, which tells you that triangle *TQS* is a $30°-60°-90°$ triangle. That makes $\angle TQS$ 60°, and you also get the length of $\overline{TS}$, which, because it's the long leg of $30°-60°-90°$ triangle *TQS*, has to be $\sqrt{3}$ times as long as its short leg, $\overline{QT}$. So $\overline{TS}$ is $3\sqrt{6}$.

Next, because $\angle PQR$ is 150°, and angles *TQS* and *SQR* are 60° and 45°, respectively, you subtract to get 45° for $\angle PQT$. That makes triangle *PQT* a $45°-45°-90°$ triangle, and thus $\overline{PT}$, like $\overline{QT}$, is $3\sqrt{2}$.

Now you have everything you need to figure the area of the quadrilateral. The area of a right triangle equals half the product of its legs, so here's the final math:

$$\begin{aligned} Area_{Quad\ PQRS} &= area_{\Delta PQT} + area_{\Delta TQS} + area_{\Delta QRS} \\ &= \frac{1}{2}(3\sqrt{2})(3\sqrt{2}) + \frac{1}{2}(3\sqrt{6})(3\sqrt{2}) + \frac{1}{2}(6)(6) \\ &= 9 + \frac{1}{2}(9\sqrt{12}) + 18 \\ &= 9 + 9\sqrt{3} + 18 \\ &= 27 + 9\sqrt{3} \end{aligned}$$

Make sure you know your $30°-60°-90°$ and $45°-45°-90°$ triangles!

32 **$10\frac{1}{3}$.**

To do this problem and the next one, you first have to establish that the two triangles are similar (the same shape). Because segments $\overline{BD}$ and $\overline{AE}$ are parallel, angles *BDC* and *AED* are corresponding angles and are therefore congruent. And the two triangles share angle *C*. Thus, by the AA (angle-angle) theorem, triangles *BCD* and *ACE* are similar.

To get the length of $\overline{BC}$, you could use similar triangle proportions, but it's a little bit quicker to use the side-splitter theorem, which tells you that $\frac{BC}{AB}=\frac{4}{8}$. Because the ratio equals $\frac{4}{8}$, you can set $\overline{BC}$ equal to $4x$ and $\overline{AB}$ equal to $8x$. They add up to 13, so you have $4x+8x=13$, or $x=\frac{13}{12}$. Plugging that into $4x$ gives you $\frac{13}{3}$ for the length of $\overline{BC}$.

Now all you need to finish is the length of $\overline{BD}$. Did you fall for the nasty trap in this problem? When you see the 4 and the 8 along the right side of triangle *ACE*, it's easy to make the mistake of thinking that $\overline{BD}$ and $\overline{AE}$ will be in the same 4-to-8 or 1-to-2 ratio and conclude that $\overline{BD}$ therefore equals 3. But $\overline{BD}$ and $\overline{AE}$ are not in a 1-to-2 ratio. To get $\overline{BD}$, you have to use a similar triangle proportion like the following:

$$\frac{\text{right side of } \Delta BCD}{\text{right side of } \Delta ACE} = \frac{\text{base of } \Delta BCD}{\text{base of } \Delta ACE}$$

$$\frac{CD}{CE} = \frac{BD}{AE}$$

$$\frac{4}{12} = \frac{BD}{6}$$

Cross multiplication gives you a length of 2 for $\overline{BD}$.

Adding up the three sides (4, $\frac{13}{3}$, and 2) gives you the perimeter.

33 $\frac{1}{9}$ **or 1:9.**

If you know the appropriate theorem for this problem, it's a snap. If you don't know the theorem, the problem's very hard. You could also get tripped up if you thought you needed the areas of the two triangles (you don't), and you could be thrown off by the trap referred to in Problem 32.

All you need is the theorem that tells you that the ratio of the areas of similar figures is equal to the square of the ratio of any of their corresponding sides. For this problem, the theorem tells you that $\frac{Area_{\Delta BCD}}{Area_{\Delta ACE}} = \left(\frac{CD}{CE}\right)^2 = \left(\frac{4}{12}\right)^2 = \left(\frac{1}{3}\right)^2 = \frac{1}{9}$.

(Note that you did not need to know the altitudes of the triangles or their areas in order to compute the ratio of their areas.)

In plain English, the idea is simply that if you take any 2-D shape and blow it up to, say, 4 times its height, its area will grow 4^2, or 16 times. By the way, if you blow up a 3-D shape, say, 4 times its height, its volume will grow 4^3, or 64 times.

34 $\frac{\sqrt{3}}{2}xy$.

When you see a 60° angle in a problem, one of the first things you should consider is the $30° - 60° - 90°$ triangle. Sure enough, that's the key to this problem.

All you need to do is to drop an altitude from *Q* straight down to base $\overline{PS}$, making a right angle with $\overline{PS}$. Call the point where the altitude meets the base point *T*. Triangle *PQT* contains a 60° angle and a 90° angle, so it has to be a $30° - 60° - 90°$ triangle. The short leg of a $30° - 60° - 90°$ triangle is half as long as its hypotenuse, so $\overline{PT}$ is half of $\overline{PQ}$, or $\frac{1}{2}y$. Then, because the long leg of a $30° - 60° - 90°$ triangle is $\sqrt{3}$ times as long as its short leg, altitude $\overline{QT}$ is $\sqrt{3} \cdot \frac{1}{2}y = \frac{\sqrt{3}}{2}y$.

Now that you have the altitude and the base of the parallelogram, you just plug them into the parallelogram area formula to get your answer: $Area_{parallelogram\ PQRS} = base \cdot height = x \cdot \frac{\sqrt{3}}{2}y$.

If you're ready to test your skills a bit more, take the following chapter quiz that incorporates all the chapter topics.

Whaddya Know? Chapter 4 Quiz

Quiz time! Complete each problem to test your knowledge on the various topics covered in this chapter. You can then find the solutions and explanations in the next section.

1. Simplify: $(x)\left(\frac{x}{2y}\right)\left(\frac{z}{2y}\right)$

2. Simplify: $\left(\frac{5}{a}\right)(xyz)$

3. Simplify: $\left(\frac{8p+8q+r}{8p^2}\right)$

4. Add the fractions: $\frac{x}{y^2}+\frac{z}{x+y}$

5. Subtract the fractions: $\frac{a+b}{a-b}-\frac{a-b}{a+b}$

6. Simplify: $\frac{x^4-16}{x^3+2x^2+4x+8}$

7. Simplify: $\frac{\left(\frac{p}{q}\right)^0\left(\frac{x-y}{z}\right)^2}{\frac{x^2-y^2}{z^2}}$

8. Solve: $\left|x^3-8\right|=-8$

9. Solve: $\left|x^3+4\right|<4$

10. Simplify: $\left(\frac{x^{1/2}x^{1/3}x^{1/4}}{x^{1/6}x^{5/12}}\right)^{-2}$ (Assume x is positive.)

11. Solve: $\sqrt[3]{x}=\sqrt[5]{x}$

12. Simplify/calculate (no calculator allowed): $\ln\frac{1}{e^2}+\log_\pi 1-\log_8 2+\log_2 8+\log\sqrt[3]{10}$

13. Solve: $x^{1/2}-5x^{1/4}+6=0$

14. Solve: $\frac{1}{x}=\frac{1}{x^2}+\frac{1}{x^3}$

15. On the coordinate plane, point A is at $(4,\ 2)$ and point B is at $(12,\ 8)$.

    (a) What's the slope of $\overline{AB}$?

    (b) What's the length of $\overline{AB}$?

    (c) What's the midpoint of $\overline{AB}$?

16. Sketch a coordinate plane, and draw $\overline{AB}$ from Problem 15. Then add the vertical line, $x=8$, that will pass through the midpoint of $\overline{AB}$. Any point on the line $x=8$ (let's call it point C) will have coordinates $(8,\ y)$. What value or values of y will make ΔABC a right triangle?

17 A right circular cone has a radius and altitude of 4. What's the surface area of a sphere that has the same volume as the cone?

18 Determine the area of the triangle shown in the following figure.

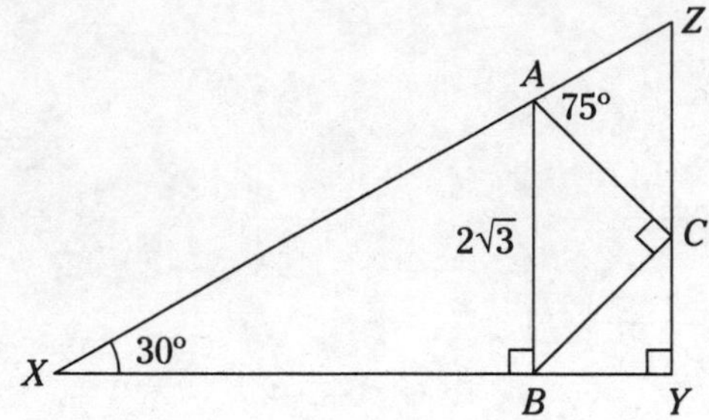

Answers to Chapter 4 Quiz

1. $\dfrac{x^2z}{4y^2}$.

$$(x)\left(\frac{x}{2y}\right)\left(\frac{z}{2y}\right)=\frac{x^2z}{4y^2}$$

Note that the 2*y* in the two denominators is *not* a common denominator (which would remain unchanged if this were a fraction addition or subtraction problem). A common denominator has no significance in a fraction multiplication problem like this (or a fraction division problem).

2. $\dfrac{5xyz}{a}$.

$$\left(\frac{5}{a}\right)(xyz)=\left(\frac{5}{a}\right)\left(\frac{xyz}{1}\right)=\frac{5xyz}{a}$$

3. **You cannot simplify this fraction.**

Tricked you. No cancellation is allowed! If you thought you could cancel one *p* from the numerator and denominator, giving you an answer of $\left(\dfrac{8+8q+r}{8p}\right)$, or that you could cancel the 8's, giving you an answer of $\left(\dfrac{p+q+r}{p^2}\right)$, make up numbers for the *p*, *q*, and *r*, plug your numbers into the original expression and into your answer, use your calculator to compute the results, and you'll see that your two numerical results are not equal.

4. $\dfrac{x^2+xy+y^2z}{y^2x+y^3}$.

First, you need a common denominator. That's $y^2(x+y)$. Now, convert the original fractions to fractions that contain the common denominator, and continue from there:

$$\frac{x(x+y)}{y^2(x+y)}+\frac{y^2z}{y^2(x+y)}=\frac{x(x+y)+y^2z}{y^2(x+y)}=\frac{x^2+xy+y^2z}{y^2x+y^3}$$

5. $\dfrac{4ab}{a^2-b^2}$.

First, rewrite the fractions with the common denominator $(a-b)(a+b)$. That gives you $\dfrac{(a+b)(a+b)}{(a-b)(a+b)}-\dfrac{(a-b)(a-b)}{(a-b)(a+b)}$. Now, simplify: $\dfrac{a^2+2ab+b^2-\left(a^2-2ab+b^2\right)}{(a-b)(a+b)}=\dfrac{4ab}{a^2-b^2}$

6. $x-2$.

First, factor the numerator, using the difference of squares factor pattern twice, and factor the denominator using the grouping technique. Then cancel for your final answer:

$$\frac{\left(x^2+4\right)\left(x^2-4\right)}{x^2(x+2)+4(x+2)}=\frac{\left(x^2+4\right)(x+2)(x-2)}{\left(x^2+4\right)(x+2)}=x-2$$

7 $\dfrac{x-y}{x+y}$.

First, you know, of course, that anything to the zero power equals 1. Next, distribute the power of 2 to the numerator and denominator of the fraction in the numerator: $=\dfrac{\frac{(x-y)^2}{z^2}}{\frac{x^2-y^2}{z^2}}$.

Flip and multiply: $=\dfrac{(x-y)^2}{z^2}\cdot\dfrac{z^2}{x^2-y^2}$. Factor the difference of squares, and then cancel for your final answer: $=\dfrac{(x-y)^2}{z^2}\cdot\dfrac{z^2}{(x+y)(x-y)}=\dfrac{x-y}{x+y}$

8 **No solution.**

STOP! The absolute value of anything can never equal a negative number. No solution.

9 $-2<x<0$.

The best way to solve an absolute value inequality is to solve the related equation and then test the regions on the number line created by the solutions to the equation.

$$|x^3+4|=4$$

$$\begin{array}{ccc} x^3+4=4 & & x^3+4=-4 \\ x^3=0 & \text{or} & x^3=-8 \\ x=0 & & x=-2 \end{array}$$

The solutions of −2 and 0 create three regions on the number line: one to the left of −2, one between −2 and 0, and one to the right of 0. Take one number (any number) from each of the three regions and test them one at a time in the original inequality. Only the middle region produces a true result, so your answer is $-2<x<0$.

10 $\dfrac{1}{x}$.

First, flip the fraction and make the −2 power positive: $=\left(\dfrac{x^{1/6}x^{5/12}}{x^{1/2}x^{1/3}x^{1/4}}\right)^2$. Multiply out the numerator and denominator by adding the powers: $=\left(\dfrac{x^{2/12}x^{5/12}}{x^{6/12}x^{4/12}x^{3/12}}\right)^2=\left(\dfrac{x^{7/12}}{x^{13/12}}\right)^2$. Subtract the powers in the numerator and denominator: $=\left(x^{-6/12}\right)^2=\left(x^{-1/2}\right)^2=x^{-1}=\dfrac{1}{x}$

11 $x=0$ **or** $x=-1$ **or** $x=1$.

The first step is optional, but many people find it easier to work with powers than with roots, so you can rewrite the equations using powers: $x^{1/3}=x^{1/5}$. Now, to get rid of those pesky fractions, raise both sides of the equation to the 15th power (because 15 is the least common multiple of 3 and 5):

$$\begin{aligned} \left(x^{1/3}\right)^{15}&=\left(x^{1/5}\right)^{15} \\ x^5&=x^3 \\ x^5-x^3&=0 \\ x^3\left(x^2-1\right)&=0 \\ x^3(x+1)(x-1)&=0,\ \text{and so on} \end{aligned}$$

(12) **1.**

First, rewrite without fractions or roots: $\ln e^{-2} + \log_{\pi} 1 - \log_8 2 + \log_2 8 + \log 10^{1/3}$. The first and last of these terms should be automatic if you know the definition of a logarithm. The other terms aren't much harder given that $\pi^0 = 1$, $8^{1/3} = 2$, and $2^3 = 8$. You've got $-2 + 0 - \frac{1}{3} + 3 + \frac{1}{3} = 1$.

(13) $\boldsymbol{x = 16}$ **or** $\boldsymbol{x = 81}$**.**

This is a disguised quadratic. Keep your eyes peeled for such problems when you see a trinomial like this. Whenever the first term (ignoring any coefficient) is the square of the middle term (ignoring any coefficient) — as here, where $x^{1/2} = \left(x^{1/4}\right)^2$ — you've got a disguised quadratic. If you set $u = x^{1/4}$, then $u^2 = x^{1/2}$. Now just rewrite the original equation in terms of u, factor, and solve for u:

$$u^2 - 5u + 6 = 0$$
$$(u-2)(u-3) = 0$$

Thus, $u = 2$ or $u = 3$. But don't forget: you want x, not u, so you've got to switch back to x and then solve for x.

$$x^{1/4} = 2 \quad \text{or} \quad x^{1/4} = 3$$
$$x = 16 \qquad\quad x = 81$$

(14) $\boldsymbol{x = \frac{1+\sqrt{5}}{2}}$ **or** $\boldsymbol{x = \frac{1-\sqrt{5}}{2}}$**.**

Multiply both sides of the equation by the least common denominator, namely, x^3, then set the resulting quadratic equation equal to zero:

$$x^3\left(\frac{1}{x}\right) = \left(\frac{1}{x^2} + \frac{1}{x^3}\right)x^3$$
$$x^2 = x + 1$$
$$x^2 - x - 1 = 0$$

You can't factor this because the discriminant $(b^2 - 4ac)$ equals 5, which is not a perfect square. So, solve with the quadratic formula:

$$x = \frac{-(-1) \pm \sqrt{(-1)^2 - (-4)}}{2}$$
$$x = \frac{1+\sqrt{5}}{2} \quad \text{or} \quad x = \frac{1-\sqrt{5}}{2}$$

(15) **Slope** $= \boldsymbol{\frac{3}{4}}$**, Distance** $= \mathbf{10}$**, Midpoint** $= \mathbf{(8,\ 5)}$**.**

(a) $\textit{Slope} = \frac{y_2 - y_1}{x_2 - x_1} = \frac{8-2}{12-4} = \frac{3}{4}$

(b) $\textit{Distance} = \sqrt{(x_2 - x_1)^2 + (y_2 - y_1)^2} = \sqrt{(12-4)^2 + (8-2)^2} = \sqrt{64 + 36} = 10$

Congrats if you noticed the shortcut: You've got a Pythagorean Triple triangle, a 6-8-10 triangle.

(c) $\textit{Midpoint} = \left(\frac{x_1 + x_2}{2},\ \frac{y_1 + y_2}{2}\right) = \left(\frac{4+12}{2},\ \frac{2+8}{2}\right) = (8,\ 5)$

(16) **The values of *y* are $\frac{40}{3}$, $-\frac{10}{3}$, 0, and 10.**

How many solutions did you find? If you found two, not too bad. Congrats if you found all four.

Let's first find the highest and lowest points. The highest point C on the line will make $\angle B$ a right angle. A right angle means that you've got perpendicular lines, and perpendicular lines have opposite reciprocal slopes. Thus, the slope of $\overline{CB}$ will have to equal the opposite reciprocal of the slope of $\overline{AB}$, which equals $\frac{3}{4}$. The opposite reciprocal of that is $-\frac{4}{3}$, so set the slope of $\overline{CB}$ equal to $-\frac{4}{3}$ and solve:

$$\frac{y-8}{8-12}=\frac{-4}{3}, \text{ etc. } y=\frac{40}{3}$$

The solution for the lowest point C works the same way except that this time, the right angle will be $\angle A$. When you do the math, you get $y=-\frac{10}{3}$.

For the final two solutions (which I'm guessing were a bit harder to find), it's $\angle C$ that will be the right angle. Thus, the slope of $\overline{CA}$, $\frac{y-2}{8-4}$, will have to equal the opposite reciprocal of the slope of $\overline{CB}$, $\frac{y-8}{8-12}$:

$$\begin{aligned}\frac{y-2}{8-4}&=\frac{-(8-12)}{y-8}\\(y-2)(y-8)&=16\\y^2-10y+16&=16\\y^2-10y&=0\\y(y-10)&=0\end{aligned}$$

Your final two solutions for *y* are 0 and 10.

(17) **$16\pi\sqrt[3]{4}$.**

First, calculate the volume of the cone using — hang onto your hat — the formula for the volume of a cone: $Volume_{Cone}=\frac{1}{3}area_{base}\cdot height=\frac{1}{3}\pi\cdot 4^2\cdot 4=\frac{4^3}{3}\pi$.

Next, set the volume of the sphere equal to that answer, and solve for *r*:

$$\begin{aligned}\frac{4}{3}\pi r^3&=\frac{4^3}{3}\pi\\r^3&=4^2=16\\r&=\sqrt[3]{16}\end{aligned}$$

Finally, use that value of *r* to determine the surface area of the sphere:

$$Surface\ Area_{Sphere}=4\pi r^2=4\pi\left(\sqrt[3]{16}\right)^2=4\pi\left(\sqrt[3]{16^2}\right)=4\pi\left(\sqrt[3]{4^4}\right)=16\pi\sqrt[3]{4}$$

(18) **$6.5\sqrt{3}+6$.**

Note that ΔXBA and ΔXYZ are $30°-60°-90°$ triangles. $\angle XAB$ is thus a $60°$ angle, and, therefore, $\angle BAC$ is $45°$, and ΔABC is thus a $45°-45°-90°$ triangle. It's then easy to see that ΔBCY is another $45°-45°-90°$ triangle. Next, $\overline{XB}$ is the long leg of a $30°-60°-90°$ triangle, so it's $\sqrt{3}$ times the length of the short leg, which is $2\sqrt{3}$. $\overline{XB}$ is therefore $\sqrt{3}\cdot 2\sqrt{3}$, or 6.

The next thing you need is $\overline{BY}$ so you can add that to $\overline{XB}$ to get base $\overline{XY}$. Because $\overline{AB}$ is the hypotenuse of a $45°-45°-90°$ triangle, you divide its length by $\sqrt{2}$ to get the length of $\overline{BC}$. But then you have to divide that answer by $\sqrt{2}$ to get $\overline{BY}$. Dividing by $\sqrt{2}$ twice is the same thing as dividing by 2, so that means that $\overline{BY}$ is half of $\overline{AB}$. $\overline{BY}$ is thus $\sqrt{3}$, and therefore $\overline{XY}$ is $6+\sqrt{3}$. We're almost done. Phew.

So, ΔXYZ (a $30°-60°-90°$ triangle) has a long leg of $6+\sqrt{3}$. You divide that by $\sqrt{3}$ to get the length of the short leg, $\overline{YZ}$. That short leg, which is the height of ΔXYZ, is thus $\frac{6+\sqrt{3}}{\sqrt{3}}$, which simplifies to $2\sqrt{3}+1$. Finally, you finish with the formula for the area of a triangle. The area of ΔXYZ equals $\frac{1}{2}base\cdot height$, or $\frac{1}{2}\cdot(6+\sqrt{3})(2\sqrt{3}+1)=\frac{1}{2}(12\sqrt{3}+6+6+\sqrt{3})=\frac{1}{2}(13\sqrt{3}+12)=$ $6.5\sqrt{3}+6$. Was that fun or what?

IN THIS CHAPTER

- Figuring out functions and relations
- Learning about lines
- Getting particular about parabolas
- Grappling with graphs
- Transforming functions and investigating inverse functions

Chapter 5

Funky Functions and Their Groovy Graphs

In Chapter 5, you continue your pre-calc warm-up that you began in Chapter 4. If algebra is the language calculus is written in, you might think of functions as the "sentences" of calculus. And they're as important to calculus as sentences are to writing. Virtually everything you do in calculus concerns functions and their graphs in one way or another. *Differential calculus* involves finding the slope or steepness of various functions, and *integral calculus* involves computing the area underneath functions. And not only is the concept of a function critical for calculus, but it's also one of the most fundamental ideas in all of mathematics.

What Is a Function?

Basically, a function is a relationship between two things in which the numerical value of one thing in some way depends on the value of the other. Examples are all around us: The average daily temperature for your city depends on, and is a function of, the time of year; the distance an object has fallen is a function of how much time has elapsed since you dropped it; the area of a circle is a function of its radius; and the pressure of an enclosed gas is a function of its temperature.

The defining characteristic of a function

REMEMBER

A function has only one output for each input.

Consider Figure 5-1.

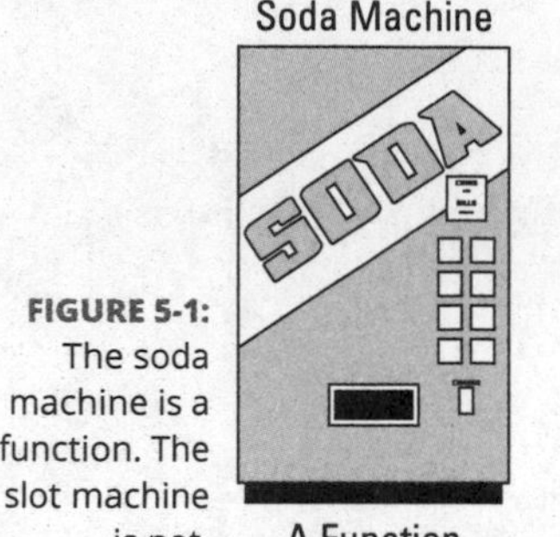

FIGURE 5-1: The soda machine is a function. The slot machine is not.

The soda machine is a function because after plugging in the inputs (your choice and your money), you know exactly what the output is. With the slot machine, on the other hand, the output is a mystery, so it's not a function. Look at Figure 5-2.

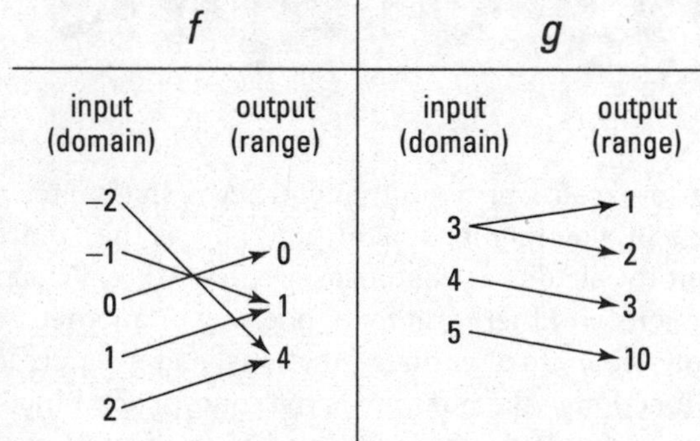

FIGURE 5-2: *f* is a function; *g* is not.

The squaring function, f, is a function because it has exactly one output assigned to each input. It doesn't matter that both 2 and –2 produce the same output of 4 because given an input, say –2, there's no mystery about the output. When you input 3 into g, however, you don't know whether the output is 1 or 2. (For now, don't worry about how the g rule turns its inputs into its outputs.) Because no output mysteries are allowed in functions, g is not a function.

REMEMBER

Good functions, unlike good literature, have predictable endings.

REMEMBER

Definitions of *domain* and *range*: The set of all inputs of a function is called the domain of the function; the set of all outputs is the range of the function.

Some people like to think of a function as a machine. Consider again the squaring function, f, from Figure 5-2. Figure 5-3 shows two of the inputs and their respective outputs.

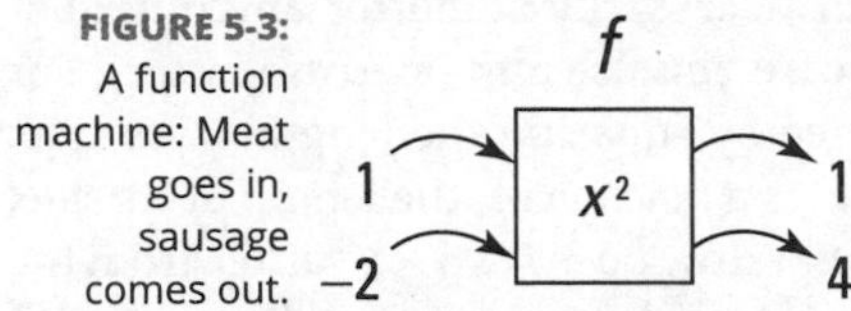

FIGURE 5-3: A function machine: Meat goes in, sausage comes out.

You pop a 1 into the function machine, and out pops a 1; you put in a −2 and a 4 comes out. A function machine takes an input, operates on it in some way, and then spits out the output.

Independent and dependent variables

REMEMBER

Definitions of *dependent variable* and *independent variable*: In a function, the thing that *depends* on the other thing is called the *dependent variable;* the other thing is the *independent variable.* Because you plug numbers into the independent variable, it's also called the *input variable.* After plugging in a number, you then calculate the output or answer for the dependent variable, so the dependent variable is also called the *output variable.* When you graph a function, the independent variable goes on the x-axis, and the dependent variable goes on the y-axis.

Sometimes the dependence between the two things is one of cause and effect — for example, raising the temperature of a gas *causes* an increase in the pressure. In this case, temperature is the *independent variable* and pressure the *dependent variable* because the pressure depends on the temperature.

Often, however, the dependence is not one of cause and effect, but just some sort of association between the two things. Usually, though, the independent variable is the thing you already know or can easily ascertain, and the dependent variable is the thing you want to figure out. For instance, you wouldn't say that time causes an object to fall (gravity is the cause), but if you know how much time has passed since you dropped an object, you can figure out how far it has fallen. So, time is the independent variable, and distance fallen is the dependent variable; and you would say that distance is a function of time.

Whatever the type of correspondence between the two variables, the dependent variable (the y-variable) is the thing you're usually more interested in. Generally, you want to know what happens to the dependent or y-variable as the independent or x-variable goes to the right: Is the y-variable (the height of the graph) rising or falling and, if so, how steeply, or is the graph level, neither going up nor down?

Function notation

A common way of writing the function $y = 5x^3 - 2x^2 + 3$ is to replace the "y" with "$f(x)$" and write $f(x) = 5x^3 - 2x^2 + 3$. It's just a different notation for the same thing. These two equations are, in every respect, mathematically identical. Students are often puzzled by function notation when they see it for the first time. They wonder what the "f" is and whether $f(x)$ means f times x. It does not. If function notation bugs you, my advice is to think of $f(x)$ as simply the way y is written in some foreign language. Don't consider the f and the x separately; just think of $f(x)$ as a single symbol for y.

You can also think of $f(x)$ (read as "f of x") as short for "a function of x." You can write $y = f(x) = 3x^2$, which is translated as "y is a function of x and that function is $3x^2$." However, sometimes other letters are used instead of f — such as $g(x)$ or $p(x)$ — often just to differentiate between functions. The function letter doesn't necessarily stand for anything, but sometimes the initial letter of a word is used (in which case you use an uppercase letter). For instance, you know that the area of a square is determined by squaring the length of its side: $Area = side^2$ or $A = s^2$. The area of a square depends on, and is a *function* of, the length of its side. With function notation, you can write $A(s) = s^2$. (Quick quiz: How does $f(x) = x^2$ differ from the area of a square function, $A(s) = s^2$? Answer: For $f(x) = x^2$, x can equal any number, but with $A(s) = s^2$, s must be positive, because the length of a side of a square cannot be negative or zero. The two functions thus have different *domains.*)

Consider, again, the squaring function $y = x^2$ or $f(x) = x^2$. When you input 3 for x, the output is 9. Function notation is convenient because you can concisely express the input and the output by writing $f(3) = 9$ (read as "f of 3 equals 9"). Remember that $f(3) = 9$ means that when x is 3, $f(3)$ is 9; or, equivalently, it tells you that when x is 3, y is 9.

Composite functions

A *composite* function is the combination of two functions. For example, the cost of the electrical energy needed to air-condition your place depends on how much electricity you use, and usage depends on the outdoor temperature. Because cost depends on usage and usage depends on temperature, cost depends on temperature. In function language, cost is a function of usage, usage is a function of temperature, and thus cost is a function of temperature. This last function, a combination of the first two, is a composite function.

Let $f(x) = x^2$ and $g(x) = 5x - 8$. Input 3 into $g(x)$: $g(3) = 5 \cdot 3 - 8$, which equals 7. Now take that output, 7, and plug it into $f(x)$: $f(7) = 7^2 = 49$. The machine metaphor shows what I did here. Look at Figure 5-4. The g machine turns the 3 into a 7, and then the f machine turns the 7 into a 49.

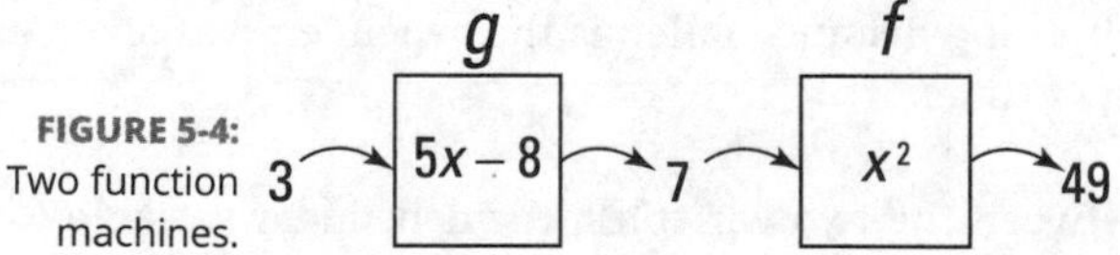

FIGURE 5-4: Two function machines.

You can express the net result of the two functions in one step with the following *composite* function:

$$f(g(3)) = 49$$

You always calculate the inside function of a composite function first: $g(3) = 7$. Then you take the output, 7, and calculate $f(7)$, which equals 49.

To determine the general composite function, $f(g(x))$, plug $g(x)$, which equals $5x-8$, into $f(x)$. In other words, you want to determine $f(5x-8)$. The f function or f machine takes an input and squares it. Thus,

$$\begin{aligned} f(5x-8) &= (5x-8)^2 \\ &= (5x-8)(5x-8) \\ &= 25x^2 - 40x - 40x + 64 \\ &= 25x^2 - 80x + 64 \end{aligned}$$

Thus, $f(g(x)) = 25x^2 - 80x + 64$.

WARNING

With composite functions, the order matters. As a general rule, $f(g(x)) \neq g(f(x))$.

What Does a Function Look Like?

I'm no math historian, but everyone seems to agree that René Descartes (1596–1650) came up with the *x-y* coordinate system shown in Figure 5-5.

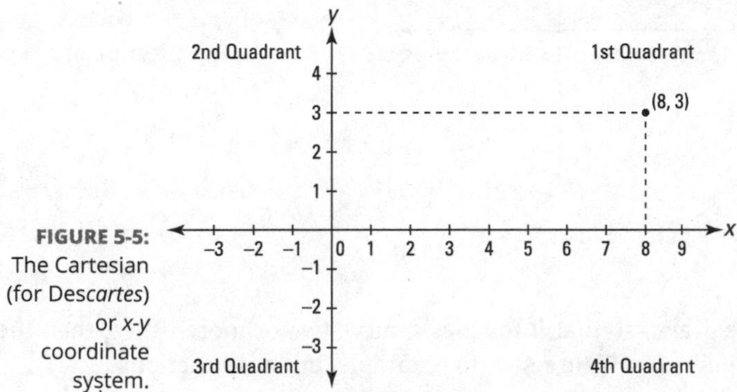

FIGURE 5-5: The Cartesian (for Des*cartes*) or *x-y* coordinate system.

Isaac Newton (1642–1727) and Gottfried Leibniz (1646–1716) are credited with inventing calculus, but it's hard to imagine that they could have done it without Descartes's contribution several decades earlier. Think of the coordinate system (or the screen on your graphing calculator) as your window into the world of calculus. Virtually everything in your calculus textbook and in this book involves (directly or indirectly) the graphs of lines or curves — usually functions — in the *x-y* coordinate system.

Consider the four graphs in Figure 5-6.

These four curves are functions because they satisfy the *vertical line test*. (*Note:* I'm using the term *curve* here to refer to any shape, whether it's curved or straight.)

REMEMBER

The vertical line test: A curve is a function if a vertical line drawn through the curve — regardless of where it's drawn — touches the curve only once. This guarantees that each input within the function's domain has exactly one output.

No matter where you draw a vertical line on any of the four graphs in Figure 5-6, the line will touch the curve at only one point. Try it.

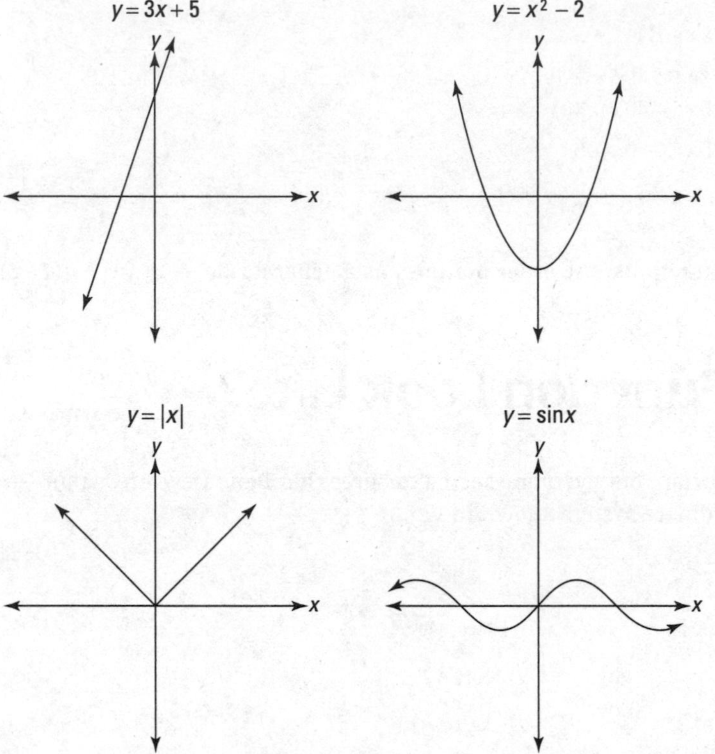

FIGURE 5-6: Four functions.

If, however, a vertical line can be drawn so that it touches a curve two or more times, then the curve is not a function. The two curves in Figure 5-7, for example, are not functions.

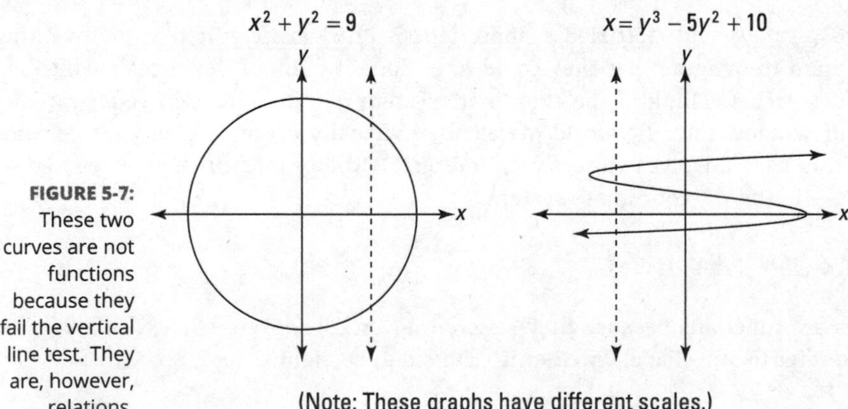

FIGURE 5-7: These two curves are not functions because they fail the vertical line test. They are, however, relations.

So, the four curves in Figure 5-6 are functions, and the two in Figure 5-7 are not, but all six of the curves are *relations.*

REMEMBER

Definition of *relation*: A relation is any collection of points on the x-y coordinate system.

You spend a little time studying some non-function relations in calculus — circles, for instance — but the vast majority of calculus problems involve functions.

EXAMPLE

Q. If $f(x)=3x^2-4x+8$, what does $f(a+b)$ equal?

A. $\mathbf{3a^2+6ab+3b^2-4a-4b+8}$**.**

$$\begin{aligned} f(x)&=3x^2-4x+8 \\ f(a+b)&=3(a+b)^2-4(a+b)+8 \\ &=3\left(a^2+2ab+b^2\right)-4a-4b+8 \\ &=3a^2+6ab+3b^2-4a-4b+8 \end{aligned}$$

YOUR TURN

1 **(a)** What's the domain of $g(x)=\sqrt{4-x}$?

(b) What's the range of g?

 What's the domain of $f(x)=\dfrac{1}{x\sqrt{x+5}}$?

---

3 For the function $f(x) = x^2$, what's $f(a+b) - f(a-b)$?

Common Functions and Their Graphs

You're going to see hundreds of functions in your study of calculus, so it wouldn't be a bad idea to familiarize yourself with the basic ones in this section: the line, the parabola, the absolute value function, the cubing and cube root functions, and the exponential and logarithmic functions.

Lines in the plane in plain English

A line is the simplest function you can graph on the coordinate plane. (Lines are important in calculus because you often study lines that are tangent to curves and because when you zoom in far enough on a curve, it looks and behaves like a line.) Figure 5-8 shows an example of a line: $y = 3x + 5$.

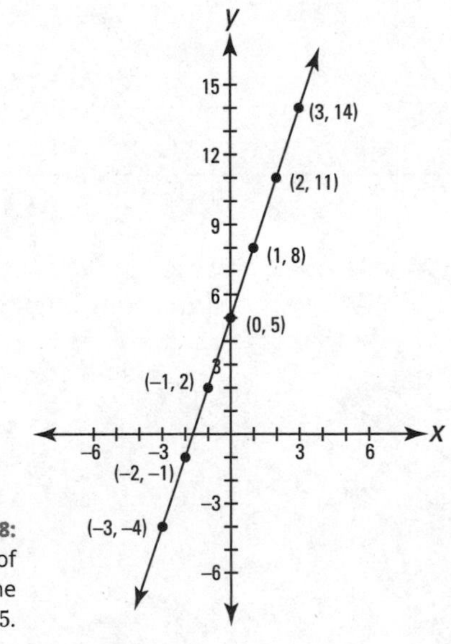

FIGURE 5-8: The graph of the line $y = 3x + 5$.

Hitting the slopes

The most important thing about the line in Figure 5-8 — at least for your study of calculus — is its slope or steepness. Notice that whenever x goes 1 to the right, y goes up by 3. A good way to visualize slope is to draw a stairway under the line (see Figure 5-9). The vertical part of the step is called the *rise*, the horizontal part is called the *run*, and the slope is defined as the ratio of the rise to the run:

$$Slope = \frac{rise}{run} = \frac{3}{1} = 3$$

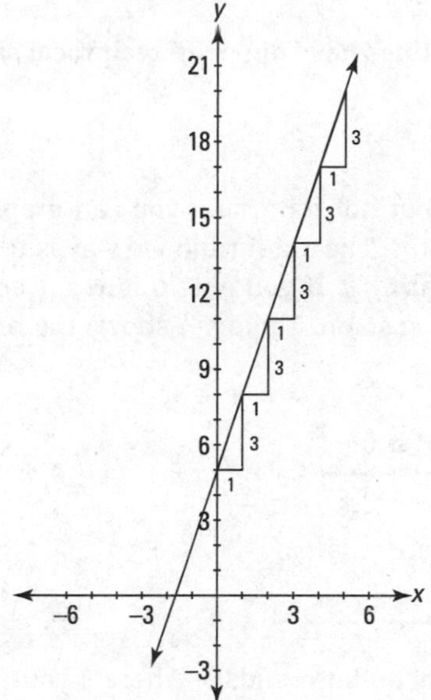

FIGURE 5-9: The line $y = 3x + 5$ has a slope of 3.

You don't have to make the run equal to 1. The ratio of rise to run, and thus the slope, always comes out the same, regardless of what size you make the steps. If you make the run equal to 1, however, the slope is the same as the rise because a number divided by 1 equals itself. This is a good way to think about slope — the slope is the amount that a line goes up (or down) as it goes 1 to the right.

Definitions of positive, negative, zero, and undefined slopes: Lines that go *up* to the right have a *positive* slope; lines that go *down* to the right have a *negative* slope. Horizontal lines have a slope of *zero*, and vertical lines do not have a slope — you say that the slope of a vertical line is *undefined*.

Here's the formula for slope:

$$Slope = \frac{y_2 - y_1}{x_2 - x_1}$$

Pick any two points on the line in Figure 5-9, say (1, 8) and (3, 14), and plug them into the formula to calculate the slope:

$$Slope = \frac{14-8}{3-1} = \frac{6}{2} = 3$$

This computation involves, in a sense, a stairway step that goes over 2 and up 6. The answer of 3 agrees with the slope you can see in Figure 5-9.

Any line parallel to this one has the same slope, and any line perpendicular to this one has a slope of $-\frac{1}{3}$, which is the *opposite reciprocal* of 3.

REMEMBER

***Parallel* lines have the same slope. *Perpendicular* lines have opposite reciprocal slopes.**

Graphing lines

If you have the equation of the line, $y = 3x + 5$, but not its graph, you can graph the line the old-fashioned way or with your graphing calculator. The old-fashioned way is to create a table of values by plugging numbers into *x* and calculating *y*. If you plug 0 into *x*, *y* equals 5; plug 1 into *x*, and *y* equals 8; plug 2 into *x*, and *y* is 11, and so on. Table 5-1 shows the results.

Table 5-1 Points on the Line $y = 3x + 5$

x	0	1	2	3	4	----→
y	5	8	11	14	17	----→

Plot the points, connect the dots, and put arrows on both ends — there's your line. This is a snap with a graphing calculator. Just enter $y = 3x + 5$ and your calculator graphs the line and produces a table like Table 5-1.

Slope-intercept and point-slope forms

You can see that the line in Figure 5-9 crosses the *y*-axis at 5 — this point is the *y-intercept* of the line. Because both the slope of 3 and the *y*-intercept of 5 appear in the equation $y = 3x + 5$, this equation is said to be in *slope-intercept* form. Here's the form written in the general way.

REMEMBER

***Slope-intercept* form:**

$$y = mx + b$$

(Where *m* is the slope and *b* is the *y*-intercept.)

(If that doesn't ring a bell — even a distant, faint bell — go directly to the registrar and drop calculus, but do *not* under any circumstances return this book.)

Using the *slope-intercept* form of the equation of a line is another way to graph the line. For example, to graph the line discussed here, $y = 3x + 5$, just start at (0, 5), the *y*-intercept, and then use the slope of 3 to go over 1 and up 3 from (0, 5) to get to (1, 8). Those two points give you your line.

All lines, except for *vertical* lines, can be written in slope-intercept form. Vertical lines are written like $x = 6$, for example. The number tells you where the vertical line crosses the x-axis.

The equation of a *horizontal* line also looks different, $y = 10$ for example. But it technically fits the form $y = mx + b$ — it's just that because the slope of a horizontal line is zero, and because zero times x is zero, there is no x-term in the equation. (But, if you felt like it, you could write $y = 10$ as $y = 0x + 10$.)

REMEMBER

Definition of a *constant function*: A line is the simplest type of function, and a horizontal line (called a constant function) is the simplest type of line. It's nonetheless fairly important in calculus, so make sure you know that a horizontal line has an equation like $y = 10$ and that its slope is zero.

If $m = 1$ and $b = 0$, you get the function $y = x$. This line goes through the *origin* $(0, 0)$ and makes a 45° angle with both coordinate axes. It's called the *identity* function because its outputs are the same as its inputs.

REMEMBER

***Point-slope* form:** In addition to the slope-intercept form for the equation of a line, you should also know the point-slope form:

$$y - y_1 = m(x - x_1)$$

To use this form, you need to know — you guessed it — a *point* on a line and the line's *slope*. You can use any point on the line. Consider the line in Figure 5-9 again. Pick any point, say $(2, 11)$, and then plug the x- and y-coordinates of the point into x_1 and y_1, and plug the slope, 3, into m:

$$y - 11 = 3(x - 2)$$

With a little algebra, you can convert this equation into the one you already know, $y = 3x + 5$. Try it.

Parabolic and absolute value functions — even-steven

You should be familiar with the two functions shown in Figure 5-10: the parabola, $f(x) = x^2$, and the absolute value function, $g(x) = |x|$.

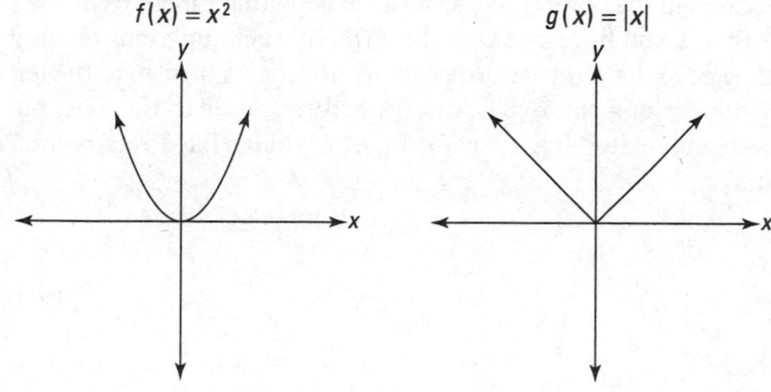

FIGURE 5-10: The graphs of $f(x) = x^2$ and $g(x) = |x|$.

Notice that both functions are symmetric with respect to the y-axis. In other words, the left and right sides of each graph are mirror images of each other. This makes them *even* functions. A *polynomial* function like $y = 9x^4 - 4x^2 + 3$, where all powers of x are even, is one type of even function. (Such an even polynomial function can contain — but need not contain — a constant term like the 3 in the preceding function. This makes sense because 3 is the same as $3x^0$ and zero is an even number.) Another even function is $y = \cos x$ (see Chapter 6).

A couple of oddball functions

Graph $f(x) = x^3$ and $g(x) = \sqrt[3]{x}$ on your graphing calculator. These two functions illustrate *odd* symmetry. Odd functions are symmetric with respect to the origin, which means that if you were to rotate them 180° about the origin, they would land on themselves. A polynomial function like $y = 4x^5 - x^3 + 2x$, where all powers of x are odd, is one type of odd function. (Unlike an even polynomial function, an odd polynomial function *cannot* contain a constant term.) Another odd function is $y = \sin x$ (see Chapter 6).

Many functions are neither even nor odd, for example $y = 3x^2 - 5x$. My high school English teacher said a paragraph should never have just one sentence, so voilà, now it's got two.

Exponential functions

An exponential function is one with a power that contains a variable, such as $f(x) = 2^x$ or $g(x) = 10^x$. Figure 5-11 shows the graphs of both these functions on the same x-y coordinate system.

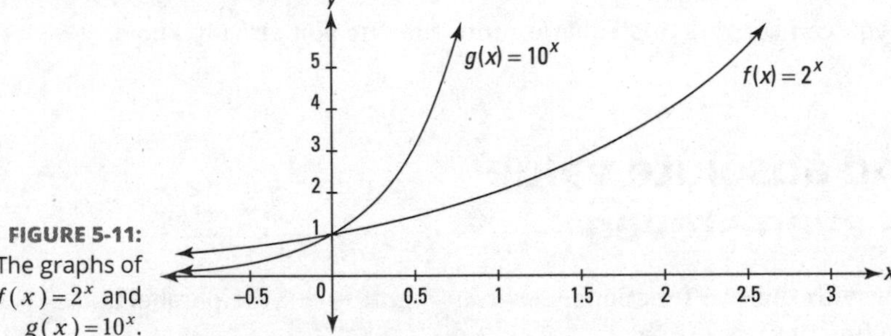

FIGURE 5-11: The graphs of $f(x) = 2^x$ and $g(x) = 10^x$.

Both functions go through the point $(0, 1)$, as do all exponential functions of the form $f(x) = b^x$. When b is greater than 1, you have *exponential growth*. All such functions get higher and higher without limit as they go to the right toward positive infinity. As they go to the left toward negative infinity, they crawl along the x-axis, always getting closer to the axis, but never touching it. You use these and related functions for figuring things like investments, inflation, and growing population.

When *b* is between 0 and 1, you have an *exponential decay* function. The graphs of such functions are like exponential growth functions in reverse. Exponential decay functions also cross the *y*-axis at $(0, 1)$, but they go up to the *left* forever, and crawl along the *x*-axis to the *right.* These functions model things that shrink over time, such as the radioactive decay of uranium.

Logarithmic functions

A logarithmic function is simply an exponential function with the *x*- and *y*-axes switched. In other words, the up-and-down direction on an exponential graph corresponds to the right-and-left direction on a logarithmic graph, and the right-and-left direction on an exponential graph corresponds to the up-and-down direction on a logarithmic graph. (If you want a refresher on logs, see Chapter 4.) You can see this relationship in Figure 5-12, in which both $f(x) = 2^x$ and $g(x) = \log_2 x$ are graphed on the same set of axes.

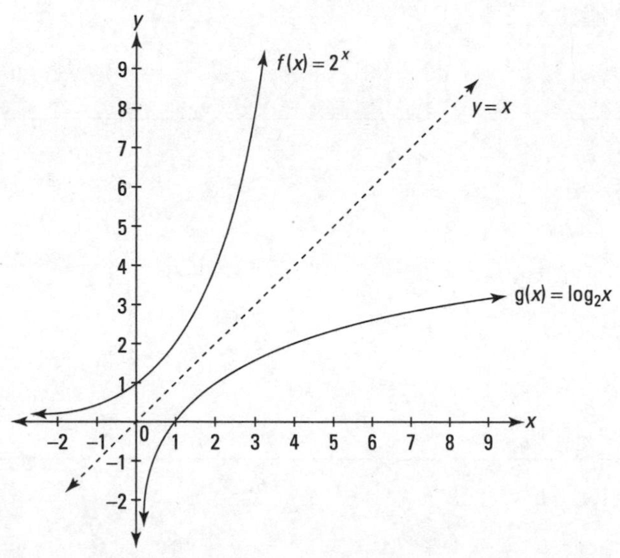

FIGURE 5-12: The graphs of $f(x) = 2^x$ and $g(x) = \log_2 x$.

Both exponential and logarithmic functions are *monotonic.* A monotonic function either goes up over its entire domain (called an *increasing* function) or goes down over its entire domain (a *decreasing* function). (I'm assuming here — as is almost always the case — that the motion along the function is from left to right.)

Notice the symmetry of the two functions in Figure 5-12 about the line $y = x$. This makes them *inverses* of each other, which brings me to the next topic.

EXAMPLE

Q. For the line $g(x) = 5 - 4x$, what's the slope and what's the y-intercept?

A. **The slope is -4 and the y-intercept is 5.** All you need to answer this question is $y = mx + b$.

YOUR TURN

4 If the slope of line l is 3,

a. What's the slope of a line parallel to l?

b. What's the slope of a line perpendicular to l?

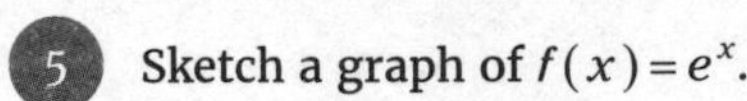

5 Sketch a graph of $f(x) = e^x$.

---

6 Sketch a graph of $g(x) = \ln x$.

Inverse Functions

The function $f(x) = x^2$ (for $x \geq 0$) and the function $f^{-1}(x) = \sqrt{x}$ (read as "f inverse of x") are inverse functions because each undoes what the other does. In other words, $f(x) = x^2$ takes an input of, say, 3 and produces an output of 9 (because $3^2 = 9$); $f^{-1}(x) = \sqrt{x}$ takes the 9 and turns it back into the 3 (because $\sqrt{9} = 3$). Notice that $f(3) = 9$ and $f^{-1}(9) = 3$. You can write all of this in one step as $f^{-1}(f(3)) = 3$. It works the same way if you start with $f^{-1}(x)$: $f^{-1}(16) = 4$ (because $\sqrt{16} = 4$), and $f(4) = 16$ (because $4^2 = 16$). If you write this in one step, you get $f\left(f^{-1}(16)\right) = 16$. (Note that while only $f^{-1}(x)$ is read as *f inverse* of x, both functions are inverses of each other.)

REMEMBER

The inverse function rule: The fancy way of summing up all of this is to say that $f(x)$ and $f^{-1}(x)$ are inverse functions if and only if $f^{-1}(f(x)) = x$ and $f\left(f^{-1}(x)\right) = x$.

WARNING

Don't confuse the superscript –1 in $f^{-1}(x)$ with the exponent –1. The exponent –1 gives you the reciprocal of something, for example $x^{-1} = \frac{1}{x}$. But $f^{-1}(x)$ is the inverse of $f(x)$. It does *not* equal $\frac{1}{f(x)}$, which is the reciprocal of $f(x)$. So why is the exact same symbol used for two different things? Beats me.

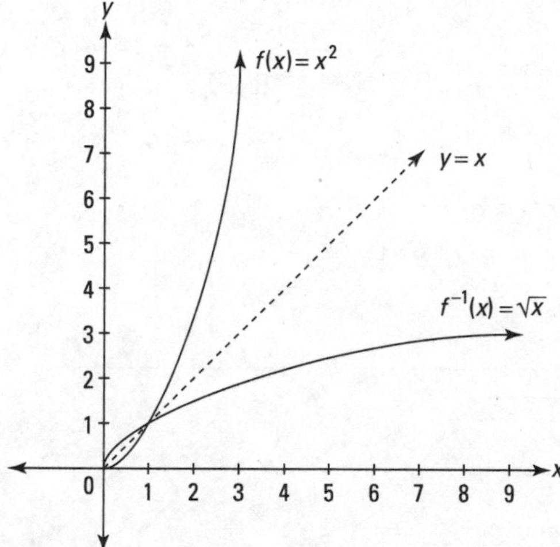

FIGURE 5-13: The graphs of $f(x) = x^2$ ($x \geq 0$), and $f^{-1}(x) = \sqrt{x}$.

When you graph inverse functions, each is the mirror image of the other, reflected over the line $y = x$. Look at Figure 5-13, which graphs the inverse functions $f(x) = x^2$ (for $x \geq 0$) and $f^{-1}(x) = \sqrt{x}$.

If you rotate the graph in Figure 5-13 counterclockwise so that the line $y = x$ is vertical, you can easily see that $f(x)$ and $f^{-1}(x)$ are mirror images of each other. One consequence of this symmetry is that if a point like $(2, 4)$ is on one of the functions, the point $(4, 2)$ will be on the other. Also, the domain of f is the range of f^{-1}, and the range of f is the domain of f^{-1}.

EXAMPLE

Q. If the inverse of f is g, and $f(p) = q$, what do $g(p)$ and $g(q)$ equal?

A. **$g(q) = p$; nothing can be said about $g(p)$.**

$g(q)$ must equal p, because since f sends p to q, f's inverse must send q to p. With regard to $g(p)$, the inverse relationship between f and g tells you nothing about the output of $g(p)$.

YOUR TURN

7 The following figure shows the graph of $f(x)$. Sketch the inverse of f, $f^{-1}(x)$.

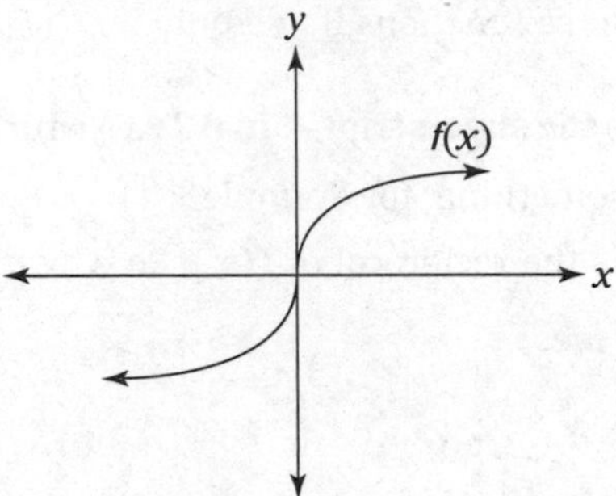

8 What's the inverse of $f(x) = \sqrt{4x + 5}$?

Shifts, Reflections, Stretches, and Shrinks

Any function can be transformed into a related function by shifting it horizontally or vertically, flipping it over horizontally or vertically, or stretching or shrinking it horizontally or vertically. I do the horizontal transformations first. Consider the exponential function $y = 2^x$, shown in Figure 5-14.

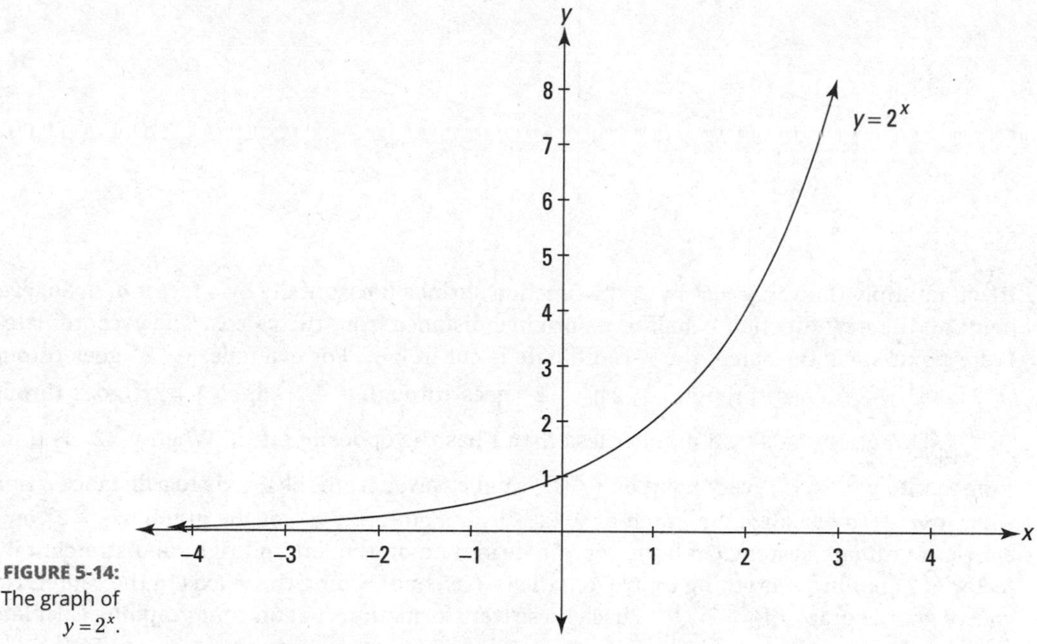

FIGURE 5-14: The graph of $y = 2^x$.

Horizontal transformations

Horizontal changes are made by adding a number to or subtracting a number from the input variable x or by multiplying x by some number. All horizontal transformations, except reflection, work the *opposite* way you'd expect: Adding to x makes the function go left, subtracting from x makes the function go right, multiplying x by a number greater than 1 shrinks the function, and multiplying x by a number less than 1 expands the function. For example, the graph of $y = 2^{x+3}$ has the same shape and orientation as the graph in Figure 5-14; it's just shifted three units to the *left*. Instead of passing through $(0, 1)$ and $(1, 2)$, the shifted function goes through $(-3, 1)$ and $(-2, 2)$. And the graph of $y = 2^{x-3}$ is three units to the *right* of $y = 2^x$. The original function and both transformations are shown in Figure 5-15.

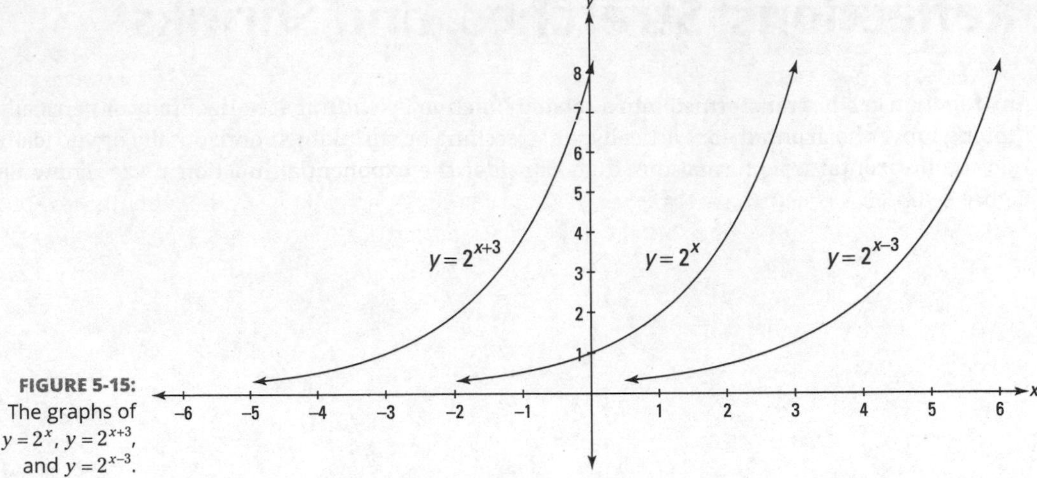

FIGURE 5-15: The graphs of $y = 2^x$, $y = 2^{x+3}$, and $y = 2^{x-3}$.

If you multiply the x in $y = 2^x$ by 2, the function shrinks horizontally by a factor of 2. So, every point on the new function is half of its original distance from the y-axis. The y-coordinate of every point stays the same; the x-coordinate is cut in half. For example, $y = 2^x$ goes through $(1, 2)$, so $y = 2^{2x}$ goes through $\left(\frac{1}{2}, 2\right)$; $y = 2^x$ goes through $\left(-4, \frac{1}{16}\right)$, so $y = 2^{2x}$ goes through $\left(-2, \frac{1}{16}\right)$. Multiplying x by a number less than 1 has the opposite effect. When $y = 2^x$ is transformed into $y = 2^{(1/4)x}$, every point on $y = 2^x$ is pulled away from the y-axis to a distance 4 times what it was. To visualize the graph of $y = 2^{(1/4)x}$, imagine you've got the graph of $y = 2^x$ on an elastic coordinate system. Grab the coordinate system on the left and right and stretch it by a factor of 4, pulling everything away from the y-axis, but keeping the y-axis in the center. Now you've got the graph of $y = 2^{(1/4)x}$. Check these transformations out on your graphing calculator.

The last horizontal transformation is a reflection over the y-axis. Multiplying the x in $y = 2^x$ by -1 reflects it over or flips it over the y-axis. For instance, the point $(1, 2)$ becomes $(-1, 2)$ and $\left(-2, \frac{1}{4}\right)$ becomes $\left(2, \frac{1}{4}\right)$. See Figure 5-16.

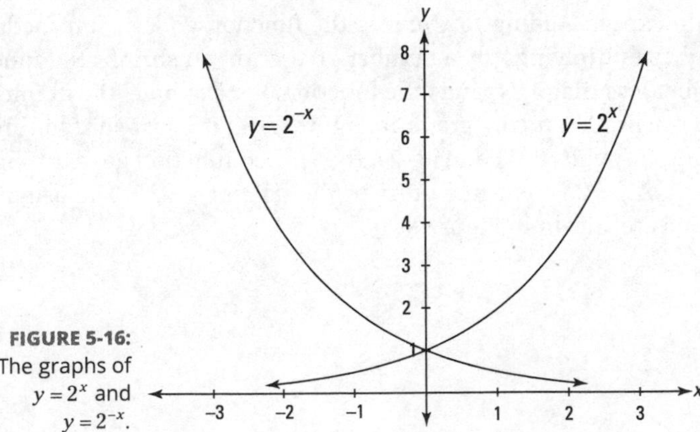

FIGURE 5-16: The graphs of $y = 2^x$ and $y = 2^{-x}$.

Vertical transformations

To transform a function vertically, you add a number to or subtract a number from the entire function or multiply the whole function by a number. To do something to an entire function, say $y=10^x$, imagine that the entire right side of the equation is inside parentheses, like $y=\left(10^x\right)$. Now, all vertical transformations are made by placing a number somewhere on the right side of the equation *outside* the parentheses. (Often, you don't actually need the parentheses, but sometimes you do.) Unlike horizontal transformations, vertical transformations work the way you expect: Adding makes the function go up, subtracting makes it go down, multiplying by a number greater than 1 stretches the function, and multiplying by a number less than 1 shrinks the function. For example, consider the following transformations of the function $y=10^x$:

$y=10^x+6$ shifts the original function up 6 units.

$y=10^x-2$ shifts the original function down 2 units.

$y=5\cdot 10^x$ stretches the original function vertically by a factor of 5.

$y=\frac{1}{3}\cdot 10^x$ shrinks the original function vertically by a factor of 3.

Multiplying the function by –1 reflects it over the x-axis, or, in other words, flips it upside down. Look at these transformations on your graphing calculator.

As you saw in the previous section, horizontal transformations change only the x-coordinates of points, leaving the y-coordinates unchanged. Conversely, vertical transformations change only the y-coordinates of points, leaving the x-coordinates unchanged.

EXAMPLE

Q. Consider the simple parabola $f(x)=x^2$. How does $g(x)=5(4x+10)^2+3$ compare to *f*?

A. ***g* makes two horizontal changes to *f*: It 1) slides it 10 to the left and 2) compresses it horizontally by a factor of 4. *g* also makes two vertical changes to *f*: It 1) stretches it vertically by a factor of 5, and 2) slides it up 3.**

Note a few things. First, I gave the horizontal changes before the vertical changes, but, when you transform a function, it doesn't matter whether you tackle the horizontal or the vertical transformations first. However, if you have more than one horizontal change, you must do those in the right order. And ditto if you have more than one vertical change. For horizontal transformations, not only do they work in the opposite way you'd expect, but you must do the PEMDAS order backwards. For this problem, the horizontal changes are multiplying by 4 and adding 10. PEMDAS tells you to do the multiplication before the addition, so you need to reverse that. That's why you *first* must slide *f* 10 to the left and *second* compress *f* by 4. Everything about vertical transformations works the way you'd expect, including PEMDAS.

YOUR TURN

9 **The figure shows the graph of $p(x) = 2^x$. Sketch the following transformation of p: $q(x) = 2^{x+3} + 5$.**

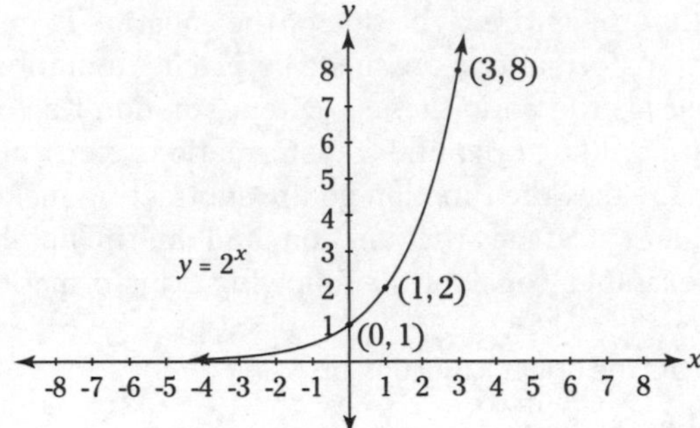

10 **Consider the parabola $f(x) = x^2$ again. The transformation $g(x) = (3x)^2$ compresses f horizontally by a factor of 3. What vertical transformation of f would achieve the same result as that horizontal transformation?**

Practice Questions Answers and Explanations

1 **(a)** $\boldsymbol{x \leq 4}$**.**

You can't take the square root of a negative (not for basic calculus, anyway, which deals with real numbers), so

$$4 - x \geq 0$$
$$4 \geq x$$

That's all there is to it. Don't forget, there's nothing wrong with the square root of zero, which equals zero. So, 4 *is* in the domain of g.

(b) $\boldsymbol{g(x) \geq 0}$**.**

Range questions are usually a bit harder than domain questions. With domain questions, you just have to figure out what x cannot be, and the domain is everything else. With range questions, there's no method quite that straightforward.

To tackle a range question, you can experiment with different input values and see what happens with the output. And, of course, you can graph the function to actually see the range, though that won't always give you the precise answer. Sometimes, like with the function in Problem 2, you can't get the precise answer without doing some calculus.

You can solve the current problem easily by just looking at the graph of the function. But it'll also come in handy to familiarize yourself with the following approach.

Consider what the graph of $y = \sqrt{x}$ looks like. If you don't remember the graph, you should graph it now on your calculator. You see the top half of a sideways parabola that begins at $(0, 0)$ and goes up and to the right forever. Because it begins at a height of zero and goes up forever, the range is $y \geq 0$.

The current function, $g(x) = \sqrt{4 - x}$, is a transformation of the parent function, $y = \sqrt{x}$. There are two transformations: the 4 and the minus sign, which is the same as multiplying x by -1. Because both transformations occur "inside" the function and change the *input* of the function, they are both *horizontal* transformations. (To transform the parent function, $y = \sqrt{x}$, into $g(x) = \sqrt{4 - x}$, you first slide it 4 to the *left* and then flip it over the y-axis.) Horizontal transformations change the domain but have no impact on the range, so the range of $g(x) = \sqrt{4 - x}$ is the same as the range of $y = \sqrt{x}$, namely, $y \geq 0$.

2. **$(-5, 0) \cup (0, \infty)$ or $x > -5$ $(x \neq 0)$.**

   Just ask yourself what x is not allowed to be. x can't equal zero because that would make the denominator zero. And x can't equal -5 because that would give you the square root of zero, which is zero, so, again, the denominator would equal zero. That takes care of the zero denominator issue. Then there's the issue of no negatives under the square root. So, x can't be less than -5. That does it. The domain is everything else — everything except what I just excluded.

3. **4ab.**

   $f(a+b)$ tells you to plug $a + b$ into the f function, x^2. Thus,

   $$f(a+b)=(a+b)^2=(a+b)(a+b)=a^2+2ab+b^2$$

   (If you thought $(a+b)^2$ was a^2+b^2, go directly to jail and do not collect $200!)

   And $f(a-b)=(a-b)^2=(a-b)(a-b)=a^2-2ab+b^2$. Finally,

   $$\begin{aligned} f(a+b)-f(a-b) &= \left(a^2+2ab+b^2\right)-\left(a^2-2ab+b^2\right) \\ &= a^2+2ab+b^2-a^2+2ab-b^2 \\ &= 4ab \end{aligned}$$

4. **(a) The slope of a line parallel to l is 3.**

   (b) The slope of a line perpendicular to l is $-\frac{1}{3}$, the *opposite reciprocal* of 3.

5. **See the following figure.**

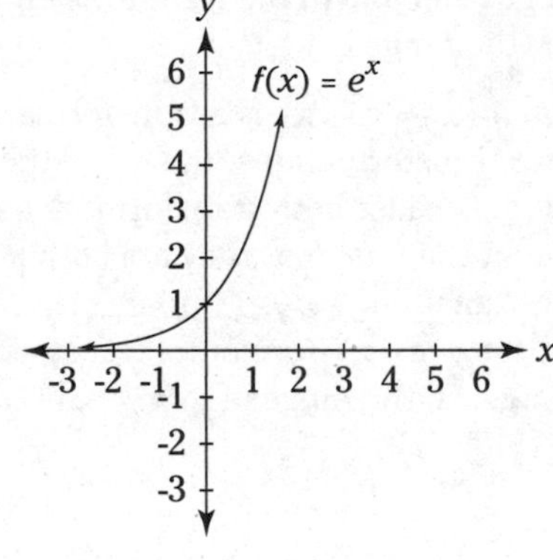

6. **See the following figure.**

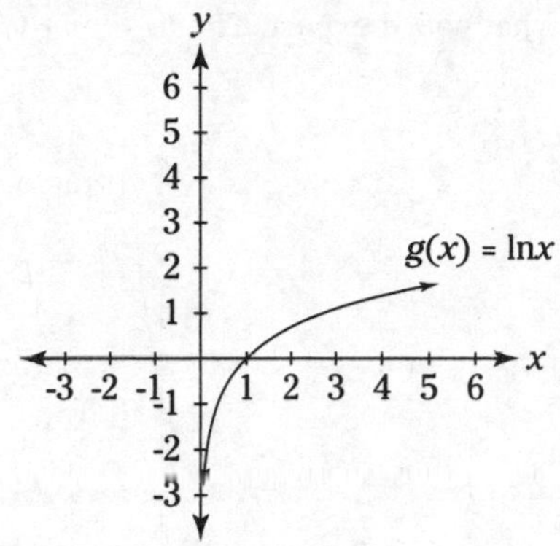

7. You obtain $f^{-1}(x)$ by reflecting $f(x)$ over the line $y = x$. **See the following figure.**

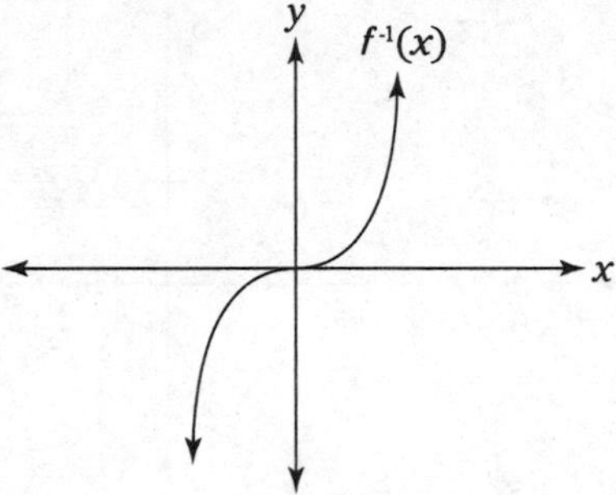

8. $\boldsymbol{f^{-1}(x) = \dfrac{x^2 - 5}{4}}$ **$(x \geq 0)$.**

First, replace $f(x)$ with y and then switch the x and y:

$$y = \sqrt{4x + 5}$$
$$x = \sqrt{4y + 5}$$

Now just solve for y:

$$x = \sqrt{4y + 5}$$
$$x^2 = 4y + 5$$
$$x^2 - 5 = 4y$$
$$\frac{x^2 - 5}{4} = y$$

That's it for the math, but one issue remains. The domain of a function equals the range of its inverse, and the range of a function equals the domain of its inverse. The range of f is $[0, \infty)$, so that must become the domain of its inverse. So, you have to restrict the domain of f^{-1} to $[0, \infty)$. That does it.

9. You obtain $q(x)$ from $p(x)$ by taking $p(x)$ and sliding it 3 to the *left* and 5 up. **See the following figure.** Note that $q(x)$ contains "*x plus* 3," but the horizontal transformation is 3 to the *left* — the opposite of what you'd expect. The "+5" in $q(x)$ tells you to go up 5.

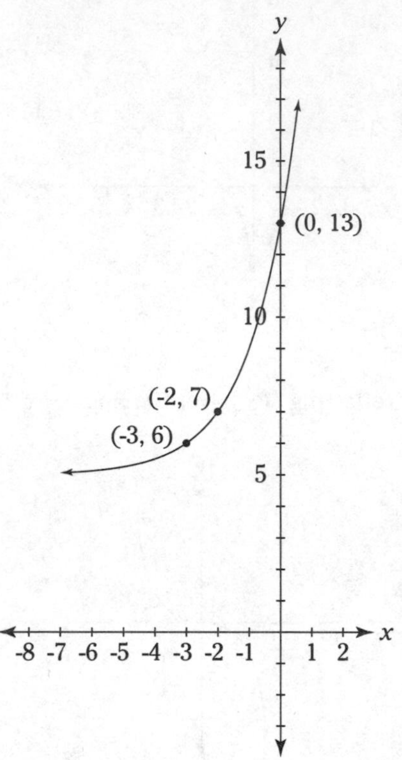

TIP

Horizontal transformations always work opposite the way you'd expect; and PEMDAS works backwards. Vertical transformations, on the other hand, go the normal way — *up* for *plus* and *down* for *minus*; and PEMDAS works the normal way.

10. **A vertical stretch by a factor of 9.**

$g(x) = (3x)^2 = 9x^2$. The multiplication by 9 you see here occurs "outside" the x^2, and is thus a vertical transformation. It's a vertical stretch by a factor of 9. (By the way, it is only with some simple functions like *f* that you can achieve the same transformed function by either a horizontal or a vertical transformation.)

If you're ready to test your skills a bit more, take the following chapter quiz that incorporates all the chapter topics.

Whaddya Know? Chapter 5 Quiz

Quiz time! Complete each problem to test your knowledge on the various topics covered in this chapter. You can then find the solutions and explanations in the next section.

1 For this question and the next, use the function $f(x)=x^3-x^2$.

What does $-f(-x)$ equal, and what's its domain and range?

2 What does $f\left(x+x^{-1}\right)$ equal?

3 Determine whether the following functions are even, odd, or neither:

(a) $f(x)=6x^2-4x$

(b) $f(x)=9x^3-5x-3$

(c) $f(x)=x^4-7$

(d) $f(x)=x^3+4x$

4 Determine whether the following functions are even, odd, or neither:

(a) $g(x)=\sin x$

(b) $g(x)=\cos x$

(c) $g(x)=\sin\left(x^2\right)$

(d) $g(x)=x\cos x$

5 What's the equation of the line that passes through the point $(-0.5,-0.5)$ and that's perpendicular to the line $f(x)=-0.5-0.5x$?

6 (a) What's the product of the slopes of the perpendicular lines from Problem 5?

(b) Give an example of two lines that are perpendicular to each other but where the product of their slopes is different from your answer to Part (a). (You should give the equations of your two example lines.)

7 Use the following functions for this and the next problem. Given $f(x)=\sqrt{x}$, what is $f^{-1}(x)$?

8 Where do $f(x)$ and $f^{-1}(x)$ intersect?

Answers to Chapter 5 Quiz

1. $\boldsymbol{x^3 + x^2}$**. The domain is** $(-\infty, \infty)$**; the range is** $(-\infty, \infty)$**.** You can avoid a bit of messiness by first determining $f(-x)$. Then your answer will just be the opposite of that:

$$\begin{aligned} f(-x) &= (-x)^3 - (-x)^2 \\ &= -x^3 - x^2 \end{aligned}$$

The answer is the opposite of that, namely, $x^3 + x^2$. (Note that you *cannot* distribute the subtraction sign in the first line of this equation.)

The domain of $-f(-x) = x^3 + x^2$ is $(-\infty, \infty)$ — or all real numbers — because there are no restrictions on what you can plug into x. The range is also $(-\infty, \infty)$ because this continuous polynomial function goes down to negative infinity and up to infinity.

2. $\boldsymbol{f\left(x + x^{-1}\right) = \left(x + x^{-1}\right)^3 - \left(x + x^{-1}\right)^2}$**.**

Now you can do the squaring and cubing of those binomials by using the squaring and cubing patterns:

$$\begin{aligned} (a+b)^2 &= a^2 + 2ab + b^2 \\ (a+b)^3 &= a^3 + 3a^2b + 3ab^2 + b^3 \end{aligned}$$

Thus,

$$\begin{aligned} \left(x + x^{-1}\right)^2 &= x^2 + 2 + x^{-2} \\ \left(x + x^{-1}\right)^3 &= x^3 + 3x + 3x^{-1} + x^{-3} \end{aligned}$$

Putting it all together gives you

$$\begin{aligned} f\left(x + x^{-1}\right) &= x^3 + 3x + 3x^{-1} + x^{-3} - \left(x^2 + 2 + x^{-2}\right) \\ &= x^3 + 3x + 3x^{-1} + x^{-3} - x^2 - 2 - x^{-2} \\ &= x^3 - x^2 + 3x - 2 + 3x^{-1} - x^{-2} + x^{-3} \end{aligned}$$

How about the nice symmetry in that final answer?

3. **(a) Neither even nor odd.**

For polynomial functions, all you need to check is whether the powers are even or odd (and since a constant, say 6, is the same as $6x^0$, and since zero is an even number, constants count as *even* terms). The coefficients are irrelevant. If all the powers are even, you have an even function; if all the powers are odd, you've got an odd function; if there's a mix, the function is neither even nor odd.

$f(x)$ has a power two term and a power one term (or, to speak more mathematically, it has a quadratic term and a linear term), so this function is neither even nor odd.

(b) Neither even nor odd.

The powers in this function are 3, 1, and 0, so this function is neither even nor odd.

(c) Even.

The powers here are 4 and 0, so this function is even.

(d) Odd.

The powers here are 3 and 1, so this function is odd.

It wouldn't be a bad idea for you to graph these four functions to confirm the answers visually.

Even functions are symmetric with respect to the y-axis (a left-right symmetry). Odd functions are symmetric with respect to the origin (that's a rotational type of symmetry — if you took the function and rotated it 180° about the origin, it would land on itself).

4 **(a) Odd.**

If you graph the sine function, you see that it's an odd function.

(b) Even.

As you can see from its graph, the cosine function is even.

(c) Even.

The ordinary sine function is odd, but this function is even. The formal way of showing that a function is even is to show that $f(-x)=f(x)$. Let's check this for $g(x)$:

$g(-x)=\sin\left((-x)^2\right)$, which equals $\sin\left(x^2\right)$, and that equals $g(x)$ so, yes, it checks.

(d) Odd.

The ordinary cosine function is even, but this function is odd. The formal way of showing that a function is odd is to show that $f(-x)=-f(x)$. Let's check this for $g(x)$:

$g(-x)=-x\cos(-x)$, which equals $-x\cos x$ from the basic trig identity, $\cos x=\cos(-x)$. $-x\cos x$ equals $-g(x)$ so, yes, it checks.

5 $\boldsymbol{y=2x+0.5}$**.**

First, note that the slope of the given line is -0.5 or $-\frac{1}{2}$. Perpendicular lines have *opposite reciprocal* slopes, and the opposite reciprocal of $-\frac{1}{2}$ is 2, so that's the slope of the line you're looking for. Finish with the *point-slope* form for the equation of a line: $y-y_1=m(x-x_1)$:

$$\begin{aligned} y-(-0.5)&=2(x-(-0.5)) \\ y+0.5&=2x+1 \\ y&=2x+0.5 \end{aligned}$$

6 **(a) −1.**

Is that a softball question or what? The slopes are $-\frac{1}{2}$ and 2, so their product is −1. The product of opposite reciprocals is always −1.

(b) Two examples are $y = -15$ and $x = 6$.

While it's true that the product of opposite reciprocals is always −1, if one of the lines is horizontal (slope of zero), and the other is vertical (undefined slope), the product of their slopes is undefined. Your answer should be any horizontal line like $y = -15$ and any vertical line like $x = 6$.

7 **$f^{-1}(x) = x^2,\ x \geq 0$.**

You've got the function $y = \sqrt{x}$. Switch the x and y then solve for y:

$$x = \sqrt{y}$$
$$y = x^2$$

Thus, $f^{-1}(x) = x^2$ — well, almost. Don't forget: You've got to check your answer's domain and range. The range of a function's inverse must be the same as the domain of the original function, and the domain of a function's inverse must be the same as the range of the original function. The range of $f^{-1}(x) = x^2$ matches the domain of $f(x) = \sqrt{x}$, but the domain of $f^{-1}(x) = x^2$, namely, all real numbers, is larger than the range of $f(x) = \sqrt{x}$, which is all nonnegative numbers. So, you've got to restrict the domain of the inverse. Your final answer is $f^{-1}(x) = x^2,\ x \geq 0$.

8 **The functions intersect at $(0,\ 0)$ and $(1,\ 1)$.**

Set the two functions equal to each other and solve:

$$\begin{aligned}\sqrt{x} &= x^2\\ \sqrt{x}^2 &= \left(x^2\right)^2\\ x &= x^4\\ x^4 - x &= 0\\ x\left(x^3 - 1\right) &= 0\end{aligned}$$

Let's pause here for a minute to observe two things. First, note that you should not cancel an x in line three — which would give you $1 = x^3$. That's generally not a good idea. That canceling will often — as it does here — cause you to miss the solution of zero. Second, if you stopped here, and sort of relied on common sense to conclude that x equals 0 or 1, you'd be correct. But if you want to make your math teacher or professor happy, you'll finish like so.

$$\begin{aligned}x\left(x^3 - 1\right) &= 0\\ x(x-1)\left(x^2 + x + 1\right) &= 0 \quad \text{(using the difference of cubes factor pattern)}\end{aligned}$$

The zero product property gives you the obvious answers of 0 and 1, and then you'd also have $x^2 + x + 1 = 0$. But since the discriminant of that quadratic is negative, its zeros are both imaginary. The function and its inverse thus intersect at $x = 0$ and $x = 1$. Plug those x-values into either function to get the y-values. The functions intersect at $(0,\ 0)$ and $(1,\ 1)$. That's a wrap.

IN THIS CHAPTER

» Socking it to 'em with SohCahToa

» Everybody's got an angle: 30°, 45°, 60°

» Circumnavigating the unit circle

» Graphing trig functions

» Investigating inverse trig functions

Chapter 6 The Trig Tango

Believe it or not, trigonometry is a very practical, real-world branch of mathematics, because it involves the measurement of lengths and angles. Surveyors use it when surveying property, making topographical maps, and so on. The Ancient Greeks and Alexandrians, among others, knew not only simple SohCahToa stuff, but a lot of sophisticated trig as well. They used it for building, navigation, and astronomy. Trigonometry comes up a lot in the study of calculus, so if you snoozed through high school trig, WAKE UP! and review the following trig basics.

Starting off with SohCahToa

The study of trig begins with the right triangle. The three main trig functions (sine, cosine, and tangent) and their reciprocals (cosecant, secant, and cotangent) all tell you something about the lengths of the sides of a right triangle that contains a given acute angle — like angle x in Figure 6-1. The longest side of this right triangle (or any right triangle), the diagonal side, is called the *hypotenuse*. The side that's 3 units long in this right triangle is referred to as the *opposite* side because it's on the opposite side of the triangle from angle x, and the side of length 4 is called the *adjacent* side because it's adjacent to, or touching, angle x.

SohCahToa is a meaningless mnemonic device that helps you remember the definitions of the sine, cosine, and tangent functions. *SohCahToa* uses the initial letters of *sine*, *cosine*, and *tangent*, and the initial letters of *hypotenuse*, *opposite*, and *adjacent* to help you remember the following definitions. (To remember how to spell *SohCahToa*, note its pronunciation and the fact that it contains three groups of three letters each.) For any angle θ,

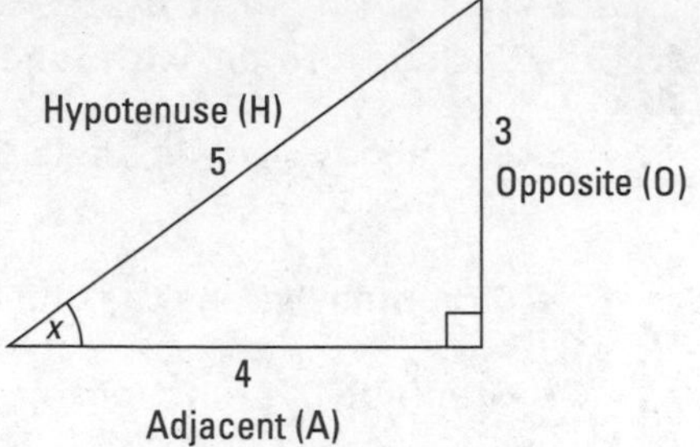

FIGURE 6-1: You're studying calculus, so maybe you thought right triangles were behind you. Guess again.

Soh $\sin\theta = \frac{O}{H}$

Cah $\cos\theta = \frac{A}{H}$

Toa $\tan\theta = \frac{O}{A}$

For the triangle in Figure 6-1,

$$\sin x = \frac{O}{H} = \frac{3}{5} \qquad \cos x = \frac{A}{H} = \frac{4}{5} \qquad \tan x = \frac{O}{A} = \frac{3}{4}$$

The other three trig functions are reciprocals of these: Cosecant (csc) is the reciprocal of sine, secant (sec) is the reciprocal of cosine, and cotangent (cot) is the reciprocal of tangent.

$$\csc\theta = \frac{1}{\sin\theta} = \frac{1}{\frac{O}{H}} = \frac{H}{O} \qquad \sec\theta = \frac{1}{\cos\theta} = \frac{1}{\frac{A}{H}} = \frac{H}{A} \qquad \cot\theta = \frac{1}{\tan\theta} = \frac{1}{\frac{O}{A}} = \frac{A}{O}$$

So, for the triangle in Figure 6-1,

$$\csc x = \frac{H}{O} = \frac{5}{3} \qquad \sec x = \frac{H}{A} = \frac{5}{4} \qquad \cot x = \frac{A}{O} = \frac{4}{3}$$

Two Important Trig Triangles

I discussed the two special right triangles shown in Figure 6-2 in the geometry section of Chapter 4. But I'm showing them to you again so you can see how *SohCahToa* works with them.

When you apply the *SohCahToa* trig functions and their reciprocals to the 45° angle in the 45°-45°-90° triangle, you get the following trig values:

$$\sin 45° = \frac{O}{H} = \frac{1}{\sqrt{2}} = \frac{\sqrt{2}}{2} \approx 0.71 \qquad \csc 45° = \frac{H}{O} = \frac{\sqrt{2}}{1} = \sqrt{2} \approx 1.41$$

$$\cos 45° = \frac{A}{H} = \frac{1}{\sqrt{2}} = \frac{\sqrt{2}}{2} \approx 0.71 \qquad \sec 45° = \frac{H}{A} = \frac{\sqrt{2}}{1} = \sqrt{2} \approx 1.41$$

$$\tan 45° = \frac{O}{A} = \frac{1}{1} = 1 \qquad \cot 45° = \frac{A}{O} = \frac{1}{1} = 1$$

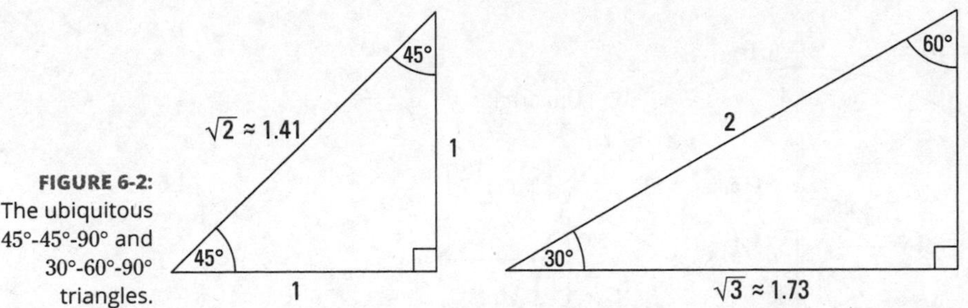

FIGURE 6-2: The ubiquitous 45°-45°-90° and 30°-60°-90° triangles.

And here's how *SahCahToa* works with the 30° angle in the 30°-60°-90° triangle:

$$\sin 30° = \frac{O}{H} = \frac{1}{2} \qquad \csc 30° = \frac{H}{O} = \frac{2}{1} = 2$$

$$\cos 30° = \frac{A}{H} = \frac{\sqrt{3}}{2} \approx 0.87 \qquad \sec 30° = \frac{H}{A} = \frac{2}{\sqrt{3}} = \frac{2\sqrt{3}}{3} \approx 1.15$$

$$\tan 30° = \frac{O}{A} = \frac{1}{\sqrt{3}} = \frac{\sqrt{3}}{3} \approx 0.58 \qquad \cot 30° = \frac{A}{O} = \frac{\sqrt{3}}{1} = \sqrt{3} \approx 1.73$$

The 30°-60°-90° triangle kills two birds with one stone because it also gives you the trig values for a 60° angle. Look at Figure 6-2 again. For the 60° angle, the $\sqrt{3}$ side of the triangle is now the *opposite* side for purposes of *SohCahToa* because it's on the opposite side of the triangle from the 60° angle. The 1-unit side becomes the *adjacent* side for the 60° angle, and the 2-unit side is still, of course, the hypotenuse. Now use *SohCahToa* again to find the trig values for the 60° angle:

$$\sin 60° = \frac{O}{H} = \frac{\sqrt{3}}{2} \approx 0.87 \qquad \csc 60° = \frac{H}{O} = \frac{2}{\sqrt{3}} = \frac{2\sqrt{3}}{3} \approx 1.15$$

$$\cos 60° = \frac{A}{H} = \frac{1}{2} \qquad \sec 60° = \frac{H}{A} = \frac{2}{1} = 2$$

$$\tan 60° = \frac{O}{A} = \frac{\sqrt{3}}{1} = \sqrt{3} \approx 1.73 \qquad \cot 60° = \frac{A}{O} = \frac{1}{\sqrt{3}} = \frac{\sqrt{3}}{3} \approx 0.58$$

The mnemonic device *SohCahToa*, along with the two oh-so-easy-to-remember right triangles in Figure 6-2, gives you the answers to 18 trig problems!

Circling the Enemy with the Unit Circle

SohCahToa only works with right triangles, and so it can only handle *acute* angles — angles less than 90°. (The angles in a triangle must add up to 180°; because a right triangle has a 90° angle, the other two angles must each be less than 90°.) With the *unit circle*, however, you can find trig values for any size angle. The *unit* circle has a radius of *one unit* and is set in an *x*-*y* coordinate system with its center at the origin. See Figure 6-3.

Figure 6-3 has quite a lot of information, but don't panic; it will all make perfect sense in a minute.

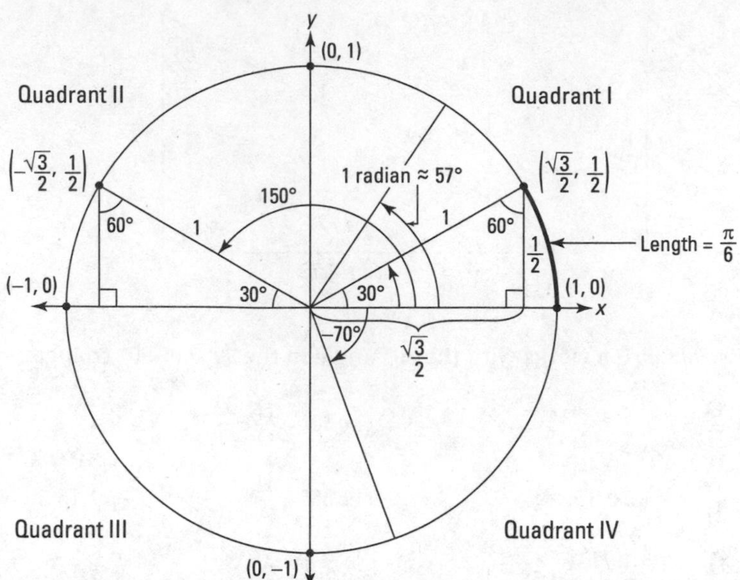

FIGURE 6-3: The so-called unit circle.

Angles in the unit circle

REMEMBER

Measuring angles: To measure an angle in the unit circle, start at the positive *x*-axis and go *counterclockwise* to the *terminal* side of the angle.

For example, the 150° angle in Figure 6-3 begins at the positive *x*-axis and ends at the segment that hits the unit circle at $\left(-\frac{\sqrt{3}}{2}, \frac{1}{2}\right)$. If you go *clockwise* instead, you get an angle with a *negative* measure (like the –70° angle in the figure).

Measuring angles with radians

You know all about *degrees*. You know what 45° and 90° angles look like; you know that *about face* means a turn of 180° and that turning all the way around till you're back to where you started is a 360° turn.

But degrees aren't the only way to measure angles. You can also use *radians*. Degrees and radians are just two different ways to measure angles, like inches and centimeters are two ways to measure length.

REMEMBER

Definition of *radian*: The radian measure of an angle is the length of the arc along the circumference of the unit circle cut off by the angle.

Look at the 30° angle in quadrant I of Figure 6-3. Do you see the bolded section of the circle's circumference that is cut off by that angle? Because a whole circle is 360°, that 30° angle is one-twelfth of the circle. So, the length of the bold arc is one-twelfth of the circle's circumference. Circumference is given by the formula $C = 2\pi r$. This circle has a radius of 1, so its circumference equals 2π. Because the bold arc is one-twelfth of that, its length is $\frac{\pi}{6}$, which is the radian measure of the 30° angle.

TIP

360° equals 2π radians. The unit circle's circumference of 2π makes it easy to remember that 360° equals 2π radians. Half the circumference has a length of π, so 180° equals π radians.

If you focus on the fact that 180° equals π radians, other angles are easy:

- 90° is half of 180°, so 90° equals half of π, or $\frac{\pi}{2}$ radians.
- 60° is a third of 180°, so 60° equals a third of π, or $\frac{\pi}{3}$ radians.
- 45° is a fourth of 180°, so 45° equals a fourth of π, or $\frac{\pi}{4}$ radians.
- 30° is a sixth of 180°, so 30° equals a sixth of π, or $\frac{\pi}{6}$ radians.

REMEMBER

Formulas for converting from degrees to radians and vice versa:

- To convert from degrees to radians, multiply the angle's measure by $\frac{\pi}{180°}$.
- To convert from radians to degrees, multiply the angle's measure by $\frac{180°}{\pi}$.

By the way, the word *radian* comes from *radius.* Look at Figure 6-3 again. An angle measuring 1 radian (about 57°) cuts off an arc along the circumference of this circle that's the same length as the circle's radius. This is true not only of unit circles, but of circles of any size. In other words, take the radius of any circle, lay it along the circle's circumference, and that arc creates an angle of 1 radian.

TIP

Radians are preferred over degrees. In this or any other calculus book, some problems use degrees and others use radians, but radians are the preferred unit. If a problem doesn't specify the unit, do the problem in radians.

Honey, I shrunk the hypotenuse

Look at the unit circle in Figure 6-3 again. See the 30°-60°-90° triangle in quadrant I? It's the same shape but half the size of the one in Figure 6-2. Each of its sides is half as long. Because its hypotenuse now has a length of 1, and because when H is 1, $\frac{O}{H}$ equals O, the sine of the 30° angle, which equals $\frac{O}{H}$, ends up equaling the length of the opposite side. The opposite side is $\frac{1}{2}$, so that's the sine of 30°. Note that the length of the opposite side is the same as the *y*-coordinate of the point $\left(\frac{\sqrt{3}}{2}, \frac{1}{2}\right)$. If you figure the cosine of 30° in this triangle, it ends up equaling the length of the adjacent side, which is the same as the *x*-coordinate of $\left(\frac{\sqrt{3}}{2}, \frac{1}{2}\right)$. Notice that these values for $\sin 30°$ and $\cos 30°$ are the same as the ones given by the 30°-60°-90° triangle in Figure 6-2. This shows you, by the way, that shrinking a right triangle down (or blowing it up) has no effect on the trigonometric values for the angles in the triangle.

Now look at the 30°-60°-90° triangle in quadrant II in Figure 6-3. Because it's the same size as the 30°-60°-90° triangle in quadrant I, which hits the circle at $\left(\frac{\sqrt{3}}{2}, \frac{1}{2}\right)$, the triangle in quadrant II hits the circle at a point that's straight across from and symmetric to $\left(\frac{\sqrt{3}}{2}, \frac{1}{2}\right)$. The coordinates of the point in quadrant II are $\left(-\frac{\sqrt{3}}{2}, \frac{1}{2}\right)$. But remember that angles on the unit circle are all measured from the positive x-axis, so the hypotenuse of this triangle indicates a 150° angle; and that's the angle, not 30°, associated with the point $\left(-\frac{\sqrt{3}}{2}, \frac{1}{2}\right)$. The cosine of 150° is given by the x-coordinate of this point, $-\frac{\sqrt{3}}{2}$, and the sine of 150° equals the y-coordinate, $\frac{1}{2}$.

REMEMBER

Coordinates on the unit circle tell you an angle's cosine and sine. The terminal side of an angle in the unit circle hits the circle at a point whose x-coordinate is the angle's cosine and whose y-coordinate is the angle's sine. Here's a mnemonic: x and y are in alphabetical order as are *cosine* and *sine*.

Putting it all together

Look at Figure 6-4. Now that you know all about the 45°-45°-90° triangle, you can easily work out — or take my word for it — that a 45°-45°-90° triangle in quadrant I hits the unit circle at $\left(\frac{\sqrt{2}}{2}, \frac{\sqrt{2}}{2}\right)$. And if you take the 30°-60°-90° triangle in quadrant I that hits the unit circle at $\left(\frac{\sqrt{3}}{2}, \frac{1}{2}\right)$ and flip it on its side, you get another 30°-60°-90° triangle with a 60° angle that hits the circle at $\left(\frac{1}{2}, \frac{\sqrt{3}}{2}\right)$. As you can see, this point has the same coordinates as those for the 30° angle, but reversed.

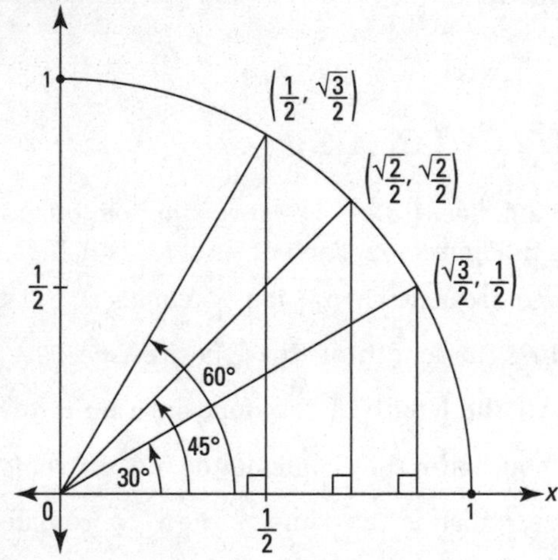

FIGURE 6-4: Quadrant I of the unit circle with three angles and their coordinates.

REMEMBER

How to draw a right triangle in the unit circle: Whenever you draw a right triangle in the unit circle, put the acute angle you care about at the origin — that's $(0, 0)$ — and then put the right angle on the x-axis — never on the y-axis.

TIP

$\frac{\sqrt{3}}{2}$ is greater than $\frac{1}{2}$. To keep from mixing up the numbers $\frac{1}{2}$ and $\frac{\sqrt{3}}{2}$ when dealing with a 30° or a 60° angle, note that because $\sqrt{3}$ is more than 1, $\frac{\sqrt{3}}{2}$ must be greater than $\frac{1}{2}$ $\left(\frac{1}{2}=0.5; \frac{\sqrt{3}}{2}\approx 0.87\right)$. Thus, because a 30° angle hits the circle further out to the right than up, the x-coordinate must be greater than the y-coordinate. So, the point must be $\left(\frac{\sqrt{3}}{2}, \frac{1}{2}\right)$, not the other way around. It's vice versa for a 60° angle.

Now for the whole enchilada. Because of the symmetry in the four quadrants, the three points in quadrant I in Figure 6-4 have counterparts in the other three quadrants, giving you 12 known points. Add to these the four points on the axes, (1, 0), (0, 1), (−1, 0), and (0, −1), and you have 16 total points, each with an associated angle, as shown in Figure 6-5.

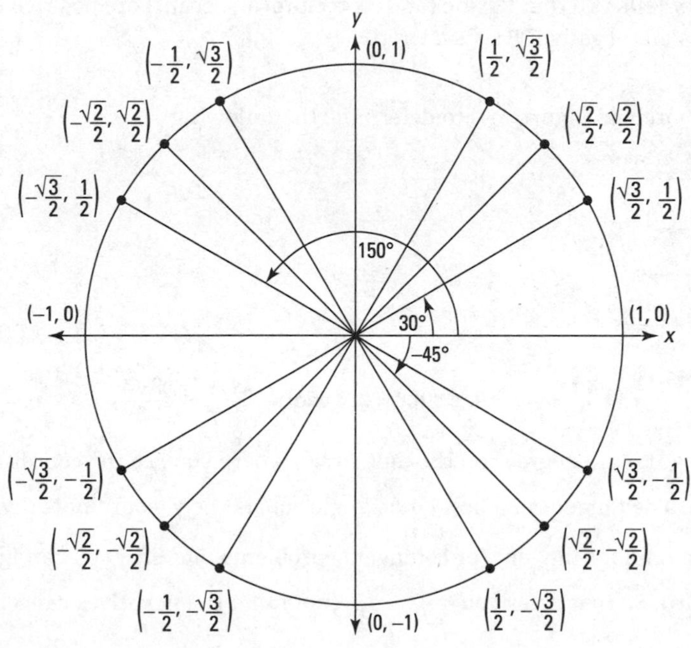

FIGURE 6-5: The unit circle with 16 angles and their coordinates.

These 16 pairs of coordinates automatically give you the cosine and sine of the 16 angles. And because $\tan\theta = \frac{\sin\theta}{\cos\theta}$, you can obtain the tangent of these 16 angles by dividing an angle's y-coordinate by its x-coordinate. (Note that when the cosine of an angle equals zero, the tangent is undefined because you can't divide by zero.) Finally, you can find the cosecant, secant, and cotangent of the 16 angles because these trig functions are just the reciprocals of sine, cosine, and tangent. (Same caution: whenever sine, cosine, or tangent equals zero, the reciprocal function is undefined.) You've now got, at your fingertips — okay, maybe that's a bit of a stretch — the answers to 96 trig questions.

TIP

Learn the unit circle. Knowing the trig values from the unit circle is quite useful in calculus. So quiz yourself. Start by memorizing the 45°-45°-90° and the 30°-60°-90° triangles. Then picture how these triangles fit into the four quadrants of the unit circle. Use the symmetry of the quadrants as an aid. With some practice, you can get pretty quick at figuring out the values for the six trig functions of all 16 angles. (Try to do this without looking at something like Figure 6-5.)

And quiz yourself with radians as well as with degrees. That would bring your total to 192 trig facts! Quick — what's the secant of 210°, and what's the cosine of $\frac{2\pi}{3}$? Here are the answers (no peeking): $-\frac{2\sqrt{3}}{3}$ and $-\frac{1}{2}$.

TIP

All Students Take Calculus. Here's a final tip to help you with the unit circle and the values of all the trig functions. Take any old unit circle (like the one in Figure 6-5) and write the initial letters of *All Students Take Calculus* in the four quadrants: Put an **A** in quadrant I, an **S** in quadrant II, a **T** in quadrant III, and a **C** in quadrant IV. These letters now tell you whether the various trig functions have positive or negative values in the different quadrants. The **A** in quadrant I tells you that **All** six trig functions have positive values in quadrant I. The **S** in quadrant II tells you that **Sine** (and its reciprocal, cosecant) are positive in quadrant II and that all other trig functions are negative there. The **T** in quadrant III tells you that **Tangent** (and its reciprocal, cotangent) are positive in quadrant III and that the other functions are negative there. Finally, the **C** in quadrant IV tells you that **Cosine** (and its reciprocal, secant) are positive there and that the other functions are negative. That's a wrap.

EXAMPLE

Q. Use the unit circle in Figure 6-5 to determine the following:

$\tan 210° =$ ______?

$\cot\frac{3\pi}{4} =$ ______?

$\sec 300° =$ ______?

$\csc\left(-\frac{2\pi}{3}\right) =$ ______?

A. $\tan 210° = \frac{\sqrt{3}}{3}$, $\cot\frac{3\pi}{4} = -1$, $\sec 300° = 2$, $\csc\left(-\frac{2\pi}{3}\right) = -\frac{2\sqrt{3}}{3}$.

For $\tan 210°$, go to 210 degrees on the unit circle, where you see the coordinates $\left(-\frac{\sqrt{3}}{2}, -\frac{1}{2}\right)$. The tangent of a unit circle angle equals the *y*-coordinate divided by the *x*-coordinate (and for tangent (or cotangent) problems like this, you can ignore the denominators). So that gives you $\frac{-1}{-\sqrt{3}} = \frac{\sqrt{3}}{3}$ (note that the negatives cancel).

For $\cot\frac{3\pi}{4}$, note the coordinates at $\frac{3\pi}{4}$ radians or 135 degrees, namely, $\left(-\frac{\sqrt{2}}{2}, \frac{\sqrt{2}}{2}\right)$. The cotangent of a unit circle angle equals the *x*-coordinate divided by the *y*-coordinate. That gives you −1.

For the third problem, note that the coordinates at 300 degrees are $\left(\frac{1}{2}, -\frac{\sqrt{3}}{2}\right)$, and that secant equals the reciprocal of cosine — which is given by the *x*-coordinate. The reciprocal of $\frac{1}{2}$ is, of course, 2.

Finally, at $-\frac{2\pi}{3}$ radians or −120 degrees, the coordinates are $\left(-\frac{1}{2}, -\frac{\sqrt{3}}{2}\right)$. Cosecant is the reciprocal of sine, which is given by the *y*-coordinate. The reciprocal of $-\frac{\sqrt{3}}{2}$ is $-\frac{2\sqrt{3}}{3}$.

1 Use the right triangle to complete the table. No peaking a few pages back!

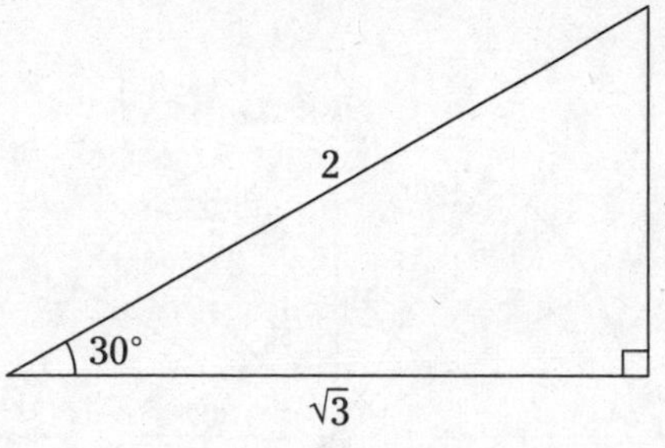

$\sin 30° =$ _____ $\csc 30° =$ _____

$\cos 30° =$ _____ $\sec 30° =$ _____

$\tan 30° =$ _____ $\cot 30° =$ _____

2 Use the triangle from Problem 1 to complete the following table.

$\sin 60° =$ _____ $\csc 60° =$ _____

$\cos 60° =$ _____ $\sec 60° =$ _____

$\tan 60° =$ _____ $\cot 60° =$ _____

3 Use the following triangle to complete the table.

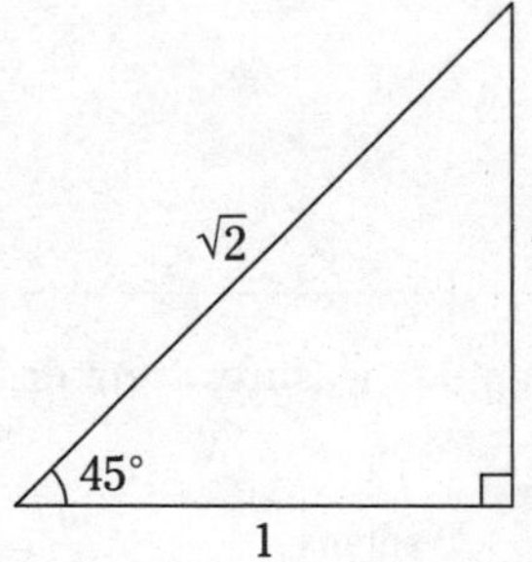

$\sin 45° =$ _____ $\csc 45° =$ _____

$\cos 45° =$ _____ $\sec 45° =$ _____

$\tan 45° =$ _____ $\cot 45° =$ _____

4 Using your results from Problems 1, 2, and 3, fill in the coordinates for the points on the unit circle.

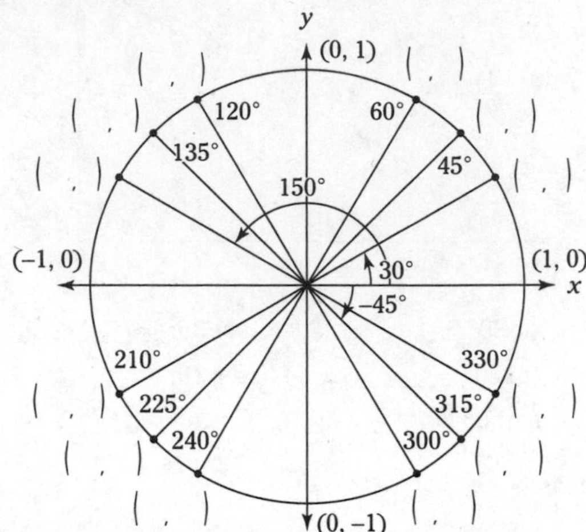

5 Complete the following table using your results from Problem 4.

$\tan 120° = \_\_\_\_\_$ $\csc 180° = \_\_\_\_\_$

$\csc 150° = \_\_\_\_\_$ $\cot 300° = \_\_\_\_\_$

$\cot 270° = \_\_\_\_\_$ $\sec 225° = \_\_\_\_\_$

6 Convert the following angle measures from degrees to radians or vice versa.

$150° = \_\_\_\_\_$ radians $\frac{4\pi}{3} = \_\_\_\_\_°$

$225° = \_\_\_\_\_$ radians $\frac{7\pi}{4} = \_\_\_\_\_°$

$300° = \_\_\_\_\_$ radians $\frac{5\pi}{2} = \_\_\_\_\_°$

$-60° = \_\_\_\_\_$ radians $-\frac{7\pi}{6} = \_\_\_\_\_°$

7 What's $\sec \frac{11\pi}{6}$?

8 What's $\csc \frac{4\pi}{3}$?

9 What's $\tan(3\pi)\cot(3\pi)$?

10 What's $\sin 30° \cdot \cos 45° \cdot \tan 60°$? Try to get the answers to the three pieces in your head — then finish the multiplication on paper.

11 Express $\frac{\sec x}{\tan^2 x}$ in terms of sines and cosines.

12 Solve $\cos x + \sin(2x) = 0$ in the interval $[0, 2\pi]$.

Graphing Sine, Cosine, and Tangent

Figure 6-6 shows the graphs of sine, cosine, and tangent, which you can, of course, produce on a graphing calculator.

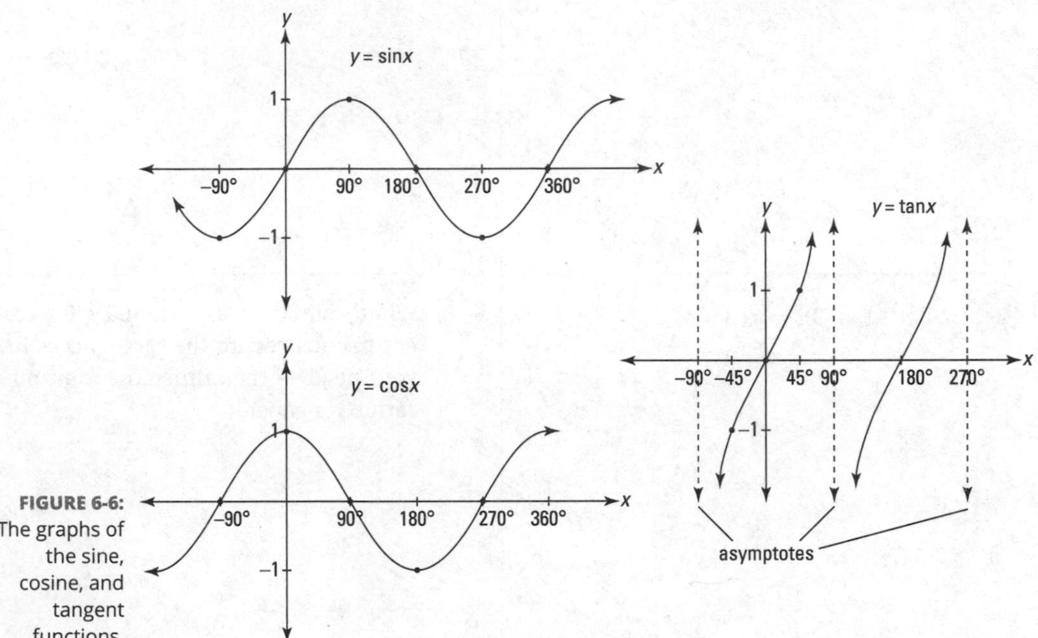

FIGURE 6-6: The graphs of the sine, cosine, and tangent functions.

REMEMBER

Definitions of *periodic* and *period*: Sine, cosine, and tangent — and their reciprocals, cosecant, secant, and cotangent — are *periodic* functions, which means that their graphs contain a basic shape that repeats over and over indefinitely to the left and the right. The *period* of such a function is the horizontal span of one of its cycles.

If you know the unit circle, you can easily reproduce these three graphs by hand. First, note that the sine and cosine graphs are the same shape — cosine is the same as sine, just slid 90° to the left. Also, notice that their simple wave shape goes as high as 1 and as low as –1 and goes on forever to the left and right, with the same shape repeating every 360°. That's the *period* of both functions, 360°. (It's no coincidence, by the way, that 360° is also once around the unit circle.) The unit circle tells you that $\sin 0° = 0$, $\sin 90° = 1$, $\sin 180° = 0$, $\sin 270° = -1$, and $\sin 360° = 0$. If you start with these five points, you can sketch one cycle. The cycle then repeats to the left and right. You can use the unit circle in the same manner to sketch the cosine function (you do this in practice problem 13).

Notice in Figure 6-6 that the period of the tangent function is 180°. If you remember that and the basic pattern of repeating backward *S*-shapes, then sketching it isn't difficult. Because $\tan\theta = \frac{y}{x}$, you can use the unit circle to determine that $\tan(-45°) = -1$, $\tan 0° = 0$, and $\tan 45° = 1$. That gives you the points $(-45°, -1)$, $(0, 0)$, and $(45°, 1)$. Since $\tan(-90°)$ and $\tan 90°$ are both undefined (because $\frac{y}{x}$ at these points gives you a zero in the denominator), you draw *vertical asymptotes* at –90° and 90°.

REMEMBER

Definition of *vertical asymptote:* A vertical asymptote is an imaginary line that a curve gets closer and closer to (but never touches) as the curve goes up toward infinity or down toward negative infinity. (In Chapters 7 and 8, you see more vertical asymptotes and also some horizontal asymptotes.)

The two asymptotes at –90° and 90° and the three points at (–45°, –1), (0, 0), and (45°, 1) show you where to sketch one backward *S*. The *S*-shapes then repeat every 180° to the left and the right.

Inverse Trig Functions

An inverse trig function, like any inverse function, reverses what the original function does. For example, $\sin 30° = \frac{1}{2}$, so the inverse sine function — written as $\sin^{-1}$ — reverses the input and output. Thus, $\sin^{-1}\frac{1}{2} = 30°$. It works the same for the other trig functions.

WARNING

The negative 1 superscript in the sine inverse function is *not* a negative 1 power, despite the fact that it looks just like it. Raising something to the negative 1 power gives you its reciprocal, so you might think that $\sin^{-1} x$ is the reciprocal of $\sin x$, but the reciprocal of sine is cosecant, *not* sine inverse. Pretty weird that the same symbol is used to mean two different things. Go figure.

The only trick with inverse trig functions is memorizing their *ranges* — that's the interval of their outputs. Consider sine inverse, for example. Because both $\sin 30° = \frac{1}{2}$ and $\sin 150° = \frac{1}{2}$, you wouldn't know whether $\sin^{-1}\frac{1}{2}$ equals 30° or 150° unless you know how the interval of sine inverse outputs is defined. And remember, in order for something to be a function, there can't be any mystery about the output for a given input. If you reflect the sine function over the line $y = x$ to create its inverse, you get a vertical wave that isn't a function because it doesn't pass the vertical line test. (See the definition of the vertical line test in Chapter 5.) To make sine inverse a function, you have to take a small piece of the vertical wave that does pass the vertical line test. The same thing goes for the other inverse trig functions. Here are their ranges:

The range of $\sin^{-1} x$ is $\left[-\frac{\pi}{2}, \frac{\pi}{2}\right]$, or $[-90°, 90°]$.

The range of $\cos^{-1} x$ is $[0, \pi]$, or $[0°, 180°]$.

The range of $\tan^{-1} x$ is $\left(-\frac{\pi}{2}, \frac{\pi}{2}\right)$, or $(-90°, 90°)$.

The range of $\cot^{-1} x$ is $(0, \pi)$, or $(0°, 180°)$.

Note the pattern: The range of $\sin^{-1} x$ is basically the same as $\tan^{-1} x$, and the range of $\cos^{-1} x$ is basically the same as $\cot^{-1} x$.

Believe it or not, calculus authors don't agree on the ranges for the secant inverse and cosecant inverse functions. You'd think they could agree on this like they do with just about everything else in mathematics. Humph. Use the ranges given in your particular textbook. If you don't have a textbook, use the $\sin^{-1} x$ range for its cousin $\csc^{-1} x$, and use the $\cos^{-1} x$ range for $\sec^{-1} x$. (By the way, I don't refer to $\csc^{-1} x$ as the reciprocal of $\sin^{-1} x$ because it's *not* its reciprocal — even though csc *x* is the reciprocal of sin *x*. Ditto for $\cos^{-1} x$ and $\sec^{-1} x$.)

Identifying with Trig Identities

Remember trig identities like $\sin^2 x + \cos^2 x = 1$ and $\sin(2x) = 2\sin x \cos x$? Tell the truth now — most people remember trig identities about as well as they remember nineteenth-century vice presidents. They come in handy in calculus, though, so a list of other useful identities is in the online Cheat Sheet. Go to `www.dummies.com` and type **Calculus All in One For Dummies Cheat Sheet** in the Search box.

EXAMPLE

Q. Graph $y = \sin^{-1} x$.

A. **Check out the following figure.** It shows the relationship between $y = \sin x$ and $y = \sin^{-1} x$. (The dashed sine curve is just like the sine graph in Figure 6-6 except that the angles on the x-axis are in radians instead of degrees.) The inverse sine graph is created by flipping the solid portion of the sine wave over the line $y = x$. The endpoints of $y = \sin^{-1} x$ are at $\left(-1, -\frac{\pi}{2}\right)$ and $\left(1, \frac{\pi}{2}\right)$.

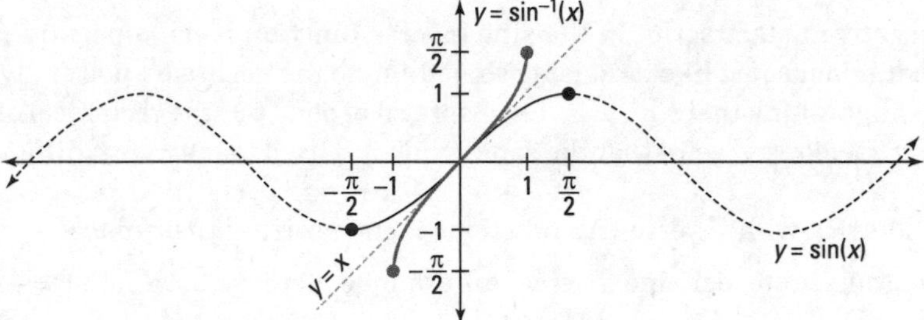

YOUR TURN

13 Using the unit circle in Figure 6-5, but not peaking at Figure 6-6, sketch $y = \cos x$.

14 Using your answers from Problem 4 and your knowledge of the ranges of the inverse trig functions, complete the following table.

$\sin^{-1}\left(\frac{1}{2}\right) =$ ____°	$\tan^{-1}\sqrt{3} =$ ____ radians
$\sin^{-1}\left(-\frac{1}{2}\right) =$ ____°	$\sin^{-1} 1 =$ ____ radians
$\cos^{-1}\left(-\frac{1}{2}\right) =$ ____°	$\cos^{-1} 1 =$ ____ radians
$\tan^{-1}(-1) =$ ____°	$\cos^{-1} 0 =$ ____ radians

Practice Questions Answers and Explanations

1. $\sin 30° = \frac{1}{2}$ $\quad \csc 30° = 2$

   $\cos 30° = \frac{\sqrt{3}}{2}$ $\quad \sec 30° = \frac{2\sqrt{3}}{3}$

   $\tan 30° = \frac{\sqrt{3}}{3}$ $\quad \cot 30° = \sqrt{3}$

2. $\sin 60° = \frac{\sqrt{3}}{2}$ $\quad \csc 60° = \frac{2\sqrt{3}}{3}$

   $\cos 60° = \frac{1}{2}$ $\quad \sec 60° = 2$

   $\tan 60° = \sqrt{3}$ $\quad \cot 60° = \frac{\sqrt{3}}{3}$

3. $\sin 45° = \frac{\sqrt{2}}{2}$ $\quad \csc 45° = \sqrt{2}$

   $\cos 45° = \frac{\sqrt{2}}{2}$ $\quad \sec 45° = \sqrt{2}$

   $\tan 45° = 1$ $\quad \cot 45° = 1$

4. **See the figure.**

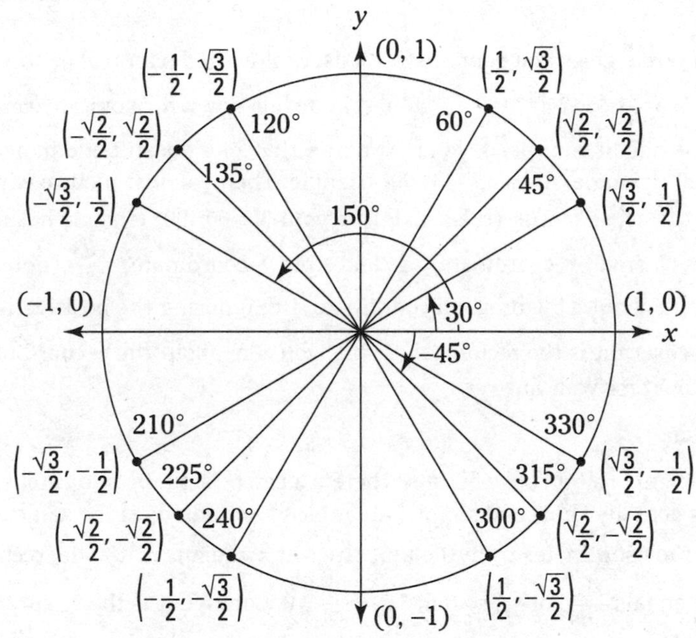

5. $\tan 120° = -\sqrt{3}$ $\quad \csc 180° = \text{undefined}$

   $\csc 150° = 2$ $\quad \cot 300° = -\frac{\sqrt{3}}{3}$

   $\cot 270° = 0$ $\quad \sec 225° = -\sqrt{2}$

6 **$150° = \frac{5\pi}{6}$ radians** **$\frac{4\pi}{3} = 240°$**

$225° = \frac{5\pi}{4}$ radians **$\frac{7\pi}{4} = 315°$**

$300° = \frac{5\pi}{3}$ radians **$\frac{5\pi}{2} = 450°$ (coterminal with 90°)**

$-60° = -\frac{\pi}{3}$ radians **$-\frac{7\pi}{6} = -210°$ (coterminal with 150°)**

7 **$\frac{2\sqrt{3}}{3}$.**

Of course, you can just look at the unit circle to get your answer. Secant is the reciprocal of cosine. The unit circle tells you that $\cos\frac{11\pi}{6}$ (or 330°) is $\frac{\sqrt{3}}{2}$. Flip that upside down for your answer: $\frac{2}{\sqrt{3}}$, or $\frac{2\sqrt{3}}{3}$.

But if you're ambitious and want to try this one in your head, you first notice that 330° doesn't end in a 5, so you have a 30°-60°-90° triangle, not a 45°-45°-90° triangle. Then you just picture where 330° is — it's in quadrant IV close to 360° (the x-axis). So, your 30°-60°-90° triangle has to be wide and short, which has a big x-coordinate, $\frac{\sqrt{3}}{2}$, and a small y-coordinate, $-\frac{1}{2}$. Because secant is the reciprocal of cosine, you care about the x-coordinate, $\frac{\sqrt{3}}{2}$. Flip it upside down for your answer.

8 **$-\frac{2\sqrt{3}}{3}$.**

The unit circle gives you your answer. Cosecant is the reciprocal of sine. The unit circle tells you that $\sin\frac{4\pi}{3}$ (or 240°) is $-\frac{\sqrt{3}}{2}$. Flip that upside down for your answer: $-\frac{2}{\sqrt{3}}$, or $-\frac{2\sqrt{3}}{3}$.

To do this one in your head, you first notice that 240° doesn't end in a 5, so you have a 30°-60°-90° triangle, not a 45°-45°-90° triangle. Then you just picture where 240° is — it's in quadrant III close to 270° (the y-axis). So your 30°-60°-90° triangle has to be narrow and tall, which has a small x-coordinate, $-\frac{1}{2}$, and a big y-coordinate, $-\frac{\sqrt{3}}{2}$ (note that in this context, when I talk about a big or small coordinate, I'm ignoring the positive/negative issue). Because cosecant is the reciprocal of sine, you care about the y-coordinate, $-\frac{\sqrt{3}}{2}$. Flip it upside down for your answer.

9 **Undefined.**

This problem is a bit tricky because there's a catch (actually two catches). But other than that, it's actually short and simple. An angle of 3π radians is the same as π radians, so you just use the coordinates from the unit circle at π radians or 180 degrees — namely, $(-1, 0)$. Tangent equals $\frac{\sin}{\cos}$, or $\frac{y}{x}$, so $\tan(3\pi) = \frac{0}{-1} = 0$. Cotangent is the reciprocal of tangent, so $\cot(3\pi) = \frac{-1}{0}$, which is undefined. (Don't forget, you can't divide by zero!) Thus, your answer for $\tan(3\pi)\cot(3\pi)$ is zero times undefined, which is undefined.

Here are the two catches: First, you might think that zero times undefined is zero because zero times anything is zero. But it doesn't work that way. If any piece of a problem is undefined, the answer is undefined. The second catch is that you could mistakenly conclude that since tangent and cotangent are reciprocals, their product would be 1. That is generally true of reciprocals, but not here because, again, one of them is undefined. The two values you get here, zero and undefined, are sort of, but not technically, reciprocals. So, you can't multiply them to get 1. No matter how you look at it, the answer is undefined.

10 $\mathbf{\frac{\sqrt{6}}{4}}$.

You should be able to picture in your head that the coordinates on the unit circle at 30°, 45°, and 60° are $\left(\frac{\sqrt{3}}{2}, \frac{1}{2}\right)$, $\left(\frac{\sqrt{2}}{2}, \frac{\sqrt{2}}{2}\right)$, and $\left(\frac{1}{2}, \frac{\sqrt{3}}{2}\right)$, respectively. So, $\sin 30° = \frac{1}{2}$ and $\cos 45° = \frac{\sqrt{2}}{2}$.

For the tangent piece of the problem, here's a tip. Tangent equals $\frac{y}{x}$, but when doing tangent problems on the unit circle, you don't have to bother dividing the y-fraction by the x-fraction. The denominators of these fractions always cancel, so you only have to put the y-numerator over the x-numerator, thus: $\tan 60° = \frac{\sqrt{3}}{1}$.

Multiply these three parts for your final answer: $\sin 30° \cdot \cos 45° \cdot \tan 60° = \frac{1}{2} \cdot \frac{\sqrt{2}}{2} \cdot \frac{\sqrt{3}}{1} = \frac{\sqrt{6}}{4}$

11 $\mathbf{\frac{\cos x}{\sin^2 x}}$.

$\frac{\sec x}{\tan^2 x} = \frac{\frac{1}{\cos x}}{\frac{\sin^2 x}{\cos^2 x}} = \frac{1}{\cos x} \cdot \frac{\cos^2 x}{\sin^2 x} = \frac{\cos^2 x}{\cos x \cdot \sin^2 x}$. Now, just cancel one of the cosines, and you're done.

12 $\mathbf{\frac{\pi}{2}, \frac{7\pi}{6}, \frac{3\pi}{2}}$, **and** $\mathbf{\frac{11\pi}{6}}$.

It's generally difficult to deal with a trig equation with two different arguments (the x and the $2x$), so you should try to do something to get rid of the $2x$. The trig identity, $\sin(2x) = 2\sin x \cos x$, is the ticket. Make the substitution:

$$\cos x + \sin(2x) = 0$$
$$\cos x + 2\sin x \cos x = 0$$

Now factor by pulling out the GCF; then use the zero product property:

$$\cos x(1 + 2\sin x) = 0$$

$$\cos x = 0 \quad \text{or} \quad 1 + 2\sin x = 0$$
$$2\sin x = -1$$
$$\sin x = -\frac{1}{2}$$

If you know the unit circle well (you should!), you know that cosine equals zero at $\frac{\pi}{2}$ and $\frac{3\pi}{2}$ and that sine equals $-\frac{1}{2}$ at $\frac{7\pi}{6}$ and $\frac{11\pi}{6}$. That's a wrap.

13. The x-coordinates of points on the unit circle tell you that $\cos 0° = 1$, $\cos 90° = 0$, $\cos 180° = -1$, $\cos 270° = 0$, and $\cos 360° = 1$. Starting with these five points, you can sketch one cycle of the sinusoidal shape of the cosine function. The cycle then repeats to the left and right. You should end up with a graph similar to the one in Figure 6-6.

14. $\sin^{-1}\frac{1}{2} = 30°$ $\tan^{-1}\sqrt{3} = \frac{\pi}{3}$ **radians**

$\sin^{-1}\left(-\frac{1}{2}\right) = -30°$ $\sin^{-1}1 = \frac{\pi}{2}$ **radians**

$\cos^{-1}\left(-\frac{1}{2}\right) = 120°$ $\cos^{-1}1 = 0$ **radians**

$\tan^{-1}(-1) = -45°$ $\cos^{-1}0 = \frac{\pi}{2}$ **radians**

Whaddya Know? Chapter 6 Quiz

Quiz time! Complete each problem to test your knowledge on the various topics covered in this chapter. You can then find the solutions and explanations in the next section.

1 For this problem and the next, study the unit circle for a few minutes, and try to commit it to memory. Then try to answer the questions without looking at it. If that's too difficult, you can use the unit circle as a "cheat sheet," but if any of your answers needs simplification, try to do that in your head.

What are the values of the six trig functions for 300°?

2 What are the values of the six trig functions for $\frac{7\pi}{6}$ radians?

3 **(a)** Express $(\cos\theta + \sin\theta)(\cos\theta - \sin\theta)$ three different ways as trig functions raised to the second power.

(b) Now use your knowledge of trig identities to express $(\cos\theta + \sin\theta)(\cos\theta - \sin\theta)$ as a single trig function raised to the first power (in other words, no squaring, cubing, etc. is allowed).

4 **(a)** $\sin\left(\sin^{-1}\left(-\frac{2}{3}\right)\right) = ?$

(b) $\cos^{-1}\left(\cos\frac{2\pi}{3}\right) = ?$

5 **(a)** $\cos^{-1}(\cos 435°) = ?$

(b) $\cos^{-1}(\cos 190°) = ?$

6 **(a)** $\cot\left(\sin^{-1}\left(-\frac{1}{2}\right)\right) = ?$

(b) $\sin^{-1}(\tan 225°) = ?$

7 How do the amplitude and period of $f(x) = 3\sin(2x)$ compare with the amplitude and period of the ordinary sine function?

8 Solve for θ: $2\sin^2\theta + \sin\theta = 1$ within the span $[0, 2\pi)$.

Answers to Chapter 6 Quiz

1. $\cos 300° = \dfrac{1}{2}$

   $\sin 300° = \dfrac{-\sqrt{3}}{2}$

   $\tan 300° = -\sqrt{3}$ **(the *y*-numerator over the *x*-numerator)**

   $\sec 300° = 2$ **(the reciprocal of cosine)**

   $\csc 300° = \dfrac{-2\sqrt{3}}{3}$ **(flip the sine answer then rationalize the denominator)**

   $\cot 300° = \dfrac{-\sqrt{3}}{3}$ **(flip the tangent answer then rationalize the denominator)**

   Can you picture where 300° is on the unit circle? The radius of the unit circle to the 300° point makes a 60° angle with the x-axis, so you've got a tall, narrow 30°-60°-90° triangle in quadrant IV. That can help you remember the coordinates at that point, because you know that the y-coordinate is "greater" than the x-coordinate (except that it's negative). The coordinates at the 300° point are $\left(\dfrac{1}{2}, \dfrac{-\sqrt{3}}{2}\right)$. Those coordinates give you the six answers you need.

2. $\cos\dfrac{7\pi}{6} = \dfrac{-\sqrt{3}}{2}$

   $\sin\dfrac{7\pi}{6} = \dfrac{-1}{2}$

   $\tan\dfrac{7\pi}{6} = \dfrac{\sqrt{3}}{3}$ **(the *y*-numerator over the *x*-numerator, then rationalize)**

   $\sec\dfrac{7\pi}{6} = \dfrac{-2\sqrt{3}}{3}$ **(the reciprocal of cosine)**

   $\csc\dfrac{7\pi}{6} = -2$ **(the reciprocal of sine)**

   $\cot\dfrac{7\pi}{6} = \sqrt{3}$ **(flip the tangent answer then rationalize)**

   At the $\dfrac{7\pi}{6}$ point, you've got a wide, short 30°-60°-90° triangle in quadrant III that makes a 30° angle with the x-axis. The coordinates at that point are $\left(\dfrac{-\sqrt{3}}{2}, \dfrac{-1}{2}\right)$. That's all you need.

3. (a) $\mathbf{\cos^2\theta - \sin^2\theta,\ 2\cos^2\theta - 1,\ \text{and}\ 1 - 2\sin^2\theta.}$

   You know the difference of squares pattern, right? You better! $(a+b)(a-b) = a^2 - b^2$. Thus, $(\cos\theta + \sin\theta)(\cos\theta - \sin\theta) = \cos^2\theta - \sin^2\theta$. That's your first answer. Next, use the Pythagorean Identity, $\sin^2\theta + \cos^2\theta = 1$, and its two other forms, $\sin^2\theta = 1 - \cos^2\theta$ and $\cos^2\theta = 1 - \sin^2\theta$, to make substitutions in your first answer.

   (b) $\mathbf{\cos(2\theta).}$

   The three answers to Part (a) are the three forms of the trig identity for $\cos(2\theta)$.

4 (a) $-\frac{2}{3}$.

When you have one of these problems with a trig function and the inverse of the same trig function — with the inverse function in the *inside* — the answer is right there. It's automatic.

(b) $\frac{2\pi}{3}$.

When the inverse function is on the outside like here, you've got to be careful. The answer to this particular problem, like with Part (a), is automatic. But this problem is automatic only because the given angle, namely $\frac{2\pi}{3}$, is in the range of the inverse cosine function, $[0, \pi]$. When that's not the case, you've got a little work to do.

5 (a) **75°.**

Okay. So here's a problem with the inverse function on the outside where the angle is *not* in the range of the inverse cosine function. What to do? First, consider where 435° is. You should be able to see that a 435° angle is in the proper location for the inverse cosine function (because it's in quadrant I). It's coterminal with a 75° angle, because $75° + 360° = 435°$. Because 75° *is* in the range of the inverse cosine function, that's your answer.

(b) **170°.**

And here's the third type of problem with the inverse function on the outside. This time, the angle, 190°, is not in the proper location for the inverse cosine function because it's in quadrant III. For this situation, you need to find an angle that *is* in the proper span for inverse cosine and that has the same cosine answer as the cosine of 190°. Consider the unit circle. The cosine of an angle in the unit circle is the x-coordinate of the point on the unit circle where the angle meets the circle. Imagine where the 190° point is on the unit circle. It's just below the x-axis in quadrant III. All you need to do is to reflect that point over the x-axis. That brings you to 170°, and that's your answer.

Another way to do this is to simply take the negative of the angle, –190°, which is in the proper location, because it's in quadrant II. Then just add 360° to –190° (170°) to produce a coterminal angle that's in the correct span.

6 (a) $-\sqrt{3}$.

From the unit circle, you can see that angles of –30° and 210° have a sine equal to $-\frac{1}{2}$. The range of inverse sine is $[-90°, 90°]$, so $\sin^{-1}\left(-\frac{1}{2}\right) = -30°$. Then the unit circle coordinates at –30°, namely $\left(\frac{\sqrt{3}}{2}, \frac{-1}{2}\right)$, give you the cotangent answer of $-\sqrt{3}$.

(b) **90°.**

From the unit circle, $\tan 225° = 1$, and also from the unit circle, $\sin^{-1} 1 = 90°$. That's it.

7. **$f(x) = 3\sin(2x)$ has an amplitude 3 times the amplitude of the ordinary sine function and a period that's half of the ordinary period. (That gives $f(x) = 3\sin(2x)$ an amplitude of 3 and a period of π.)**

   From your knowledge of the transformation of functions, you should know that the 3 is a vertical stretch and that the 2 is a horizontal *shrink*.

8. **$\frac{3\pi}{2}, \frac{\pi}{6}, \frac{5\pi}{6}$.**

   This is a quadratic equation in $\sin\theta$. So, like with most quadratic equations, you should set this equal to zero: $2\sin^2\theta + \sin\theta - 1 = 0$. Now factor: $(2\sin\theta - 1)(\sin\theta + 1) = 0$. The zero product property then gives you the two answers of $\sin\theta = \frac{1}{2}$ or $\sin\theta = -1$. You can then find your final answers on the unit circle: you might notice that sine equals −1 at $-\frac{\pi}{2}$, but that's not in the given span, so you've got to be sure to give your answer as $\frac{3\pi}{2}$. Sine equals $\frac{1}{2}$ at $\frac{\pi}{6}$, so that's another answer. And don't forget about the third answer of $\frac{5\pi}{6}$.

3 Limits

In This Unit . . .

CHAPTER 7: **Limits and Continuity**

Take It to the Limit — NOT
Linking Limits and Continuity
The 33333 Limit Mnemonic
Practice Questions Answers and Explanations
Whaddya Know? Chapter 7 Quiz
Answers to Chapter 7 Quiz

CHAPTER 8: **Evaluating Limits**

Easy Does It — Easy Limits
The "Real Deal" Limit Problems
Evaluating Limits at ±Infinity
Practice Questions Answers and Explanations
Whaddya Know? Chapter 8 Quiz
Answers to Chapter 8 Quiz

IN THIS CHAPTER

» Taking a look at limits

» Evaluating functions with holes — break out the mothballs

» Exploring continuity and discontinuity

Chapter 7
Limits and Continuity

As discussed in Chapter 3, you can use ordinary algebra and geometry when the things in a math or science problem *aren't* changing (sort of) and when the graph of the function in question is a *straight* line. But you need calculus when things *are* changing — these changing things show up on graphs as *curves*. Calculus can handle such things by zooming in on the curves till they become straight (zooming in infinitely far, sort of). At that point, ordinary algebra and geometry can be used. *Limits* are the seemingly magical trick or tool that does this zooming-in process. It's the mathematics of limits that makes calculus work.

Limits are fundamental for both differential and integral calculus. The formal definition of a derivative involves a limit, as does the definition of a definite integral. (If you're a real go-getter and can't wait to read the actual definitions, check out Chapters 9 and 14.) Now, it turns out that after you learn the shortcuts for calculating derivatives and integrals, you won't need to use the longer limit methods anymore. But understanding the mathematics of limits is nonetheless important because it forms the foundation upon which the vast architecture of calculus is built (okay, so I got a bit carried away there). In this chapter, I lay the groundwork for differentiation and integration by exploring limits and the closely related topic, continuity.

Take It to the Limit — NOT

Limits can be tricky. Don't worry if you don't grasp the concept right away.

MATH RULES

Informal definition of *limit* (the formal definition is in a few pages): The limit of a function (if it exists) for some x-value c is the height the function gets closer and closer to as x gets closer and closer to c from the left and the right. (*Note:* This definition does not apply to limits where x approaches infinity or negative infinity. More about those limits later in this chapter and in Chapter 8.)

Got it? You're kidding! Let me say it another way. A function has a limit for a given x-value c if the function zeros in on some height as x gets closer and closer to the given value c from the left and the right. Did that help? I didn't think so. It's much easier to understand limits through examples than through this sort of mumbo jumbo, so let's take a look at some.

Using three functions to illustrate the same limit

Consider the function $f(x)=3x+1$, shown on the left in Figure 7-1. When we say that the limit of $f(x)$ as x approaches 2 is 7, written as $\lim_{x\to 2} f(x)=7$, we mean that as x gets closer and closer to 2 from the left and the right, $f(x)$ gets closer and closer to a height of 7. By the way, as far as I know, the number 2 in this example doesn't have a formal name, but I call it the *arrow-number*. The arrow-number gives you a horizontal location in the x direction. Don't confuse it with the *answer* to the limit problem, called the *limit*, which refers to a y-value or *height* of the function (7 in this example). Now, look at Table 7-1.

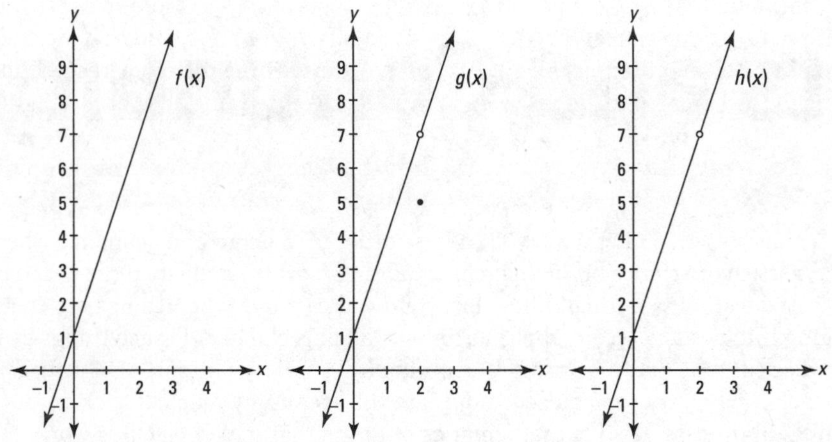

FIGURE 7-1: The graphs of the functions of *f*, *g*, and *h*.

Table 7-1 Input and Output Values of $f(x)=3x+1$ as x Approaches 2

x approaches 2 from the left → ← *x* approaches 2 from the right

x	1	1.5	1.9	1.99	1.999	2.001	2.01	2.1	2.5	3
$f(x)$	4	5.5	6.7	6.97	6.997	7.003	7.03	7.3	8.5	10

y approaches 7 → ← *y* approaches 7

Table 7-1 shows that y is approaching 7 as x approaches 2 from both the left and the right, and thus the limit is 7. If you're wondering what all the fuss is about — why not just plug the number 2 into x in $f(x)=3x+1$ and obtain the answer of 7 — I'm sure you've got a lot of company. In fact, if all functions were *continuous* (without gaps) like *f*, you *could* just plug in the arrow-number to get the answer, and this type of limit problem would basically be pointless. You need to use limits in calculus because of *discontinuous* functions like *g* and *h* that have holes.

Function g in the middle of Figure 7-1 is identical to f except for the hole at $(2, 7)$ and the point at $(2, 5)$. Actually, this function, $g(x)$, would never come up in an ordinary calculus problem — I only use it to illustrate how limits work. (Keep reading. I have a bit more groundwork to lay before you see why I include it.)

The important functions for calculus are the functions like h on the right in Figure 7-1, which come up frequently in the study of derivatives. This third function is identical to $f(x)$ except that the point $(2, 7)$ has been plucked out, leaving a hole at $(2, 7)$ and no other point where x equals 2.

Imagine what the table of input and output values would look like for $g(x)$ and $h(x)$. Can you see that the values would be identical to the values in Table 7-1 for $f(x)$? For both g and h, as x gets closer and closer to 2 from the left and the right, y gets closer and closer to a height of 7. For all three functions, the limit as x approaches 2 is 7.

This brings me to a critical point: When determining the limit of a function as x approaches, say, 2, the value of $f(2)$ — or even whether $f(2)$ exists at all — is totally irrelevant. Take a look at all three functions again where $x = 2$: $f(2)$ equals 7, $g(2)$ is 5, and $h(2)$ doesn't exist (or, as mathematicians say, it's *undefined*). But, again, those three different results don't affect the answers to the three limit problems — which all have the same answer.

You never get to the arrow-number. In a limit problem, x gets closer and closer to the arrow-number c, but technically *never gets there,* and what happens to the function when x equals the arrow-number c has *no effect* on the answer to the limit problem (though for continuous functions like $f(x)$, the function value is the same as the limit answer).

Sidling up to one-sided limits

One-sided limits work like regular, two-sided limits except that x approaches the arrow-number c from just the left or just the right. The most important purpose for such limits is that they're used in the formal definition of a regular limit (see the next section on the formal definition of a limit).

To indicate a one-sided limit, you put a little superscript subtraction sign on the arrow-number when x approaches the arrow-number from the left, or a superscript addition sign when x approaches the arrow-number from the right. It looks like this:

$$\lim_{x \to 5^-} f(x) \text{ or } \lim_{x \to 0^+} g(x)$$

Look at Figure 7-2. As x approaches 3 from the left, $p(x)$ zeros in on a height of 6, and when x approaches 3 from the right, $p(x)$ zeros in on a height of 2. As with regular limits, the value of $p(3)$ has no effect on the answer to either of these one-sided limit problems. Thus,

$$\lim_{x \to 3^-} p(x) = 6 \text{ and } \lim_{x \to 3^+} p(x) = 2$$

The answer to the regular, two-sided limit problem, $\lim_{x \to 3} p(x)$, is that the limit does not exist because $p(x)$ is zeroing in on different heights as x approaches 3 from the left and from the right.

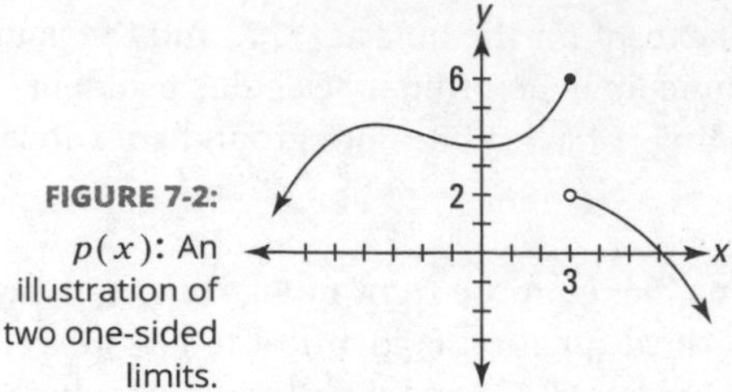

FIGURE 7-2: $p(x)$: An illustration of two one-sided limits.

A function like $p(x)$ in Figure 7-2 is called a *piecewise function* because it's got separate pieces. Each part of a piecewise function has its own equation — like, for example, the following three-piece function:

$$y = \begin{cases} x^2 & \text{for } x \le 1 \\ 3x - 2 & \text{for } 1 < x \le 10 \\ x + 5 & \text{for } x > 10 \end{cases}$$

Sometimes a chunk of a piecewise function connects with its neighboring chunk, in which case the function is continuous there. And sometimes, like with $p(x)$, a piece does not connect with the adjacent piece; this results in a discontinuity.

The formal definition of a limit — just what you've been waiting for

Now that you know about one-sided limits, I can give you the formal mathematical definition of a limit. Here goes:

MATH RULES

Formal definition of limit: Let f be a function and let c be a real number.

$\lim\limits_{x \to c} f(x)$ exists if and only if

1. $\lim\limits_{x \to c^-} f(x)$ exists,
2. $\lim\limits_{x \to c^+} f(x)$ exists, and
3. $\lim\limits_{x \to c^-} f(x) = \lim\limits_{x \to c^+} f(x)$.

Calculus books always present this as a three-part test for the existence of a limit, but condition 3 is the only one you need to worry about because 1 and 2 are built into 3. You just have to remember that you can't satisfy condition 3 if the left and right sides of the equation are both undefined or nonexistent; in other words, it is *not* true that *undefined = undefined* or that *nonexistent = nonexistent.* (I think this is why calc texts use the three-part definition.) As long as you've got that straight, condition 3 is all you need to check.

MATH RULES

When we say a limit exists, it means that the limit equals a *finite* number. Some limits equal infinity or negative infinity, but you nevertheless say that they *do not exist.* That may seem strange, but take my word for it. (More about infinite limits in the next section.)

Limits and vertical asymptotes

A *rational function* like $f(x)=\frac{(x+2)(x-5)}{(x-3)(x+1)}$ has vertical asymptotes at $x=3$ and $x=-1$. Remember asymptotes? They're imaginary lines that the graph of a function gets closer and closer to as it goes up, down, left, or right toward infinity or negative infinity. The function $f(x)$ is shown in Figure 7-3.

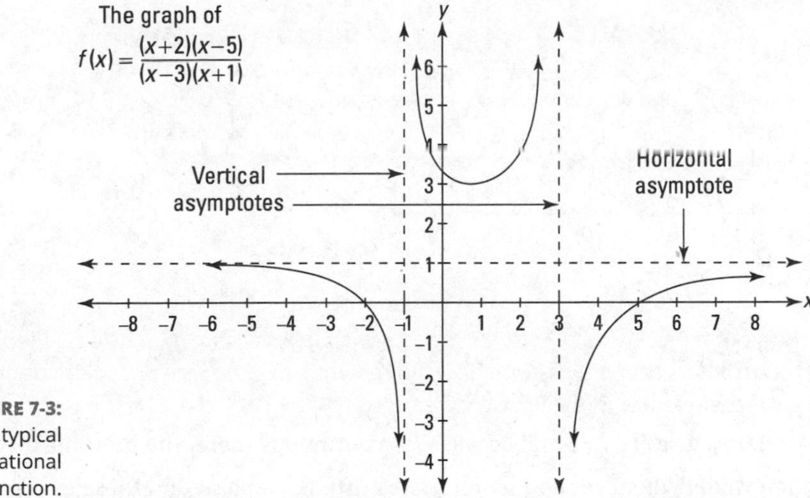

FIGURE 7-3: A typical rational function.

Consider the limit of the function in Figure 7-3 as x approaches 3. As x approaches 3 from the left, $f(x)$ goes up to infinity, and as x approaches 3 from the right, $f(x)$ goes down to negative infinity. Sometimes it's informative to indicate this by writing the following:

$$\lim_{x\to 3^-} f(x)=\infty \text{ and } \lim_{x\to 3^+} f(x)=-\infty$$

But it's also correct to say that both of these limits *do not exist* because infinity is not a real number. If you're asked to determine the regular, two-sided limit, $\lim_{x\to 3} f(x)$, you have no choice but to say that it does not exist because the limits from the left and from the right are not equal.

Limits and horizontal asymptotes

Up till now, I've been looking at limits where x approaches a regular, finite number. But x can also approach infinity or negative infinity. Limits at infinity exist when a function has a horizontal asymptote. For example, the function in Figure 7-3 has a horizontal asymptote at $y=1$, which the function gets closer and closer to as it goes toward infinity to the right and negative infinity to the left. (Going left, the function crosses and goes above the horizontal asymptote at $x=-7$ (not shown in Figure 7-3) and then gradually comes down toward the asymptote. Going right, the function stays below the asymptote and gradually rises up toward it.) The limits equal the height of the horizontal asymptote and are written as

$$\lim_{x\to\infty} f(x)=1 \text{ and } \lim_{x\to-\infty} f(x)=1$$

You see more limits at infinity in Chapter 8.

Use Figure 7-4 for the example problems and for Problems 1 through 6.

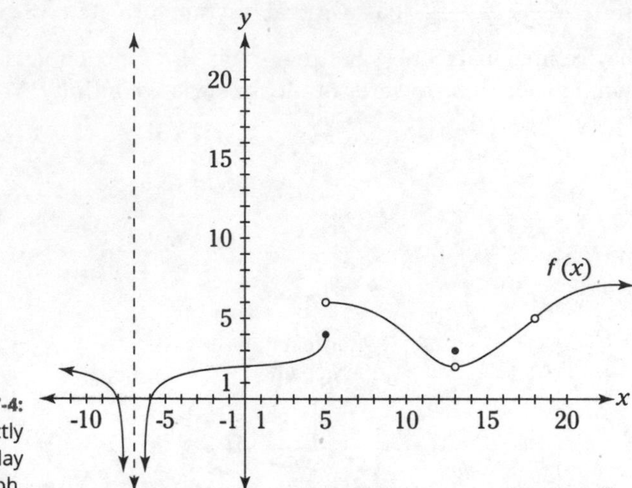

FIGURE 7-4: Not exactly your everyday graph.

EXAMPLE

Q. $\lim_{x \to 0} f(x) = ?$

A. $\boldsymbol{\lim_{x \to 0} f(x) = 2}$**.** Because $f(0) = 2$ and because f is continuous there, the limit must equal the function value. Whenever a function passes through a point and there's no discontinuity at that point, the limit equals the function value.

Q. $\lim_{x \to 13} f(x) = ?$

A. $\boldsymbol{\lim_{x \to 13} f(x) = 2}$**.** There's a hole at (13, 2), and the limit at a hole is the height of the hole.

YOUR TURN

1 $\lim_{x \to -7} f(x) = ?$

2 **(a)** $f(5) = ?$

(b) $f(18) = ?$

3 $\lim_{x \to 5} f(x) = ?$

4 $\lim_{x \to 18} f(x) = ?$

5 $\lim_{x \to 5^-} f(x) = ?$

6 $\lim_{x \to 5^+} f(x) = ?$

7 $\lim_{x \to \infty} \frac{1}{x} = ?$ See the following graph of $y = \frac{1}{x}$.

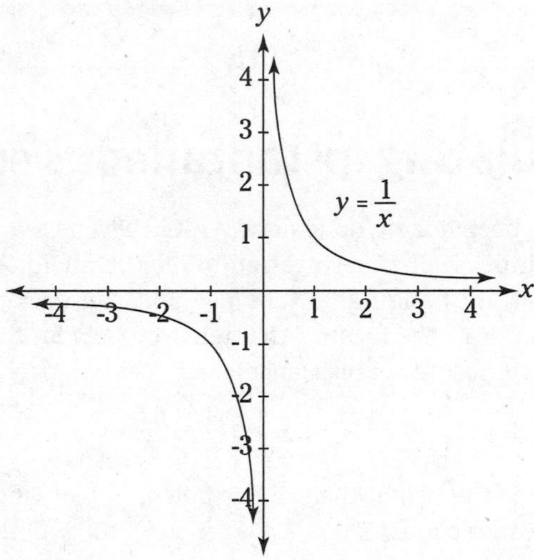

Sketch *by hand* the function $f(x) = \frac{|x|}{x}$; then refer to your sketch for Problems 8, 9, and 10.

8 $\lim_{x \to 0^-} f(x) = ?$

9 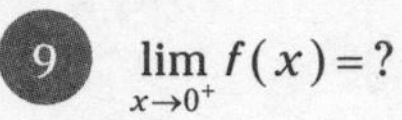 $\lim_{x \to 0^+} f(x) = ?$

10 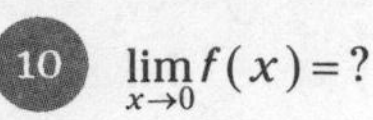 $\lim_{x \to 0} f(x) = ?$

Calculating instantaneous speed with limits

If you've been dozing up to now, WAKE UP! The following problem, which eventually turns out to be a limit problem, brings you to the threshold of real calculus. Say you and your calculus-loving cat are hanging out one day and you decide to drop a ball out of your second-story window. Here's the formula that tells you how far the ball has dropped after a given number of seconds (ignoring air resistance):

$$h(t) = 16t^2$$

(where h is the height the ball has fallen, in feet, and t is the amount of time since the ball was dropped, in seconds)

If you plug 1 into t, h is 16; so, the ball falls 16 feet during the first second. During the first 2 seconds, it falls a total of $16 \cdot 2^2$, or 64 feet, and so on. Now, what if you wanted to determine the ball's speed exactly 1 second after you dropped it? You can start by whipping out this trusty ol' formula:

REMEMBER

$$\textbf{\textit{Distance}} = \textbf{\textit{rate}} \cdot \textbf{\textit{time}} \quad \textbf{so} \quad \textbf{\textit{Rate}} = \frac{\textbf{\textit{distance}}}{\textbf{\textit{time}}}.$$

Using the *rate* (or *speed*) formula, you can easily figure out the ball's average speed during the 2nd second of its fall. Because it dropped 16 feet after 1 second and a total of 64 feet after 2 seconds, it fell $64 - 16$, or 48 feet from $t = 1$ second to $t = 2$ seconds. The following formula gives you the average speed:

$$\begin{aligned} \textit{Average speed} &= \frac{\textit{total distance}}{\textit{total time}} \\ &= \frac{64 - 16}{2 - 1} \\ &= 48 \text{ feet per second} \end{aligned}$$

But this isn't the answer you want because the ball falls faster and faster as it drops, and you want to know its speed exactly 1 second after you drop it. The ball speeds up between 1 and 2 seconds, so this *average* speed of 48 feet per second during the 2nd second is certain to be faster than the ball's *instantaneous* speed at the end of the 1st second. For a better approximation, calculate the average speed between $t = 1$ second and $t = 1.5$ seconds. After 1.5 seconds, the ball has fallen $16 \cdot 1.5^2$, or 36 feet, so from $t = 1$ to $t = 1.5$, it falls $36 - 16$, or 20 feet. Its average speed is thus

$$\begin{aligned} \textit{Average speed} &= \frac{36-16}{1.5-1} \\ &= 40 \text{ feet per second} \end{aligned}$$

If you continue this process for elapsed times of a quarter of a second, a tenth of a second, then a hundredth, a thousandth, and a ten thousandth of a second, you arrive at the list of average speeds shown in Table 7-2.

Table 7-2 Average Speeds from 1 Second to *t* Seconds

t seconds	2	$1\frac{1}{2}$	$1\frac{1}{4}$	$1\frac{1}{10}$	$1\frac{1}{100}$	$1\frac{1}{1{,}000}$	$1\frac{1}{10{,}000}$
Avg. speed from 1 sec. to *t* sec.	48	40	36	33.6	32.16	32.016	32.0016

As *t* gets closer and closer to 1 second, the average speeds appear to get closer and closer to 32 feet per second.

Here's the formula I used to generate the numbers in Table 7-2. It gives you the average speed between 1 second and *t* seconds:

$$\begin{aligned} \textit{Average speed} &= \frac{16t^2 - 16 \cdot 1^2}{t-1} \\ &= \frac{16\left(t^2-1\right)}{t-1} \\ &= \frac{16(t-1)(t+1)}{t-1} \\ &= 16t + 16 \quad (\text{where } t \neq 1) \end{aligned}$$

(In the last line of the solution, recall that *t* cannot equal 1 because that would result in a zero in the denominator of the original equation. This restriction remains in effect even after you cancel the $t - 1$.)

Figure 7-5 shows the graph of this function.

This graph is identical to the graph of the line $y = 16t + 16$ except for the hole at $(1, 32)$. There's a hole there because if you plug 1 into *t* in the average speed function, you get

$$\textit{Average speed} = \frac{16\left(1^2-1\right)}{1-1} = \frac{0}{0}$$

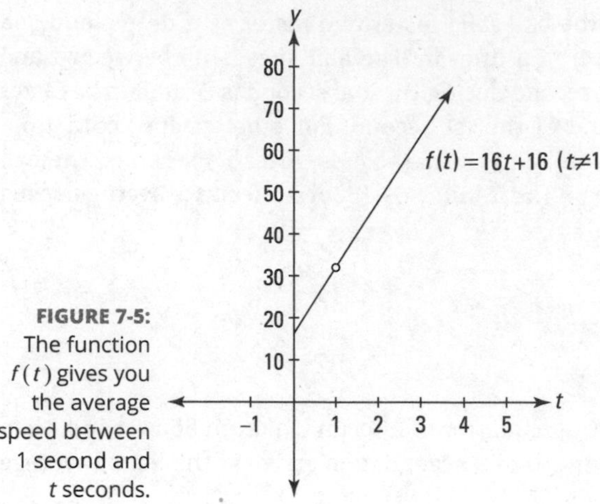

FIGURE 7-5: The function $f(t)$ gives you the average speed between 1 second and t seconds.

which is undefined. And why do you get $\frac{0}{0}$? Because you're trying to determine an average speed — which equals *total distance* divided by *elapsed time* — from $t = 1$ to $t = 1$. But from $t = 1$ to $t = 1$ is, of course, *no* time, and "during" this point in time, the ball doesn't travel any distance, so you get $\frac{\text{zero feet}}{\text{zero seconds}}$ as the average speed from $t = 1$ to $t = 1$.

Obviously, there's a problem here. Hold on to your hat, you've arrived at one of the big "Ah ha!" moments in the development of differential calculus.

MATH RULES

Definition of *instantaneous speed:* Instantaneous speed is defined as the limit of the average speed as the elapsed time approaches zero.

For the falling-ball problem, you'd have

$$\begin{aligned}\underset{\text{at } t=1 \text{ second}}{\textit{Instantaneous speed}} &= \lim_{t\to1}\frac{16\left(t^2-1\right)}{t-1}\\ &= \lim_{t\to1}\frac{16(t-1)(t+1)}{t-1}\\ &= \lim_{t\to1}(16t+16)\\ &= 32 \text{ feet per second}\end{aligned}$$

The fact that the elapsed time never gets to zero doesn't affect the precision of the answer to this limit problem — the answer is exactly 32 feet per second, the height of the hole in Figure 7-4. What's remarkable about limits is that they enable you to calculate the precise, instantaneous speed at a *single* point in time by taking the limit of a function that's based on an *elapsed* time, a period between *two* points of time.

Linking Limits and Continuity

Before I expand on the material on limits from the earlier sections of this chapter, I want to introduce a related idea — *continuity.* This is such a simple concept. A *continuous* function is simply a function with no gaps — a function that you can draw without taking your pencil off the paper. Consider the four functions shown in Figure 7-6.

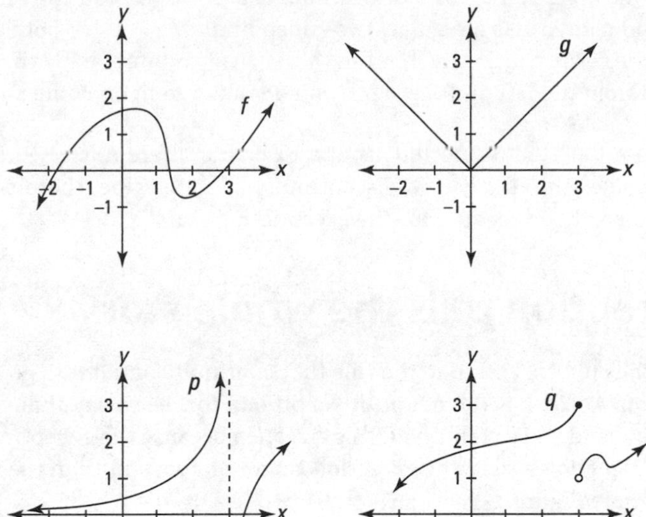

FIGURE 7-6: The graphs of the functions *f*, *g*, *p*, and *q*.

Whether or not a function is continuous is almost always obvious. The first two functions in Figure 7-6, $f(x)$ and $g(x)$, have no gaps, so they're continuous. The next two, $p(x)$ and $q(x)$, have gaps at $x = 3$, so they're not continuous. That's all there is to it. Well, not quite. The two functions with gaps are not continuous everywhere, but because you can draw sections of them without taking your pencil off the paper, you can say that parts of those functions are continuous. And sometimes a function is continuous everywhere it's defined. Such a function is described as being *continuous over its entire domain,* which means that its gap or gaps occur at x-values where the function is undefined. The function $p(x)$ is continuous over its entire domain; $q(x)$, on the other hand, is not continuous over its entire domain because it's not continuous at $x = 3$, which is in the function's domain. Often, the important issue is whether a function is continuous at a particular x-value. It is unless there's a gap there.

MATH RULES

Continuity of polynomial functions: All polynomial functions are continuous everywhere.

MATH RULES

Continuity of rational functions: All rational functions (a *rational function* is the quotient of two polynomial functions) are continuous over their entire domains. They are discontinuous at x-values not in their domains — that is, x-values where the denominator is zero.

Continuity and limits usually go hand in hand

Look at the four functions in Figure 7-6 where $x = 3$. Consider whether each function is continuous there and whether a limit exists at that x-value. The first two, f and g, have no gaps at $x = 3$, so they're continuous there. Both functions also have limits at $x = 3$, and in both cases, the limit equals the height of the function at $x = 3$, because as x gets closer and closer to 3 from the left and the right, y gets closer and closer to $f(3)$ and $g(3)$, respectively.

Functions p and q, on the other hand, are not continuous at $x = 3$ (or you can say that they're *discontinuous* there), and neither has a regular, two-sided limit at $x = 3$. For both functions, the gaps at $x = 3$ not only break the continuity, but they also cause no limits to be there because, as you move toward $x = 3$ from the left and the right, you do not zero in on some single y-value.

So there you have it. If a function is continuous at an x-value, there must be a regular, two-sided limit for that x-value. And if there's a discontinuity at an x-value, there's no two-sided limit there . . . well, almost. Keep reading for the exception.

The hole exception tells the whole story

The hole exception is the only exception to the rule that continuity and limits go hand in hand, but it's a *huge* exception. And, I have to admit, it's a bit odd for me to say that continuity and limits *usually* go hand in hand and to talk about this *exception* because the exception is the whole point. When you come right down to it, the exception is more important than the rule. Consider the two functions shown in Figure 7-7.

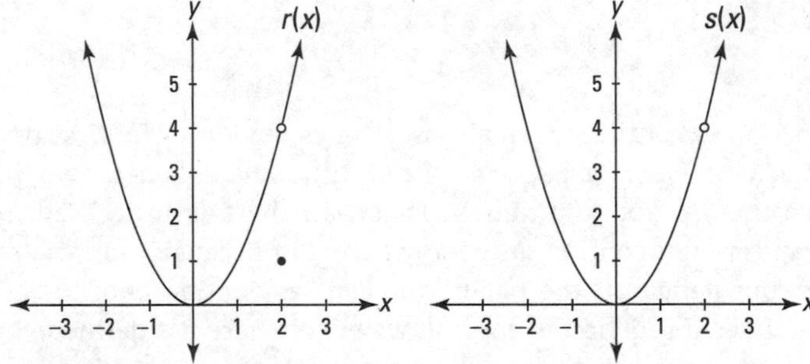

FIGURE 7-7: The graphs of the functions r and s.

These functions have gaps at $x = 2$ and are obviously not continuous there, but they *do* have limits as x approaches 2. In each case, the limit equals the height of the hole.

MATH RULES

The hole exception: The only way a function can have a regular, two-sided limit where it is not continuous is where the discontinuity is an infinitesimal hole in the function.

So, both functions in Figure 7-7 have the same limit as x approaches 2; the limit is 4, and the facts that $r(2) = 1$ and that $s(2)$ is undefined are irrelevant. For both functions, as x zeros in on 2 from either side, the height of the function zeros in on the height of the hole — that's the limit. This bears repeating:

MATH RULES

The limit at a hole: The limit at a hole is the height of the hole.

"That's great," you may be thinking. "But why should I care?" Well, stick with me for just a minute. In the falling ball example in the section, "Calculating instantaneous speed with limits," earlier in this chapter, I tried to calculate the average speed during zero elapsed time. This gave me $\frac{\text{zero distance}}{\text{zero time}}$.

Because $\frac{0}{0}$ is undefined, the result was a hole in the function. Function holes often come about from the impossibility of dividing zero by zero. It's these functions where the limit process is critical, and such functions are at the heart of the meaning of a derivative, and derivatives are at the heart of differential calculus.

MATH RULES

The derivative-hole connection: A derivative always involves the undefined fraction $\frac{0}{0}$ and always involves the limit of a function with a hole. (If you're curious, all the limits in Chapter 9 — where the derivative is formally defined — are limits of functions with holes.)

Sorting out the mathematical mumbo jumbo of continuity

All you need to know to fully *understand* the idea of continuity is that a function is continuous at some particular x-value if there is no gap there. However, because you might be tested on the following formal definition, I suppose you'll want to know it.

MATH RULES

Definition of continuity: A function $f(x)$ is *continuous* at a point $x = a$ if the following three conditions are satisfied:

1. $f(a)$ is defined,
2. $\lim_{x \to a} f(x)$ exists, and
3. $f(a) = \lim_{x \to a} f(x)$.

Just like with the formal definition of a limit, the definition of continuity is always presented as a three-part test, but condition 3 is the only one you really need to worry about because conditions 1 and 2 are built into 3. You must remember, however, that condition 3 is *not* satisfied when the left and right sides of the equation are both undefined or nonexistent.

The 33333 Limit Mnemonic

Here's a great memory device that pulls a lot of information together in one fell swoop. It may seem contrived or silly, but with mnemonic devices, contrived and silly work. The 33333 limit mnemonic helps you remember five groups of three things: two groups involving limits, two involving continuity, and one about derivatives. (I realize I haven't gotten to derivatives yet, but this is the best place to present this mnemonic. Take my word for it — nothing's perfect.)

First, note that the word *limit* has five letters and that there are five 3's in this mnemonic. Next, write *limit* with a lowercase "l" and uncross the "t" so it becomes another "l" — like this:

limil

Now, the two "l"s are for limits, the two "i"s are for continuity (notice that the letter "i" has a gap in it, thus it's not continuous), and the "m" is for slope (remember $y = mx + b$?), which is what derivatives are all about (you'll see that in Chapter 9 in just a few pages).

Each of the five letters helps you remember three things — like this:

```
l i m i l
3 3 3 3 3
```

- 3 parts to the definition of a limit:

  Look back to the definition of a limit in the section, "The formal definition of a limit — just what you've been waiting for." Remembering that it has three parts helps you remember the parts — trust me.

- 3 cases where a limit fails to exist:
  - At a vertical asymptote — called an *infinite discontinuity* — like at $x = 3$ on function p in Figure 7-6.
  - At a jump discontinuity, like where $x = 3$ on function q in Figure 7-6.
  - With a limit at infinity of an *oscillating function* like $\sin x$, which goes up and down forever, never zeroing in on a single height.

- 3 parts to the definition of continuity:

  Just as with the definition of a limit, remembering that the definition of continuity has 3 parts helps you remember the 3 parts (see the section, "Sorting out the mathematical mumbo jumbo of continuity").

- 3 types of discontinuity:
  - A *removable discontinuity* — that's a fancy term for a hole — like the holes in functions r and s in Figure 7-7.
  - An infinite discontinuity like at $x = 3$ on function p in Figure 7-6.
  - A jump discontinuity like at $x = 3$ on function q in Figure 7-6.

  Note that the three types of discontinuity (hole, infinite, and jump) begin with three consecutive letters of the alphabet. Since they're consecutive, there are no gaps between *h*, *i*, and *j*, so they're "continuous" letters. Hey, was this book worth the price or what?

- 3 cases where a derivative fails to exist (I explain this in Chapter 9):
  - At any type of *discontinuity*.
  - At a sharp point on a function, namely, at a *cusp* or a *corner*.
  - At a *vertical tangent* (because the slope is undefined there).

Well, there you have it. Did you notice that another way this mnemonic works is that it gives you 3 cases where a limit fails to exist, 3 cases where continuity fails to exist, and 3 cases where a derivative fails to exist? *Holy triple trio of nonexistence, Batman, that's yet another 3 — the 3 topics of the mnemonic: limits, continuity, and derivatives!*

Use Figure 7-8 for the example problem and for Problems 11 through 15.

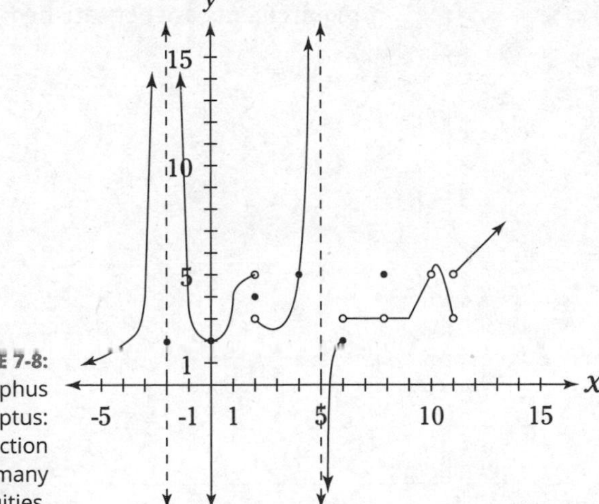

FIGURE 7-8: Graphus interruptus: A function with many discontinuities.

EXAMPLE

Q. List the x-coordinates of all discontinuities of the function, state whether the discontinuities are removable or nonremovable, and give the type of discontinuity — hole, jump, or infinite.

A. **At $x = -2$ and $x = 5$, the vertical asymptotes are nonremovable, infinite discontinuities.**

At $x = 2$, $x = 6$, and $x = 11$, there are nonremovable, jump discontinuities.

At $x = 8$ and $x = 10$, there are holes; holes are removable discontinuities. Though infinitely small, these are nevertheless discontinuities. They're "removable" discontinuities because you can "fix" the function by plugging the holes.

YOUR TURN

11 At which of the following x-values are all three requirements for the existence of a limit satisfied, and what is the limit at those x-values? $x = -2$, 0, 2, 4, 5, 6, 8, 10, and 11.

12 For the x-values at which all three limit requirements are not met, state which of the three requirements are not satisfied. If one or both one-sided limits exist at any of these x-values, give the value of the one-sided limit.

 At which of the x-values are all three requirements for continuity satisfied?

 For the rest of the x-values, state which of the three continuity requirements are not satisfied.

15 $\lim_{x\to\infty} \sin x = ?$ See the following graph.

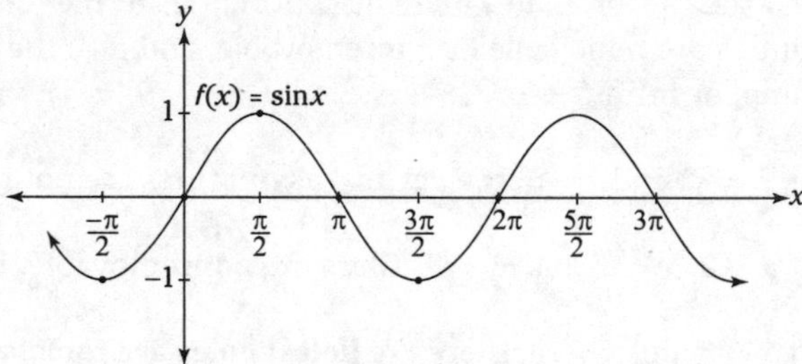

Practice Questions Answers and Explanations

1. **$\lim_{x \to -7} f(x)$ does not exist (DNE) because there's a vertical asymptote at −7.** Or, because $f(x)$ approaches negative infinity both from the left and from the right, you could say the limit equals negative infinity.

2. **(a) $f(5) = 4$, the height of the solid dot at $x = 5$.**

   (b) $f(18)$ is undefined because f has no y-value corresponding to the x-value of 18.

After reviewing the following solutions to Problems 3 through 6, reflect on how the answers to those problems compare to the answers to Problem 2.

3. **$\lim_{x \to 5} f(x)$ does not exist.** The limit does not exist because the limit from the left does not equal the limit from the right. Or, you could say that the limit DNE because there's a jump discontinuity at $x = 5$.

4. **$\lim_{x \to 18} f(x) = 5$.** Like the second example problem, the limit at a hole is the height of the hole. The fact that $f(18)$ is undefined is irrelevant to this limit question.

5. **$\lim_{x \to 5^-} f(x) = 4$.** The limit is 4 because $f(5) = 4$ and f is continuous from the left at $(5, 4)$.

6. **$\lim_{x \to 5^+} f(x) = 6$.** This question is just like Problem 5, except that there's a hollow dot — instead of a solid one — when you arrive at the gap. But the hollow dot at $(5, 6)$ is irrelevant to the limit question — just as in Problem 4, where the hole was irrelevant.

7. **$\lim_{x \to \infty} \frac{1}{x} = 0$.** As you go out farther and farther to the right, the function gets closer and closer to zero, so that's the limit.

8. **$\lim_{x \to 0^-} f(x) = -1$.**

   Of course, you can graph f with your graphing calculator, but it's a good idea to graph functions by hand now and then. It helps you understand why the function looks the way it does. All you need to do to sketch this one by hand is to plug a few negative and positive numbers into x. You'll soon see that whenever the input is negative, the output is −1, and whenever the input is positive, the output is 1. And you need the hollow dots on the y-axis at −1 and 1 because $f(0)$ is undefined. Your sketch should look something like the following figure.

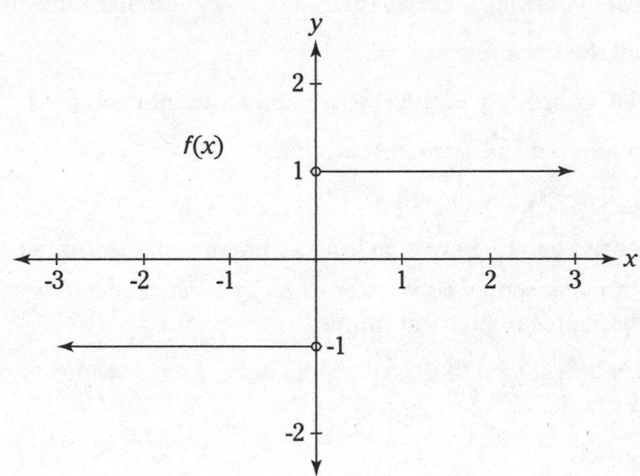

For the one-sided limit, $\lim_{x \to 0^-} f(x)$, nothing to the right of zero is relevant. And, as with all limit problems, what actually happens to the function (namely, whether it exists and, if it exists, what it equals) when x gets to the limit number doesn't affect the limit answer. All that matters is what's happening to the function as x gets closer and closer to the limit number. As x gets closer and closer to zero from the left, y is staying precisely at -1, so that's the limit.

9 **$\lim_{x \to 0^+} f(x) = 1$.** See the solution to Problem 8. The limit in this problem works exactly the same way.

10 **$\lim_{x \to 0} f(x)$ does not exist.** As you see in the solutions to Problems 8 and 9, $\lim_{x \to 0^-} f(x) \neq \lim_{x \to 0^+} f(x)$, and, therefore, the ordinary, two-sided limit does not exist.

11 **At 0, the limit is 2; at 4, the limit is 5; at 8, the limit is 3; at 10, the limit is 5.**

MATH RULES

A limit exists at a particular x-value of a curve when the curve is *heading toward* some particular y-value and keeps *heading toward* that y-value as you continue to zoom in on the curve at the x-value. The curve must head toward that y-value (that height) as you move along the curve both from the right and from the left (unless the limit is one where x approaches infinity). I emphasize *heading toward* because what happens precisely at the given x-value isn't relevant to this limit inquiry. That's why there is a limit at a hole like the ones at $x = 8$ and $x = 10$.

12 **At −2 and 5, all three conditions fail.**

At 2, 6, and 11, only the third requirement is not satisfied.

At 2, the limit from the left equals 5 and the limit from the right equals 3.

At 6, the limit from the left is 2 and the limit from the right is 3.

Finally, at 11, the limit from the left equals 3 and the limit from the right equals 5.

13 **The function in Figure 7-8 is continuous at 0 and 4.** The common-sense way of thinking about continuity is that a curve is continuous wherever you can draw the curve without taking your pen off the paper. It should be obvious that that's true at 0 and 4, but not at any of the other listed x-values.

14 **All listed x-values other than 0 and 4 are points of discontinuity.** A *discontinuity* is just a highfalutin calculus way of saying a gap. If you have to take your pen off the paper at some point when drawing a curve, then the curve has a discontinuity there.

At 5 and 11, all three conditions fail.

At −2, 2, and 6, continuity requirements 2 and 3 are not satisfied.

At 10, requirements 1 and 3 are not satisfied.

At 8, requirement 3 is not satisfied.

15 **$\lim_{x \to \infty} \sin x$ does not exist.** There's no limit as x approaches infinity because the curve oscillates — it never settles down to one precise y-value. (The three-part definition of a limit does not apply to limits at infinity.)

If you're ready to test your skills a bit more, take the following chapter quiz that incorporates all the chapter topics.

Whaddya Know? Chapter 7 Quiz

Quiz time! Complete each problem to test your knowledge on the various topics covered in this chapter. You can then find the solutions and explanations in the next section.

Use the following figure for the questions in this quiz.

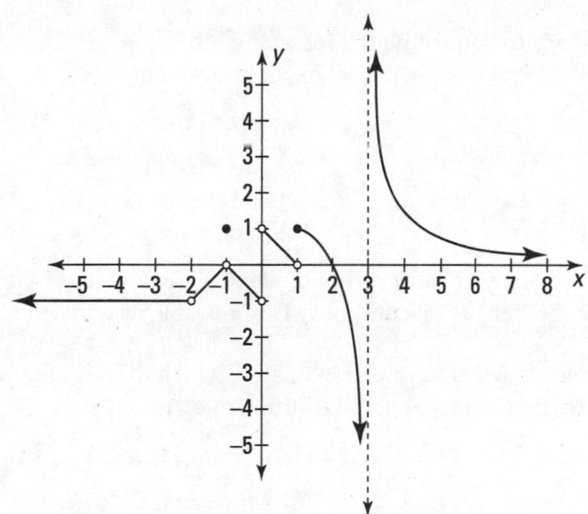

1 Use the graph of f to determine the following function values:

$f(-3)=?$

$f(-2)=?$

$f(-1)=?$

$f(0)=?$

$f(1)=?$

$f(2)=?$

$f(3)=?$

2 Use the graph of f to determine the following limits:

$\lim_{x\to -3} f(x)=?$

$\lim_{x\to -2} f(x)=?$

$\lim_{x\to -1} f(x)=?$

$\lim_{x\to 0} f(x)=?$

$\lim_{x\to 1} f(x)=?$

$\lim_{x\to 2} f(x)=?$

$\lim_{x\to 3} f(x)=?$

3 Use the graph of f to determine the following left-sided limits:

$\lim_{x\to -3^-} f(x)=?$

$\lim_{x\to -2^-} f(x)=?$

$\lim_{x\to -1^-} f(x)=?$

$\lim_{x\to 0^-} f(x)=?$

$\lim_{x\to 1^-} f(x)=?$

$\lim_{x\to 2^-} f(x)=?$

$\lim_{x\to 3^-} f(x)=?$

4. Use the graph of f to determine the following right-sided limits:

$\lim_{x \to -3^+} f(x) = ?$

$\lim_{x \to -2^+} f(x) = ?$

$\lim_{x \to -1^+} f(x) = ?$

$\lim_{x \to 0^+} f(x) = ?$

$\lim_{x \to 1^+} f(x) = ?$

$\lim_{x \to 2^+} f(x) = ?$

$\lim_{x \to 3^+} f(x) = ?$

5. Use the graph of f to determine the following limits:

$\lim_{x \to -\infty} f(x) = ?$

$\lim_{x \to \infty} f(x) = ?$

6. At what x-values is f not continuous?

7. For each of your answers to Question 6, identify the type of discontinuity (hole, jump, or infinite) and whether the discontinuity is removable or nonremovable.

8. For each of your answers to Question 6, state which of the three continuity requirements are not satisfied (from the three-part definition of continuity).

Answers to Chapter 7 Quiz

1. Thought I'd start you off with a softball question:

   $f(-3) = -1$

   $f(-2)$ **is undefined** $\qquad f(1) = 1$

   $f(-1) = 1$ $\qquad f(2) = 0$

   $f(0)$ **is undefined** $\qquad f(3)$ **is undefined**

2. $\lim\limits_{x \to -3} f(x) = -1$. $f(-3)$ is −1 and the graph is continuous there, so that's the limit.

   $\lim\limits_{x \to -2} f(x) = -1$. The limit at a hole is the height of the hole.

   $\lim\limits_{x \to -1} f(x) = 0$. Ditto.

   $\lim\limits_{x \to 0} f(x)$ **is undefined.** The one-sided limits are unequal, so this two-sided limit is undefined.

   $\lim\limits_{x \to 1} f(x)$ **is undefined.** Ditto.

   $\lim\limits_{x \to 2} f(x) = 0$. $f(2)$ equals zero and the graph is continuous there, so that's the limit.

   $\lim\limits_{x \to 3} f(x)$ **is undefined.** The left-sided limit is negative infinity, and the right-sided limit is positive infinity, so this two-sided limit is undefined.

3. You would often want to consider the one-sided limits (like the ones in this problem and Problem 4) before doing the two-sided limits (like the ones in Problem 2), but I wanted to see if you could correctly do Questions 4, 5, and 7 in Problem 2 without the hints that the related questions in Problems 3 and 4 would provide.

   $\lim\limits_{x \to -3^-} f(x) = -1$

   $\lim\limits_{x \to -2^-} f(x) = -1$ $\qquad \lim\limits_{x \to 1^-} f(x) = 0$

   $\lim\limits_{x \to -1^-} f(x) = 0$ $\qquad \lim\limits_{x \to 2^-} f(x) = 0$

   $\lim\limits_{x \to 0^-} f(x) = -1$ $\qquad \lim\limits_{x \to 3^-} f(x) = -\infty$

4. $\lim\limits_{x \to -3^+} f(x) = -1$

   $\lim\limits_{x \to -2^+} f(x) = -1$ $\qquad \lim\limits_{x \to 1^+} f(x) = 1$

   $\lim\limits_{x \to -1^+} f(x) = 0$ $\qquad \lim\limits_{x \to 2^+} f(x) = 0$

   $\lim\limits_{x \to 0^+} f(x) = 1$ $\qquad \lim\limits_{x \to 3^+} f(x) = \infty$

5. $\lim\limits_{x \to -\infty} f(x) = -1$

   $\lim\limits_{x \to \infty} f(x) = 0$. The limit at infinity is the height of the horizontal asymptote.

6. **−2, −1, 0, 1, and 3.**

7. **At −2 and −1, there are holes, which are removable discontinuities.**

   At 0 and 1, there are jump discontinuities, which are nonremovable.

   At 3, there's a nonremovable infinite discontinuity.

8. **At −2, conditions 1 and 3 fail.**

   At −1, only condition 3 fails.

   At 0, all three conditions fail.

   At 1, conditions 2 and 3 fail.

   At 3, all three conditions fail.

IN THIS CHAPTER

» Algebra, schmalgebra

» Calculators — taking the easy way out

» Making limit sandwiches

» Infinity — "Are we there yet?"

» Conjugate multiplication — sounds R rated, but it's strictly PG

Chapter 8 Evaluating Limits

Chapter 7 introduces the concept of a limit and goes over the theoretical limit stuff. This chapter gets down to the nitty-gritty techniques for calculating the answers to limit problems. Here you practice two main methods for solving limit problems: using your calculator and using algebra.

The calculator techniques are useful for a few reasons: 1) You can solve some limit problems on your calculator that are either impossible or just very difficult to do with algebra, 2) You can check your algebraic answers with your calculator, and 3) Limit problems can be solved with a calculator when you're not required to show your work — like maybe on a multiple-choice test. (Of course, depending on your teacher and depending on the situation, you may not be allowed to use your calculator.)

Learning the algebraic techniques is valuable for a couple reasons: The first, *incredibly* important reason is that the mathematics involved in the algebraic methods is beautiful, pure, and rigorous; and, second — something so trivial that perhaps I shouldn't mention it — you'll be tested on it. Do I have my priorities straight or what?

Easy Does It — Easy Limits

Before getting to the algebraic and calculator techniques, let's dispense with some really easy limit problems.

A few limit problems are *very* easy. They're so easy that I don't have to waste your time with unnecessary introductory remarks and unneeded words that take up space and do nothing to further your knowledge of the subject — instead, I can just cut to the chase and give you only the critical facts and get to the point and get down to business and . . . Okay, so are you ready?

Limits to memorize

You should memorize the following limits. If you fail to memorize the limits in the last four bullets, you could waste *a lot* of time trying to figure them out.

- $\lim_{x\to a} c = c$

  ($y = c$ is a horizontal line, so the limit — which is the function height — must equal c regardless of the arrow-number.)
- $\lim_{x\to 0^+} \frac{1}{x} = \infty$
- $\lim_{x\to 0^-} \frac{1}{x} = -\infty$
- $\lim_{x\to \infty} \frac{1}{x} = 0$
- $\lim_{x\to -\infty} \frac{1}{x} = 0$
- $\lim_{x\to 0} \frac{\sin x}{x} = \lim_{x\to 0} \frac{x}{\sin x} = 1$
- $\lim_{x\to 0} \frac{\cos x - 1}{x} = 0$
- $\lim_{x\to \infty} \left(1 + \frac{1}{x}\right)^x = e \approx 2.718$
- $\lim_{x\to \infty} \left(1 - \frac{1}{x}\right)^x = \frac{1}{e} \approx 0.368$

Plugging and chugging

Plug-and-chug problems make up the second category of easy limits. Just plug the arrow-number into the limit function, and if the computation results in a number, that's your answer (but see the following warning). For example,

$$\lim_{x\to 3}\left(x^2 - 10\right) = -1$$

(Don't forget that for this method to work, the result you get after plugging in must be an ordinary number, not infinity or negative infinity or something that's undefined.)

If you're dealing with a function that's continuous everywhere (like the one in this example) or a function that's continuous over its entire domain, this method will always work. These are *well-duh* limit problems, and, to be perfectly frank, there's really no point to them. The limit is simply the function value. If you're dealing with any other type of function, this method will only work sometimes — read on.

Beware of discontinuities. The plug-and-chug method works for any type of function, including piecewise functions, *unless* there's a discontinuity at the arrow-number you plug in. In that case, if you get a number after plugging in, that number is *not* the limit; the limit might equal some other number or it might not exist. (See Chapter 7 for a description of piecewise functions.)

What happens when plugging in gives you a non-zero number over zero? If you plug the arrow-number into a limit like $\lim_{x \to 5} \frac{10}{x-5}$ and you get any number (other than zero) divided by zero — like $\frac{10}{0}$ — then you know that the limit does not exist; in other words, the limit does not equal a finite number. (The answer might be infinity or negative infinity or just a plain old "does not exist.")

The "Real Deal" Limit Problems

When you tackle any limit problem, you should always begin by plugging the arrow-number into the limit expression. Sometimes (not often), your answer will be an ordinary number or a non-zero number over zero, and you'll be done — as explained in the preceding section (though make sure you heed the warning). But if you plug in the arrow-number and the result is undefined (excluding the case covered in the previous tip), you've got a "for real" limit problem — and a bit of work to do. This is the main focus of this chapter. These are the interesting limit problems, the ones that likely have infinitesimal holes, and the ones that are important for differential calculus — you see more of them in Chapter 9.

With these real-deal limit problems (where plugging in often gives you $\frac{0}{0}$), you can try four things: your calculator, general algebraic techniques, making a limit sandwich (a special algebraic technique), and L'Hôpital's rule (which is covered in Chapter 18).

Figuring a limit with your calculator

Your calculator is a great tool for understanding limits. It can often give you a better feel for how a limit works than the algebraic techniques can. A limit problem asks you to determine what the y-value of a function is zeroing in on as the x-value approaches a particular number. With your calculator, you can actually witness the process and the result. Say you want to evaluate the following limit: $\lim_{x \to 5} \frac{x^2-25}{x-5}$. The plug-and-chug method doesn't work because plugging 5 into x produces the undefined result of $\frac{0}{0}$. Let's solve this limit problem with a calculator. I'll go over two basic methods.

A note about calculators and other technology: With every passing year, there are more and more powerful calculators and more and more resources on the Internet that can do calculus for you. One thing that allows these technologies to do calculus is that they can handle algebra (using CAS, a Computer Algebra System). Say you input $(x+3)(2x-5)$. These technologies can FOIL that expression and give you the algebraic answer of $2x^2+x-15$. A calculator like the TI-Nspire, or any other calculator with CAS, or websites like Wolfram Alpha (`www.wolframalpha.com`), can actually do the above limit problem, and all sorts of more difficult calculus problems, and give you the exact numerical or algebraic answer.

Older calculator models can't do algebra or calculus in the real, precise, algebraic way. They can, however, give you approximate answers to calc problems — which will often suffice.

Different calculus teachers have different policies on what technology they allow in their classes. Many do not allow the use of CAS calculators and comparable technologies because they basically do all the calculus work for you. So, the following discussion (and the rest of this book) assumes you're using a more basic calculator (like the TI-84) without CAS capability.

Method one

The first calculator method is to test the limit function with two numbers: one slightly less than the arrow-number and one slightly more than it. So, here's what you do for the earlier problem, $\lim_{x\to 5}\frac{x^2-25}{x-5}$. If you have a calculator like a Texas Instruments TI-84, enter the first number, say 4.9999, on the home screen, and press the Sto (store) button, then the *x* button, and then the Enter button (this stores the number into *x*). Then enter the function, $\frac{x^2-25}{x-5}$, and hit Enter. The result, 9.9999, is extremely close to a round number, 10, so 10 is likely your answer. Now take a number a little higher than the arrow-number, like 5.0001, and repeat the process. Since the result, 10.0001, is also very close to 10, that clinches it. The answer is 10 (almost certainly). (By the way, if you're using a different calculator model, you can likely achieve the same result with the same technique or something very close to it.) This method can be effective, but it often doesn't give you a good feel for how the *y*-values zero in on the result. To get a better picture of this process, you can store three or four numbers into *x* (one after another), each a bit closer to the arrow-number, and look at the sequence of results.

Method two

The second calculator method is to set *y* equal to the limit expression and enter it in your calculator's graphing mode. You can then investigate the limit problem on your calculator by 1) looking at the table of values and 2) viewing the graph of the function. Let's do it. Enter $y=\frac{x^2-25}{x-5}$ in graphing mode on your calculator. Then go to Table Setup and enter the arrow-number, 5, as the TblStart number; then, enter a small number, say 0.001, for ΔTbl — that's the size of the *x*-increments in the table. Hit the Table button to produce the table. Now scroll up until you can see a couple of numbers less than 5, and you should see a table of values similar to the one in Table 8-1.

Table 8-1 TI-84 Table for $y=\frac{x^2-25}{x-5}$ after Scrolling Up to 4.998

x	y
4.998	9.998
4.999	9.999
5	error
5.001	10.001
5.002	10.002
5.003	10.003

Because y gets very close to 10 as x zeros in on 5 from above and below, 10 is the limit (almost certainly . . . you can't be absolutely positive with these calculator methods, but they almost always work).

Next, take a look at the graph of this function. Go into the window and tweak the Xmin, Xmax, Ymin, and Ymax settings, if necessary, so that the part of the function corresponding to the arrow-number is within the viewing window. Now, use the trace feature to trace along the function until you get *close* to the arrow-number. You can't trace exactly *onto* the arrow-number because there's a little hole in the function there, the height of which, by the way, is your answer. When you trace close to the arrow-number, the y-value will get close to the limit answer. Use the ZoomBox feature to draw a little box around the part of the graph containing the arrow-number, and zoom in until you see that the y-values are getting very close to a round number — that's your answer.

These calculator techniques are useful for a number of reasons. Your calculator can give you the answers to limit problems that are impossible to do algebraically. And it can solve limit problems that you could do with paper and pencil except that you're stumped. Also, for problems that you do solve on paper, you can use your calculator to check your answers. And even when you choose to solve a limit algebraically — or are required to do so — it's a good idea to create a table like Table 8-1 not just to confirm your answer, but to see how the function behaves near the arrow-number. This gives you a *numerical* grasp on the problem, which enhances your *algebraic* understanding of it. If you then look at the graph of the function on your calculator, you have a third, *graphical* or *visual* way of thinking about the problem.

TIP

Many calculus problems can be done *algebraically, graphically,* and *numerically*. When possible, use two or three of the approaches. Each approach gives you a different perspective on a problem and enhances your grasp of the relevant concepts.

Use the calculator methods to supplement algebraic methods, but don't rely too much on them. First of all, the non-CAS-calculator techniques won't allow you to deduce an exact answer unless the numbers your calculator gives you are getting close to a number you recognize. If your calculator gives you 9.999, for example, you can be pretty sure the exact answer is 10, and you can be quite confident that the exact answer is $\frac{1}{3}$ if your calculator says 0.333332. And perhaps you recognize that 1.414211 is very close to $\sqrt{2}$. But if the exact answer to a limit problem is something like $\frac{1}{2\sqrt{3}}$, you probably won't recognize it. The number $\frac{1}{2\sqrt{3}}$ is approximately equal to 0.288675. When you see numbers on your calculator close to that decimal, you won't recognize $\frac{1}{2\sqrt{3}}$ as the limit — unless you're an Archimedes, a Gauss, or a Ramanujan (members of the mathematics hall of fame). However, even when you don't recognize the *exact* answer in such cases, you can still learn an approximate answer, in decimal form, to the limit question, and that can be helpful.

WARNING

Gnarly functions may stump your calculator. Another calculator limitation is that it won't work at all with some peculiar functions. Consider, for example, $\lim_{x\to 5}\left(\sqrt[25]{x-5}\cdot\sin\left(\frac{1}{x-5}\right)\right)$. This limit equals zero, but you can't get that result with your calculator.

By the way, even when the non-CAS-calculator methods work, these calculators can do some quirky things from time to time. For example, if you're solving a limit problem where x approaches 3, and you put numbers in your calculator that are *too* close to 3 (like 3.0000000001), you can get too close to the calculator's maximum decimal length. This can result in answers that get *further* from the limit answer, even as you input numbers closer and closer to the arrow-number.

The moral of the story is that you should think of your calculator as one of several tools at your disposal for solving limits — not as a foolproof substitute for algebraic techniques.

EXAMPLE

Q. Evaluate $\lim_{x \to 6} \frac{x^2 - 5x - 6}{\sin(x-6)}$.

A. **The answer is 7.**

Method One:

1. **Use the Sto button to store 5.99 into x.**
2. **Enter $\frac{x^2 - 5x - 6}{\sin(x-6)}$ on the home screen and hit Enter. (*Note:* You must be in radian mode.)**

   This gives you a result of ~ 6.99, suggesting that the answer is 7.

3. **Repeat Steps 1 and 2 with 5.999 stored into x.**

   This gives you a result of ~ 6.999 (even closer to 7).

4. **Repeat Steps 1 and 2 with 6.01 stored into x.**

   This gives you a result of ~ 7.01 (close to 7 again).

5. **Repeat Steps 1 and 2 with 6.001.**

   This gives you a result of ~ 7.001 (even closer).

Because the results are obviously homing in on the round number of 7, that's your answer.

Method Two:

1. **Enter $\frac{x^2 - 5x - 6}{\sin(x-6)}$ in graphing or "y =" mode.**
2. **Go to Table Setup and set TblStart to the arrow-number, 6, and ΔTbl to 0.01.**
3. **Go to the Table, and you'll see the y-values getting closer and closer to 7 as you scroll toward $x = 6$ from above and below 6.**

   So 7 is your answer.

4. **Graph the function. For expressions containing trig functions, ZoomStd, ZoomFit, and ZoomTrig are good windows to try for your first viewing.**

   For this funny function, none of these three window options works very well, but ZoomStd is the best.

5. **Trace close to $x = 6$ and you'll see that y is near 7. Use ZoomBox to draw a little box around the point $(6, 7)$; then hit Enter.**
6. **Trace near $x = 6$ on this zoomed-in graph until you get very near to $x = 6$.**
7. **Repeat the Zoombox process maybe two more times and you should be able to trace extremely close to $x = 6$.**

   (When I did this, I could trace to $x = 6.0000022$, $y = 7.0000023$.)

The answer is 7.

YOUR TURN

1 Use your calculator to evaluate $\lim_{x\to -3}\frac{x^2-5x-24}{x+3}$. Try both methods.

2 Use your calculator to determine $\lim_{x\to 0}\frac{\sin x}{\tan^{-1}x}$. Use both methods.

Solving limit problems with algebra

MATH RULES

Don't forget to plug in! You can solve limit problems with several algebraic techniques (discussed next). But before you try any algebra, your first step should always be to plug the arrow-number into the limit expression. If the function is continuous at the arrow-number and if plugging in results in an ordinary number, then that's the answer. You're done. For example, to evaluate $\lim_{x\to 5}\frac{x^2-10}{x}$, just plug in the arrow-number. You get $\frac{5^2-10}{5}=3$. That's all there is to it. You're also done if plugging in the arrow-number gives you a number or infinity or negative infinity over zero, like $\frac{3}{0}$, or $\frac{\pm\infty}{0}$; in these cases the limit does not exist (DNE).

When plugging in fails because it gives you $\frac{0}{0}$, you've got a nontrivial limit problem and a bit of work to do. You have to convert the fraction into some expression where plugging in *does* work (it works because the algebra you do plugs an infinitesimal hole in the function where x equals the arrow-number). Here are some algebraic methods you can try:

- FOILing
- Factoring
- Finding the least common denominator
- Canceling
- Simplification
- Conjugate multiplication

A couple of these methods are illustrated in the following examples. You'll practice all the methods in the practice problems.

Here's a factoring example.

EXAMPLE

Q. Evaluate $\lim_{x\to 5}\frac{x^2-25}{x-5}$, the same problem you did with a calculator in the preceding section.

A. **The limit is 10.**

1. **Try plugging 5 into x — you should *always* try substitution first.**

   You get $\frac{0}{0}$ — no good, on to plan B.

2. **x^2-25 can be factored, so do it.**

   $$\lim_{x\to 5}\frac{x^2-25}{x-5}=\lim_{x\to 5}\frac{(x-5)(x+5)}{x-5}$$

3. **Cancel the $(x-5)$ from the numerator and denominator.**

   $$=\lim_{x\to 5}(x+5)$$

4. **Now substitution will work.**

   $$=5+5=10$$

   So, $\lim_{x\to 5}\frac{x^2-25}{x-5}=10$, confirming the calculator answer.

   By the way, the function you got after canceling the $(x-5)$, namely $y=(x+5)$, is identical to the original function, $y=\frac{x^2-25}{x-5}$, except that the hole in the original function at $(5, 10)$ has been plugged. And note that the limit as x approaches 5 is 10, which is the height of the hole at $(5, 10)$.

Try *conjugate multiplication* for fraction functions that contain square roots. Conjugate multiplication *rationalizes* the numerator or denominator of a fraction, which means getting rid of square roots.

EXAMPLE

Q. Evaluate $\lim_{x\to 4}\frac{\sqrt{x}-2}{x-4}$.

A. **The limit is $\frac{1}{4}$.**

1. **Try substitution.**

   Plug in 4: That gives you $\frac{0}{0}$ — time for plan B.

2. **Multiply the numerator *and* denominator by the conjugate of $\sqrt{x}-2$, which is $\sqrt{x}+2$.**

Definition of *conjugate*: The conjugate of a two-term expression is just the same expression with subtraction switched to addition or vice versa. The product of conjugates always equals the first term squared minus the second term squared.

Now do the rationalizing.

$$\begin{aligned}&\lim_{x\to 4}\frac{\sqrt{x}-2}{x-4}\\ &=\lim_{x\to 4}\left(\frac{\sqrt{x}-2}{x-4}\cdot\frac{\sqrt{x}+2}{\sqrt{x}+2}\right)\\ &=\lim_{x\to 4}\frac{\sqrt{x}^2-2^2}{(x-4)(\sqrt{x}+2)}\\ &=\lim_{x\to 4}\frac{(x-4)}{(x-4)(\sqrt{x}+2)}\end{aligned}$$

3. **Cancel the $(x-4)$ from the numerator and denominator.**

$$=\lim_{x\to 4}\frac{1}{\sqrt{x}+2}$$

4. **Now substitution works.**

$$=\frac{1}{\sqrt{4}+2}=\frac{1}{4}$$

So, $\lim_{x\to 4}\frac{\sqrt{x}-2}{x-4}=\frac{1}{4}$.

As with the factoring example, this rationalizing process plugged the hole in the original function. In this example, 4 is the arrow-number, $\frac{1}{4}$ is the limit answer, and the function $\frac{\sqrt{x}-2}{x-4}$ has a hole at $\left(4,\ \frac{1}{4}\right)$.

EXAMPLE

When factoring and conjugate multiplication don't work, try some other basic algebra, like adding or subtracting fractions, multiplying or dividing fractions, canceling, or some other form of simplification. Here's an example:

Q. Evaluate $\lim_{x\to 0}\frac{\frac{1}{x+4}-\frac{1}{4}}{x}$.

A. **The limit is $-\frac{1}{16}$.**

1. **Try substitution.**

Plug in 0: That gives you $\frac{0}{0}$ — no good.

2. **Simplify the complex fraction (that's a big fraction that contains little fractions) by multiplying the numerator and denominator by the least common denominator of the little fractions, namely $4(x+4)$.**

Note: You can also simplify a complex fraction by adding or subtracting the little fractions in the numerator and/or denominator, but the method described here is a bit quicker.

$$\lim_{x\to 0}\frac{\frac{1}{x+4}-\frac{1}{4}}{x}$$

$$=\lim_{x\to 0}\left(\frac{\frac{1}{x+4}-\frac{1}{4}}{x}\cdot\frac{4(x+4)}{4(x+4)}\right)$$

$$=\lim_{x\to 0}\frac{4-(x+4)}{4x(x+4)}$$

$$=\lim_{x\to 0}\frac{-x}{4x(x+4)}$$

$$=\lim_{x\to 0}\frac{-1}{4(x+4)}$$

3. **Now substitution works.**

$=\frac{-1}{4(0+4)}=-\frac{1}{16}$. That's the limit.

YOUR TURN

 $\lim_{x\to 3}\frac{x^2-9}{x-3}$

4 $\lim_{x\to 1}\frac{x-1}{x^2+x-2}$

 5 $\lim_{x\to -2}\frac{x+2}{x^3+8}$

 6 $\lim_{x\to 2}\frac{x^2-4}{4x^2+5x-6}$

7 $\lim_{x \to 9} \frac{x-9}{3-\sqrt{x}}$

8 $\lim_{x \to 10} \frac{\sqrt{x-5}-\sqrt{5}}{x-10}$

9 $\lim_{x \to 0} \frac{\cos x - 1}{x}$

10 $\lim_{x \to 2} \frac{\frac{1}{x}-\frac{1}{2}}{x-2}$

11 $\lim_{x \to 0} \frac{x}{\frac{1}{6}+\frac{1}{x-6}}$

12 $\lim_{x \to 0} \frac{\sin x}{x}$

13 $\lim_{x \to 0} \frac{x}{\sin(3x)}$

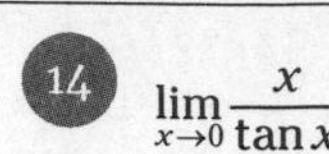

14 $\lim_{x \to 0} \frac{x}{\tan x}$

15 $\lim_{x \to 6} \frac{x-6}{\sqrt{6}-\sqrt{x}}$

16 $\lim_{x \to 5} 8$

17 $\lim_{x \to 0} k$ (k is a constant)

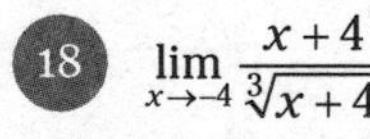

18 $\lim_{x \to -4} \frac{x+4}{\sqrt[3]{x+4}}$

Take a break and make yourself a limit sandwich

The *sandwich* or *squeeze* method is a special algebraic technique you can try when you can't solve a limit problem with ordinary algebra. The basic idea is to find one function that's always less than or equal to the limit function (at least near the arrow-number) and another function that's always greater than or equal to the limit function. Both of your new functions must have the same limit as *x* approaches the arrow-number. Then, because the limit function is "sandwiched" between the other two, like salami between slices of bread, it must have that same limit as well. Consider Figure 8-1. Function *f* is sandwiched between *B* (for *bottom*) and *T* (for *top*). Because near the arrow-number of 2, *f* is always higher than or the same height as *B* and always lower than or the same height as *T*, and because $\lim_{x\to 2} B(x) = \lim_{x\to 2} T(x)$, $f(x)$ must have the same limit as *x* approaches 2 because *f* is squeezed between *B* and *T*. The limit of both *B* and *T* as *x* approaches 2 is 3. So, 3 has to be the limit of *f* as well. It's got nowhere else to go.

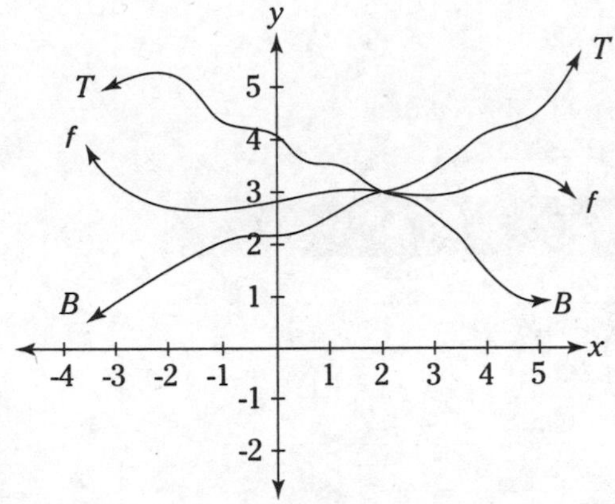

FIGURE 8-1: A limit sandwich — functions *B* and *T* are the bread and *f* is the salami.

EXAMPLE

Q. Evaluate $\lim_{x\to 0}\left(x\sin\frac{1}{x}\right)$.

A. The limit is zero.

1. **Try substitution.**

   Plug 0 into *x*. That gives you $0\cdot\sin\frac{1}{0}$ — no good; this is undefined. On to plan B.

2. **Try the ordinary algebraic methods from the last section.**

   Knock yourself out. You can't do it. Plan C.

3. **Try your calculator.**

   It's always a good idea to see what your calculator tells you even if this is a "show your work" problem. To graph this function, set your graphing calculator's mode to *radian* and the window to

   Xmin = −0.4

   Xmax = 0.4

   Ymin = −0.3

   Ymax = 0.3

Figure 8-2 shows what the graph looks like.

It definitely looks like the limit of f is zero as x approaches zero from the left and the right. Now, check the table of values on your calculator (set TblStart to zero and ΔTbl to 0.001). Table 8-2 gives some of the values from the calculator table.

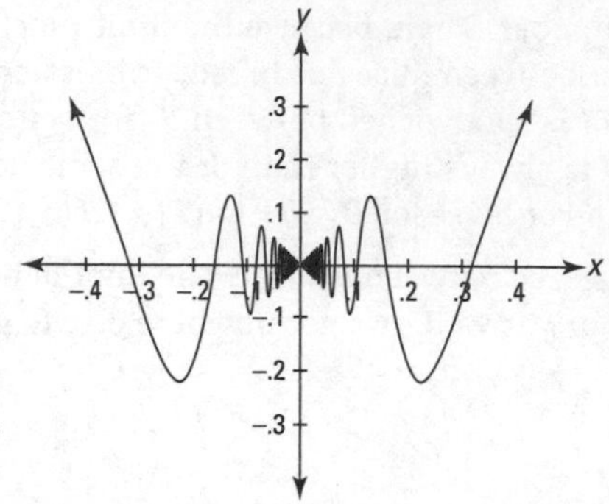

FIGURE 8-2: The graph of $f(x) = x\sin\frac{1}{x}$.

Table 8-2 Table of Values for $f(x) = x\sin\frac{1}{x}$

x	$f(x)$
0	Error
0.001	0.0008269
0.002	−0.000936
0.003	0.0009565
0.004	−0.003882
0.005	−0.004366
0.006	−0.000969
0.007	−0.006975
0.008	−0.004928
0.009	−0.008234

These numbers sort of look like they're getting closer and closer to zero as x gets close to zero, but they're not convincing. This type of table doesn't work so great for oscillating functions like sine or cosine. (Some function values on the table, for example -0.000969 for $x = 0.006$, are closer to zero than other values higher on the table where x is smaller. That's the opposite of what you want to see. And you'll run into the same problem if you look at small negative values of x.)

A better way of seeing that the limit of f is zero as x approaches zero is to use the first calculator method I discuss in the section "Figuring a limit with your calculator." Enter the function on the home screen and successively plug in the x-values listed in Table 8-3 to obtain the corresponding function values. (*Note:* Don't be confused. Table 8-3 is called a "table," but the values in that table were not generated by a calculator's table function. Get it?)

Table 8-3 Another Table of Values for $f(x) = x \sin \frac{1}{x}$

x	$f(x)$
0.1	–0.054
0.01	–0.0051
0.001	0.00083
0.0001	–0.000031
0.00001	0.00000036

Now you can definitely see that f is headed toward zero. (You should also look at the values of $f(x)$ as x approaches zero from the negative side. These values will confirm that g is headed to zero.)

4. **Now you should prove the limit mathematically even though you've already solved it on your calculator. One way to do this is to make a limit sandwich.**

For some sandwich method problems, it can be a challenge to come up with the "bread" functions B and T. For such challenging sandwich problems, you've got to think about and/or look at the shape of the salami function f, and then use your knowledge of functions and your imagination to come up with some good prospects for the bread functions.

However, the above sandwich problem is a snap. It's an example of a common type of sandwich problem of the form $f(x) = x^n \sin(\ )$ or $f(x) = x^n \cos(\ )$ — basically, x to a positive integer power times the sine or cosine of just about anything. For this class of functions, there are two cases: If n is even, you can use $B(x) = -x^n$ and $T(x) = x^n$ for the bread functions; if n is odd, you can use $B(x) = -|x^n|$ and $T(x) = |x^n|$ for the bread functions. Piece o' cake.

Following this rule for the current problem, you can use $-|x|$ for the bottom function and $|x|$ for the top function. You should graph $B(x) = -|x|$ and $T(x) = |x|$ along with $f(x) = x \sin \frac{1}{x}$ to confirm that B and T make adequate bread functions for f. Figure 8-3 shows that they do. We've shown — though perhaps not to a mathematician's satisfaction, *egad!* — that $B(x) \le f(x) \le T(x)$. And because $\lim_{x \to 0} B(x) = \lim_{x \to 0} T(x) = 0$, it follows that $f(x)$ must have the same limit: voilà — $\lim_{x \to 0} f(x) = 0$.

5. **Taking the easy way out.**

I hope you won't be too upset with me when I only now show you the easy way to solve this limit problem. (I needed to show you the important ideas in Steps 3 and 4 because you really should learn the theory of the sandwich theorem and also how to squeeze a function between two others. And on top of that, I've heard that some teachers and professors may not like the following approach. I'm not sure why because it's perfectly valid.)

All you need to do to solve the current limit problem is to observe 1) that $\sin\frac{1}{x}$ never gets below −1 or above 1 (in other words, it's *bounded* by −1 and 1), and 2) that $\lim_{x\to 0} x = 0$. Thus, basically, for $\lim_{x\to 0}\left(x\sin\frac{1}{x}\right)$, you've got zero times bounded, which equals zero. That's all there is to it.

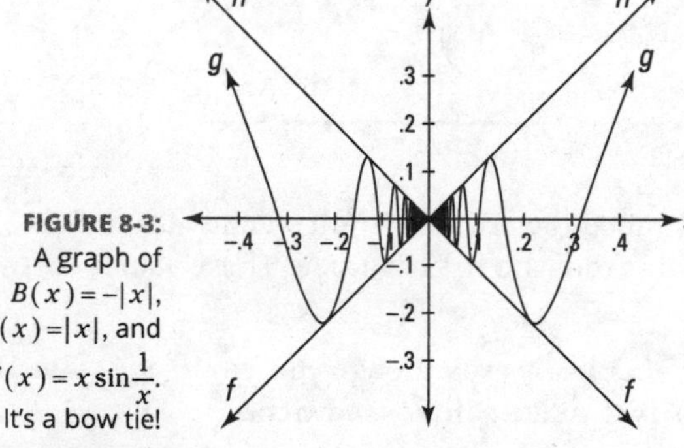

FIGURE 8-3: A graph of $B(x) = -|x|$, $T(x) = |x|$, and $f(x) = x\sin\frac{1}{x}$. It's a bow tie!

YOUR TURN

19 Evaluate $\lim_{x\to 0}\left(x\sin\frac{1}{x^2}\right)$.

20 Evaluate $\lim_{x\to 0}\left(x^2\cos\frac{1}{x}\right)$.

THE LONG AND WINDING ROAD

Consider the function $f(x) = x\sin\frac{1}{x}$, shown in Figures 8-2 and 8-3 and discussed in the section about making a limit sandwich. It's defined everywhere except at zero. If you now alter it slightly — by renaming it $g(x)$ and then defining $g(0)$ to be 0 — you create a new function with bizarre properties. The function is now continuous everywhere; in other words, it has no gaps. But at $(0, 0)$, it seems to contradict the basic idea of continuity that says you can trace the function without taking your pencil off the paper.

Imagine starting anywhere on $g(x)$ — which looks exactly like $f(x)$ in Figures 8-2 and 8-3 — to the left of the *y*-axis and driving along the winding road toward the origin, $(0, 0)$. Get this: You can start your drive as close to the origin as you like — how about the width of a proton away from $(0, 0)$ — and the length of road between you and $(0, 0)$ is *infinitely* long! That's right. It winds up and down with such increasing frequency as you get closer and closer to $(0, 0)$ that the length of your drive is actually infinite, despite the fact that each "straightaway" is getting shorter and shorter. On this long and winding road, you'll never get to their door.

This altered function is clearly continuous at every point — with the possible exception of $(0, 0)$ — because it's a smooth, connected, winding road. And because $\lim_{x\to 0}\left(x\sin\frac{1}{x}\right) = 0$ (see the limit sandwich section for proof), and because $g(0)$ is defined to be 0, the three-part test for continuity at 0 is satisfied. The function is thus continuous everywhere.

But tell me, how can the curve ever reach $(0, 0)$ or connect to $(0, 0)$ from the left (or the right)? Assuming you can traverse an infinite distance by driving infinitely fast, when you finally drive through the origin, are you on one of the up legs of the road or one of the down legs? Neither seems possible because no matter how close you are to the origin, you have an infinite number of legs and an infinite number of turns ahead of you. There is no last turn before you reach $(0, 0)$. So, it seems that the function can't connect to the origin and that, therefore, it can't be continuous there — despite the fact that the math tells you that it is.

Here's another way of looking at it. Imagine a vertical line drawn on top of the function at $x = -0.2$. Now, keeping the line vertical, slowly slide the line to the right over the function until you pass over $(0, 0)$. There are no gaps in the function, so at every instance, the vertical line crosses the function somewhere. Think about the point where the line intersects with the function. As you drag the line to the right, that point travels along the function, winding up and down along the road, and, as you drag the line over the origin, the point reaches and then passes $(0, 0)$. Now tell me this: When the point hits $(0, 0)$, is it on its way up or down? How can you reconcile all this? I wish I knew.

Stuff like this really messes with your mind.

Evaluating Limits at ±Infinity

In the previous sections, I look at limits as *x* approaches a finite number, but you can also have limits where *x* approaches infinity or negative infinity. Consider the function $f(x) = \frac{1}{x}$ and check out its graph in Figure 8-4.

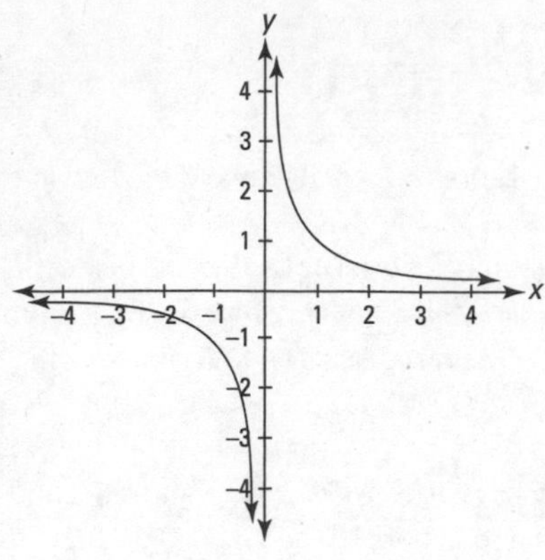

FIGURE 8-4: The graph of $f(x) = \frac{1}{x}$.

You can see on the graph (in the first quadrant) that as *x* gets bigger and bigger — in other words, as *x* approaches infinity — the height of the function gets lower and lower but never gets to zero. This is confirmed by considering what happens when you plug bigger and bigger numbers into $\frac{1}{x}$: The outputs get smaller and smaller and approach zero. This graph thus has a horizontal asymptote of $y = 0$ (the *x*-axis), and you say that $\lim_{x\to\infty} \frac{1}{x} = 0$. The fact that *x* never actually reaches infinity and that *f* never gets to zero has no relevance. When we say that $\lim_{x\to\infty} \frac{1}{x} = 0$, we mean that as *x* gets bigger and bigger without end, *f* is closing in on a height of zero (or *f* is ultimately getting infinitely close to a height of zero). If you look at the third quadrant, you can see that the function *f* also approaches zero as *x* approaches negative infinity, which is written as $\lim_{x\to-\infty} \frac{1}{x} = 0$.

Like with limits where *x* approaches a finite number, to solve limit problems where *x* approaches infinity or negative infinity, you can use your calculator or algebra. But before I go through those techniques, let's first take care of a special class of limits at ±infinity where no calculus is needed: rational function limits.

Limits of rational functions at ±infinity

This section deals with the horizontal asymptotes of *rational functions*. A rational function is a fraction function with polynomials in the numerator and denominator. (You probably studied rational functions and their asymptotes, both vertical and horizontal, in an algebra and/or a precalculus class.) Determining the limit of a function as *x* approaches infinity or negative infinity is the same as finding the height of its horizontal asymptote.

Say you've got a rational function like $f(x) = \frac{4x^2 - x + 8}{5x^3 + 10x + 1}$. First, note the degree of the numerator, namely, the highest power of *x* in the numerator (2 in this function) and the degree of the denominator (3 in this function). There are three cases:

- If the degree of the numerator is greater than the degree of the denominator, for example, $f(x) = \frac{6x^3 + x^2 - 7}{2x^2 + 8}$, then there's no horizontal asymptote, and the limit of the function as *x* approaches infinity (or negative infinity) does not exist (the limit will be positive or negative infinity).

» If the degree of the denominator is greater than the degree of the numerator, for example, $g(x)=\frac{4x^2-9}{x^4+12}$, then the x-axis (that's the line $y=0$) is the horizontal asymptote, and $\lim_{x\to\infty} g(x) = \lim_{x\to-\infty} g(x)=0$.

» If the degrees of the numerator and denominator are equal, take the coefficient of the highest power of x in the numerator and divide it by the coefficient of the highest power of x in the denominator. That quotient gives you the answer to the limit problem and the height of the asymptote — for example, if $h(x)=\frac{4x^3-10x+1}{5x^3+2x^2-x}$, $\lim_{x\to\infty} h(x) = \lim_{x\to-\infty} h(x)=\frac{4}{5}$, and h has a horizontal asymptote at $y=\frac{4}{5}$.

TIP

Talk like a professor. To impress your friends, point your index finger upward, raise one eyebrow, and say in a professorial tone, "In a rational function where the numerator and denominator are of equal degrees, the limit of the function as x approaches infinity or negative infinity equals the quotient of the coefficients of the leading terms. A horizontal asymptote occurs at this same value."

WARNING

$\frac{\infty}{\infty}$ does *not* equal 1. Substitution doesn't work for the problems in this section. If you try plugging infinity into x in any of the rational functions in this section, you get $\frac{\infty}{\infty}$ but that does *not* necessarily equal 1 ($\frac{\infty}{\infty}$ sometimes equals 1, but it often does not). A result of $\frac{\infty}{\infty}$ tells you nothing about the answer to a limit problem.

EXAMPLE

Examples? EXAMPLES?! We don't need no stinking examples. (My apologies to the late, great director John Huston). See the bullet points earlier in this section and then try the following practice problems.

YOUR TURN

21 What's $\lim_{x\to\infty}\frac{5x^3-x^2+10}{2x^4+x+3}$? Explain your answer.

22 What's $\lim_{x\to-\infty}\frac{3x^4+100x^3+4}{8x^4+1}$? Explain your answer.

Solving limits at ±infinity with a calculator

In many situations in calculus (and math in general), your calculator can be a great tool. Sometimes you can use it to solve problems you can't solve the pure, algebraic way. It can also give you ideas that you can use to solve a problem the "right" way. And you can use it to check your answers. Your calculator is a good thing to try with some limits at plus or minus infinity.

Here's a limit problem that can't be done by the method in the previous section because it doesn't involve a rational function: $\lim_{x\to\infty}\left(\sqrt{x^2+x}-x\right)$. But it's a snap with a calculator. Enter the function in graphing mode, then go to Table Setup and set TblStart to 100,000 and ΔTbl to 100,000. Table 8-4 shows the results.

Table 8-4 Table of Values for $\sqrt{x^2+x}-x$

x	y
100,000	0.4999988
200,000	0.4999994
300,000	0.4999996
400,000	0.4999997
500,000	0.4999998
600,000	0.4999998
700,000	0.4999998
800,000	0.4999998
900,000	0.4999999

You can see that *y* is getting extremely close to 0.5 as *x* gets larger and larger. So, 0.5 is the limit of the function as *x* approaches infinity. (The function is not a rational function, but it has a horizontal asymptote at $y = 0.5$. All functions, rational or not, that have a finite limit answer as *x* approaches infinity or negative infinity have a horizontal asymptote at the limit answer height.) If you have any doubts that the limit equals 0.5, go back to Table Setup and put in a humongous TblStart and ΔTbl, say 1,000,000,000, and check the table results again. All you see is a column of 0.5's. That's the limit.

By the way, unlike with the rational functions in the previous section, the limit of this function as *x* approaches negative infinity doesn't equal the limit as *x* approaches positive infinity. For this function, here's what happens when *x* approaches negative infinity: $\lim_{x\to-\infty}\left(\sqrt{x^2+x}-x\right)=\lim_{x\to-\infty}\left(\sqrt{x(x+1)}-x\right)$. Now plug in: $=\sqrt{-\infty(-\infty+1)}-(-\infty)=\sqrt{(-\infty)(-\infty)}+\infty$ $=\sqrt{\infty}+\infty=\infty+\infty=\infty$. One more thing: Just as with regular limits, using a non-CAS (Computer Algebra System) calculator for infinite limits won't allow you to deduce the exact answer unless the numbers in the table are getting close to a number you recognize, like 0.5. If the exact limit answer is something like $\frac{\sqrt{3}}{5}$, on the other hand, you won't recognize that exact answer from its decimal approximation.

WARNING

$\infty - \infty$ **does *not* equal zero.** Substitution does not work for the problem, $\lim_{x\to\infty}\left(\sqrt{x^2+x}-x\right)$. If you plug ∞ into *x*, you get $\infty - \infty$, which does *not* necessarily equal zero ($\infty - \infty$ sometimes equals zero, but it often does not). A result of $\infty - \infty$ tells you nothing about the answer to a limit problem.

You'll get some practice using your calculator in some of the Your Turn problems in the next section.

Solving limits at ±infinity with algebra

In general, to find a limit at infinity ($\lim_{x\to\infty}$ or $\lim_{x\to-\infty}$) with algebra, you can use the same techniques from the bulleted list in the section, "Solving limit problems with algebra," earlier in this chapter. And, if you're dealing with a fraction function that's not a rational function, a good algebraic technique is to divide the numerator and denominator of the fraction by the highest power of *x* that appears anywhere in the fraction. Also, for some limits at positive infinity, the following tip may come in handy.

TIP

Evaluating infinity/infinity. Consider the following four types of expressions: x^{10}, 5^x, $x!$, and x^x. If a limit at positive infinity involves a fraction with one of them over another, you can apply this handy little tip. These four expressions are listed from "smallest" to "largest." The limit will equal zero if you have a "smaller" expression over a "larger" one, and the limit will equal infinity if you have a "larger" expression over a "smaller" one. (This isn't a true ordering; it's only for problems of this type.) And note the following things:

1. Coefficients don't change the order — for example, $1000x^{10}$, $12\cdot 5^x$, $300x!$, and $0.065x^x$.
2. For the first expression (x^{10}), the power can be any number, while for the second expression (5^x), the number must be greater than one.
3. Replacing *x* with a multiple of *x* (like $5^{(10x)}$ or $(6x)^{10}$) doesn't change the order — with one important exception: with the factorial expression, replacing the *x* with *kx* (if $k \geq 1$ like with $(4x)!$ or $(1.8x)!$) makes the expression the largest of the four; if $0 < k < 1$, the order doesn't change.

EXAMPLE

Q. Find $\lim_{x\to\infty}\frac{x^3}{1.01^x}$.

A. **The limit is zero.**

This is an example of a "smaller" expression over a "larger" one, so the answer is 0. Perhaps this result surprises you. You may think that this fraction will keep getting bigger and bigger because it seems that no matter what power 1.01 is raised to, it will never grow very large. And, in fact, if you plug 1000 into *x*, the quotient is big — over 47,000. But if you enter $\frac{x^3}{1.01^x}$ in graphing mode and then set both TblStart and ΔTbl to 1000, the table values show quite convincingly that the limit is 0. By the time $x = 3000$, the answer is about 0.00293, and when $x = 10{,}000$, the answer is roughly 6×10^{-32}.

Q. Find $\lim_{x\to\infty}\frac{100x^2}{5x^3-\cos(x^2)}$.

A. **The limit is zero.**

1. **Divide the numerator and denominator by the highest power of x appearing in the fraction — that's x^3.**

$$=\lim_{x\to\infty}\frac{\frac{100x^2}{x^3}}{\frac{5x^3}{x^3}-\frac{\cos(x^2)}{x^3}}=\lim_{x\to\infty}\frac{\frac{100}{x}}{5-\frac{\cos(x^2)}{x^3}}$$

2. **Now you can plug in.**

$$=\frac{\frac{100}{\infty}}{5-\frac{\cos(\infty)}{\infty}}=\frac{0}{5-0}=0$$

Here's the justification for $\frac{\cos(\infty)}{\infty}=0$: Values of the cosine function are always between −1 and 1 inclusive. You say that the cosine function is *bounded* by −1 and 1, or, symbolically, that $-1\le\cos x\le 1$ for all x. For problems like this one, you can use the rule that $\frac{\text{bounded}}{\infty}=0$. For other limit problems, a related rule may come in handy: zero · bounded = zero.

Note: While the rational function rule discussed earlier in this chapter will give you the correct answer for this problem, you should not use the rule here because this function is not a rational function.

EXAMPLE

Q. Find $\lim_{x\to\infty}\left(\sqrt{x^2+x}-x\right)$ with algebra. You got the answer earlier in this chapter with a calculator, but all things being equal, it's better to solve the problem algebraically because then you have a mathematically airtight answer. The calculator answer for this limit is *very* convincing, but it's not mathematically rigorous, so if you stop there, the math police might get you.

A. **The limit is $\frac{1}{2}$.**

1. **Try substitution — always a good idea.**

No good. You get $\infty-\infty$, which tells you nothing — see the Warning in the previous section. On to plan B.

Because $\left(\sqrt{x^2+x}-x\right)$ contains a square root, the conjugate multiplication method would be a natural choice, except that that method is used for fraction functions. Well, just put $\left(\sqrt{x^2+x}-x\right)$ over the number 1 and, voilà, you've got a fraction: $\frac{\sqrt{x^2+x}-x}{1}$. Now do the conjugate multiplication.

2. Multiply the numerator and denominator by the conjugate of $\left(\sqrt{x^2+x}-x\right)$ and simplify.

$$\lim_{x\to\infty}\frac{\sqrt{x^2+x}-x}{1}$$

$$=\lim_{x\to\infty}\left(\frac{\sqrt{x^2+x}-x}{1}\cdot\frac{\sqrt{x^2+x}+x}{\sqrt{x^2+x}+x}\right)$$

$$=\lim_{x\to\infty}\frac{x^2+x-x^2}{\sqrt{x^2+x}+x}$$

(First, cancel the x^2s in the numerator. Then factor x out of the denominator. Yes, you heard that right.)

$$=\lim_{x\to\infty}\frac{x}{x\left(\sqrt{1+\frac{1}{x}}+1\right)}$$

(Now, cancel the x's.)

$$=\lim_{x\to\infty}\frac{1}{\sqrt{1+\frac{1}{x}}+1}$$

3. Now substitution works.

$$=\frac{1}{\sqrt{1+\frac{1}{\infty}}+1}$$

$$=\frac{1}{\sqrt{1+0}+1}$$

(Recall that $\lim_{x\to\infty}\frac{1}{x}=0$ from the section, "Limits to memorize.")

$$=\frac{1}{1+1}=\frac{1}{2}$$

Thus, $\lim_{x\to\infty}\left(\sqrt{x^2+x}-x\right)=\frac{1}{2}$, which confirms the calculator answer.

YOUR TURN

23 Use your calculator to figure $\lim_{x\to\infty}\frac{x^x}{x!}$.

24 Determine $\lim_{x\to\infty}\frac{5x+2}{\sqrt{4x^2-1}}$.

25 Evaluate $\lim_{x\to-\infty}\left(4x+\sqrt{16x^2-3x}\right)$.

26 Evaluate 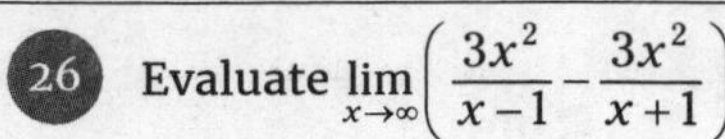 $\lim_{x\to\infty}\left(\frac{3x^2}{x-1}-\frac{3x^2}{x+1}\right)$.

27 $\lim_{x\to-\infty}\cos\left(x^2\right)$

28 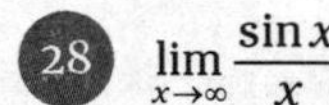 $\lim_{x\to\infty}\frac{\sin x}{x}$

29 $\lim_{x\to\infty}\left(1+\frac{1}{x}\right)^x$

30 $\lim_{x\to\infty}\left(1+\frac{1}{3x}\right)^x$

Practice Questions Answers and Explanations

1. $\lim_{x \to -3} \frac{x^2 - 5x - 24}{x + 3} = -11$

   You want the limit as x approaches −3, so pick a number really close to −3, like −3.0001, plug that into x in your function $\frac{x^2 - 5x - 24}{x + 3}$, and enter that into your calculator. (If you've got a calculator like a Texas Instruments TI-84, a good way to do this is to use the Sto button to store −3.0001 into x, then enter $\frac{x^2 - 5x - 24}{x + 3}$ into the home screen and hit Enter.)

   The calculator's answer is −11.0001. Because this is near the round number −11, your answer is −11. By the way, you can do this problem easily with algebra as well.

2. $\lim_{x \to 0} \frac{\sin x}{\tan^{-1} x} = 1$

   Enter the function in graphing mode like this: $\frac{\sin x}{\tan^{-1} x}$. Then go to Table Setup and enter a small increment into ΔTbl (try 0.01 for this problem), and enter the arrow-number, 0, into tblStart. When you scroll through the table near $x = 0$, you'll see the y-values getting closer and closer to the round number 1. That's your answer. This problem, unlike Problem 1, is *not* easy to do with algebra — until, that is, you learn L'Hôpital's rule in Chapter 18.

3. $\lim_{x \to 3} = \frac{x^2 - 9}{x - 3} = 6$

   Factor, cancel, and plug in.

   $$= \lim_{x \to 3} \frac{(x-3)(x+3)}{(x-3)}$$
   $$= \lim_{x \to 3} (x+3) = 6$$

4. $\lim_{x \to 1} \frac{x - 1}{x^2 + x - 2} = \frac{1}{3}$

   Factor, cancel, and plug in.

   $$= \lim_{x \to 1} \frac{(x-1)}{(x-1)(x+2)}$$
   $$= \lim_{x \to 1} \frac{1}{x+2} = \frac{1}{3}$$

5. $\lim_{x \to -2} \frac{x + 2}{x^3 + 8} = \frac{1}{12}$

   Factor, cancel, and plug in.

   $$= \lim_{x \to -2} \frac{(x+2)}{(x+2)\left(x^2 - 2x + 4\right)}$$
   $$= \lim_{x \to -2} \frac{1}{x^2 - 2x + 4}$$
   $$= \frac{1}{(-2)^2 - 2(-2) + 4} = \frac{1}{12}$$

6. $\lim_{x \to 2} \frac{x^2 - 4}{4x^2 + 5x - 6} = 0$

MATH RULES

Did you waste your time factoring the numerator and denominator? Gotcha! **Always plug in first!** When you plug 2 into the limit expression, you get $\frac{0}{20}$, which equals zero — that's your answer.

7 $\lim\limits_{x\to 9}\frac{x-9}{3-\sqrt{x}}=-6$

1. **Multiply the numerator and denominator by the conjugate of the denominator — namely, $3+\sqrt{x}$.**

$$=\lim_{x\to 9}\left(\frac{x-9}{3-\sqrt{x}}\cdot\frac{3+\sqrt{x}}{3+\sqrt{x}}\right)$$

2. **Multiply out the part of the fraction containing the conjugate pair (the denominator in this problem).**

$$=\lim_{x\to 9}\frac{(x-9)(3+\sqrt{x})}{(9-x)}$$

3. **Cancel.**

$$=\lim_{x\to 9}\left(-1(3+\sqrt{x})\right)$$

TIP

Any fraction of the form $\frac{a-b}{b-a}$ equals −1.

4. **Plug in.**

$$=-1(3+\sqrt{9})=-6$$

8 $\lim\limits_{x\to 10}\frac{\sqrt{x-5}-\sqrt{5}}{x-10}=\frac{\sqrt{5}}{10}$

Multiply the numerator and denominator by the conjugate, FOIL, cancel, and plug in.

$$=\lim_{x\to 10}\left(\frac{\sqrt{x-5}-\sqrt{5}}{x-10}\cdot\frac{\sqrt{x-5}+\sqrt{5}}{\sqrt{x-5}+\sqrt{5}}\right)$$

$$=\lim_{x\to 10}\frac{(x-5)-5}{(x-10)(\sqrt{x-5}+\sqrt{5})}$$

$$=\lim_{x\to 10}\frac{(x-10)}{(x-10)(\sqrt{x-5}+\sqrt{5})}$$

$$=\lim_{x\to 10}\frac{1}{\sqrt{x-5}+\sqrt{5}}$$

$$=\frac{1}{\sqrt{10-5}+\sqrt{5}}=\frac{1}{2\sqrt{5}}=\frac{\sqrt{5}}{10}$$

9 $\lim\limits_{x\to 0}\frac{\cos x-1}{x}=0$

Did you try multiplying the numerator and denominator by the conjugate of $\cos x-1$? Gotcha again! That method doesn't work here. The answer to this limit is zero, something you just have to memorize.

10 $\lim\limits_{x\to 2}\frac{\frac{1}{x}-\frac{1}{2}}{x-2}=-\frac{1}{4}$

1. **Multiply the numerator and denominator by the least common denominator of the little fractions inside the big fraction — namely, 2x.**

$$= \lim_{x\to 2}\left(\frac{\frac{1}{x}-\frac{1}{2}}{x-2}\cdot\frac{2x}{2x}\right)$$

2. **Multiply out the numerator.**

$$= \lim_{x\to 2}\frac{(2-x)}{(x-2)(2x)}$$

3. **Cancel.**

$$= \lim_{x\to 2}\frac{-1}{2x}$$

4. **Plug in.**

$$= \frac{-1}{2\cdot 2} = -\frac{1}{4}$$

(11) $\lim\limits_{x\to 0}\dfrac{x}{\frac{1}{6}+\frac{1}{x-6}} = -36$

Multiply by the least common denominator, multiply out, cancel, and plug in.

$$= \lim_{x\to 0}\left(\frac{x}{\frac{1}{6}+\frac{1}{x-6}}\cdot\frac{6(x-6)}{6(x-6)}\right)$$

$$= \lim_{x\to 0}\frac{6x(x-6)}{(x-6)+6}$$

$$= \lim_{x\to 0}\frac{6x(x-6)}{x}$$

$$= \lim_{x\to 0}(6(x-6)) = -36$$

(12) $\lim\limits_{x\to 0}\dfrac{\sin x}{x} = 1$

No work required — except for the memorization, that is.

(13) $\lim\limits_{x\to 0}\dfrac{x}{\sin(3x)} = \dfrac{1}{3}$

Did you get it? If not, try the following hint before you read the solution: This fraction sort of resembles the one in Problem 12. Still stuck? Okay, here you go:

1. **Multiply the numerator and denominator by 3.**

   You have a $3x$ in the denominator, so you need $3x$ in the numerator as well (to make the fraction look more like the one in Problem 12).

$$= \lim_{x\to 0}\left(\frac{x}{\sin(3x)}\cdot\frac{3}{3}\right)$$

$$= \lim_{x\to 0}\frac{3x}{3\sin(3x)}$$

2. **Pull the $\frac{1}{3}$ through the lim symbol (the 3 in the denominator is really a $\frac{1}{3}$, right?).**

$$= \frac{1}{3} \lim_{x \to 0} \frac{3x}{\sin(3x)}$$

Now, if your calc teacher lets you, you can just stop here (since it's "obvious" that $\lim_{x \to 0} \frac{3x}{\sin(3x)} = 1$) and put down your final answer of $\frac{1}{3} \cdot 1$, or $\frac{1}{3}$. But if your teacher's a stickler for showing work, you'll have to do a couple more steps.

3. **Set $u = 3x$.**
4. **Substitute u for $3x$ in the numerator and denominator.** And, as x approaches 0 in $u = 3x$, u also approaches 0, so you can substitute u for x under the lim symbol.

$$= \frac{1}{3} \lim_{u \to 0} \frac{u}{\sin u}$$
$$= \frac{1}{3} \cdot 1 = \frac{1}{3}$$

TIP

Because $\lim_{x \to 0} \frac{\sin x}{x} = 1$, the limit of the reciprocal of $\frac{\sin x}{x}$, namely $\frac{x}{\sin x}$, must equal the reciprocal of 1 — which is, of course, 1.

(14) $\lim_{x \to 0} \frac{x}{\tan x} = 1$

1. **Use the fact that $\lim_{x \to 0} \frac{\sin x}{x} = 1$ and replace $\tan x$ with $\frac{\sin x}{\cos x}$.**

$$= \lim_{x \to 0} \frac{x}{\frac{\sin x}{\cos x}}$$

2. **Multiply the numerator and denominator by $\cos x$.**

$$= \lim_{x \to 0} \left(\frac{x}{\frac{\sin x}{\cos x}} \cdot \frac{\cos x}{\cos x} \right)$$
$$= \lim_{x \to 0} \frac{x \cos x}{\sin x}$$

3. **Rewrite the expression as the product of two functions.**

$$= \lim_{x \to 0} \left(\frac{x}{\sin x} \cdot \frac{\cos x}{1} \right)$$

4. **Break this into two limits, using the fact that $\lim_{x \to c} (f(x) \cdot g(x)) = \lim_{x \to c} f(x) \cdot \lim_{x \to c} g(x)$ (provided that both limits on the right exist).**

$$= \lim_{x \to 0} \frac{x}{\sin x} \cdot \lim_{x \to 0} \cos x$$
$$= 1 \cdot 1 = 1$$

(15) $\lim_{x \to 6} \frac{x - 6}{\sqrt{6} - \sqrt{x}} = -2\sqrt{6}$

Plugging in 6 produces $\frac{0}{0}$. Check. Your work begins.

Multiply the numerator and denominator by the conjugate of the denominator, simplify, and cancel:

$$\begin{aligned}\lim_{x\to6}\frac{x-6}{\sqrt{6}-\sqrt{x}}&=\lim_{x\to6}\left(\frac{x-6}{\sqrt{6}-\sqrt{x}}\cdot\frac{\sqrt{6}+\sqrt{x}}{\sqrt{6}+\sqrt{x}}\right)\\&=\lim_{x\to6}\frac{(x-6)(\sqrt{6}+\sqrt{x})}{\sqrt{6}^2-\sqrt{x}^2}\\&=\lim_{x\to6}\frac{(x-6)(\sqrt{6}+\sqrt{x})}{6-x}\\&=\lim_{x\to6}\left(-1(\sqrt{6}+\sqrt{x})\right)\\&=\lim_{x\to6}\left(-\sqrt{6}-\sqrt{x}\right)=-\sqrt{6}-\sqrt{6}=-2\sqrt{6}\end{aligned}$$

(16) $\lim_{x\to5}8=8$

This probably seems like an odd problem, because there's no x in the limit expression for you to plug the 5 into. Think of it this way. The 8 represents the function $y=8$, which is a horizontal line at a height of 8. The limit problem asks you to determine what y is getting closer and closer to along the function as x gets closer and closer to 5. But, since the function is a horizontal line, y is always equal to 8 regardless of the value of x. Thus, $\lim_{x\to5}8=\lim_{x\to-3}8=\lim_{x\to2023}8=\lim_{x\to-\infty}8=\lim_{x\to\infty}8=8$.

(17) $\lim_{x\to0}k=k$ **(k is a constant)**

Don't forget that for all calculus problems, constants behave like ordinary numbers. In Problem 16, the 8 represented the horizontal line $y=8$, so in this problem, the k represents the horizontal line $y=k$. So, y is always at a height of k regardless of the value of x. Thus, the limit equals k.

(18) $\lim_{x\to-4}\frac{x+4}{\sqrt[3]{x+4}}=0$

Plug in the arrow-number: You get $\frac{0}{0}$, so keep going and try some basic algebra.

$$\begin{aligned}&\lim_{x\to-4}\frac{x+4}{\sqrt[3]{x+4}}\\&=\lim_{x\to-4}\frac{(x+4)^1}{(x+4)^{1/3}}\\&=\lim_{x\to-4}(x+4)^{2/3}\end{aligned}$$

Now you can plug in:

$$=(-4+4)^{2/3}=0^{2/3}=0$$

(19) $\lim_{x\to0}\left(x\sin\frac{1}{x^2}\right)=0$

There are three ways to do this. The easiest method is to note that $\lim_{x\to0}x=0$ and that $\sin\frac{1}{x^2}$ never gets greater than 1 or less than −1 (it's bounded by −1 and 1). Then, because $\text{zero}\times\text{bounded}=\text{zero}$, the limit is zero. *Warning:* There's a chance that your calc teacher may not like this logic.

Second, you can use your calculator: Store something small like 0.1 into x and then input $x\sin\frac{1}{x^2}$ into your home screen and hit Enter. You should get a result of ~-0.05. Now store 0.01 into x and use the Entry button to get back to $x\sin\frac{1}{x^2}$ and hit Enter again. The result is

~ 0.003. Now try 0.001, then 0.0001 (giving you ~ –0.00035 and ~ 0.00009), and so on. It's pretty clear — though probably not to the satisfaction of your professor — that the limit is 0.

The third way will definitely satisfy a persnickety professor. You've got to *sandwich* (or *squeeze*) your *salami* function, $x\sin\frac{1}{x^2}$, between two *bread* functions that have identical limits as *x* approaches the same arrow-number that it approaches in the salami function, namely the arrow-number of zero. Use the rule explained in the example problem. The current function, $x\sin\frac{1}{x^2}$, is the same thing as $x^1\sin\frac{1}{x^2}$. Because the power on the first *x* is 1, an odd number, you've got the second case where you can use $-|x|$ for the bottom function and $|x|$ for the top function. Graph $B(x)=-|x|$, $f(x)=x\sin\frac{1}{x^2}$, and $T(x)=|x|$ at the same time on your graphing calculator, and you can see that $x\sin\frac{1}{x^2}$ is always greater than or equal to $-|x|$ and always less than or equal to $|x|$. Because $\lim_{x\to 0}(-|x|)=0$ and $\lim_{x\to 0}|x|=0$, and because $x\sin\frac{1}{x^2}$ is sandwiched between them, $\lim_{x\to 0}\left(x\sin\frac{1}{x^2}\right)$ must also be zero.

(20) $\lim_{x\to 0}\left(x^2\cos\frac{1}{x}\right)=0$

For $\lim_{x\to 0}\left(x^2\cos\frac{1}{x}\right)$, the power on *x* is an even number (namely 2), so you've got the first case explained in the text where you can use $B(x)=-x^2$ and $T(x)=x^2$ for the bread functions. The cosine of anything is always between –1 and 1, so $x^2\cos\frac{1}{x}$ is sandwiched between those two bread functions. (You should confirm this by looking at their graphs. Use the following window on your graphing calculator: radian mode, Xmin = –0.15625, Xmax = 0.15625, Xscl = 0.05, Ymin = –0.0125, Ymax = 0.0125, Yscl = 0.005.) Because $\lim_{x\to 0}(-x^2)=0$ and $\lim_{x\to 0}x^2=0$, $\lim_{x\to 0}\left(x^2\cos\frac{1}{x}\right)$ must also be zero.

(21) $\lim_{x\to\infty}\frac{5x^3-x^2+10}{2x^4+x+3}=0$

Because the degree of the numerator is less than the degree of the denominator, this is a Case I problem. So, the limit as *x* approaches infinity is zero.

(22) $\lim_{x\to-\infty}\frac{3x^4+100x^3+4}{8x^4+1}=\frac{3}{8}$

$\lim_{x\to-\infty}\frac{3x^4+100x^3+4}{8x^4+1}$ is a Case II example because the degrees of the numerator and denominator are both 4. The limit is thus the quotient of the coefficients of the leading terms in the numerator and denominator, namely, $\frac{3}{8}$.

(23) $\lim_{x\to\infty}\frac{x^x}{x!}=\infty$

According to the "larger" over "smaller" tip, this answer must be infinity. Or you can get this result with your calculator. If you set the table (don't forget: fork on the left, spoon on the right) with something like TblStart = 100 and ΔTbl = 100, and then look at the table, you may see "undef" for some or all of the *y*-values, depending on your calculator model. You have to be careful when trying to interpret what "undef" (for "undefined") means on your calculator. It often means infinity, but not always, so don't just jump to that conclusion. Instead, make TblStart and ΔTbl smaller, say, 10. Sure enough, the *y*-values grow huge very fast, and you can safely conclude that the limit is infinity.

24 $\lim_{x\to\infty} \frac{5x+2}{\sqrt{4x^2-1}} = \frac{5}{2}$

1. **Divide the numerator and denominator by the highest power of x appearing in the fraction — that's x or x^1.** (The x^2 isn't a higher power of x because the x^2 gets square rooted. Square rooting something is the same as raising it to the $\frac{1}{2}$ power, so, you've basically got $\left(x^2\right)^{1/2} = x^1$.)

$$= \lim_{x\to\infty} \frac{\frac{5x+2}{x}}{\frac{\sqrt{4x^2-1}}{x}}$$

2. **Put the x into the square root (it becomes x^2).**

$$= \lim_{x\to\infty} \frac{\frac{5x+2}{x}}{\sqrt{\frac{4x^2-1}{x^2}}}$$

3. **Distribute the division in the numerator and denominator.**

$$= \lim_{x\to\infty} \frac{5+\frac{2}{x}}{\sqrt{4-\frac{1}{x^2}}}$$

4. **Plug in and simplify.**

$$= \frac{5+\frac{2}{\infty}}{\sqrt{4-\frac{1}{\infty^2}}} = \frac{5+0}{\sqrt{4-0}} = \frac{5}{2}$$

Here's a calculator solution. Plug 100 then 1000 then 10,000 then 100,000 into x. The results are 2.51, 2.501, 2.5001, and 2.50001. Since these numbers are getting closer and closer to the round number 2.5, that's the limit. It's great when you can do a problem two completely different ways and obtain the same result. That should give you a great deal of confidence that your answer is correct.

25 $\lim_{x\to-\infty} \left(4x + \sqrt{16x^2 - 3x}\right) = \frac{3}{8}$

1. **Put the entire expression over 1 so you can use the conjugate trick.**

$$= \lim_{x\to-\infty} \left(\frac{4x+\sqrt{16x^2-3x}}{1} \cdot \frac{4x-\sqrt{16x^2-3x}}{4x-\sqrt{16x^2-3x}} \right)$$

2. **FOIL the numerator.**

$$= \lim_{x\to-\infty} \frac{16x^2-\left(16x^2-3x\right)}{4x-\sqrt{16x^2-3x}}$$

3. **Simplify the numerator and factor out $16x^2$ inside the radicand.**

$$= \lim_{x \to -\infty} \frac{3x}{4x - \sqrt{16x^2\left(1 - \frac{3}{16x}\right)}}$$

4. **Pull the $16x^2$ out of the square root; it becomes $-4x$.**

You have to pull a *positive* out of the radicand (as always), so you pull out *negative* $4x$ because when x is negative (which it is as it approaches negative infinity), $-4x$ is positive. Got it?

$$= \lim_{x \to -\infty} \frac{3x}{4x - (-4x)\sqrt{1 - \frac{3}{16x}}}$$

$$= \lim_{x \to -\infty} \frac{3x}{4x\left(1 + \sqrt{1 - \frac{3}{16x}}\right)}$$

5. **Cancel, then plug in.**

$$= \lim_{x \to -\infty} \frac{3}{4\left(1 + \sqrt{1 - \frac{3}{16x}}\right)}$$

$$= \frac{3}{4\left(1 + \sqrt{1 - \frac{3}{16(-\infty)}}\right)}$$

$$= \frac{3}{4\left(1 + \sqrt{1 - 0}\right)}$$

$= \frac{3}{8}$ Piece o' cake.

Here's a calculator solution: plug −100 then −1000 then −10,000 then −100,000 into x in the limit expression. Here's what you get: 0.374824, 0.374982, 0.374998, 0.375. Bingo. Do you recognize 0.375?

26 $\lim_{x \to \infty} \left(\frac{3x^2}{x - 1} - \frac{3x^2}{x + 1}\right) = 6$

1. **Subtract the fractions using the LCD of $(x - 1)(x + 1) = x^2 - 1$.**

$$= \lim_{x \to \infty} \frac{3x^2(x + 1) - 3x^2(x - 1)}{x^2 - 1}$$

2. **Simplify.**

$$= \lim_{x \to \infty} \frac{3x^3 + 3x^2 - 3x^3 + 3x^2}{x^2 - 1}$$

$$= \lim_{x \to \infty} \frac{6x^2}{x^2 - 1}$$

3. **Your answer is the quotient of the coefficients of x^2 in the numerator and the denominator.** See Case 3 in the section, "Limits of rational functions at ±infinity."

$= 6$

Note that had you plugged infinity into the original problem, you would have

$$\frac{3\infty^2}{\infty-1}-\frac{3\infty^2}{\infty+1}$$
$$=\infty-\infty$$
$$=0?$$

WARNING

It may seem strange, but **infinity minus infinity does *not* equal zero.**

Here's a calculator confirmation of your algebraic solution. Plug 100 then 1000 then 10,000 into x in the limit expression. That gives you 6.0006, 6.000006, 6.00000006. That's clearly homing in on 6.

(27) $\lim\limits_{x\to-\infty}\cos\left(x^2\right)$ **does not exist (DNE)**

The best approach to this limit problem is to simply sketch or picture the graph of the cosine function (or graph it on your graphing calculator). As x moves left toward negative infinity, the cosine curve oscillates between heights of –1 and 1. The curve never approaches a single height; the oscillation goes on forever. This tells you that $\lim\limits_{x\to-\infty}\cos x$ does not exist (and, by the same reasoning, $\lim\limits_{x\to\infty}\cos x$ DNE). The function in this problem, $\lim\limits_{x\to-\infty}\cos\left(x^2\right)$, has a different shape than $\cos x$, but it oscillates forever in the same way between heights of –1 and 1 (it oscillates faster and faster the further out you go toward infinity or negative infinity). Thus, $\cos\left(x^2\right)$ does not exist (DNE).

(28) $\lim\limits_{x\to\infty}\frac{\sin x}{x}=0$

Like $\lim\limits_{x\to\infty}\cos x$, $\lim\limits_{x\to\infty}\sin x$ DNE because the sine function oscillates forever between heights of –1 and 1 as x gets larger and larger. But it doesn't follow that the answer to the current problem is also DNE. The function, $\frac{\sin x}{x}$, does oscillate forever as x gets larger and larger, but the amplitude of the oscillation gets damped more and more as x gets larger. Near $x=100$, for example, the amplitude of the oscillation gets divided by about 100, so $\frac{\sin x}{x}$ oscillates between heights of about –0.01 and 0.01. Near $x=1000$, $\frac{\sin x}{x}$ oscillates between about –0.001 and 0.001, and so on. The crests and troughs of the oscillating wave get smaller and smaller and closer and closer to a height of zero. That's the limit: zero.

(29) $\lim\limits_{x\to\infty}\left(1+\frac{1}{x}\right)^x=e\approx 2.718$

No work required here. This is one of the handful of limits you should just memorize.

Since the number e came up here, I can't resist mentioning what some say is the most elegant equation in mathematics — one short, simple equation that contains the five most important numbers in mathematics: 0, 1, π, e, and i (the square root of –1). Here it is: $e^{i\pi}+1=0$.

(30) $\lim_{x\to\infty}\left(1+\frac{1}{3x}\right)^x = \sqrt[3]{e}$

For this problem, keep in mind the solution to Problem 29: $\lim_{x\to\infty}\left(1+\frac{1}{x}\right)^x = e$. The idea for the current problem is to manipulate the limit with the $3x$ in it until you get something that resembles the solution from Problem 29. Here's what you do:

First, set the limit in question equal to *y*; then cube both sides:

$$\lim_{x\to\infty}\left(1+\frac{1}{3x}\right)^x = y$$

$$\left(\lim_{x\to\infty}\left(1+\frac{1}{3x}\right)^x\right)^3 = y^3$$

On the left, you can pull the lim symbol to the outside of the parentheses (just take my word for it):

$$\lim_{x\to\infty}\left(\left(1+\frac{1}{3x}\right)^x\right)^3 = y^3$$

Now, use the power-to-a-power rule:

$$\lim_{x\to\infty}\left(1+\frac{1}{3x}\right)^{3x} = y^3$$

See how this limit resembles the limit from Problem 29? You're almost there. The next step is to set *u* equal to $3x$ so you can replace each $3x$ with a *u*. And, because $u = 3x$, as *x* approaches infinity, so does *u*; thus, you can replace the *x* below the lim symbol with a *u*:

$$\lim_{u\to\infty}\left(1+\frac{1}{u}\right)^u = y^3$$

Finally, this limit is mathematically identical to the one from Problem 29, which equals *e*. Therefore,

$$e = y^3$$

But you need *y*, not y^3, because you set the limit you want equal to *y*. So, solve for *y*, and you're done:

$y = \sqrt[3]{e}$, thus $\lim_{x\to\infty}\left(1+\frac{1}{3x}\right)^x = \sqrt[3]{e}$

If you're ready to test your skills a bit more, take the following chapter quiz that incorporates all the chapter topics.

Whaddya Know? Chapter 8 Quiz

Quiz time! Complete each problem to test your knowledge on the various topics covered in this chapter. You can then find the solutions and explanations in the next section.

1. $\lim_{x\to 0^+} \frac{x^2}{\sqrt{3x}-\sqrt{2x}} = ?$

2. $\lim_{x\to 1} \frac{x^3-1}{x^6-1} = ?$

3. $\lim_{x\to 1} \frac{x^6-1}{x^4-1} = ?$

4. $\lim_{x\to -1} \frac{x^3+1}{x+1} = ?$

5. $\lim_{x\to 1} \frac{x^3+1}{x^6+1} = ?$

6. $\lim_{x\to \frac{1}{\pi}} \pi = ?$

7. $\lim_{x\to 0^+} \frac{|2x+1|-|x|-1}{|2x|-|x|} = ?$

8. $\lim_{x\to 0^-} \frac{|2x+1|-|x|-1}{|2x|-|x|} = ?$

9. $\lim_{x\to 0} \frac{|2x+1|-|x|-1}{|2x|-|x|} = ?$

10. $\lim_{x\to \frac{\pi}{2}} \frac{2\pi \cos x}{\sin(2x)} = ?$

11. $\lim_{x\to -\pi} \frac{(\sec x+1)\cos^2 x}{\sin^2 x} = ?$

12. $\lim_{x\to -\pi} \frac{821\sin^2 x}{\cos x+1} = ?$

    Bonus trivia: What's the significance of the answer in the history of mathematics?

13. $\lim_{x\to \infty} \frac{3k-2x-x^2}{3x^2-2x-1} = ?$ (k is a constant)

14. $\lim_{x\to \infty} \left(\cos\frac{1}{x}+\cos\frac{-1}{x}\right) = ?$

15. $\lim_{x\to \infty} \sin k = ?$ (k is a constant)

16. $\lim_{x\to \infty} \left(x-\sqrt[3]{x}\right) = ?$

17. $\lim_{x\to \infty} \frac{1492^x}{x!} = ?$

Answers to Chapter 8 Quiz

1. $\lim\limits_{x\to 0^+} \dfrac{x^2}{\sqrt{3x}-\sqrt{2x}} = 0$

Plugging in gives you zero over zero, so you've got some work to do. Multiply the top and bottom by the conjugate of the denominator (always a good thing to try):

$$= \lim_{x\to 0^+}\left(\frac{x^2}{\sqrt{3x}-\sqrt{2x}}\cdot\frac{\sqrt{3x}+\sqrt{2x}}{\sqrt{3x}+\sqrt{2x}}\right)$$

$$= \lim_{x\to 0^+}\frac{x^2\left(\sqrt{3x}+\sqrt{2x}\right)}{\sqrt{3x}^2-\sqrt{2x}^2}$$

$$= \lim_{x\to 0^+}\frac{x^2\left(\sqrt{3x}+\sqrt{2x}\right)}{x}$$

Now, cancel and then plug in:

$$= \lim_{x\to 0^+}\left(x\left(\sqrt{3x}+\sqrt{2x}\right)\right) = 0$$

That's it. And don't forget: while $\frac{0}{0}$ will sometimes equal zero (like here), it very often does not.

2. $\lim\limits_{x\to 1}\dfrac{x^3-1}{x^6-1} = \dfrac{1}{2}$

The first thing you see (and read) when you see this fraction is *x cubed*. That might make you think of factoring the numerator and denominator using the difference of *cubes* pattern. Your first step would look like this:

$$= \lim_{x\to 1}\frac{(x-1)\left(x^2+x+1\right)}{\left(x^2-1\right)\left(x^4+x^2+1\right)}$$

You could keep going from there, and you'd get the correct answer. However, you can save some effort if you notice that you can factor the denominator using the difference of *squares* pattern instead (nothing needs to be done to the numerator):

$$= \lim_{x\to 1}\frac{x^3-1}{\left(x^3-1\right)\left(x^3+1\right)}$$

$$= \lim_{x\to 1}\frac{1}{x^3+1} = \frac{1}{2}$$

3. $\lim\limits_{x\to 1}\dfrac{x^6-1}{x^4-1} = \dfrac{3}{2}$

You can factor the top with the difference of squares pattern or the difference of cubes pattern. The difference of squares is the only option for the denominator, so you might think that'd be the way to go for the entire fraction. You can get the correct answer that way, but, here, the quickest solution is to use the difference of cubes pattern on the top and the difference of squares pattern on the bottom, because then you can immediately cancel:

$$= \lim_{x \to 1} \frac{(x^2-1)(x^4+x^2+1)}{(x^2-1)(x^2+1)}$$
$$= \lim_{x \to 1} \frac{(x^4+x^2+1)}{(x^2+1)}$$

Plug in and you're done: $\frac{3}{2}$. (I hope it goes without saying that *no cancelling* can be done in the original fraction until you factor! Ditto for the original fraction in Problem 2.)

4 $\lim_{x \to -1} \frac{x^3+1}{x+1} = 3$

Here's a quickie. It's another difference of cubes problem:

$$= \lim_{x \to -1} \frac{(x+1)(x^2-x+1)}{x+1}$$
$$= \lim_{x \to -1} (x^2-x+1)$$
$$= (-1)^2 - (-1) + 1 = 3$$

5 $\lim_{x \to 1} \frac{x^3+1}{x^6+1} = 1$

Did you begin by using the difference of cubes pattern on the top and either the difference of cubes or difference of squares pattern on the bottom? Tricked you! All you do is plug in for your answer: $\frac{1+1}{1+1} = 1$. Don't forget: *Always plug in first!*

6 $\lim_{x \to \frac{1}{\pi}} \pi = \pi$

This question concerns a limit of the function $f(x) = \pi$ or $y = \pi$. But that function is a horizontal line with a constant height of $\pi \approx 3.14159....$ The arrow-number is irrelevant. For any arrow-number, the answer is π.

7 $\lim_{x \to 0^+} \frac{|2x+1| - |x| - 1}{|2x| - |x|} = 1$

Absolute value bars can make it difficult to do ordinary algebra, so you should always consider whether there's some way to get rid of them. Here's how you do that in this problem. For the first absolute value expression $|2x+1|$, when x is near zero (whether x is approaching zero from the left or the right), $2x+1$ will be close to 1, and, therefore, the absolute value bars are doing nothing and can be removed. For $|x|$ and $|2x|$, since x is approaching zero from the right, x and $2x$ will always be positive, and, again, the bars are doing nothing and can be removed. So now you've got the following:

$$= \lim_{x \to 0^+} \frac{2x+1-x-1}{2x-x}$$

Simplify and you're done:

$$= \lim_{x\to 0^+} \frac{x}{x}$$
$$= \lim_{x\to 0^+} 1 = 1$$

(8) $\lim\limits_{x\to 0^-} \dfrac{|2x+1|-|x|-1}{|2x|-|x|} = -3$

The first absolute expression works the same, of course, as in Problem 7, so just get rid of the bars. But in this problem, since x is approaching zero from the left, $|x|$ and $|2x|$ will always be negative, so you can't just eliminate the bars. Absolute value bars turn a negative into a positive, and you can achieve that by multiplying by −1. Thus, you can get rid of the absolute value bars around the x and the $2x$ by multiplying by −1 instead. Like so.

$$= \lim_{x\to 0^-} \frac{2x+1-(-1)\cdot x-1}{-1\cdot 2x-(-1)\cdot x}$$

Simplify and you're done:

$$= \lim_{x\to 0^-} \frac{2x+1+x-1}{-2x+x}$$
$$= \lim_{x\to 0^-} \frac{3x}{-x}$$
$$= \lim_{x\to 0^-} (-3) = -3$$

(9) $\lim\limits_{x\to 0} \dfrac{|2x+1|-|x|-1}{|2x|-|x|} =$ **undefined**

The limit from the left does not equal the limit from the right, so this two-sided limit is undefined.

(10) $\lim\limits_{x\to \frac{\pi}{2}} \dfrac{2\pi \cos x}{\sin(2x)} = \pi$

For trig limit problems, always be ready to use your trusty trig identities. Just use the identity for $\sin(2x)$, then cancel and plug in. Piece o' cake:

$$= \lim_{x\to \frac{\pi}{2}} \frac{2\pi \cos x}{2\sin x \cos x}$$
$$= \lim_{x\to \frac{\pi}{2}} \frac{\pi}{\sin x} = \frac{\pi}{1} = \pi$$

(11) $\lim\limits_{x\to -\pi} \dfrac{(\sec x+1)\cos^2 x}{\sin^2 x} = -\dfrac{1}{2}$

If you noticed that $\dfrac{\cos^2 x}{\sin^2 x} = \cot^2 x$, and that, therefore, you could rewrite the original problem like $\lim\limits_{x\to -\pi} \left((\sec x+1)\cot^2 x\right)$, that's a good thing to notice. Unfortunately, however, that doesn't help much in this particular problem. You could finish the problem from there, but, actually, the original version is a bit better, because you want to be dealing with a fraction. The key is to multiply the top and bottom by the conjugate of $\sec x+1$, namely, $\sec x-1$. Like so.

$$= \lim_{x \to -\pi} \frac{(\sec x - 1)(\sec x + 1)\cos^2 x}{(\sec x - 1)\sin^2 x}$$

$$= \lim_{x \to -\pi} \frac{(\sec^2 x - 1)\cos^2 x}{(\sec x - 1)\sin^2 x}$$

Next, you use the Pythagorean Identity for tangent and secant, $\tan^2 x = \sec^2 x - 1$:

$$= \lim_{x \to -\pi} \frac{\tan^2 x \cdot \cos^2 x}{(\sec x - 1)\sin^2 x}$$

And then use the fact that $\tan^2 x = \dfrac{\sin^2 x}{\cos^2 x}$:

$$= \lim_{x \to -\pi} \frac{\dfrac{\sin^2 x}{\cos^2 x} \cdot \cos^2 x}{(\sec x - 1)\sin^2 x}$$

Cancel everything, plug in, and you're done:

$$= \lim_{x \to -\pi} \frac{1}{\sec x - 1} = \frac{1}{-1-1} = -\frac{1}{2}$$

12 $\displaystyle \lim_{x \to -\pi} \frac{821 \sin^2 x}{\cos x + 1} = 1642$

Multiply the top and bottom by the conjugate of $\cos x + 1$:

$$= \lim_{x \to -\pi} \frac{(821 \sin^2 x)(\cos x - 1)}{(\cos x + 1)(\cos x - 1)}$$

$$= \lim_{x \to -\pi} \frac{(821 \sin^2 x)(\cos x - 1)}{\cos^2 x - 1}$$

Now you can use the Pythagorean Identity to replace $\cos^2 x - 1$ with $-\sin^2 x$:

$$= \lim_{x \to -\pi} \frac{(821 \sin^2 x)(\cos x - 1)}{-\sin^2 x}$$

Finally, cancel and plug in for your answer:

$$= \lim_{x \to -\pi} (-821(\cos x - 1))$$

$$= -821(\cos(-\pi) - 1) = -821(-1-1) = 1642$$

Bonus trivia: What's the significance of the answer in the history of mathematics?

Isaac Newton was born in 1642. He and Gottfried Leibniz were the co-inventors of calculus (or at least, that's the one-sentence story). For a fuller account, check out the Wikipedia article, "History of calculus."

13 $\lim_{x\to\infty}\frac{3k-2x-x^2}{3x^2-2x-1}=-\frac{1}{3}$

Don't forget: *Constants behave exactly like ordinary numbers.* So, you could ask yourself how the problem would work if the k were, say, a 5 instead of a k. Then you'd have $\lim_{x\to\infty}\frac{15-2x-x^2}{3x^2-2x-1}$. This is a straightforward horizontal asymptote problem. It's the third case discussed in the text where the polynomial in the numerator is of the same degree as the polynomial in the denominator. The horizontal asymptote in that case is given by the quotient of the coefficients of the highest-power terms in the numerator and denominator. The answer is thus $\frac{-1}{3}$ or $-\frac{1}{3}$.

14 $\lim_{x\to\infty}\left(\cos\frac{1}{x}+\cos\frac{-1}{x}\right)=2$

Just plug in infinity. Both $\frac{1}{\infty}$ and $\frac{-1}{\infty}$ "equal" zero. (This is not a precise mathematical equality, but don't sweat it.) That leaves you with $\cos 0+\cos 0$, or 2.

15 $\lim_{x\to\infty}\sin k=\sin k$

This is a tricky little problem. First, don't forget, again, that constants behave exactly like ordinary numbers. So, you might ask yourself how the problem would work if the k were, say, a 10. Then you'd have $\lim_{x\to\infty}\sin 10$. Since $\sin 10$ is just a number, $\lim_{x\to\infty}\sin 10$ would be asking for the height of the horizontal line $y=\sin 10$ as you go out toward infinity. But that height is the same everywhere, so the answer would be $\sin 10$. Since constants behave exactly like numbers, $\lim_{x\to\infty}\sin k=\sin k$. The other tricky thing is that even if you got this far, you might be tempted to say that the limit is undefined because you don't know what k is. But imagine that I asked you to determine the product $(k+1)(k+2)$. That product is k^2+3k+2, and that would be your answer. You wouldn't say that the product was undefined. It works the same in the current problem.

16 $\lim_{x\to\infty}\left(x-\sqrt[3]{x}\right)=\infty$

Don't forget: You cannot plug in infinity, which would give you $\infty-\infty$, and conclude that the answer is zero. *Infinity minus infinity does not equal zero.*

First, turn the limit expression into a fraction (a good strategy to keep in mind for limit problems). So, you've got $\lim_{x\to\infty}\left(\frac{x-\sqrt[3]{x}}{1}\right)$. Now, divide the numerator and denominator by the highest power of x appearing in the fraction, namely x^1. That gives you $\lim_{x\to\infty}\left(\frac{\frac{x}{x}-\frac{\sqrt[3]{x}}{x}}{\frac{1}{x}}\right)=\lim_{x\to\infty}\left(\frac{1-\frac{1}{x^{2/3}}}{\frac{1}{x}}\right)$. Now you can plug in infinity: $=\frac{1-\frac{1}{\infty}}{\frac{1}{\infty}}=\frac{1-0}{0}=\frac{1}{0}=\infty$.

17 $\lim_{x\to\infty}\frac{1492^x}{x!}=0$

You can solve this with the tip earlier in this chapter (under the section "Solving limits at infinity with algebra") involving the four types of expressions: x^{10}, 5^x, $x!$, and x^x (these are ordered from small to big). The current problem has a smaller expression type over a larger one, so the answer is a zero. The large number 1492 has no impact on the result.

4 Differentiation

In This Unit . . .

CHAPTER 9: **Differentiation Orientation**

Differentiating: It's Just Finding the Slope
The Derivative: It's Just a Rate
The Derivative of a Curve
The Difference Quotient
Average Rate and Instantaneous Rate
To Be or Not to Be? Three Cases Where the Derivative Does Not Exist
Practice Questions Answers and Explanations
Whaddya Know? Chapter 9 Quiz
Answers to Chapter 9 Quiz

CHAPTER 10: **Differentiation Rules — Yeah, Man, It Rules**

Basic Differentiation Rules
Differentiation Rules for Experts — Oh, Yeah, I'm a Calculus Wonk
Differentiating Implicitly
Differentiating Inverse Functions
Scaling the Heights of Higher-Order Derivatives
Practice Questions Answers and Explanations
Whaddya Know? Chapter 10 Quiz
Answers to Chapter 10 Quiz

CHAPTER 11: **Differentiation and the Shape of Curves**

Taking a Calculus Road Trip
Finding Local Extrema — My Ma, She's Like, Totally Extreme
Finding Absolute Extrema on a Closed Interval
Finding Absolute Extrema over a Function's Entire Domain
Locating Concavity and Inflection Points
Looking at Graphs of Derivatives Till They Derive You Crazy
The Mean Value Theorem — Go Ahead, Make My Day
Practice Questions Answers and Explanations
Whaddya Know? Chapter 11 Quiz
Answers to Chapter 11 Quiz

CHAPTER 12: **Your Problems Are Solved: Differentiation to the Rescue!**

Getting the Most (or Least) Out of Life: Optimization Problems
Yo-Yo a Go-Go: Position, Velocity, and Acceleration
Related Rates — They Rate, Relatively
Practice Questions Answers and Explanations
Whaddya Know? Chapter 12 Quiz
Answers to Chapter 12 Quiz

CHAPTER 13: **More Differentiation Problems: Going Off on a Tangent**

Tangents and Normals: Joined at the Hip
Straight Shooting with Linear Approximations
Business and Economics Problems
Practice Questions Answers and Explanations
Whaddya Know? Chapter 13 Quiz
Answers to Chapter 13 Quiz

IN THIS CHAPTER

» Discovering the simple algebra behind calculus

» Getting a grip on weird calculus symbols

» Differentiating with Laurel and Hardy

» Finding the derivatives of lines and curves

» Tackling the tangent line problem and the difference quotient

Chapter 9 Differentiation Orientation

Differential calculus is the mathematics of *change* and the mathematics of *infinitesimals*. You might say that it's the mathematics of infinitesimal changes — changes that occur every gazillionth of a second.

Without differential calculus — if you've got only algebra, geometry, and trigonometry — you're limited to the mathematics of things that either don't change or that change or move at an *unchanging* rate. Remember those problems from algebra? "One train leaves the station at 3 p.m. going west at 80 mph. Two hours later, another train leaves going east at 50 mph . . ." You can handle such a problem with algebra because the speeds or rates are unchanging. Our world, however, isn't one of unchanging rates — rates are in constant flux.

Think about putting an astronaut on the moon. Apollo 11 took off from a *moving* launch pad (the earth is both rotating on its axis and revolving around the sun). As the rocket flew higher and higher, the friction caused by the atmosphere and the effect of the earth's gravity were changing not just every second, not just every millionth of a second, but every *infinitesimal* fraction of a second. The spacecraft's weight was also constantly changing as it burned fuel. All of these things influenced the rocket's changing speed. On top of all that, the rocket had to hit a *moving* target, the moon. All of these things were changing, and their rates of change were changing. Say the rocket was going 1000 mph one second and 1020 mph a second later — during that one second, the rocket's speed literally passed through the *infinite* number of different speeds between 1000 and 1020 mph. How can you do the math for these ephemeral things that change every *infinitesimal* part of a second? You can't do it without differential calculus.

Differential calculus is used for all sorts of terrestrial things as well. Much of modern economic theory, for example, relies on differentiation. In economics, everything is in constant flux. Prices go up and down, supply and demand fluctuate, and inflation is constantly changing. Because these things are constantly changing, the ways they affect each other are constantly changing. You need calculus for this.

Differential calculus is one of the most practical and powerful inventions in the history of mathematics. So let's get started already.

Differentiating: It's Just Finding the Slope

Differentiation is the first of the two major ideas in calculus (the other is integration, which I cover in Unit 5). Differentiation is the process of finding the derivative of a function like $y = x^2$. The *derivative* is just a fancy calculus term for a simple idea you know from algebra: slope. *Slope,* as you know, is the fancy algebra term for steepness. And *steepness* is the fancy word for . . . No! Steepness is the *ordinary* word you've known since you were a kid, as in, "Hey, this road sure is steep." Everything you study in differential calculus all relates back to the simple idea of steepness.

MATH RULES

In *differential* calculus, you study *differentiation,* which is the process of *deriving* — that's finding — *derivatives.* These are big words for a simple idea: finding the *steepness* or *slope* of a line or curve. Throw some of these terms around to impress your friends. By the way, the root of the words *differential* and *differentiation* is *difference* — I explain the connection at the end of this chapter in the section on the *difference quotient.*

Consider Figure 9-1. A steepness of $\frac{1}{2}$ means that as the stick man walks one foot to the right, he goes up $\frac{1}{2}$ foot; where the steepness is 3, he goes up 3 feet as he walks 1 foot to the right. Where the steepness is zero, he's at the top, going neither up nor down; and where the steepness is negative, he's going down. A steepness of –2, for example, means he goes *down* 2 feet for every foot he goes to the right. This is shown more precisely in Figure 9-2.

TIP

Negative slope: To remember that going down to the right (or up to the left) is a *negative* slope, picture an uppercase *N*, as shown in Figure 9-3.

WARNING

Don't be among the legions of students who mix up the slopes of vertical and horizontal lines. How steep is a flat, horizontal road? Not steep at all, of course. Zero steepness. So, a horizontal line has a slope of *zero.* (Like where the stick man is at the top of the hill in Figure 9-1.) What's it like to drive up a vertical road? You can't do it. And you can't get the slope of a vertical line — it doesn't exist, or, as mathematicians say, it's *undefined.*

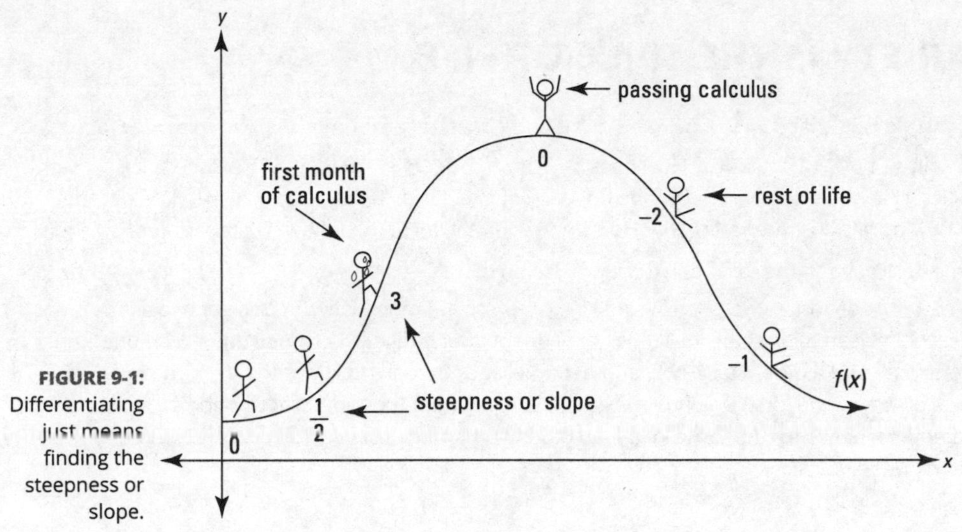

FIGURE 9-1: Differentiating just means finding the steepness or slope.

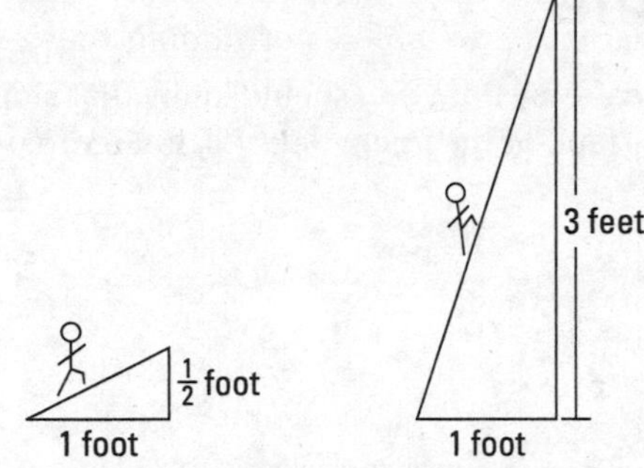

English:	steepness = $\frac{1}{2}$	steepness = 3
Algebra:	slope = $\frac{1}{2}$	slope = 3
Calculus:	$\frac{dy}{dx} = \frac{1}{2}$	$\frac{dy}{dx} = 3$

($\frac{dy}{dx}$, pronounced *dee y dee x*, is one of the many symbols for the derivative — see sidebar.)

FIGURE 9-2: The derivative = slope = steepness.

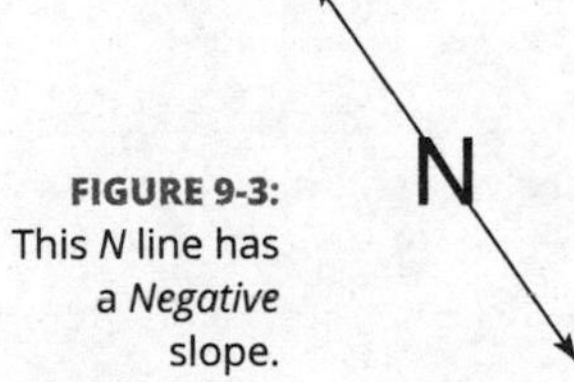

FIGURE 9-3: This *N* line has a *Negative* slope.

VARIETY IS THE SPICE OF LIFE

Everyone knows that $3^2 = 9$. Now, wouldn't it be weird if the next time you read this math fact, it was written as ${}^{2}3 = 9$ or ${}_{2}3 = 9$? How does $\overset{2}{3} = 9$ grab you? Or $\underset{2}{3} = 9$? Variety is *not* the spice of mathematics. When mathematicians decide on a way of expressing an idea, they stick to it — except, that is, with calculus. Are you ready? Hold on to your hat. All of the following are different symbols for the derivative — they all mean *exactly* the same thing: $\frac{dy}{dx}$ or $\frac{df}{dx}$ or $\frac{df(x)}{dx}$ or $D_x y$ or $\frac{d}{dx}f(x)$ or $f'(x)$ or y' or $\dot{f}$ or $\dot{y}$ or $D_x f$ or Df or $D_x f(x)$. There are more. Now, you've got two alternatives: 1) Beat your head against the wall trying to figure out things like why some author uses one symbol one time and a different symbol another time, and what exactly does the *d* or *f* mean anyway, and so on and so on, or 2) Don't try to figure it out; just treat these different symbols like words in different languages for the same idea — in other words, *don't sweat it.* I strongly recommend the second option.

The slope of a line

Keep going with the slope idea — by now, you should know that slope is what differentiation is all about. Take a look at the graph of the line, $y = 2x + 3$, in Figure 9-4.

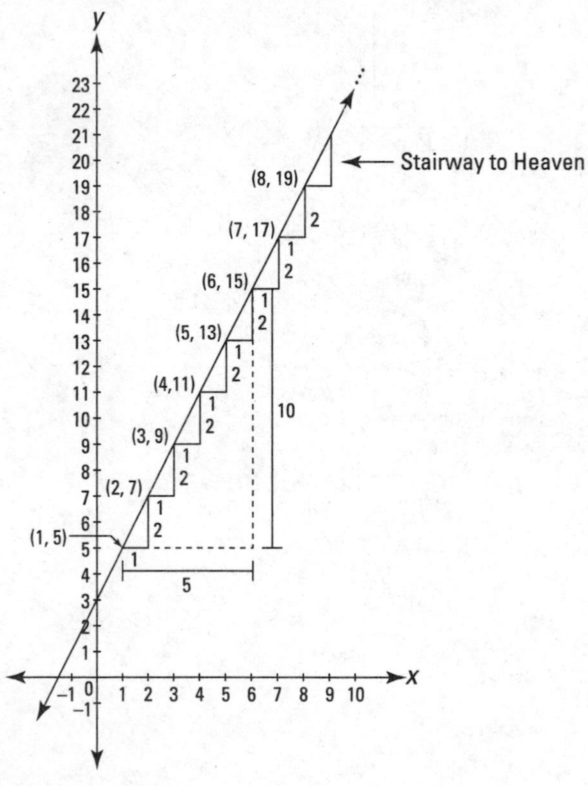

FIGURE 9-4: The graph of $y = 2x + 3$.

You remember from algebra — I'm *totally confident* of this — that you can find points on this line by plugging numbers into *x* and calculating *y*: plug 1 into *x* and *y* equals 5, which gives you the point located at $(1, 5)$; plug 4 into *x* and *y* equals 11, giving you the point $(4, 11)$, and so on.

I'm sure you also remember how to calculate the slope of this line. I realize that no calculation is necessary here — you go up 2 as you go over 1, so the slope is automatically 2. You can also simply note that $y = 2x + 3$ is in slope-intercept form $(y = mx + b)$ and that, since $m = 2$, the slope is 2. (See Chapter 5 if you want to review $y = mx + b$.) But bear with me because you need to know what follows. First, recall that

REMEMBER

$$\textbf{Slope} = \frac{\textbf{rise}}{\textbf{run}} = \frac{y_2 - y_1}{x_2 - x_1}$$

The *rise* is the distance you go up (the vertical part of a stair step), and the *run* is the distance you go across (the horizontal part of a stair step). Now, take any two points on the line, say, $(1, 5)$ and $(6, 15)$, and figure the rise and the run. You *rise* up 10 from $(1, 5)$ to $(6, 15)$ because 5 plus 10 is 15 (or you could say that 15 minus 5 is 10). And you *run* across 5 from $(1, 5)$ to $(6, 15)$ because 1 plus 5 is 6 (or in other words, 6 minus 1 is 5). Next, you divide to get the slope:

$$Slope = \frac{rise}{run} = \frac{10}{5} = 2$$

Here's how you do the same problem using the slope formula:

$$Slope = \frac{y_2 - y_1}{x_2 - x_1}$$

Plug in the points $(1, 5)$ and $(6, 15)$:

$$Slope = \frac{15 - 5}{6 - 1} = \frac{10}{5} = 2$$

Okay, let's summarize what you know about this line. Table 9-1 shows six points on the line and the unchanging slope of 2.

Table 9-1 Points on the Line $y = 2x + 3$ and the Slope at Those Points

x (horizontal position)	1	2	3	4	5	6	etc.
y (height)	5	7	9	11	13	15	etc.
slope	2	2	2	2	2	2	etc.

The derivative of a line

The preceding section showed you the algebra of slope. Now, here's the calculus. The derivative (the slope) of the line in Figure 9-4 is always 2, so you write

$$\frac{dy}{dx} = 2 \quad \text{(Read } \textit{dee y dee x equals } 2.)$$

Another common way of writing the same thing is

$$y' = 2 \quad \text{(Read } \textit{y prime equals } 2.)$$

And you say,

> The derivative of the function, $y = 2x + 3$, is 2.
>
> (Read *The derivative of the function, $y = 2x + 3$, is 2.* That was a joke.)

The Derivative: It's Just a Rate

Here's another way to understand the idea of a derivative that's even more fundamental than the concept of slope: A derivative is a *rate*. So why did I start the chapter with *slope*? Because slope is in some respects the easier of the two concepts, and slope is the idea you return to again and again in this book and any other calculus textbook as you look at the graphs of dozens and dozens of functions. But before you've got a slope, you've got a rate. A slope is, in a sense, a picture of a rate; the rate comes first, the picture of it comes second. Just like you can have a function before you see its graph, you can have a rate before you see it as a slope.

Calculus on the playground

Imagine Laurel and Hardy on a teeter-totter — check out Figure 9-5. (You don't remember Laurel and Hardy? Shocking! They were the internationally famous comedy duo from the late 1920s till the mid-1950s. Many movies were made about them; most recently was the 2018 film, *Stan & Ollie.* Check it out.)

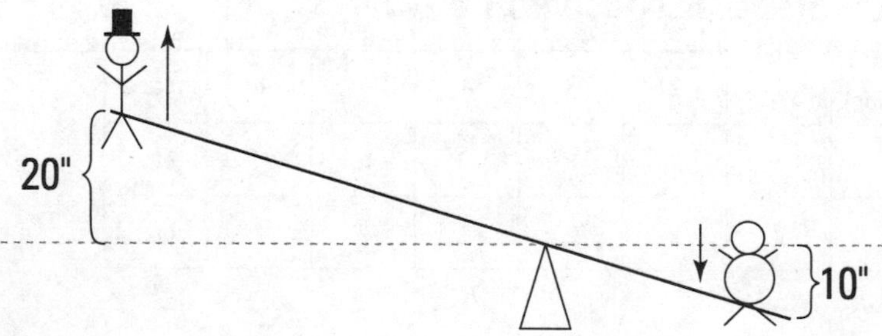

FIGURE 9-5: Laurel and Hardy — blithely unaware of the calculus implications.

Assuming Hardy weighs twice as much as Laurel, Hardy has to sit twice as close to the center as Laurel for them to balance. And for every inch that Hardy goes down, Laurel goes up two inches. So Laurel moves twice as much as Hardy. Voilà, you've got a derivative!

MATH RULES

A derivative is a rate. A *derivative* is simply a measure of how much one thing changes compared to another — and that's a *rate.*

Laurel moves twice as much as Hardy, so with calculus symbols, you write

$$dL = 2dH$$

Loosely speaking, dL can be thought of as the change in Laurel's position and dH as the change in Hardy's position. You can see that if Hardy goes down 10 inches, then dH is 10, and because dL equals 2 times dH, dL is 20 — so Laurel goes up 20 inches. Dividing both sides of this equation by dH gives you

$$\frac{dL}{dH} = 2$$

And that's the derivative of Laurel with respect to Hardy. (It's read as, "dee L, dee H," or as, "the derivative of L with respect to H.") The fact that $\frac{dL}{dH} = 2$ simply means that Laurel is moving 2 times as much as Hardy. Laurel's *rate* of movement is 2 inches per inch of Hardy's movement.

Now let's look at it from Hardy's point of view. Hardy moves half as much as Laurel, so you can also write

$$dH = \frac{1}{2}dL$$

Dividing by dL gives you

$$\frac{dH}{dL} = \frac{1}{2}$$

This is the derivative of Hardy with respect to Laurel, and it means that Hardy moves $\frac{1}{2}$ inch for every inch that Laurel moves. Thus, Hardy's rate is $\frac{1}{2}$ inch per inch of Laurel's movement.

By the way, you can also get this derivative by taking $\frac{dL}{dH} = 2$, which is the same as $\frac{dL}{dH} = \frac{2}{1}$, and flipping it upside down to get $\frac{dH}{dL} = \frac{1}{2}$.

These rates of 2 *inches per inch* and $\frac{1}{2}$ *inch per inch* may seem a bit odd because we often think of rates as referring to something per unit of time, like *miles per hour.* But a rate can be *anything per anything.* So, whenever you've got a *this per that,* you've got a rate; and if you've got a rate, you've got a derivative.

Speed — the most familiar rate

Speaking of *miles per hour*, say you're driving at a constant speed of 60 miles per hour. That's your car's *rate*, and 60 miles per hour is the derivative of your car's position, *p*, with respect to time, *t*. With calculus symbols, you write

$$\frac{dp}{dt} = 60\,\frac{\text{miles}}{\text{hour}}$$

This tells you that your car's position changes 60 miles for each hour that the time changes. Or you can say that your car's position (in miles) changes 60 times as much as the time changes (in hours). Again, a derivative just tells you how much one thing changes compared to another.

And just like the Laurel and Hardy example, this derivative, like all derivatives, can be flipped upside down:

$$\frac{dt}{dp} = \frac{1}{60}\,\frac{\text{hours}}{\text{mile}}$$

This *hours-per-mile* rate is certainly much less familiar than the ordinary *miles-per-hour* rate, but it's nevertheless a perfectly legitimate rate. It tells you that for each mile you go, the time changes $\frac{1}{60}$ of an hour. And it tells you that the time (in hours) changes $\frac{1}{60}$ as much as the car's position (in miles).

MATH RULES

There's no end to the different rates you might see. You just saw *miles per hour* and *hours per mile*. Then there's *miles per gallon* (for gas mileage), *gallons per minute* (for water draining out of a pool), *output per employee* (for a factory's productivity), and so on. Rates can be constant or changing. In either case, every rate is a derivative, and every derivative is a rate.

The rate-slope connection

Rates and slopes have a simple connection. All of the previous rate examples can be graphed on an *x-y* coordinate system, where each rate appears as a slope. Consider the Laurel and Hardy example again. Laurel moves twice as much as Hardy. This can be represented by the following equation:

$$L = 2H$$

Figure 9-6 shows the graph of this function.

The inches on the *H-axis* indicate how far Hardy has moved up or down from the teeter-totter's starting position; the inches on the *L-axis* show how far Laurel has moved up or down. The line goes up 2 inches for each inch it goes to the right, and its slope is thus $\frac{2}{1}$, or 2. This is the visual depiction of $\frac{dL}{dH} = 2$, showing that Laurel's position changes 2 times as much as Hardy's.

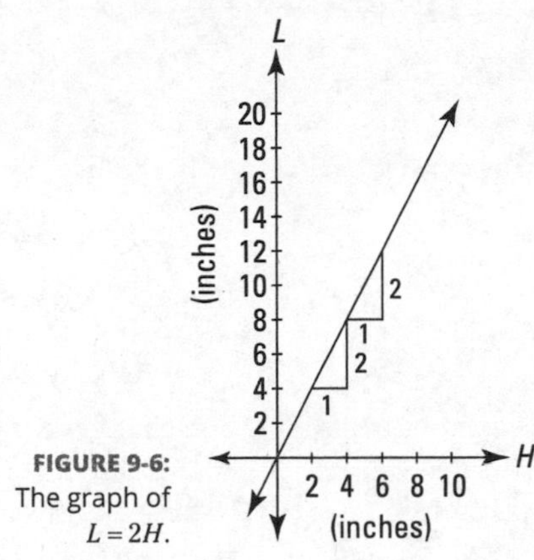

FIGURE 9-6: The graph of $L = 2H$.

One last comment. You know that $slope = \frac{rise}{run}$. Well, you can think of dL as the *rise* and dH as the *run*. That ties everything together quite nicely.

$$slope = \frac{rise}{run} = \frac{dL}{dH} = rate$$

A derivative is just a slope, and a derivative is also just a rate.

The Derivative of a Curve

The sections so far in this chapter have involved *linear* functions — straight lines with *unchanging* slopes. But if all functions and graphs were lines with unchanging slopes, there'd be no need for calculus. The derivative of the Laurel and Hardy function graphed previously is 2, but you don't need calculus to determine the slope of a line. Calculus is the mathematics of change, so now is a good time to move on to *parabolas*: curves with *changing* slopes. Figure 9-7 is the graph of the parabola, $y = \frac{1}{4}x^2$.

Notice how the parabola gets steeper and steeper as you go to the right. You can see from the graph that at the point $(2, 1)$, the slope is 1; at $(4, 4)$, the slope is 2; at $(6, 9)$, the slope is 3, and so on. Unlike the unchanging slope of a line, the slope of a parabola depends on where you are; it depends on the x-coordinate of wherever you are on the parabola. So, the derivative (or slope) of the function $y = \frac{1}{4}x^2$ is itself a function of x — namely, $\frac{1}{2}x$ (I'll show you how I got that in a minute). To find the slope of the curve at any point, you just plug the x-coordinate of the point into the derivative, $\frac{1}{2}x$, and you've got the slope. For instance, if you want the slope at the point $(3, 2.25)$, plug 3 into the x, and the slope is $\frac{1}{2}$ times 3, or 1.5. Table 9-2 shows some points on the parabola and the steepness at those points.

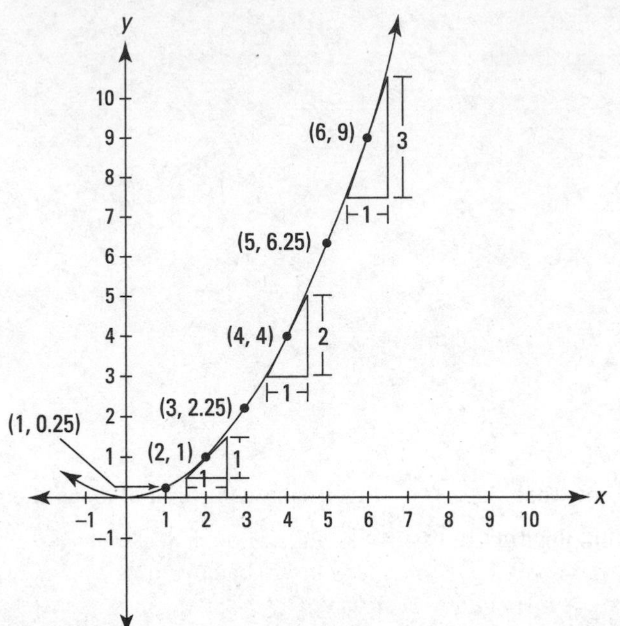

FIGURE 9-7: The graph of $y = \frac{1}{4}x^2$.

Table 9-2 Points on the Parabola $y = \frac{1}{4}x^2$ and the Slopes at Those Points

x (horizontal position)	1	2	3	4	5	6	etc.
y (height)	0.25	1	2.25	4	6.25	9	etc.
$\frac{1}{2}x$ (slope)	0.5	1	1.5	2	2.5	3	etc.

Here's the calculus. You write

$$\frac{dy}{dx} = \frac{1}{2}x \quad \text{or} \quad y' = \frac{1}{2}x$$

And you say,

The derivative of the function $y = \frac{1}{4}x^2$ is $\frac{1}{2}x$.

Or you can say,

The derivative of $\frac{1}{4}x^2$ is $\frac{1}{2}x$.

I promised to tell you how to *derive* this derivative of $y = \frac{1}{4}x^2$, so here you go:

1. **Beginning with the original function, $\frac{1}{4}x^2$, take the power and put it in front of the coefficient.**

   $2 \cdot \frac{1}{4}x^{2}$

2. **Multiply.**

   2 times $\frac{1}{4}$ is $\frac{1}{2}$, so that gives you $\frac{1}{2}x^2$.

3. **Reduce the power by 1.**

   In this example, the 2 becomes a 1. So, the derivative is

   $\frac{1}{2}x^1$ or just $\frac{1}{2}x$.

This and many other differentiation techniques are discussed in Chapter 10.

The following problems emphasize the fact that a derivative is basically just a rate or a slope. So to solve these problems, all you have to do is answer the questions as if they had asked you to determine a rate or a slope instead of a derivative.

EXAMPLE

Q. What's the derivative of $y = 5 - 4x$?

A. **The answer is −4.** $y = 5 - 4x$ is the same as $y = -4x + 5$. And you know, of course, that the slope of $y = -4x + 5$ is −4, right? No? Egad! Any line of the form $y = mx + b$ has a slope equal to m. The derivative of a line or curve is the same thing as its slope, so the derivative of this line is −4.

Remember: You can think of the derivative $\frac{dy}{dx}$ as basically $\frac{rise}{run}$.

YOUR TURN

1 If you leave your home at time $= 0$, and speed away in your car at $60\frac{\text{miles}}{\text{hour}}$, what's $\frac{dp}{dt}$, the derivative of your position with respect to time?

2 Using the information from Problem 1, write a function that gives your position as a function of time.

3 What's the slope of the parabola $y=-\frac{1}{3}x^2+\frac{23}{3}x-\frac{85}{3}$ at the point $(7, 9)$? See the following figure.

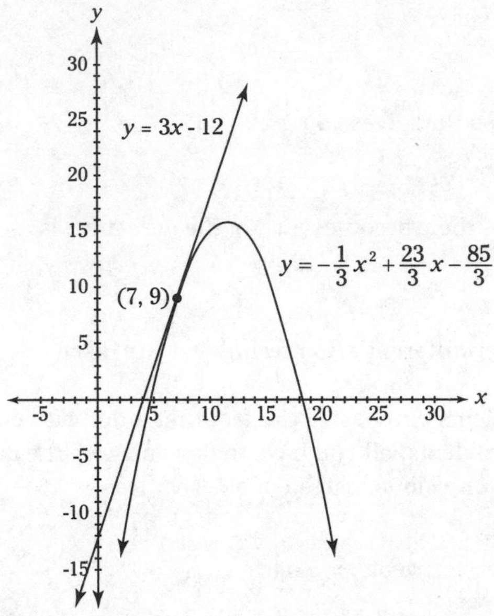

---

4 What's the derivative of the parabola $y=-x^2+5$ at the point $(0, 5)$? *Hint:* Graph the parabola.

---

5 With your graphing calculator, graph both the line $y=-4x+9$ and the parabola $y=5-x^2$. You'll see that they're tangent at the point $(2, 1)$.

(a) What is the derivative of $y=5-x^2$ when $x=2$?

(b) On the parabola, how fast is y changing compared to x when $x=2$?

The Difference Quotient

Sound the trumpets! You come now to what is perhaps the cornerstone of differential calculus: the difference quotient, the bridge between limits and the derivative. (But you're going to have to be patient here, because it's going to take me a few pages to explain the logic behind the difference quotient before I can show you what it is.) Okay, so here goes. I keep repeating — have you noticed? — the important fact that a derivative is just a slope. You learned how to find the slope of a line in algebra. In Figure 9-7, I gave you the slope of the parabola at several points, and then I showed you the short-cut method for finding the derivative — but I left out the important math in the middle. That math involves limits, and it takes you to the threshold of calculus. Hold on to your hat.

REMEMBER

***Slope* is defined as $\frac{rise}{run}$, and *slope* $= \frac{y_2 - y_1}{x_2 - x_1}$.**

To compute a slope, you need two points to plug into this formula. For a line, this is easy. You just pick any two points on the line and plug them in. But it's not so simple if you want, say, the slope of the parabola $f(x) = x^2$ at the point $(2, 4)$. Check out Figure 9-8.

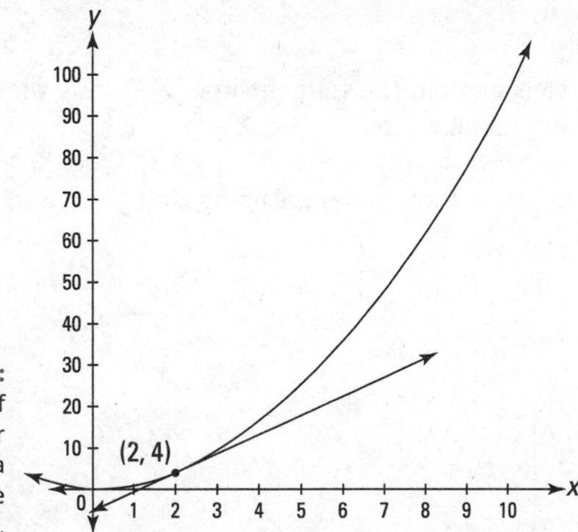

FIGURE 9-8: The graph of $f(x) = x^2$ or $y = x^2$ with a tangent line at $(2, 4)$.

You can see the line drawn tangent to the curve at $(2, 4)$. Because the slope of the tangent line is the same as the slope of the parabola at $(2, 4)$, all you need is the slope of the tangent line to give you the slope of the parabola. But you don't know the equation of the tangent line, so you can't get the second point — in addition to $(2, 4)$ — that you need for the slope formula.

Here's how the inventors of calculus got around this roadblock. Figure 9-9 shows the tangent line again and a secant line intersecting the parabola at $(2, 4)$ and at $(10, 100)$.

REMEMBER

Definition of *secant line:* A secant line is a line that intersects a curve at two points. This is a bit oversimplified, but it'll do.

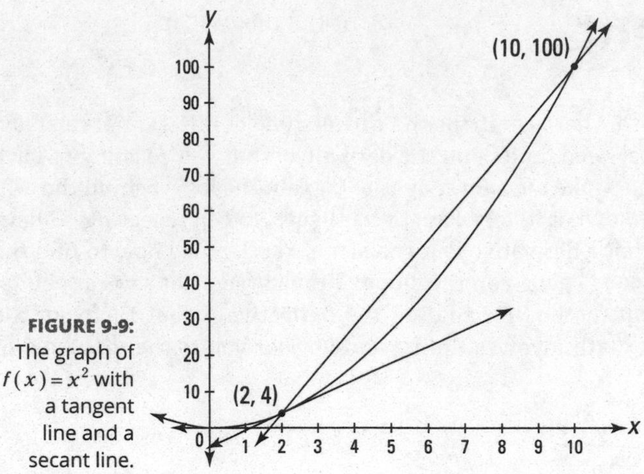

FIGURE 9-9: The graph of $f(x) = x^2$ with a tangent line and a secant line.

The slope of this secant line is given by the slope formula:

$$Slope = \frac{rise}{run} = \frac{y_2 - y_1}{x_2 - x_1} = \frac{100 - 4}{10 - 2} = \frac{96}{8} = 12$$

You can see that this secant line is steeper than the tangent line, and thus the slope of the secant, 12, is higher than the slope you're looking for.

Now add one more point at $(6, 36)$ and draw another secant using that point and $(2, 4)$ again. See Figure 9-10.

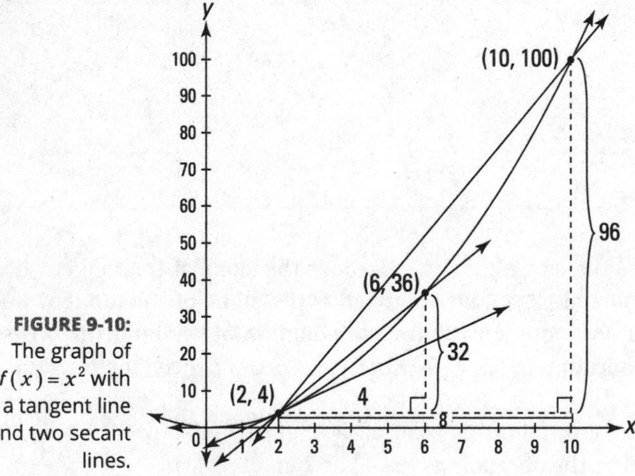

FIGURE 9-10: The graph of $f(x) = x^2$ with a tangent line and two secant lines.

Calculate the slope of this second secant:

$$Slope = \frac{36 - 4}{6 - 2} = \frac{32}{4} = 8$$

You can see that this secant line is a better approximation of the tangent line than the first secant.

Now, imagine what would happen if you grabbed the point at $(6, 36)$ and slid it down the parabola toward $(2, 4)$, dragging the secant line along with it. Can you see that as the point gets closer and closer to $(2, 4)$, the secant line gets closer and closer to the tangent line, and that the slope of this secant thus gets closer and closer to the slope of the tangent?

So, you can get the slope of the tangent if you take the *limit* of the slopes of this moving secant. Let's give the moving point the coordinates (x_2, y_2). As this point (x_2, y_2) slides closer and closer to (x_1, y_1), namely $(2, 4)$, the *run*, which equals $x_2 - x_1$, gets closer and closer to zero. So, here's the limit you need:

$$\begin{aligned} Slope_{of\ tangent} &= \lim_{\substack{as\ point\ slides\\ toward\ (2,\ 4)}} \left(slope_{of\ moving\ secant}\right) \\ &= \lim_{run \to 0} \frac{rise}{run} \\ &= \lim_{x_2 \to x_1} \frac{y_2 - y_1}{x_2 - x_1} \\ &= \lim_{x_2 \to 2} \frac{y_2 - 4}{x_2 - 2} \end{aligned}$$

Watch what happens to this limit when you plug in four more points on the parabola that are closer and closer to $(2, 4)$:

When the point (x_2, y_2) slides to $(3, 9)$, the slope is $\frac{9-4}{3-2}$, or 5.

When the point slides to $(2.1, 4.41)$, the slope is $\frac{4.41-4}{2.1-2}$, or 4.1.

When the point slides to $(2.01, 4.0401)$, the slope is 4.01.

When the point slides to $(2.001, 4.004001)$, the slope is 4.001.

Sure looks like the slope is headed toward 4. (By the way, the fact that the slope at $(2, 4)$ — which you'll see in a minute does turn out to be 4 — is the same as the y-coordinate of the point is a meaningless coincidence, as is the pattern you may have noticed in these numbers between the y-coordinates and the slopes.)

As with all limit problems, the variable in this problem, x_2, *approaches* but never actually gets to the arrow-number (2 in this case). If it got to 2 — which would happen if you slid the point you grabbed along the parabola until it was actually on top of $(2, 4)$ — you'd get $\frac{4-4}{2-2} = \frac{0}{0}$, which is undefined. But, of course, the slope at $(2, 4)$ is precisely the slope you want: the slope of the line when the point *does* land on top of $(2, 4)$. Herein lies the beauty of the limit process. With this limit, you get the *exact* slope of the *tangent* line at $(2, 4)$ even though the limit function, $\frac{y_2 - 4}{x_2 - 2}$, generates slopes of *secant* lines.

Here again is the equation for the slope of the tangent line:

$$Slope = \lim_{x_2 \to 2} \frac{y_2 - 4}{x_2 - 2}$$

And the slope of the tangent line is — you guessed it — the derivative.

MATH RULES

Meaning of the *derivative*: The derivative of a function $f(x)$ at some number $x = c$, written as $f'(c)$, is the slope of the tangent line to f drawn at c.

The slope fraction $\frac{y_2 - 4}{x_2 - 2}$ is expressed with algebra terminology. Now let's rewrite it to give it that highfalutin calculus look. But first (finally!), the definition you've been waiting for.

MATH RULES

Definition of the *difference quotient*: There's a fancy calculus term for the general slope fraction, $\frac{rise}{run}$ or $\frac{y_2 - y_1}{x_2 - x_1}$, when you write it in the fancy calculus way. A fraction is a *quotient*, right? And both $y_2 - y_1$ and $x_2 - x_1$ are *differences*, right? So, voilà, it's called the *difference quotient*. Here it is:

$$\frac{f(x+h)-f(x)}{h}$$

(This is the most common way of writing the difference quotient. You may run across other, equivalent ways.) In the following pages, I show you how $\frac{y_2 - y_1}{x_2 - x_1}$ morphs into the difference quotient.

Okay, let's lay out this morphing process. First, the *run*, $x_2 - x_1$ (in this example, $x_2 - 2$), is called — don't ask me why — h. Next, because $x_1 = 2$ and the *run* equals h, x_2 equals $2 + h$. You then write y_1 as $f(2)$ and y_2 as $f(2+h)$. Making all the substitutions gives you the derivative of x^2 at $x = 2$:

$$\begin{aligned} f'(2) &= \lim_{run \to 0} \frac{rise}{run} \\ &= \lim_{x_2 \to 2} \frac{y_2 - 4}{x_2 - 2} \\ &= \lim_{h \to 0} \frac{f(2+h)-f(2)}{(2+h)-2} \\ &= \lim_{h \to 0} \frac{f(2+h)-f(2)}{h} \end{aligned}$$

MATH RULES

$\lim_{h \to 0} \frac{f(2+h)-f(2)}{(2+h)-2}$ **is simply the shrinking $\frac{rise}{run}$ stair-step** that you can see in Figure 9-10 as the point slides down the parabola toward $(2, 4)$.

Figure 9-11 is basically the same as Figure 9-10 except that, instead of exact points like $(6, 36)$ and $(10, 100)$, the sliding point has the general coordinates of $(2+h, f(2+h))$, and the *rise* and the *run* are expressed in terms of h. Figure 9-11 is the ultimate figure for $f'(2) = \lim_{h \to 0} \frac{f(2+h)-f(2)}{h}$.

Have I confused you with these two figures? Don't sweat it. They both show the same thing. Both figures are visual representations of $f'(2) = \lim_{h \to 0} \frac{f(2+h)-f(2)}{h}$. I just thought it'd be a good idea to show you a figure with exact coordinates before showing you Figure 9-11 with all that strange-looking f and h stuff in it.

Doing the math gives you, at last, the slope of the tangent line at $(2, 4)$:

$$
\begin{aligned}
f'(2) &= \lim_{h\to 0}\frac{f(2+h)-f(2)}{h} \\
&= \lim_{h\to 0}\frac{(2+h)^2-(2)^2}{h} \quad \text{(The function is } f(x)=x^2, \text{ so } f(2+h)=(2+h)^2\text{, right?)} \\
&= \lim_{h\to 0}\frac{\left(4+4h+h^2\right)-4}{h} \\
&= \lim_{h\to 0}\frac{4h+h^2}{h} \\
&= \lim_{h\to 0}\frac{h(4+h)}{h} \\
&= \lim_{h\to 0}(4+h) \\
&= 4+0=4
\end{aligned}
$$

So, the slope at the point $(2, 4)$ is 4.

MATH RULES

Main definition of the *derivative*: If you replace the point $(2, f(2))$ — in Figure 9-11 and the limit math that follows it — with the general point $(x, f(x))$, you get the general definition of the derivative as a function of *x*:

$$f'(x)=\lim_{h\to 0}\frac{f(x+h)-f(x)}{h}$$

So, at last you see that the derivative is defined as the limit of the difference quotient.

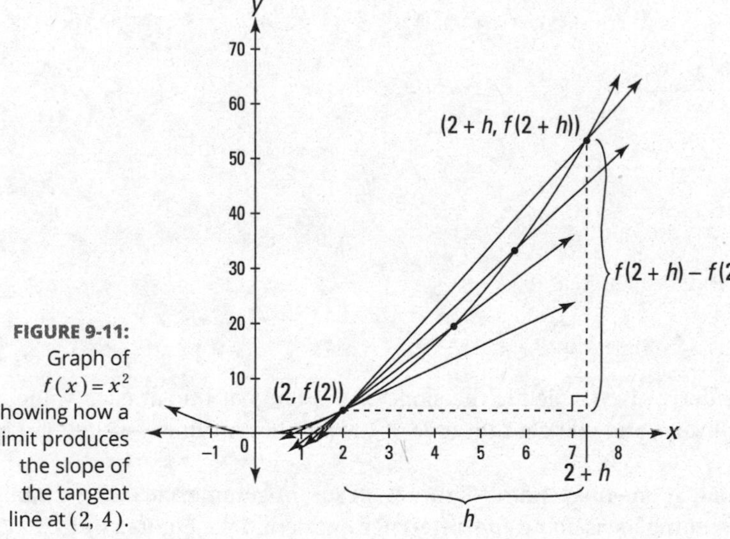

FIGURE 9-11: Graph of $f(x)=x^2$ showing how a limit produces the slope of the tangent line at $(2, 4)$.

Figure 9-12 shows this general definition graphically. Note that Figure 9-12 is virtually identical to Figure 9-11, except that x's replace the 2's in Figure 9-11 and that the moving point in Figure 9-12 slides down toward any old point $(x, f(x))$ instead of toward the specific point $(2, f(2))$.

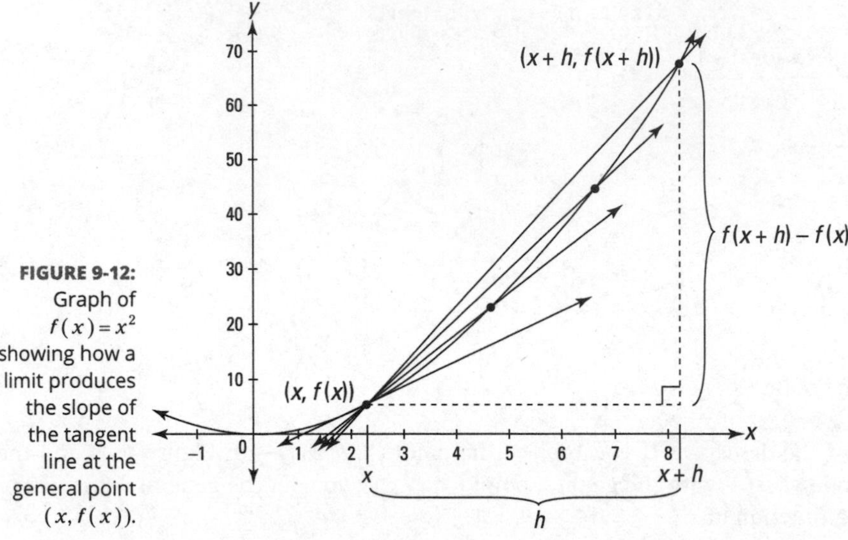

FIGURE 9-12: Graph of $f(x) = x^2$ showing how a limit produces the slope of the tangent line at the general point $(x, f(x))$.

Now work out this limit and get the derivative for the parabola $f(x) = x^2$:

$$
\begin{aligned}
f'(x) &= \lim_{h \to 0} \frac{f(x+h) - f(x)}{h} \\
&= \lim_{h \to 0} \frac{(x+h)^2 - x^2}{h} \qquad \begin{array}{l}\text{(The function is } f(x) = x^2, \\ \quad \text{so, } f(x+h) = (x+h)^2.\text{)}\end{array} \\
&= \lim_{h \to 0} \frac{\left(x^2 + 2xh + h^2\right) - x^2}{h} \\
&= \lim_{h \to 0} \frac{2xh + h^2}{h} \\
&= \lim_{h \to 0} \frac{h(2x+h)}{h} \\
&= \lim_{h \to 0} (2x + h) \\
&= 2x + 0 = 2x
\end{aligned}
$$

Thus, for this parabola, the derivative (which is the slope of the tangent line at each value x) equals $2x$. Plug any number into x, and you get the slope of the parabola at that x-value. Try it.

To close this section, let's look at one final figure. Figure 9-13 sort of summarizes (in a simplified way) all the difficult preceding ideas about the difference quotient. Like Figures 9-10, 9-11, and 9-12, Figure 9-13 contains a basic slope stair-step, a secant line, and a tangent line. The slope of the secant line is $\frac{rise}{run}$, or $\frac{\Delta y}{\Delta x}$. The slope of the tangent line is $\frac{dy}{dx}$. You can think of $\frac{dy}{dx}$

as $\frac{a \text{ little (ultimately infinitesimal) bit of } y}{a \text{ little (ultimately infinitesimal) bit of } x}$, and you can see why this is one of the symbols used for the derivative. As the secant line stair-step shrinks down to nothing, or, in other words, in the limit as Δx and Δy go to zero,

$$\frac{dy}{dx}\text{(the slope of the tangent line)} = \frac{\Delta y}{\Delta x}\text{(the slope of the secant line).}$$

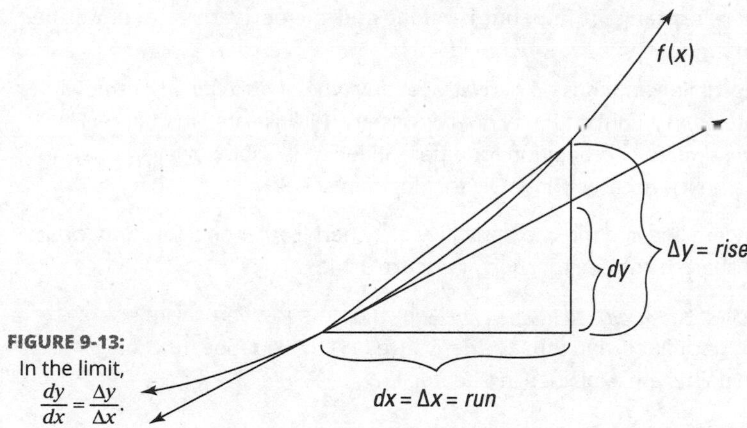

FIGURE 9-13: In the limit, $\frac{dy}{dx} = \frac{\Delta y}{\Delta x}$.

Average Rate and Instantaneous Rate

Returning once again to the connection between slopes and rates, a slope is just the visual depiction of a rate: The slope, $\frac{rise}{run}$, just tells you the rate at which y changes compared to x. If, for example, the y-coordinate tells you distance traveled (in miles), and the x-coordinate tells you elapsed time (in hours), you get the familiar rate of *miles per hour.*

Each secant line in Figures 9-9 and 9-10 has a slope given by the formula $\frac{y_2 - y_1}{x_2 - x_1}$. That slope is the *average* rate over the interval from x_1 to x_2. If y is in miles and x is in hours, you get the *average* speed in *miles per hour* during the time interval from x_1 to x_2.

When you take the limit using the difference quotient and get the slope of the tangent line, you get the *instantaneous* rate at the point (x_1, y_1). Again, if y is in miles and x is in hours, you get the *instantaneous* speed at the single point in time, x_1. Because the slope of the tangent line is the derivative, this gives you another definition of the derivative.

MATH RULES

Another definition of the *derivative*: The derivative of a function $f(x)$ at some x-value is the *instantaneous* rate of change of f with respect to x at that value.

To Be or Not to Be? Three Cases Where the Derivative Does Not Exist

To close this chapter, I want to discuss the three situations where a derivative fails to exist (see the section "33333 Limit Mnemonic," in Chapter 7). By now, you certainly know that the derivative of a function at a given point is the slope of the tangent line at that point. So, if you can't draw a tangent line, there's no derivative — that happens in the first two cases discussed here. In the third case, there's a tangent line, but its slope and the derivative are undefined.

- **Case I:** There's no tangent line and thus no derivative at any type of *discontinuity:* removable, infinite, or jump. (These types of discontinuity are discussed and illustrated in Chapter 7.) Continuity is, therefore, a *necessary* condition for differentiability. It's not, however, a *sufficient* condition, as the next two cases show. Dig that logician-speak.
- **Case II:** There's no tangent line and thus no derivative at a sharp *corner* on a function (or at a *cusp*, a really pointy, sharp turn). See function *f* in Figure 9-14.
- **Case III:** Where a function has a *vertical tangent line* (which occurs at a vertical inflection point), the slope is undefined, and thus the derivative fails to exist. See function *g* in Figure 9-14. (Inflection points are explained in Chapter 11.)

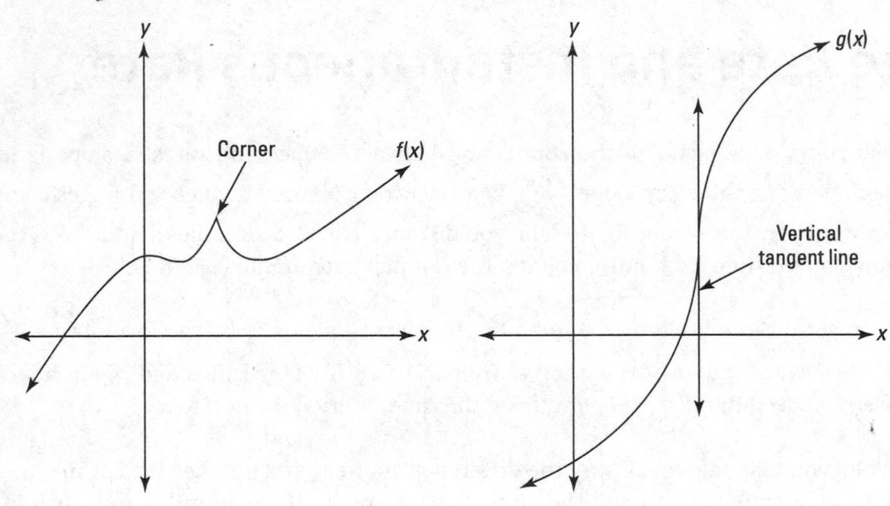

FIGURE 9-14: Cases II and III where there's no derivative.

EXAMPLE

Q. What's the slope of the parabola $f(x) = 10 - x^2$ at $x = 3$?

A. **The slope is −6.**

Here's the definition of the derivative again:

$$f'(x) = \lim_{h \to 0} \frac{f(x+h) - f(x)}{h}$$

1. **Because $f(x)=10-x^2$, $f(x+h)=10-(x+h)^2$, so the derivative is**

$$f'(x)=\lim_{h\to 0}\frac{\left(10-(x+h)^2\right)-\left(10-x^2\right)}{h}$$

2. **Simplify.**

$$=\lim_{h\to 0}\frac{10-\left(x^2+2xh+h^2\right)-10+x^2}{h}$$
$$=\lim_{h\to 0}\frac{10-x^2-2xh-h^2-10+x^2}{h}$$
$$=\lim_{h\to 0}\frac{-2xh-h^2}{h}$$

3. **Factor out *h*.**

$$=\lim_{h\to 0}\frac{h(-2x-h)}{h}$$

4. **Cancel.**

$$=\lim_{h\to 0}(-2x-h)$$

5. **Plug the arrow-number into *h*.**

$$=-2x-0$$
$$f'(x)=-2x$$

6. **You want the slope or derivative at $x=3$, so plug in 3.**

$$f'(3)=-2\cdot 3$$
$$=-6$$

YOUR TURN

Use the difference quotient to determine the derivative of the line $y=4x-3$.

Use the difference quotient to find the derivative of the parabola $f(x)=3x^2$.

8 Use the difference quotient to find the derivative of the parabola from Problem 4, $y = -x^2 + 5$.

---

9 (a) Determine the derivative of $g(x) = \sqrt{4x + 5}$ using the difference quotient.

(b) What's the slope or derivative of g at $x = 5$?

---

10 Use the parabola from Problem 7, but make it a position function, $s(t) = 3t^2$, where t is in hours and $s(t)$ is in miles.

(a) What's the average velocity from $t = 4$ to $t = 5$?

(b) What's the average velocity from $t = 4$ to $t = 4.1$?

(c) What's the average velocity from $t = 4$ to $t = 4.01$?

---

11 For the position function in Problem 10, what's the *instantaneous* velocity at $t = 4$? *Hint:* Use the derivative.

Practice Questions Answers and Explanations

1 $\frac{dp}{dt} = 60$

A derivative is always a rate, and a rate is always a derivative. So, if your speed, or rate, is $60\frac{\text{miles}}{\text{hour}}$, the derivative, $\frac{dp}{dt}$, is also 60.

2 **The function is $p(t) = 60t$ or $p = 60t$, where t is in hours and p is in miles.**

If you plug 1 into t, your position is 60 miles; plug 2 into t and your position is 120 miles. The function $p = 60t$ is a line, of course, in the form $y = mx + b$ (where $b = 0$ because you started your trip at your home where your position is zero). So, the slope is 60 and the derivative is thus 60. And, again, you see that a derivative is a slope and a rate.

3 **The slope is 3.**

You can see that the line, $y = 3x - 12$, is tangent to the parabola, $y = -\frac{1}{3}x^2 + \frac{23}{3}x - \frac{85}{3}$, at the point $(7, 9)$. You know from $y = mx + b$ that the slope of $y = 3x - 12$ is 3. At the point $(7, 9)$, the parabola is exactly as steep as the line, so the derivative (that's the slope) of the parabola at $(7, 9)$ is also 3.

REMEMBER

The slope of the parabloa is different at every point. The slope of the line stays constant, but the slope of the parabola changes as you climb up from $(7, 9)$, getting less and less steep. Even if you go to the right just 0.001 to $x = 7.001$, the slope will no longer be exactly 3.

4 **Zero**

The point $(0, 5)$ is the very top of the parabola, $y = -x^2 + 5$. At the top, the parabola is going neither up nor down — just like you're going neither up nor down at the moment when you walk across the crest of a hill. The top of the parabola is flat or level in this sense, and thus the slope and derivative both equal zero.

MATH RULES

The fact that **the derivative is zero at the top of a hill (and at the bottom of a valley)** is a critically important point that you'll return to time and time again.

5 **(a) −4**

The derivative of a curve tells you its slope or steepness. Because the line and the parabola are equally steep at $(2, 1)$, and because you know the slope of the line is −4, the slope of the parabola at $(2, 1)$ is also −4 and so is its derivative.

(b) It's decreasing 4 times as fast as *x* increases.

A derivative is a rate as well as a slope. Because the derivative of the parabola is −4 at $(2, 1)$, that tells you that y is changing 4 times as fast as x, but because the 4 is *negative*, *y decreases* 4 times as fast as *x increases*. This is the rate of y compared to x only for the one instant at $(2, 1)$ — and thus it's called an *instantaneous* rate. A split second later — say at $x = 2.000001$ — y will be decreasing a bit faster.

(6) $\boldsymbol{y' = 4}$

$$y' = \lim_{h\to 0}\frac{(4(x+h)-3)-(4x-3)}{h}$$
$$= \lim_{h\to 0}\frac{4x+4h-3-4x+3}{h}$$
$$= \lim_{h\to 0}\frac{4h}{h}$$
$$= \lim_{h\to 0}4$$
$$y' = 4$$

You can also figure this out because the slope of $y = 4x-3$ is 4.

(7) $\boldsymbol{f'(x) = 6x}$

$$f'(x) = \lim_{h\to 0}\frac{3(x+h)^2-3x^2}{h}$$
$$= \lim_{h\to 0}\frac{3\left(x^2+2xh+h^2\right)-3x^2}{h}$$
$$= \lim_{h\to 0}\frac{3x^2+6xh+3h^2-3x^2}{h}$$
$$= \lim_{h\to 0}\frac{6xh+3h^2}{h} \qquad \text{(Now, factor out the } h.)$$
$$= \lim_{h\to 0}\frac{h(6x+3h)}{h} \qquad \text{(Cancel the } h.)$$
$$= \lim_{h\to 0}(6x+3h) \qquad \text{(Now plug 0 into } h.)$$
$$= 6x+3\cdot 0$$
$$f'(x) = 6x$$

(8) $\boldsymbol{y' = -2x}$

$$y' = \lim_{h\to 0}\frac{\left(-(x+h)^2+5\right)-\left(-x^2+5\right)}{h}$$
$$= \lim_{h\to 0}\frac{\left(-\left(x^2+2xh+h^2\right)+5\right)-\left(-x^2+5\right)}{h}$$
$$= \lim_{h\to 0}\frac{-x^2-2xh-h^2+5+x^2-5}{h}$$
$$= \lim_{h\to 0}\frac{-2xh-h^2}{h} \qquad \text{(Now factor.)}$$
$$= \lim_{h\to 0}\frac{h(-2x-h)}{h} \qquad \text{(And cancel.)}$$
$$= \lim_{h\to 0}(-2x-h)$$
$$y' = -2x$$

In Problem 4, you see that the top of this parabola ($y = -x^2+5$) is at the point $(0, 5)$ and that the derivative is zero there because the parabola is going neither up nor down at its peak. That explanation was based on common sense. But now, with the result given by the difference quotient, namely $y' = -2x$, you have a rigorous confirmation of the derivative's value at $(0, 5)$. Just plug 0 in for x in $y' = -2x$, and you get $y' = 0$.

9 **(a)** $\mathbf{g'(x) = \frac{2}{\sqrt{4x+5}}}$

If you got this one, give yourself a pat on the back. It's a bit tricky.

$$g(x) = \sqrt{4x+5}$$

$$g'(x) = \lim_{h\to 0}\frac{\sqrt{4(x+h)+5}-\sqrt{4x+5}}{h}$$

$$= \lim_{h\to 0}\frac{\sqrt{4x+4h+5}-\sqrt{4x+5}}{h}$$

$$= \lim_{h\to 0}\frac{\left(\sqrt{4x+4h+5}-\sqrt{4x+5}\right)}{h}\cdot\frac{\left(\sqrt{4x+4h+5}+\sqrt{4x+5}\right)}{\left(\sqrt{4x+4h+5}+\sqrt{4x+5}\right)} \quad \text{(Conjugate multiplication)}$$

$$= \lim_{h\to 0}\frac{(4x+4h+5)-(4x+5)}{h\left(\sqrt{4x+4h+5}+\sqrt{4x+5}\right)} \quad \text{(Because } (a-b)(a+b) = a^2-b^2\text{)}$$

$$= \lim_{h\to 0}\frac{4h}{h\left(\sqrt{4x+4h+5}+\sqrt{4x+5}\right)}$$

$$= \lim_{h\to 0}\frac{4}{\sqrt{4x+4h+5}+\sqrt{4x+5}} \quad \text{(Now you can plug in.)}$$

$$= \frac{4}{\sqrt{4x+4\cdot 0+5}+\sqrt{4x+5}}$$

$$= \frac{4}{2\sqrt{4x+5}}$$

$$g'(x) = \frac{2}{\sqrt{4x+5}}$$

(b) $\mathbf{g'(5) = \frac{2}{5}}$

$$g'(5) = \frac{2}{\sqrt{4\cdot 5+5}} = \frac{2}{5}$$

10 Use this formula: *Average velocity* $= \frac{\textit{total displacement}}{\textit{total time}}$.

(a) 27 miles/hour

$$\textit{Average velocity}_{4\text{ to }5} = \frac{s(5)-s(4)}{5-4}$$

$$= \frac{3\cdot 5^2 - 3\cdot 4^2}{1}$$

$$= 27\ \frac{\text{miles}}{\text{hour}}$$

(b) 24.3 miles/hour

$$\textit{Average velocity}_{4\text{ to }4.1} = \frac{s(4.1)-s(4)}{4.1-4}$$

$$= \frac{3\cdot 4.1^2 - 3\cdot 4^2}{0.1}$$

$$= 24.3\ \frac{\text{miles}}{\text{hour}}$$

(c) 24.03 miles/hour

$$\begin{aligned} \textit{Average velocity}_{4 \text{ to } 4.01} &= \frac{s(4.01)-s(4)}{4.01-4} \\ &= \frac{3\cdot 4.01^2 - 3\cdot 4^2}{0.01} \\ &= 24.03\ \frac{\text{miles}}{\text{hour}} \end{aligned}$$

11 **24 miles/hour**

Problem 7 gives you the derivative of this parabola, $f'(x)=6x$. The position function in this problem is the same except for different variables, so its derivative is $s'(t)=6t$.

Plug in 4 for t, and you get $s'(4)=24\frac{\text{miles}}{\text{hour}}$. Notice how the average velocities in Problem 10 get closer and closer to $24\frac{\text{miles}}{\text{hour}}$ as the total travel time gets less and less and the ending time homes in on $t=4$. That's precisely how the difference quotient works as h shrinks to zero.

If you're ready to test your skills a bit more, take the following chapter quiz that incorporates all the chapter topics.

Whaddya Know? Chapter 9 Quiz

Quiz time! Complete each problem to test your knowledge on the various topics covered in this chapter. You can then find the solutions and explanations in the next section.

1 For Problems 1 and 2, refer to the following figure. The figure shows a graph of $f(x)=x^3$ and a line tangent to f at (1, 1) that crosses f at (−2, −8).

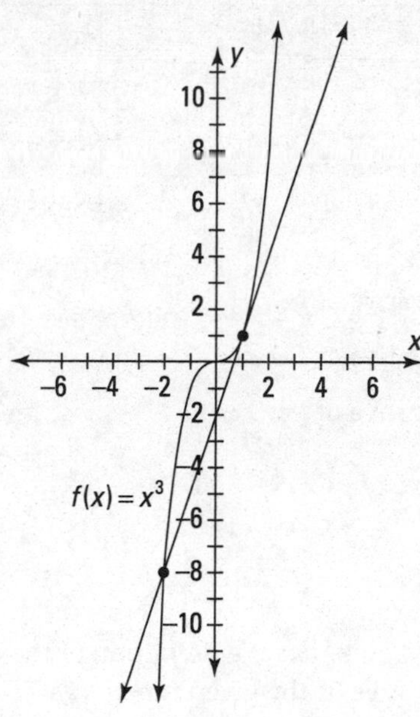

(a) What's the derivative of f at $x = 0$?

(b) What's the derivative of f at $x = 1$?

2 What's the derivative of f at $x = -1$ and why?

3 For this problem, refer to the following figure. The figure shows a graph of $f(x)=\sqrt[3]{x}$ and a line tangent to f at $(1, 1)$ that crosses f at $(-8, -2)$.

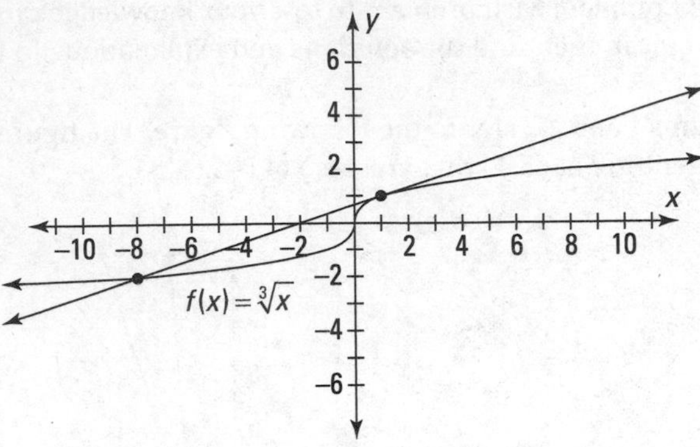

(a) What's the derivative of f at $x=1$?

(b) What's the derivative of f at $x=-1$?

(c) What's the derivative of f at $x=0$?

4 What's the derivative of $g(x)=-1$ at $(-1, -1)$?

For Problems 5 through 8, use the definition of the derivative, $f'(x)=\lim_{h\to 0}\frac{f(x+h)-f(x)}{h}$, to determine the derivative of the given function.

5 $f(x)=-8$. (You should know the derivative here without using this definition, but you should use the definition for practice.)

6 $g(x)=-x$. (You should be able to get this derivative in your head without using the difference quotient, but, again, it's a good idea to practice using the difference quotient to learn how to apply it to all sorts of functions.)

7 $f(x)=-x^3$

8 $g(x)=10-\sqrt{x}$

Answers to Chapter 9 Quiz

1 **(a) The derivative is zero.**

A derivative is a slope, so to determine the derivative of $f(x) = x^3$ at $x = 0$, you need merely to notice that f is horizontal for one infinitesimal point as it crosses the origin. The slope of a horizontal line is zero, so the derivative of $f(x) = x^3$ at $x = 0$ is zero.

(b) The derivative at $x = 1$ equals 3.

A derivative tells you the slope or steepness of a curve at a given point, and that's the same thing as the slope of a tangent line drawn at that point. So, all you need is the slope of the line tangent to f at (1, 1). That line crosses f at $(-2, -8)$. The slope formula gives you its slope: $slope = \frac{y_2 - y_1}{x_2 - x_1} = \frac{1-(-8)}{1-(-2)} = \frac{9}{3} = 3.$

2 **The derivative at $x = -1$ (like the derivative at $x = 1$) equals 3.**

The key to this problem is to note that f is an odd function (see Chapter 5 for a discussion of even and odd functions). Odd functions are symmetrical about the origin. One thing this means is that if you take an odd function like f and rotate it 180° about the origin, it lands on itself. You can see this by turning the figure for Question 2 upside down. You see the very same shape except that the tangent line now touches the curve at $(-1, -1)$ and crosses the curve at $(2, 8)$. That tangent line (on the upside-down graph) has the same slope as the original tangent line.

By the way, a feature of all odd functions is that if you know the derivative of the function at any x-value, the derivative at the opposite x-value will be the same. (And a feature of all even functions is that if you know the derivative of the function at any x-value, the derivative at the opposite x-value will be the opposite of the original derivative.)

3 **(a) The derivative at $x = 1$ is $\frac{1}{3}$.**

The line tangent to f at (1, 1) crosses f at $(-8, -2)$. The slope formula gives you a slope of $\frac{1}{3}$.

(b) The derivative at $x = -1$ (like the derivative at $x = 1$) is $\frac{1}{3}$.

This problem works just like Problem 2. Refer back to that solution. The function $f(x) = \sqrt[3]{x}$, like $f(x) = x^3$, is an odd function, and thus the derivative of $\sqrt[3]{x}$ at $x = -1$ is the same as it is at $x = 1$.

(c) The derivative at $x = 0$ is undefined.

As $f(x) = \sqrt[3]{x}$ crosses the origin, it's vertical for one infinitesimal point. In other words, a tangent line drawn to f at the origin would be vertical. The slope of a vertical line is undefined, and thus the derivative of f at $x = 0$ is undefined.

4 **The derivative is zero.**

The function $g(x)=-1$ is a horizontal line, so its slope is zero. And, thus, the derivative of g at $(-1,-1)$ (or at any other point on g) equals zero. End of story.

5 **The derivative is zero.**

Of course, like with Problem 4, this function, $f(x)=-8$, is a horizontal line, so its slope and derivative equal zero.

Here's how you do it with the limit of the difference quotient: $f'(x)=\lim_{h\to 0}\frac{f(x+h)-f(x)}{h}$.

For $f(x)=-8$, f of anything equals -8, so $f(x+h)=-8$. Plug in and finish:

$f'(x)=\lim_{h\to 0}\frac{-8-(-8)}{h}=\lim_{h\to 0}\frac{0}{h}=\lim_{h\to 0}0=0.$

6 **The derivative equals −1.**

The function $g(x)=-x$ is a line (in $y=mx+b$ form, g would be written as $y=-1x+0$). So, its slope and derivative equal -1.

Now let's get the derivative with the limit of the difference quotient. To compute $g'(x)=\lim_{h\to 0}\frac{g(x+h)-g(x)}{h}$, you need $g(x+h)$; that equals $-(x+h)$. Now you can plug in and finish: $g'(x)=\lim_{h\to 0}\frac{-(x+h)-(-x)}{h}=\lim_{h\to 0}\frac{-x-h+x}{h}=\lim_{h\to 0}\frac{-h}{h}=\lim_{h\to 0}(-1)=-1.$

7 **The derivative equals $-3x^2$.**

You need $f(x+h)$, so you can plug that into $f'(x)=\lim_{h\to 0}\frac{f(x+h)-f(x)}{h}$. The function is $f(x)=-x^3$, so $f(x+h)$ equals $-(x+h)^3$. Expanding that gives you $-\left(x^3+3x^2h+3xh^2+h^3\right)$. Plug in and finish:

$$\begin{aligned}
f'(x)&=\lim_{h\to 0}\frac{-\left(x^3+3x^2h+3xh^2+h^3\right)-\left(-x^3\right)}{h}\\
&=\lim_{h\to 0}\frac{-x^3-3x^2h-3xh^2-h^3+x^3}{h}\\
&=\lim_{h\to 0}\frac{-3x^2h-3xh^2-h^3}{h}\\
&=\lim_{h\to 0}\frac{h\left(-3x^2-3xh-h^2\right)}{h}\\
&=\lim_{h\to 0}\left(-3x^2-3xh-h^2\right)\\
&=-3x^2
\end{aligned}$$

(8) **The derivative equals $-\frac{1}{2\sqrt{x}}$.**

You are given that $g(x)=10-\sqrt{x}$, so $g(x+h)=10-\sqrt{x+h}$. Plugging those expressions into the definition of the derivative gives you $g'(x)=\lim_{h\to 0}\frac{10-\sqrt{x+h}-(10-\sqrt{x})}{h}$. Now cancel the 10's, then reverse the order of the two radicals in the numerator to make the numerator more ordinary-looking. Then multiply the numerator and denominator by the conjugate of the numerator, and finish:

$$
\begin{aligned}
g'(x) &= \lim_{h\to 0}\frac{-\sqrt{x+h}+\sqrt{x}}{h} \\
&= \lim_{h\to 0}\frac{\sqrt{x}-\sqrt{x+h}}{h} \\
&= \lim_{h\to 0}\left(\frac{\sqrt{x}-\sqrt{x+h}}{h}\cdot\frac{\sqrt{x}+\sqrt{x+h}}{\sqrt{x}+\sqrt{x+h}}\right) \\
&= \lim_{h\to 0}\frac{x-(x+h)}{h\left(\sqrt{x}+\sqrt{x+h}\right)} \\
&= \lim_{h\to 0}\frac{-h}{h\left(\sqrt{x}+\sqrt{x+h}\right)} \\
&= \lim_{h\to 0}\frac{-1}{\sqrt{x}+\sqrt{x+h}} \\
&= -\frac{1}{2\sqrt{x}}
\end{aligned}
$$

IN THIS CHAPTER

» Learning the rules whether you like it or not — sorry, buddy, but those are the rules

» Mastering the basic differentiation rules and graduating to expert rules

» Figuring out implicit differentiation

» Using logarithms in differentiation

» Differentiating inverse functions

Chapter 10

Differentiation Rules — Yeah, Man, It Rules

Chapter 9 gives you the basic idea of what a derivative is — it's just a rate like speed and it's simply the slope of a function. It's important that you have a solid, intuitive grasp of these fundamental ideas.

You also now know the mathematical foundation of the derivative and its technical definition involving the limit of the difference quotient. Now, I'm going to be forever banned from the Royal Order of Pythagoras for saying this, but, to be perfectly candid, you can basically forget that limit stuff — except that you need to know it for your final — because in this chapter I give you shortcut techniques for finding derivatives that avoid the difficulties of limits and the difference quotient.

Some of this material is unavoidably dry. If you have trouble staying awake while slogging through these rules, look back to the last chapter and take a peek at the next three chapters to see why you should care about mastering these differentiation rules. Countless problems in business, economics, medicine, engineering, and physics, as well as other disciplines, deal with how fast a function is rising or falling, and that's what a derivative tells you. And it's often important to know where a function is rising or falling the fastest (the largest and smallest slopes) and where its peaks and valleys are (where the slope is zero). Before you can do these interesting problems, you've got to learn how to find derivatives. If Chapters 11, 12, and 13 are like playing the piano, then this chapter is like learning your scales — it's dull, but you've got to do it. You may want to order up a latte with an extra shot.

Basic Differentiation Rules

Calculus can be difficult, but you'd never know it judging by this section alone. Learning these first half dozen or so rules is a snap. If you get tired of this easy stuff, however, I promise plenty of challenges in the next section.

The constant rule

This is simple: $f(x)=5$ is a horizontal line with a slope of zero, and thus its derivative is also zero. So, for any number c, if $f(x)=c$, then $f'(x)=0$. Or you can write $\frac{d}{dx}c=0$. End of story.

The power rule

Say $f(x)=x^5$. To find its derivative, take the power, 5, bring it in front of the x, and then reduce the power by 1 (in this example, the power becomes a 4). That gives you $f'(x)=5x^4$. To repeat, bring the power in front, then reduce the power by 1. That's all there is to it.

In Chapter 9, I differentiated $y=x^2$ with the difference quotient:

$$\begin{aligned}
y &= x^2 \\
y' &= \lim_{h\to 0}\frac{(x+h)^2-x^2}{h} \\
&= \lim_{h\to 0}\frac{x^2+2xh+h^2-x^2}{h} \\
&= \lim_{h\to 0}\frac{2xh+h^2}{h} \\
&= \lim_{h\to 0}(2x+h) \\
&= 2x
\end{aligned}$$

That takes some doing. Instead of all that, just use the power rule: Bring the 2 in front, and reduce the power by 1; this leaves you with a power of 1 that you can drop (because a power of 1 does nothing). Thus,

$$\begin{aligned}
y &= x^2 \\
y' &= 2x
\end{aligned}$$

Because this is so simple, you may be wondering why I didn't skip the complicated difference quotient stuff and just go straight to the shortcut method. Well, admittedly, that would have saved some time, especially considering the fact that once you know this and other shortcut methods, you'll never need the difference quotient again — except for your final exam. But the difference quotient is included in every calculus book and course because it gives you a fuller, richer understanding of calculus and its foundations — think of it as a mathematical character builder. *Or* because math teachers are sadists. You be the judge.

The power rule works for any power — a positive, a negative, or a fraction:

$$\text{If } f(x)=x^{-2} \quad \text{then} \quad f'(x)=-2x^{-3}$$

$$\text{If } g(x)=x^{2/3} \quad \text{then} \quad g'(x)=\frac{2}{3}x^{-1/3}$$

$$\text{If } h(x)=x \quad \text{then} \quad h'(x)=1$$

WARNING

The derivative of x is 1. Make sure you remember how to do the derivative of the last function in this list. It's the simplest of these functions, yet the easiest one to miss.

The best way to understand this derivative is to realize that $h(x)=x$ is a line that fits the form $y=mx+b$, because $h(x)=x$ is the same as $h(x)=1x+0$ (or $y=1x+0$). The slope (m) of this line is 1, so the derivative equals 1. Or you can just memorize that the derivative of x is 1. But if you forget both of these ideas, you can always use the power rule. Rewrite $h(x)=x$ as $h(x)=x^1$, then apply the rule: Bring the 1 in front and reduce the power by 1 to zero, giving you $h'(x)=1x^0$. Because x^0 equals 1, you've got $h'(x)=1$.

TIP

Rewrite functions so you can use the power rule. You can differentiate radical functions by rewriting them as power functions and then using the power rule. For example, if $f(x)=\sqrt[3]{x^2}$, rewrite it as $f(x)=x^{2/3}$ and use the power rule. You can also use the power rule to differentiate functions like $f(x)=\frac{1}{x^3}$. Rewrite this as $f(x)=x^{-3}$, then use the power rule.

The constant multiple rule

What if the function you're differentiating begins with a coefficient? Makes no difference. A coefficient has no effect on the process of differentiation. You just ignore it and differentiate according to the appropriate rule. The coefficient stays where it is until the final step, when you simplify your answer by multiplying by the coefficient. Here's how it works.

Differentiate $y=4x^3$.

Solution: You know by the power rule that the derivative of x^3 is $3x^2$, so the derivative of $4\left(x^3\right)$ is $4\left(3x^2\right)$. The 4 just sits there doing nothing. Then, as a final step, you simplify: $4\left(3x^2\right)$ equals $12x^2$. So $y'=12x^2$. (By the way, most people just bring the 3 to the front, like this: $y'=3\cdot 4x^2$, which gives you the same result.)

Differentiate $y=5x$.

Solution: This is a line of the form $y=mx+b$ with $m=5$, so the slope is 5, and thus the derivative is 5: $y'=5$. (It's important to think graphically like this from time to time.) But you can also solve the problem with the power rule: $\frac{d}{dx}x^1=1x^0=1$; so $\frac{d}{dx}5\left(x^1\right)=5(1)=5$.

One final example: Differentiate $y=\frac{5x^{1/3}}{4}$.

Solution: The coefficient here is $\frac{5}{4}$. So, because $\frac{d}{dx}x^{1/3}=\frac{1}{3}x^{-2/3}$ (by the power rule), $\frac{d}{dx}\frac{5}{4}\left(x^{1/3}\right)=\frac{5}{4}\left(\frac{1}{3}x^{-2/3}\right)=\frac{5}{12}x^{-2/3}$.

WARNING

Pi, e, c, k, and so on, are *not* variables! Don't forget that π (which equals, of course, about 3.14) and e (which equals about 2.72) are numbers, not variables, so they behave like ordinary numbers. Constants in problems, like c and k, also behave like ordinary numbers. (By the way, the number e, named for the great mathematician Leonhard Euler, is perhaps the most important number in all of mathematics, but I don't get into that here.)

Thus, if $y = \pi x$, $y' = \pi$ — this works exactly like differentiating $y = 5x$. And because π^3 is just a number, if $y = \pi^3$ then $y' = 0$ — this works exactly like differentiating $y = 10$. You'll also see problems containing constants like c and k. Be sure to treat them like regular numbers. For example, the derivative of $y = 5x + 2k^3$ (where k is a constant) is 5, not $5 + 6k^2$.

The sum rule — hey, that's some rule you got there

When you want the derivative of a sum of terms, take the derivative of each term separately.

What's $f'(x)$ if $f(x) = x^6 + x^3 + x^2 + x + 10$?

Solution: Just use the power rule for each of the first four terms and the constant rule for the final term. Thus, $f'(x) = 6x^5 + 3x^2 + 2x + 1$.

The difference rule — it makes no difference

If you've got a difference (that's subtraction) instead of a sum, it makes no difference. You still differentiate each term separately. Thus, if $y = 3x^5 - x^4 - 2x^3 + 6x^2 + 5x$, then $y' = 15x^4 - 4x^3 - 6x^2 + 12x + 5$. The addition and subtraction signs are unaffected by the differentiation.

EXAMPLE

Q. What's the derivative of $5x^3$?

A. $15x^2$

1. **Bring the power in front, multiplying it by the coefficient.**

   This first step gives you $15x^3$. (Note that this does not equal the original function or its derivative, so you should not put an equal sign in front of it. In fact, there's no reason to write this interim step down at all. I do it simply to make the process clear.)

2. **Reduce the power by one.**

   This gives you the final answer of $15x^2$.

YOUR TURN

1. What's the derivative of $f(x) = 8$?

2. What's the derivative of $g(x) = \pi^3$?

3. What's the derivative of $g(x) = k \sin\frac{\pi}{2} \cos(2\pi)$, where k is a constant?

4. For $f(x) = 5x^4$, $f'(x) = ?$

5. For $g(x) = \frac{-x^3}{10}$, what's $g'(x)$?

6. Find y' if $y = \sqrt{x^{-5}}$ $(x > 0)$.

7. What's the derivative of $s(t) = 7t^6 + t + 10$?

8. Find the derivative for $y = \left(x^3 - 6\right)^2$.

Differentiating trig functions

Ladies and gentlemen, I have the high honor and distinct privilege of introducing you to the derivatives of the six trig functions:

$$\frac{d}{dx}\sin x = \cos x \qquad \frac{d}{dx}\cos x = -\sin x$$

$$\frac{d}{dx}\tan x = \sec^2 x \qquad \frac{d}{dx}\cot x = -\csc^2 x$$

$$\frac{d}{dx}\sec x = \sec x \tan x \qquad \frac{d}{dx}\csc x = -\csc x \cot x$$

Make sure you memorize the derivatives of sine and cosine. They're a snap, and I've never known anyone to forget them. If you're good at rote memorization, memorize the other four as well. Or, if you're not wild about memorization or are afraid that this knowledge will crowd out the date of the Battle of Hastings (1066) — which is much more likely to come up in a board game than trig derivatives — you can figure out the last four derivatives from scratch by using the quotient rule (see the section, "The quotient rule," later in this chapter). A third option is to use the following mnemonic trick.

TIP

Psst, what's the derivative of cosecant? Imagine you're taking a test and can't remember those four last trig derivatives. You lean over to the guy next to you and whisper, "Psst, what's the derivative of $\csc x$?" Now, the last three letters of *psst* (sst) are the initial letters of sec, sec, tan. Write these three down, and below them, write their cofunctions: csc, csc, cot. Put a negative sign on the csc in the middle. Finally, add arrows like in the following diagram:

$$\begin{array}{ccccc} \sec & \rightarrow & \sec & \leftarrow & \tan \\ \csc & \rightarrow & -\csc & \leftarrow & \cot \end{array}$$

(This may seem complicated, but take my word for it, you'll remember the word *psst,* and after that, the diagram is very easy to remember.)

Look at the top row. The *sec* on the left has an arrow pointing to *sec tan* — so the derivative of $\sec x$ is $\sec x \tan x$. The *tan* on the right has an arrow pointing to *sec sec,* so the derivative of $\tan x$ is $\sec^2 x$. The bottom row works the same way, except that both derivatives are negative.

Differentiating exponential and logarithmic functions

Caution: Memorization ahead.

Exponential functions

If you can't memorize the next rule, hang up your calculator.

$$\frac{d}{dx}e^x = e^x$$

That's right — break out the smelling salts — the derivative of e^x is itself! This is a special function: e^x and its multiples, like $5e^x$, are the only functions that are their own derivatives. Think about what this means. Look at the graph of $y = e^x$ in Figure 10-1.

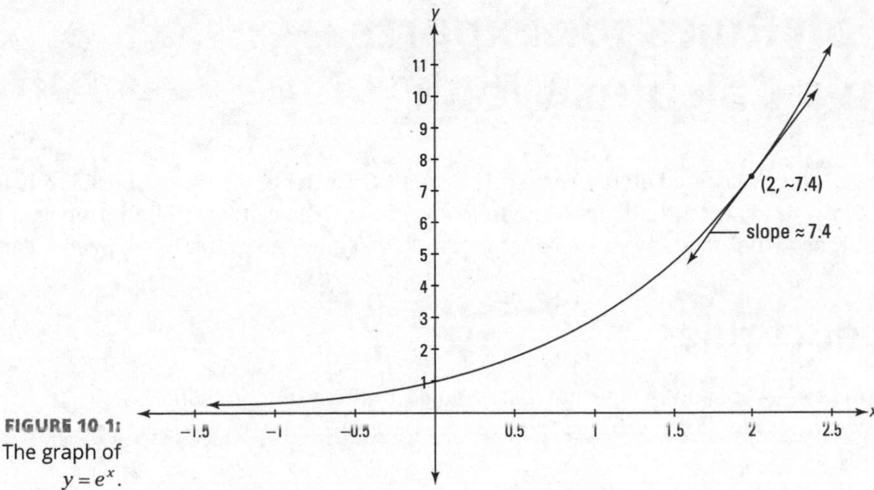

FIGURE 10-1: The graph of $y = e^x$.

Pick any point on this function, say $(2, \sim 7.4)$, and the height of the function at that point, ~ 7.4, is the same as the slope at that point.

If the base is a number other than e, you have to tweak the derivative by multiplying it by the natural log of the base:

$$\text{If } y = 2^x \quad \text{then} \quad y' = 2^x \ln 2$$
$$\text{If } y = 10^x \quad \text{then} \quad y' = 10^x \ln 10$$

Logarithmic functions

And now — what you've all been waiting for — the derivatives of logarithmic functions. (See Chapter 4 if you want to brush up on logs.) Here's the derivative of the *natural* log — that's the log with base *e:*

$$\frac{d}{dx}\ln x = \frac{1}{x}$$

If the log base is a number other than e, you tweak this derivative — like with exponential functions — except that you *divide* by the natural log of the base instead of multiplying. Thus,

$$\frac{d}{dx}\log_2 x = \frac{\frac{1}{x}}{\ln 2} = \frac{1}{x \ln 2}, \quad \text{and}$$

$$\frac{d}{dx}\log x = \frac{1}{x \ln 10} \quad \text{(Recall that } \log x \text{ means } \log_{10} x\text{, so the base is 10.)}$$

Good news: You get to practice finding derivatives of trig, exponential, and log functions in the next section.

Differentiation Rules for Experts — Oh, Yeah, I'm a Calculus Wonk

Now that you've *totally* mastered all the basic rules, take a breather and rest on your laurels for a minute. . . . Okay, ready for a challenge? The following rules, especially the chain rule, can be tough. But you know what they say: "No pain, no gain," "No guts, no glory," yada, yada, yada.

The product rule

You use this rule for — hold on to your hat — the *product* of two functions like

$$y = x^3 \cdot \sin x$$

The product rule:

$$\begin{aligned} &\text{If } y = \textit{this} \cdot \textit{that}, \\ &\text{then } y' = \textit{this}' \cdot \textit{that} + \textit{this} \cdot \textit{that}' \end{aligned}$$

So, for $y = x^3 \cdot \sin x$,

$$\begin{aligned} y' &= \left(x^3\right)' \cdot \sin x + x^3 \cdot (\sin x)' \\ &= 3x^2 \sin x + x^3 \cos x \end{aligned}$$

The quotient rule

I have a feeling that you can guess what this rule is for — the *quotient* of two functions like

$$y = \frac{\sin x}{x^4}$$

The quotient rule:

$$\begin{aligned} &\text{If } y = \frac{\textit{top}}{\textit{bottom}}, \\ &\text{then } y' = \frac{\textit{top}' \cdot \textit{bottom} - \textit{top} \cdot \textit{bottom}'}{\textit{bottom}^2} \end{aligned}$$

Many calculus books give the quotient rule in a slightly different form that's harder to remember. And some books give a "mnemonic" involving the words *lodeehi* and *hideelo* or *hodeehi* and *hideeho*, which is very easy to get mixed up — great, thanks a lot.

Here's a good way to remember the quotient rule. When you read a product, you read from left to right, and when you read a quotient, you read from top to bottom. So just remember that the quotient rule, like the product rule, works in the natural order in which you read. *Both rules begin with the derivative of the first function you read.* Also note that the numerator of the quotient rule works just like the product rule, except for the minus sign (which no one forgets). Another thing you won't forget is the denominator of the quotient rule — take my word for it. Focus on these points and you'll remember the quotient rule ten years from now — oh, sure.

So, here's the derivative of $y = \frac{\sin x}{x^4}$:

$$\begin{aligned} y' &= \frac{(\sin x)' \cdot x^4 - \sin x \cdot (x^4)'}{(x^4)^2} \\ &= \frac{x^4 \cos x - 4x^3 \sin x}{x^8} \\ &= \frac{x^3(x\cos x - 4\sin x)}{x^8} \\ &= \frac{x\cos x - 4\sin x}{x^5} \end{aligned}$$

In the section "Differentiating trig functions," I promised to show you how to find the derivatives of four trig functions — *tangent, cotangent, secant,* and *cosecant* — with the quotient rule. I'm a man of my word, so here goes. All four of these functions can be written in terms of *sine* and *cosine,* right? (See Chapter 6.) For instance, $\tan x = \frac{\sin x}{\cos x}$. Now, if you want the derivative of $\tan x$, you can use the quotient rule:

$$\begin{aligned} \tan x &= \frac{\sin x}{\cos x} \\ (\tan x)' &= \frac{(\sin x)' \cdot \cos x - \sin x \cdot (\cos x)'}{\cos^2 x} \\ &= \frac{\cos x \cdot \cos x - \sin x \cdot (-\sin x)}{\cos^2 x} \\ &= \frac{\cos^2 x + \sin^2 x}{\cos^2 x} \\ &= \frac{1}{\cos^2 x} \quad \text{(The Pythagorean Identity tells you that } \cos^2 x + \sin^2 x = 1.) \\ &= \sec^2 x \end{aligned}$$

Granted, this is quite a bit of work compared to just memorizing the answer or using the mnemonic device presented several pages back, but it's nice to know that you can get the answer this way as a last resort. The other three functions are no harder. Give them a try.

EXAMPLE

Q. $\frac{d}{dx}(x^2 \sin x) = ?$

A. $\frac{d}{dx}(x^2 \sin x) = (x^2)'(\sin x) + (x^2)(\sin x)'$

$= 2x \sin x + x^2 \cos x$

Q. $\frac{d}{dx}\frac{x^2}{(\sin x)}=?$

A. $$\frac{d}{dx}\frac{x^2}{(\sin x)}=\frac{(x^2)'(\sin x)-(x^2)(\sin x)'}{(\sin x)^2}$$
$$=\frac{2x\sin x-x^2\cos x}{\sin^2 x}$$

(I purposely designed this example to resemble the product rule example, so you can see the similarity between the quotient rule numerator and the product rule.)

YOUR TURN

9 $\frac{d}{dx}\left(x^3\cos x\right)=?$

10 $\frac{d}{dx}(\sin x\tan x)=?$

11 $\frac{d}{dx}\left(5x^3\ln x\right)=?$

12 $\frac{d}{dx}\left(x^2e^x\ln x\right)=?$

13 $\frac{d}{dx}\frac{x^3}{\cos x} = ?$

14 $\frac{d}{dx}\frac{\cos x}{e^x} = ?$

15 $\frac{d}{dx}\frac{\sin x}{x^3 \ln x} = ?$

Linking up with the chain rule

The chain rule is by far the trickiest derivative rule, but it's not really that bad if you carefully focus on a few important points. Let's begin by differentiating $y = \sqrt{4x^3 - 5}$. You use the chain rule here because you've got a *composite* function, that's one function $(4x^3 - 5)$ inside another function (the square root function).

TIP

How to spot a *composite* function: $y = \sqrt{x}$ is *not* a composite function because the *argument* of the square root function — that's the thing you take the square root of — is simply *x*. Whenever the argument of a function is anything other than a plain old *x*, you've got a composite function. Be careful to distinguish a composite function from something like $y = \sqrt{x} \cdot \sin x$, which is the *product* of two functions, $\sqrt{x}$ and $\sin x$, each of which *does* have just a plain old *x* as its argument.

Okay, so you've got this composite function, $y=\sqrt{4x^3-5}$. Here's how to differentiate it with the chain rule:

1. **You start with the *outside* function, $\sqrt{}$, and differentiate that, *IGNORING* what's inside. To make sure you ignore the inside, temporarily replace the inside function with the word *stuff*.**

   So, you've got $y=\sqrt{stuff}$. Okay, now differentiate $y=\sqrt{stuff}$ the same way you'd differentiate $y=\sqrt{x}$. Because $y=\sqrt{x}$ is the same as $y=x^{1/2}$, the power rule gives you $y'=\frac{1}{2}x^{-1/2}$. So, for this problem, you begin with $\frac{1}{2}stuff^{-1/2}$.

2. **Multiply the result from Step 1 by the derivative of the inside function, *stuff'*.**

   $$y'=\frac{1}{2}stuff^{-1/2}\cdot stuff'$$

   Take a good look at this. *All* chain rule problems follow this basic idea. You do the derivative rule for the outside function, ignoring the inside *stuff*, then multiply that by the derivative of the *stuff*.

3. **Differentiate the inside *stuff*.**

   The inside *stuff* in this problem is $4x^3-5$ and its derivative is $12x^2$ by the power rule.

4. **Now put the real *stuff* and its derivative back where they belong.**

   $$y'=\frac{1}{2}\left(4x^3-5\right)^{-1/2}\cdot\left(12x^2\right)$$

5. **Simplify.**

   $$y'=6x^2\left(4x^3-5\right)^{-1/2}$$

   Or, if you've got something against negative powers, $y'=\dfrac{6x^2}{\left(4x^3-5\right)^{1/2}}$.

   Or, if you've got something against fraction powers, $y'=\dfrac{6x^2}{\sqrt{4x^3-5}}$.

Try differentiating another composite function, $y=\sin\left(x^2\right)$:

1. **The outside function is the sine function, so you start there, taking the derivative of sine and ignoring the inside *stuff*, x^2. The derivative of $\sin x$ is $\cos x$, so for this problem, you begin with**

   $$\cos(stuff)$$

2. **Multiply the derivative of the outside function by the derivative of the *stuff*.**

   $$y'=\cos(stuff)\cdot stuff'$$

3. **The *stuff* in this problem is x^2, so *stuff'* is $2x$. When you plug these terms back in, you get**

   $$\begin{aligned}y'&=\cos\left(x^2\right)\cdot 2x\\&=2x\cos\left(x^2\right)\end{aligned}$$

Sometimes figuring out which function is inside which can be a bit tricky — especially when a function is inside another and then both of them are inside a *third* function (you can have four or more nested functions, but three is probably the most you'll see). Here's a tip.

TIP

Parentheses are your friend. For chain rule problems, rewrite a composite function with a set of parentheses around each inside function, and rewrite trig functions like $\sin^2 x$ with the power outside a set of parentheses: $(\sin x)^2$.

Let's try a problem — this is a tough one, gird your loins: differentiate $y = \sin^3(5x^2 - 4x)$. First, rewrite the cubed sine function: $y = \left(\sin\left(5x^2 - 4x\right)\right)^3$. Now it's easy to see the order in which the functions are nested. The innermost function is inside the innermost parentheses — that's $5x^2 - 4x$. Next, the sine function is inside the next set of parentheses — that's $\sin(\textit{stuff})$. Last, the cubing function is on the outside of everything — that's $\textit{stuff}^3$. (Because I'm a math teacher, I'm honor-bound to point out that the *stuff* in $\textit{stuff}^3$ is different from the *stuff* in $\sin(\textit{stuff})$. It's quite unmathematical of me to use the same term to refer to different things, but don't sweat it. I'm just using the term *stuff* to refer to whatever is inside any function.) Okay, now that you know the order of the functions, you can differentiate from *outside in:*

1. **The outermost function is $\textit{stuff}^3$ and its derivative is given by the power rule.**

   $3\textit{stuff}^2$

2. **As with all chain rule problems, you multiply that by *stuff'*.**

   $3\textit{stuff}^2 \cdot \textit{stuff}'$

3. **Now put the *stuff*, $\sin\left(5x^2 - 4x\right)$, back where it belongs.**

   $3\left(\sin\left(5x^2 - 4x\right)\right)^2 \cdot \left(\sin\left(5x^2 - 4x\right)\right)'$

4. **Use the chain rule again.**

   You can't finish this problem quickly by just taking a simple derivative because you have to differentiate another composite function, $\sin\left(5x^2 - 4x\right)$. Just treat $\sin\left(5x^2 - 4x\right)$ as if it were the original problem and take its derivative. The derivative of $\sin x$ is $\cos x$, so the derivative of $\sin(\textit{stuff})$ begins with $\cos(\textit{stuff})$. Multiply that by *stuff'*. Thus, the derivative of $\sin(\textit{stuff})$ is

   $\cos(\textit{stuff}) \cdot \textit{stuff}'$

5. **The *stuff* for this step is $5x^2 - 4x$ and its derivative is $10x - 4$. Plug those things back in.**

   $\cos\left(5x^2 - 4x\right) \cdot (10x - 4)$

6. **Now that you've got the derivative of $\sin\left(5x^2 - 4x\right)$, plug this result into the result from Step 3, giving you the whole enchilada.**

   $3\left(\sin\left(5x^2 - 4x\right)\right)^2 \cdot \cos\left(5x^2 - 4x\right) \cdot (10x - 4)$

7. **This can be simplified a bit.**

   $(30x - 12)\sin^2\left(5x^2 - 4x\right)\cos\left(5x^2 - 4x\right)$

I told you it was a tough one.

It may have occurred to you that you can save some time by not switching to the word *stuff* and then switching back. That's true, but some people like to use the technique because it forces them to leave the *stuff* alone during each step of a problem. That's the critical point.

WARNING

Make sure you . . . DON'T TOUCH THE *STUFF*.

As long as you remember this, you don't need to actually use the word *stuff* when doing a chain rule problem. You've just got to be sure you don't change the inside function while differentiating the outside function. Say you want to differentiate $f(x)=\ln(x^3)$. The argument of this natural logarithm function is x^3. Don't touch it during the first step of the solution, which is to use the natural log rule: $\frac{d}{dx}\ln x=\frac{1}{x}$. This rule tells you to put the argument of the function in the denominator under the number 1. So, after the first step in differentiating $\ln(x^3)$, you've got $\frac{1}{x^3}$. You then finish the problem by multiplying that by the derivative of x^3, which is $3x^2$. Final answer: $\frac{3}{x}$.

TIP

With the chain rule, don't use two derivative rules at the same time. Another way to make sure you've got the chain rule straight is to remember that you never use more than one derivative rule at a time.

In the preceding example, $\ln(x^3)$, you first use the natural log rule, then, as a *separate step,* you use the power rule to differentiate x^3. At no point in any chain rule problem do you use both rules at the same time. For example, with $\ln(x^3)$, you do not use the natural log rule and the power rule at the same time to come up with $\frac{1}{3x^2}$.

Here's the chain rule mumbo jumbo.

MATH RULES

The chain rule (for differentiating a composite function):

$$\text{If } y=f(g(x)),$$
$$\text{then } y'=f'(g(x))\cdot g'(x)$$

Or, equivalently,

$$\text{If } y=f(u) \text{ and } u=g(x),$$
$$\text{then } \frac{dy}{dx}=\frac{dy}{du}\cdot\frac{du}{dx} \quad \text{(Notice how the } du\text{'s cancel.)}$$

See the sidebar, "Why the chain rule works," for a plain-English explanation of this mumbo jumbo.

One final example and one last tip. Differentiate $4x^2\sin(x^3)$. This problem has a new twist — it involves the chain rule *and* the product rule. How should you begin?

TIP

Where do I begin? If you're not sure where to begin differentiating a complex expression, imagine plugging a number into x and then evaluating the expression on your calculator one step at a time. Your *last* computation tells you the *first* thing to do.

WHY THE CHAIN RULE WORKS

You wouldn't know it from the difficult math in this section or the fancy chain rule mumbo jumbo, but the chain rule is based on a *very* simple idea. Say one person is walking, another jogging, and a third is riding a bike. If the biker goes four times as fast as the jogger, and the jogger goes twice as fast as the walker, then the biker goes 4 times 2, or 8 times as fast as the walker, right? That's the chain rule in a nutshell — you just multiply the relative rates.

Flip back to Chapter 9 and take a look at Figure 9-5 showing Laurel and Hardy on a teeter-totter. For every inch Hardy goes down, Laurel goes up 2 inches. So, Laurel's rate of movement is twice Hardy's rate, and therefore $\frac{dL}{dH}=2$. Now imagine that Laurel has one of those party favors in his mouth (the kind that unrolls as you blow into it) and that for every inch he goes up, he blows the noisemaker out 3 inches. The rate of movement of the noisemaker (*N*) is thus 3 times Laurel's rate of movement. In calculus symbols, $\frac{dN}{dL}=3$. So, how fast is the noisemaker moving compared to Hardy? This is just common sense. The noisemaker is moving 3 times as fast as Laurel, and Laurel is moving 2 times as fast as Hardy, so the noisemaker is moving 3 times 2, or 6 times as fast as Hardy. Here it is in symbols (note that this is the same as the formal definition of the chain rule next to the Mumbo Jumbo icon):

$$\frac{dN}{dH}=\frac{dN}{dL}\cdot\frac{dL}{dH}=3\cdot 2=6$$

Mere child's play.

Say you plug the number 5 into the *x*'s in $4x^2\sin(x^3)$. You evaluate $4\cdot 5^2$ — that's 100; then, after getting $5^3=125$, you do $\sin(125)$, which is about -0.616. Finally, you multiply 100 by -0.616. Because your *last* computation is *multiplication*, your *first* step in differentiating is to use the *product* rule. (Had your last computation been instead something like $\sin(125)$, then you'd begin with the chain rule.) Okay, so for this problem, you begin with the product rule.

REMEMBER

The product rule:

If $y = this\cdot that$, then $y' = this'\cdot that + this\cdot that'$.

So, for $f(x)=4x^2\sin(x^3)$,

$$f'(x)=\left(4x^2\right)'\sin\left(x^3\right)+4x^2\left(\sin\left(x^3\right)\right)'$$

You finish the problem by taking the derivative of $4x^2$ with the power rule and the derivative of $\sin(x^3)$ with the chain rule:

$$f'(x)=8x\sin\left(x^3\right)+4x^2\cos\left(x^3\right)\cdot 3x^2$$

And now simplify:

$$f'(x)=8x\sin\left(x^3\right)+12x^4\cos\left(x^3\right)$$

16 $f(x)=\cos\left(x^4\right)$

$f'(x)=?$

17 $g(x)=\sin^3 x$

$g'(x)=?$

18 $s(t)=\tan(\ln t)$

$s'(t)=?$

19 $y=e^{4x^3}$

$y'=?$

20 $f(x)=x^4\sin^3 x$

$f'(x)=?$

21 $g(x)=\dfrac{\ln^2 x}{5x-4}$

$g'(x)=?$

22 $y=\cos^3\left(4x^2\right)$

$y'=?$

23 $\dfrac{d}{dx}\tan^3\left(e^{x^2}\right)=?$

24 $p = \sqrt{\cos x}$ $p' = ?$	**25** $p = \dfrac{1}{\ln q}$ $\dfrac{dp}{dq} = ?$
26 $f(x) = \ln\frac{1}{x}$. What's $f'(x)$ at the point $(5, -\ln 5)$?	**27** $y = x - \cos(1 - x)$ $y' = ?$

Differentiating Implicitly

All the differentiation problems presented in previous sections of this chapter are functions like $y = x^2 + 5x$ or $y = \sin x$. In such cases, y is written *explicitly* as a function of x. This means that the equation is solved for y; in other words, y is by itself on one side of the equation. (Note that y was sometimes written as $f(x)$ as in $f(x) = x^3 - 4x^2$, but remember that that's the same thing as $y = x^3 - 4x^2$.)

Sometimes, however, you are asked to differentiate an equation that's not solved for y, like $y^5 + 3x^2 = \sin x - \cos y$. This equation defines y *implicitly* as a function of x, and you can't write it as an explicit function because it can't be solved for y. For such a problem, you need *implicit differentiation*. (If you can solve for y, implicit differentiation will still work, but it's not necessary.) When differentiating implicitly, all the derivative rules work the same, with one exception: When you differentiate a term with a y in it, you use the chain rule with a little twist.

Remember using the chain rule to differentiate something like $\sin(x^3)$ with the *stuff* technique? The derivative of sine is cosine, so the derivative of $\sin(stuff)$ is $\cos(stuff) \cdot stuff'$. You finish the problem by finding the derivative of the *stuff*, x^3, which is $3x^2$, and then making the substitutions to give you $\cos(x^3) \cdot 3x^2$. With implicit differentiation, a *y* works like the word *stuff*. Thus, because

$$(\sin(stuff))' = \cos(stuff) \cdot stuff',$$

$$(\sin y)' = \cos y \cdot y'$$

The twist is that while the word *stuff* is temporarily taking the place of some *known* function of *x* (x^3 in this example), *y* is some *unknown* function of *x* (you don't know what the *y* equals in terms of *x*). And because you don't know what *y* equals, the *y* and the *y'* — unlike the *stuff* and the *stuff'* — must remain in the final answer. But the concept is exactly the same, and you treat *y* just like the *stuff*. You just can't make the switch back to *x*'s at the end of the problem like you can with a regular chain rule problem.

I suppose you're wondering whether I'm ever going to get around to actually doing the problem. Here goes:

EXAMPLE

Q. Differentiate $y^5 + 3x^2 = \sin x - \cos y$.

A. $y' = \dfrac{\cos x - 6x}{5y^4 - \sin y}$

1. **Differentiate each term on *both* sides of the equation.**

   For the first and fourth terms, you use the power rule and the cosine rule respectively, and, because these terms contain *y*'s, you also use the chain rule. For the second term, you use the regular power rule. And for the third term, you use the regular sine rule.

   $$5y^4 \cdot y' + 6x = \cos x - (-\sin y \cdot y')$$
   $$= \cos x + \sin y \cdot y'$$

2. **Collect all terms containing a *y'* on the left side of the equation and all other terms on the right side.**

   $$5y^4 \cdot y' - \sin y \cdot y' = \cos x - 6x$$

3. **Factor out *y'*.**

   $$y'(5y^4 - \sin y) = \cos x - 6x$$

4. **Divide for the final answer.**

   $$y' = \frac{\cos x - 6x}{5y^4 - \sin y}$$

Note that this derivative, unlike the others you've done so far, is expressed in terms of x and y instead of just x. So, if you want to evaluate the derivative to get the slope at a particular point, you need to have values for both x and y to plug into the derivative.

Also note that in many textbooks, the symbol $\frac{dy}{dx}$ is used instead of y' in every step of solutions like the one shown here. I find y' easier and less cumbersome to work with. But $\frac{dy}{dx}$ does have the advantage of reminding you that you're finding the derivative of y with respect to x. Either way is fine. Take your pick.

YOUR TURN

28 If $y^3 - x^2 = x + y$, find $\frac{dy}{dx}$ by implicit differentiation.

29 If $3y + \ln y = 4e^x$, find y'.

30 For $x^2y = y^3x + 5y + x$, find $\frac{dy}{dx}$ by implicit differentiation.

31 If $y + \cos^2(y^3) = \sin(5x^2)$, find the slope of the curve at $\left(\sqrt{\frac{\pi}{10}},\ 0\right)$.

32 If $8y + 5x^2 = \tan y$, find $\frac{dy}{dx}$.

33 Find the slope of the line tangent to the circle $x^2 + y^2 = 5$ at the point $(2,\ 1)$.

34 If $3y^4 + 5x = x^3 + y^{-3}$, find $\frac{dy}{dx}$.

35 Find the slope of the normal line to the ellipse $3x^2 + y^2 = 19$ at the point $(1,\ -4)$.

Differentiating Inverse Functions

There's a difficult-looking formula involving the derivatives of inverse functions, but before we get to it, look at Figure 10-2, which nicely sums up the whole idea.

Figure 10-2 shows a pair of inverse functions, f and g. Recall that inverse functions are symmetric with respect to the line, $y = x$. As with any pair of inverse functions, if the point $(10,\ 4)$ is on one function, $(4,\ 10)$ is on its inverse. And, because of the symmetry of the graphs, you can see that the slopes at those points are reciprocals: At $(10,\ 4)$ the slope is $\frac{1}{3}$, and at $(4,\ 10)$ the slope is $\frac{3}{1}$. That's how the idea works graphically, and if you're with me so far, you've got it down at least visually.

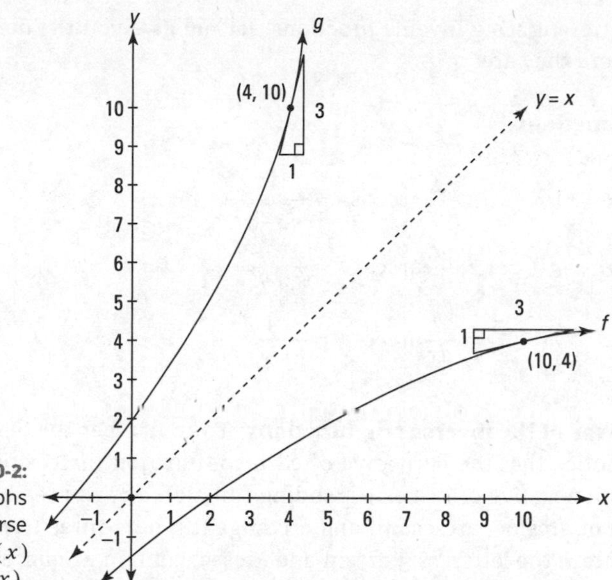

FIGURE 10-2: The graphs of inverse functions, $f(x)$ and $g(x)$.

The algebraic explanation is a bit trickier, however. The point $(10, 4)$ on f can be written as $(10, f(10))$ and the slope at this point — and thus the derivative — can be expressed as $f'(10)$. The point $(4, 10)$ on g can be written as $(4, g(4))$. Then, because $f(10) = 4$, you can replace the 4's in $(4, g(4))$ with $f(10)$s giving you $(f(10), g(f(10)))$. The slope and derivative at this point can be expressed as $g'(f(10))$. These two slopes are reciprocals, so that gives you the equation

$$f'(10) = \frac{1}{g'(f(10))}$$

This difficult equation expresses nothing more and nothing less than the two triangles on the two functions in Figure 10-2.

Using x instead of 10 gives you the general formula:

MATH RULES

The derivative of an inverse function: If f and g are inverse functions, then

$$f'(x) = \frac{1}{g'(f(x))}$$

In words, this formula says that the derivative of a function, f, with respect to x, is the reciprocal of the derivative of its inverse function with respect to f.

Okay, so maybe it was *a lot* trickier.

While I'm on the topic of differentiating inverse functions, let me give you the derivatives of the inverse trig functions. Here they are:

MATH RULES

Derivatives of inverse trig functions.

$$\frac{d}{dx}\arcsin x = \frac{1}{\sqrt{1-x^2}},\quad x \neq \pm 1 \qquad \frac{d}{dx}\arccos x = \frac{-1}{\sqrt{1-x^2}},\quad x \neq \pm 1$$

$$\frac{d}{dx}\operatorname{arcsec} x = \frac{1}{|x|\sqrt{x^2-1}},\quad x \neq 0,\ \pm 1 \qquad \frac{d}{dx}\operatorname{arccsc} x = \frac{-1}{|x|\sqrt{x^2-1}},\quad x \neq 0,\ \pm 1$$

$$\frac{d}{dx}\arctan x = \frac{1}{1+x^2} \qquad \frac{d}{dx}\operatorname{arccot} x = \frac{-1}{1+x^2}$$

TIP

Memory aid for the derivatives of the inverse trig functions. To remember the derivatives of the inverse trig functions, notice that the derivative of each co-function (arccosine, arccosecant, and arccotangent) is the negative of its corresponding function. So, you really only need to memorize the derivatives of arcsine, arcsecant, and arctangent. These three have a 1 in the numerator. The two that contain the letter "S," arcsin and arcsec, contain a Square root in the denominator and also a Subtraction sign. Arctan has no "S," so no square root and no subtraction sign (it has an addition sign instead).

EXAMPLE

Q. Use the rule for the derivative of an inverse function to compute $\arcsin'\left(\frac{1}{2}\right)$, then confirm your answer using the derivative rule for arcsine.

A. $\frac{2}{\sqrt{3}}$ **or** $\frac{2\sqrt{3}}{3}$

For inverse functions f and g,

$$f'(x) = \frac{1}{g'(f(x))}, \text{ so}$$

$$\begin{aligned}\arcsin'\frac{1}{2} &= \frac{1}{\sin'\left(\arcsin\frac{1}{2}\right)}\\ &= \frac{1}{\sin'\frac{\pi}{6}}\\ &= \frac{1}{\cos\frac{\pi}{6}}\\ &= \frac{1}{\frac{\sqrt{3}}{2}}\\ &= \frac{2}{\sqrt{3}} \text{ or } \frac{2\sqrt{3}}{3}\end{aligned}$$

Use the derivative rule for arcsine to confirm this answer.

$$\arcsin'\frac{1}{2} = \frac{1}{\sqrt{1-\left(\frac{1}{2}\right)^2}} = \frac{1}{\sqrt{\frac{3}{4}}} = \sqrt{\frac{4}{3}} = \frac{2}{\sqrt{3}}$$

It checks.

36 Use the formula for the derivative of an inverse function to compute the derivative of the arctangent function at $x = 1$, then confirm your answer using the rule for the derivative of arctangent.

37 Given that f and g are inverse functions, that $f(2) = 5$, and that the line tangent to f at $(2, 5)$ has a slope of 4, find the equation of the line tangent to g at $x = 5$.

Scaling the Heights of Higher-Order Derivatives

Finding a second, third, fourth, or higher derivative is incredibly simple. The second derivative of a function is just the derivative of its first derivative. The third derivative is the derivative of the second derivative, the fourth derivative is the derivative of the third, and so on. For example, here's a function and its first, second, third, and subsequent derivatives. In this problem, all the derivatives are obtained by the power rule:

$$\begin{aligned} f(x) &= x^4 - 5x^2 + 12x - 3 \\ f'(x) &= 4x^3 - 10x + 12 \\ f''(x) &= 12x^2 - 10 \\ f'''(x) &= 24x \\ f^{(4)}(x) &= 24 \\ f^{(5)}(x) &= 0 \\ f^{(6)}(x) &= 0 \\ \text{etc.} &= 0 \\ \text{etc.} &= 0 \end{aligned}$$

All polynomial functions like this one eventually go to zero when you differentiate repeatedly. Rational functions like $f(x)=\frac{x^2-5}{x+8}$, on the other hand, get messier and messier as you take higher and higher derivatives. And, as shown in the following problem, the higher derivatives of sine and cosine are cyclical.

$$\begin{aligned} y &= \sin x \\ y' &= \cos x \\ y'' &= -\sin x \\ y''' &= -\cos x \\ y^{(4)} &= \sin x \end{aligned}$$

The cycle repeats indefinitely with every multiple of four.

In Chapters 11 and 12, I show you several uses of higher derivatives — mainly second derivatives. (Here's a sneak preview: The first derivative of position is velocity, and the second derivative of position is acceleration.) But for now, let me give you just one of the main ideas in a nutshell. A first derivative, as you know, tells you how fast a function is changing — how fast it's going up or down — that's its slope. A second derivative tells you how fast the first derivative is changing — or, in other words, how fast the slope is changing. A third derivative tells you how fast the second derivative is changing, which tells you how fast the rate of change of the slope is changing. If you're getting a bit lost here, don't worry about it — I'm getting lost myself. It gets increasingly difficult to get a handle on what higher derivatives tell you as you go past the second derivative, because you start getting into a rate of change of a rate of change of a rate of change, and so on.

YOUR TURN

38 For $y=x^5$, what's the 1000th derivative?

39 For $y=x^3-x^{-3}$, find the first through fifth derivatives.

40 For $y = \sin x + \cos x$, find the first through sixth derivatives.

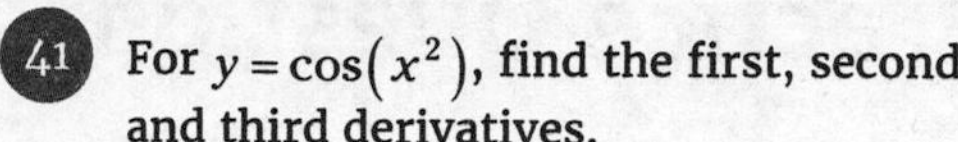

41 For $y = \cos\left(x^2\right)$, find the first, second, and third derivatives.

42 For $y = \dfrac{8^x}{(\ln 8)^3}$, find the sixth derivative.

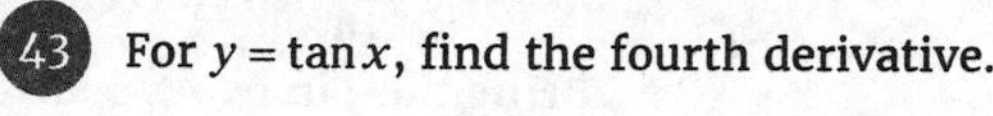

43 For $y = \tan x$, find the fourth derivative.

Practice Questions Answers and Explanations

(1) $f'(x) = 0$

The derivative of any constant is zero.

(2) $g'(x) = 0$

Because π is just a number, π^3 is also just a number. Therefore, $g(x) = \pi^3$ is a horizontal line with a slope and a derivative of zero.

(3) $g'(x) = 0$

If you feel bored because the first few problems were so easy, just enjoy it; it won't last.

(4) $f'(x) = 20x^3$

Bring the 4 in front and multiply it by the 5; at the same time reduce the power by 1, from 4 to 3.

(5) $g'(x) = -\frac{3}{10}x^2$ or $-\frac{3x^2}{10}$

You can just write down the derivative without showing any work (bring the 3 in front of the x, reduce the power 3 to a 2, and the 10 sits there doing nothing): $g'(x) = \frac{-3x^2}{10}$.

But if you want to do it more methodically, it works like this:

1. **Rewrite $\frac{-x^3}{10}$ so you can see an ordinary coefficient: $-\frac{1}{10}x^3$.**
2. **Bring the 3 in front, multiply, and reduce the power by 1.**

   $g'(x) = -\frac{3}{10}x^2$. This is the same, of course, as $-\frac{3x^2}{10}$.

(6) $y' = -\frac{5}{2}x^{-7/2}$

Rewrite with an exponent $\left(\sqrt{x^{-5}} = x^{-5/2}\right)$ and finish like Problem 5: Bring the power in front and reduce the power by one: $-\frac{5}{2}x^{-7/2}$.

To write your answer without a negative power, you write $y' = -\frac{5}{2x^{7/2}}$ or $y' = \frac{-5}{2x^{7/2}}$. Or you can write your answer without a fraction power, to wit: $y' = -\frac{5}{2\sqrt{x^7}}$ or $\frac{-5}{2\sqrt{x^7}}$ or $-\frac{5}{2\sqrt{x}^7}$ or $\frac{-5}{2\sqrt{x}^7}$. You say "po-tay-to," I say "po-tah-to."

(7) $s'(t) = 42t^5 + 1$

REMEMBER

Note that the derivative of plain old *t* or plain old *x* (or any other variable) is simply 1. This is one of the simplest of all derivative rules, yet for some reason, many people get it wrong.

8 **$y' = 6x^5 - 36x^2$**

FOIL and then take the derivative.

$$\begin{aligned} y &= (x^3 - 6)(x^3 - 6) \\ &= x^6 - 12x^3 + 36 \\ y' &= 6x^5 - 36x^2 \end{aligned}$$

9 **$3x^2 \cos x - x^3 \sin x$**

$$\begin{aligned} \frac{d}{dx}(x^3 \cos x) &= (x^3)'(\cos x) + (x^3)(\cos x)' \\ &= 3x^2 \cos x + x^3(-\sin x) \\ &= 3x^2 \cos x - x^3 \sin x \end{aligned}$$

10 **$\sin x + \sec x \tan x$**

$$\begin{aligned} \frac{d}{dx}(\sin x \tan x) &= (\sin x)' \tan x + \sin x (\tan x)' \\ &= \cos x \tan x + \sin x \sec^2 x \\ &= \sin x + \sec x \tan x \end{aligned}$$

11 **$5x^2(3\ln x + 1)$**

When doing this derivative, you can deal with the "5" in two ways. First, you can ignore it temporarily, do the differentiating, then multiply your answer by 5. (If you do it this way, don't forget that the "5" multiplies the entire derivative, not just the first term.) The second way is probably easier and better: Just make the "5" part of the first function. To wit:

$$\begin{aligned} \frac{d}{dx}(5x^3 \ln x) &= (5x^3)' \ln x + 5x^3 (\ln x)' \\ &= 15x^2 \ln x + 5x^3 \cdot \frac{1}{x} \\ &= 15x^2 \ln x + 5x^2 \quad \text{or} \quad 5x^2(3\ln x + 1) \end{aligned}$$

12 **$e^x \ln x(x^2 + 2x) + xe^x$**

This is a challenging problem because, as you've probably noticed, there are three functions in this product instead of two. But it's a piece o' cake. Just make it two functions: either $(x^2e^x)(\ln x)$ or $(x^2)(e^x \ln x)$. Take your pick.

1. **Rewrite this "triple function" as the product of two functions.**

$$= \frac{d}{dx}((x^2e^x)(\ln x))$$

2. **Apply the product rule.**

$$\frac{d}{dx}((x^2e^x)(\ln x)) = (x^2e^x)'(\ln x) + (x^2e^x)(\ln x)'$$

3. **Apply the product rule to the first chunk in the Step 2 answer: $(x^2e^x)'$.**

$$\begin{aligned} (x^2e^x)' &= (x^2)'e^x + x^2(e^x)' \\ &= 2xe^x + x^2e^x \end{aligned}$$

4. **Plug that answer where it belongs in the Step 2 equation, do the derivative of ln x, and then simplify.**

$$\left(x^2e^x\right)'(\ln x)+\left(x^2e^x\right)(\ln x)'$$
$$=\left(2xe^x+x^2e^x\right)(\ln x)+\left(x^2e^x\right)\cdot\frac{1}{x}$$
$$=2xe^x\ln x+x^2e^x\ln x+xe^x \quad \text{or}$$
$$=x^2e^x\ln x+2xe^x\ln x+xe^x \quad \text{or}$$
$$=xe^x(x\ln x+2\ln x+1) \quad \text{or}$$
$$=e^x\ln x\left(x^2+2x\right)+xe^x$$

You say "pa-jam-ahs," I say "pa-jah-mas."

(13) $\dfrac{\mathbf{3x^2\cos x+x^3\sin x}}{\mathbf{\cos^2 x}}$

$$\frac{d}{dx}\frac{x^3}{\cos x}=\frac{\left(x^3\right)'\cos x-x^3(\cos x)'}{(\cos x)^2}$$
$$=\frac{3x^2\cos x-x^3(-\sin x)}{\cos^2 x}$$
$$=\frac{3x^2\cos x+x^3\sin x}{\cos^2 x}$$

(14) $\dfrac{\mathbf{-\sin x-\cos x}}{\mathbf{e^x}}$

$$\frac{d}{dx}\frac{\cos x}{e^x}=\frac{(\cos x)'e^x-\cos x\left(e^x\right)'}{\left(e^x\right)^2}$$
$$=\frac{-e^x\sin x-e^x\cos x}{e^{2x}}$$
$$=\frac{-\sin x-\cos x}{e^x}$$

(15) $\dfrac{\mathbf{x\cos x\ln x-3\sin x\ln x-\sin x}}{\mathbf{x^4\ln^2 x}}$

$$\frac{d}{dx}\frac{\sin x}{x^3\ln x}=\frac{(\sin x)'x^3\ln x-\sin x\left(x^3\ln x\right)'}{\left(x^3\ln x\right)^2}$$
$$=\frac{\cos x\cdot x^3\ln x-\sin x\left(\left(x^3\right)'\ln x+x^3(\ln x)'\right)}{x^6\ln^2 x}$$
$$=\frac{x^3\cos x\ln x-\sin x\left(3x^2\ln x+x^3\left(\frac{1}{x}\right)\right)}{x^6\ln^2 x}$$
$$=\frac{x^3\cos x\ln x-3x^2\sin x\ln x-x^2\sin x}{x^6\ln^2 x}$$
$$=\frac{x\cos x\ln x-3\sin x\ln x-\sin x}{x^4\ln^2 x}$$

16 $f'(x) = -4x^3 \sin(x^4)$

Because the argument of the cosine function is something other than a plain old x, this is a chain rule problem.

1. **Temporarily think of the argument, x^4, as a *glob*.**

   So, you've got $f(x) = \cos(glob)$.

2. **Use the regular derivative rule.**

   $f(x) = \cos(glob)$, so

   $f'(x) = -\sin(glob)$

   (This is only a provisional answer, so the "=" sign is false — egad! The math police are going to pull me over.)

3. **Multiply this by the derivative of the argument.**

   $f'(x) = -\sin(glob) \cdot glob'$

4. **Get rid of the *glob*.**

   The *glob* equals x^4, so *glob'* equals $4x^3$.

   $$\begin{aligned} f'(x) &= -\sin(x^4) \cdot 4x^3 \\ &= -4x^3 \sin(x^4) \end{aligned}$$

17 $g'(x) = 3\sin^2 x \cos x$

Rewrite $\sin^3 x$ as $(\sin x)^3$ so that it's clear that the outermost function is the cubing function. By the chain rule, the derivative of $stuff^3$ is $3stuff^2 \cdot stuff'$. The stuff here is $\sin x$ and thus $stuff'$ is $\cos x$. So your final answer is $3(\sin x)^2 \cdot \cos x$, or $3\sin^2 x \cos x$.

18 $s'(t) = \sec^2(\ln t) \cdot \frac{1}{t}$

The derivative of $\tan x$ is $\sec^2 x$, so the derivative of $\tan(lump)$ is $\sec^2(lump) \cdot lump'$. You better know by now that the derivative of $\ln t$ is $\frac{1}{t}$, so your final result is $\sec^2(\ln t) \cdot \frac{1}{t}$.

19 $y' = 12x^2 e^{4x^3}$

The derivative of e^x is e^x, so by the chain rule, the derivative of e^{glob} is $e^{glob} \cdot glob'$. So, $y' = e^{4x^3} \cdot 12x^2$ or $12x^2 e^{4x^3}$.

20 $f'(x) = 4x^3 \sin^3 x + 3x^4 \sin^2 x \cos x$

$$f(x) = x^4 \sin^3 x$$

$$f'(x) = (x^4)' \sin^3 x + x^4 (\sin^3 x)'$$

Use the chain rule to solve $\left(\sin^3 x\right)'$, then go back and finish the problem. $\sin^3 x$ means $(\sin x)^3$ and that's *stuff*^{3}. The derivative of *stuff*^{3} is 3*stuff*$^2 \cdot$ *stuff*$'$, so the derivative of $(\sin x)^3$ is $3(\sin x)^2 \cdot \cos x$. Plug that in where it belongs, and then finish.

$$\begin{aligned} f'(x) &= \left(x^4\right)' \sin^3 x + x^4 \cdot 3(\sin x)^2 \cos x \\ &= 4x^3 \sin^3 x + 3x^4 \sin^2 x \cos x \end{aligned}$$

21 $\boldsymbol{g'(x) = \dfrac{2\ln x}{x(5x-4)} - \dfrac{5\ln^2 x}{(5x-4)^2}}$

Here you have the chain rule inside the quotient rule. Start with the quotient rule:

$$g'(x) = \frac{\left((\ln x)^2\right)'(5x-4) - (\ln x)^2(5x-4)'}{(5x-4)^2}$$

Next, take care of the chain rule solution for $\left((\ln x)^2\right)'$. You want the derivative of *glob*^{2} — that's 2*glob* $\cdot$ *glob*$'$. So, the derivative of $(\ln x)^2$ is $2(\ln x)\left(\frac{1}{x}\right)$. Now you can finish:

$$\begin{aligned} g'(x) &= \frac{2(\ln x)\left(\frac{1}{x}\right)(5x-4) - (\ln x)^2(5x-4)'}{(5x-4)^2} \\ &= \frac{(10x-8)(\ln x)\left(\frac{1}{x}\right) - 5(\ln x)^2}{(5x-4)^2} \\ &= \frac{(10x-8)\ln x - 5x(\ln x)^2}{x(5x-4)^2} \\ &= \frac{2\ln x}{x(5x-4)} - \frac{5(\ln x)^2}{(5x-4)^2} \end{aligned}$$

22 $\boldsymbol{y' = -24x\cos^2\left(4x^2\right)\sin\left(4x^2\right)}$

Triply nested!

$$y = \left(\cos\left(4x^2\right)\right)^3$$

The derivative of *stuff*^{3} is 3*stuff*$^2 \cdot$ *stuff*$'$, so you have

$$y' = 3\left(\cos\left(4x^2\right)\right)^2 \cdot \left(\cos\left(4x^2\right)\right)'$$

Now you do the derivative of $\cos($*glob*$)$, which is $-\sin($*glob*$) \cdot$ *glob*$'$. Two down, one to go:

$$\begin{aligned} y' &= 3\left(\cos\left(4x^2\right)\right)^2\left(-\sin\left(4x^2\right)\right) \cdot \left(4x^2\right)' \\ &= 3\cos^2\left(4x^2\right)\left(-\sin\left(4x^2\right)\right) \cdot 8x \\ &= -24x\cos^2\left(4x^2\right)\sin\left(4x^2\right) \end{aligned}$$

(23) $\frac{d}{dx}\tan^3\left(e^{x^2}\right) = 6xe^{x^2}\tan^2\left(e^{x^2}\right)\sec^2\left(e^{x^2}\right)$

Holy quadrupely nested quadruple nestedness, Batman! This is one for the Riddler.

$$\begin{aligned}
&\frac{d}{dx}\tan^3\left(e^{x^2}\right)\\
&=\frac{d}{dx}\left(\tan\left(e^{x^2}\right)\right)^3\\
&=3\left(\tan\left(e^{x^2}\right)\right)^2\cdot\left(\tan\left(e^{x^2}\right)\right)' \qquad \left(\text{because } \frac{d}{dx}\mathit{stuff}^3 = 3\mathit{stuff}^2\cdot\mathit{stuff}'\right)\\
&=3\tan^2\left(e^{x^2}\right)\sec^2\left(e^{x^2}\right)\cdot\left(e^{x^2}\right)' \qquad \left(\text{because } \frac{d}{dx}\tan(\mathit{glob}) = \sec^2(\mathit{glob})\cdot\mathit{glob}'\right)\\
&=3\tan^2\left(e^{x^2}\right)\sec^2\left(e^{x^2}\right)\cdot e^{x^2}\cdot\left(x^2\right)' \qquad \left(\text{because } \frac{d}{dx}e^{\mathit{lump}} = e^{\mathit{lump}}\cdot\mathit{lump}'\right)\\
&=3\tan^2\left(e^{x^2}\right)\sec^2\left(e^{x^2}\right)\cdot e^{x^2}\cdot 2x\\
&=6xe^{x^2}\tan^2\left(e^{x^2}\right)\sec^2\left(e^{x^2}\right)
\end{aligned}$$

POW!!
CRUNCH!
BAM!!!

(24) $p' = -\frac{\sin x}{2\sqrt{\cos x}}$

First, rewrite the original function with a power: $p = \sqrt{\cos x} = (\cos x)^{1/2}$.

This works like $\mathit{stuff}^{1/2}$, so you use the power rule and then finish, as with all chain rule problems, by multiplying by *stuff′*.

$$\begin{aligned}
&p = \mathit{stuff}^{1/2}\\
&p' = \frac{1}{2}\mathit{stuff}^{-1/2}\cdot\mathit{stuff}'
\end{aligned}$$

The stuff is $\cos x$, and the derivative of the stuff is thus $-\sin x$. Just plug those in for your final answer and then simplify:

$$\begin{aligned}
p' &= \frac{1}{2}(\cos x)^{-1/2}\cdot(-\sin x)\\
&= -\frac{\sin x}{2\cos^{1/2}x} \quad \text{or} \quad -\frac{\sin x}{2\sqrt{\cos x}}
\end{aligned}$$

(25) $\frac{dp}{dq} = -\frac{1}{q(\ln q)^2}$

You could use the quotient rule for this problem, but you were asked to use the chain rule. To do that, rewrite the original function as a power: $p = (\ln q)^{-1}$.

This works like $stuff^{-1}$, so you use the power rule and then finish by multiplying by *stuff'*:

$$p = stuff^{-1}$$
$$\frac{dp}{dq} = -1 stuff^{-2} \cdot stuff'$$

The stuff is $\ln q$, and, thus, $stuff' = \frac{1}{q}$. Plug those in and you're done:

$$\frac{dp}{dq} = -1(\ln q)^{-2} \cdot \frac{1}{q}$$
$$= -\frac{1}{q(\ln q)^2}$$

(26) **On $f(x) = \ln\frac{1}{x}$ at $(5, -\ln 5)$, $f'(x) = -\frac{1}{5}$**

This function can be modeled by $\ln(blob)$, so you use the natural log rule and then finish by multiplying by *blob'*.

$$f(x) = \ln(blob)$$
$$f'(x) = \frac{1}{blob} \cdot blob'$$

The blob is $\frac{1}{x}$, or x^{-1}, so $blob' = -x^{-2}$. Now just plug in and simplify:

$$f'(x) = \frac{1}{\frac{1}{x}} \cdot (-x^{-2}) = -\frac{1}{x}$$

Thus, $f'(5) = -\frac{1}{5}$.

(27) $y' = 1 - \sin(1 - x)$

The derivative of $\cos(stuff)$ is $-\sin(stuff) \cdot stuff'$, so you have

$$y = x - \cos(1 - x)$$
$$y' = 1 - (-\sin(1 - x)(-1))$$
$$= 1 - \sin(1 - x)$$

This equals $1 + \sin(x - 1)$, by the way, which is just slightly easier on the eyes. Do you see why they're equivalent?

(28) $y' = \frac{2x + 1}{3y^2 - 1}$

1. **Take the derivative of all four terms, using the chain rule (sort of) for all terms containing a *y*.**

   $$3y^2y' - 2x = 1 + y'$$

2. **Move all terms containing y' to the left, move all other terms to the right, and factor out y'.**

$$3y^2y' - y' = 1 + 2x$$
$$y'(3y^2 - 1) = 1 + 2x$$

3. **Divide and voilà!**

$$y' = \frac{2x+1}{3y^2-1}$$

(29) $\mathbf{y' = \dfrac{4ye^x}{3y+1}}$

Follow the steps for Problem 28.

$$3y' + \frac{1}{y}y' = 4e^x$$
$$y'\left(3 + \frac{1}{y}\right) = 4e^x$$
$$y' = \frac{4e^x}{3+\frac{1}{y}} = \frac{4ye^x}{3y+1}$$

(30) $\mathbf{y' = \dfrac{y^3 - 2xy + 1}{-3y^2x + x^2 - 5}}$

This time, you have two products to deal with, so use the product rule for the two products and the regular rules for the other two terms.

$$(x^2)'y + x^2y' = (y^3)'x + y^3x' + 5y' + 1$$
$$2xy + x^2y' = 3y^2y'x + y^3 + 5y' + 1$$
$$x^2y' - 3y^2y'x - 5y' = y^3 + 1 - 2xy$$
$$y'(x^2 - 3y^2x - 5) = y^3 + 1 - 2xy$$
$$y' = \frac{y^3 - 2xy + 1}{-3y^2x + x^2 - 5}$$

(31) **The slope is zero.**

You need a slope, so you need the derivative.

$$\underbrace{y'}_{\substack{\text{Implicit}\\ \text{Differentiation}}} + \underbrace{2\cos(y^3)\cdot(-\sin(y^3))(y^3)'}_{\substack{\text{Chain Rule}\\ \text{(twice nested)}}} = \underbrace{\cos(5x^2)(10x)}_{\text{Chain Rule}}$$

$$y' + 2\cos(y^3)\cdot(-\sin(y^3))(3y^2y') = 10x\cos(5x^2)$$
$$y'(1 - 6y^2\cos(y^3)\sin(y^3)) = 10x\cos(5x^2)$$
$$y' = \frac{10x\cos(5x^2)}{1 - 6y^2\cos(y^3)\sin(y^3)}$$

You need the slope at $x=\sqrt{\frac{\pi}{10}}$, $y=0$, so plug those numbers into the derivative. Actually, you can save yourself some work if you notice that the numerator will equal zero (because $\cos\left(5\sqrt{\frac{\pi}{10}}^{2}\right)=0$) and the denominator will equal 1 (because $y=0$). Thus, the slope of the curve at this point is zero. (A tangent line with a zero slope is horizontal, and because this tangent line touches the curve where $y=0$, the tangent line is the x-axis.)

(32) $\boldsymbol{\frac{dy}{dx}=\frac{10x}{\sec^2 y-8}}$

$$\begin{aligned}8y+5x^2&=\tan y\\8y'+10x&=\sec^2 y\cdot y'\\8y'-\sec^2 y\cdot y'&=-10x\\y'\left(8-\sec^2 y\right)&=-10x\\y'&=\frac{-10x}{8-\sec^2 y}=\frac{10x}{\sec^2 y-8}\end{aligned}$$

(33) **The slope is –2.**

For the slope of the tangent line, you need the derivative, of course, so take the derivative with implicit differentiation:

$$\begin{aligned}x^2+y^2&=5\\2x+2yy'&=0\\2yy'&=-2x\\y'&=\frac{-2x}{2y}=-\frac{x}{y}\end{aligned}$$

To finish, just plug the x- and y-coordinates of the point into this derivative:

$$y'_{(2,\,1)}=-\frac{2}{1}=-2$$

That's a wrap. By the way, if you know that $x^2+y^2=5$ is the equation of a circle with center at $(0, 0)$, you can solve this problem with high school geometry. Do you see how?

(34) $\boldsymbol{\frac{dy}{dx}=\frac{3x^2-5}{12y^3+3y^{-4}}}$

$$\begin{aligned}3y^4+5x&=x^3+y^{-3}\\12y^3y'+5&=3x^2-3y^{-4}y'\\12y^3y'+3y^{-4}y'&=3x^2-5\\y'\left(12y^3+3y^{-4}\right)&=3x^2-5\\y'&=\frac{3x^2-5}{12y^3+3y^{-4}}\end{aligned}$$

(35) **The slope is $-\frac{4}{3}$.**

When you see "normal line," think "tangent line," and when you see "tangent line" and/or "slope," think "derivative"!

So, get the derivative with implicit differentiation:

$$3x^2 + y^2 = 19$$
$$6x + 2yy' = 0$$
$$2yy' = -6x$$
$$y' = \frac{-6x}{2y} = -\frac{3x}{y}$$

Plug in the point to get the slope of the tangent line:

$$y'_{(1,\,-4)} = -\frac{3 \cdot 1}{-4} = \frac{3}{4}$$

Finally, the slope of the normal line is the opposite reciprocal of that, namely, $-\frac{4}{3}$.

(36) $\mathbf{\frac{1}{2}}$

$$\begin{aligned} \arctan' 1 &= \frac{1}{\tan'(\arctan 1)} \\ &= \frac{1}{\tan' \frac{\pi}{4}} \\ &= \frac{1}{\sec^2 \frac{\pi}{4}} \\ &= \cos^2 \frac{\pi}{4} \\ &= \left(\frac{\sqrt{2}}{2}\right)^2 = \frac{1}{2} \end{aligned}$$

Confirm this answer with the derivative of arctan:

$\arctan' x = \frac{1}{1+x^2}$, so, $\arctan' 1 = \frac{1}{1+1^2} = \frac{1}{2}$. It checks.

(37) $\boldsymbol{y = \frac{1}{4}x + \frac{3}{4}}$

You're given that f and g are inverses and that $f(2) = 5$, so $g(5) = 2$. Thus, the problem is asking for the equation of the line tangent to g at the point $(5, 2)$. All you need, then, is the slope of g at that point, namely, $g'(5)$, and you'll be able to write the equation of the tangent line using the point-slope form of the equation of a line: $y - y_1 = m(x - x_1)$.

The line tangent to f at $(2, 5)$ has a slope of 4, so that tells you that $f'(2) = 4$. Now you can use the formula for the derivative of an inverse function to get $g'(5)$:

$$\begin{aligned} g'(5) &= \frac{1}{f'(g(5))} \\ &= \frac{1}{f'(2)} \\ &= \frac{1}{4} \end{aligned}$$

Finally, plug everything into point-slope form:

$$y-2=\frac{1}{4}(x-5)$$
$$y=\frac{1}{4}x+\frac{3}{4}$$

38 **The answer is zero.**

$$y'=5x^4$$
$$y''=20x^3$$
$$y'''=60x^2$$
$$y^{(4)}=180x$$
$$y^{(5)}=180$$
$$y^{(6)}=0$$
$$y^{(7)}=0$$

And so on.

39 $$\boldsymbol{y=x^3-x^{-3}}$$
$$\boldsymbol{y'=3x^2+3x^{-4}}$$
$$\boldsymbol{y''=6x-12x^{-5}}$$
$$\boldsymbol{y'''=6+60x^{-6}}$$
$$\boldsymbol{y^{(4)}=-360x^{-7}}$$
$$\boldsymbol{y^{(5)}=2520x^{-8}}$$

40 $$\boldsymbol{y'=\cos x-\sin x}$$
$$\boldsymbol{y''=-\sin x-\cos x}$$
$$\boldsymbol{y'''=-\cos x+\sin x}$$
$$\boldsymbol{y^{(4)}=\sin x+\cos x}$$
$$\boldsymbol{y^{(5)}=\cos x-\sin x}$$
$$\boldsymbol{y^{(6)}=-\sin x-\cos x}$$

Notice that the fourth derivative equals the original function, the fifth derivative equals the first, and so on. This cycle of four functions repeats ad infinitum.

41 $$\boldsymbol{y'=-2x\sin\left(x^2\right)}$$
$$\boldsymbol{y''=-2\sin\left(x^2\right)-4x^2\cos\left(x^2\right)}$$
$$\boldsymbol{y'''=8x^3\sin\left(x^2\right)-12x\cos\left(x^2\right)}$$

$$y=\cos\left(x^2\right)$$
$$y'=-2x\sin\left(x^2\right) \quad \text{(chain rule)}$$
$$y''=(-2x)'\sin\left(x^2\right)+(-2x)\left(\sin\left(x^2\right)\right)' \quad \text{(product rule)}$$
$$=-2\sin\left(x^2\right)-2x\cos\left(x^2\right)2x \quad \text{(chain rule)}$$
$$=-2\sin\left(x^2\right)-4x^2\cos\left(x^2\right)$$

$$\begin{aligned} y''' &= -2\cos(x^2)2x - \left((4x^2)' \cos(x^2) + 4x^2 (\cos(x^2))' \right) \\ &= -4x\cos(x^2) - (8x\cos(x^2) + 4x^2(-\sin(x^2)2x)) \\ &= -4x\cos(x^2) - 8x\cos(x^2) + 8x^3\sin(x^2) \\ &= 8x^3\sin(x^2) - 12x\cos(x^2) \end{aligned}$$

(42) $(\ln 8)^3 8^x$

$$y = \frac{8^x}{(\ln 8)^3} = \frac{1}{(\ln 8)^3} 8^x$$

I've rewritten the function this way simply to emphasize that while the $\frac{1}{(\ln 8)^3}$ may look a bit advanced, it's just a number and just a coefficient. As such, it just sits there and has no effect on how you differentiate. The derivative of 8^x is $8^x \ln 8$, so

$$\begin{aligned} y &= \frac{1}{(\ln 8)^3} 8^x \\ y' &= \frac{1}{(\ln 8)^3} 8^x \ln 8 = \frac{1}{(\ln 8)^2} 8^x \\ y'' &= \frac{1}{(\ln 8)^2} 8^x \ln 8 = \frac{1}{\ln 8} 8^x \\ y''' &= \frac{1}{\ln 8} 8^x \ln 8 = 8^x \\ y^{(4)} &= 8^x \ln 8 = (\ln 8) 8^x \\ y^{(5)} &= (\ln 8) 8^x \ln 8 = (\ln 8)^2 8^x \\ y^{(6)} &= (\ln 8)^2 8^x \ln 8 = (\ln 8)^3 8^x \end{aligned}$$

Is that a thing of beauty or what? (To best see the pattern of this series of derivatives, look at the far right of each line.)

(43) $8\sec^2 x \tan^3 x + 16\sec^4 x \tan x$

$y = \tan x$

The first derivative is a memorized rule:

$y' = \sec^2 x$

For the second derivative, you use the chain rule:

$y'' = 2\sec x \cdot \sec x \tan x = 2\sec^2 x \tan x$

The third derivative is a product rule problem where you use the chain rule for one of the product rule derivatives:

$$y'' = 2\sec^2 x \tan x$$
$$y''' = 2\left(2\sec x \cdot \sec x \tan x \cdot \tan x + \sec^2 x \cdot \sec^2 x\right)$$
$$= 4\sec^2 x \tan^2 x + 2\sec^4 x$$

Finally, for the fourth derivative, you have a product rule piece with two chain rules inside of it plus another chain rule piece!

$$y''' = 4\sec^2 x \tan^2 x + 2\sec^4 x$$
$$y^{(4)} = 4\left(2\sec x \cdot \sec x \tan x \cdot \tan^2 x + \sec^2 x \cdot 2\tan x \cdot \sec^2 x\right) + 8\sec^3 x \cdot \sec x \tan x$$
$$= 8\sec^2 x \tan^3 x + 8\sec^4 x \tan x + 8\sec^4 x \tan x$$
$$= 8\sec^2 x \tan^3 x + 16\sec^4 x \tan x$$

Wasn't that fun?

If you're ready to test your skills a bit more, take the following chapter quiz that incorporates all the chapter topics.

Whaddya Know? Chapter 10 Quiz

Quiz time! Complete each problem to test your knowledge on the various topics covered in this chapter. You can then find the solutions and explanations in the next section.

For problems 1 through 15, find $\frac{dy}{dx}$.

1. $y = \ln(\sin x)$
2. $y = \ln\left(\sin\frac{\pi}{4}\right)$
3. $y = \cos(\ln x)$
4. $y = \cos x \sin x$
5. $y = \cos x \tan x$
6. $y = \tan^2 x - \sec^2 x$
7. $y = \tan^2(\tan x)$
8. $y = \frac{\sin^3 x}{1 - \sin x}$
9. $y = \pi x^\pi$
10. $xe^y = ye^x$
11. $y = k \cdot e \cdot x$ (k is a constant)
12. $y = \frac{\ln x}{e^x}$
13. $y = \ln(e^x)$
14. $x = \sin^2 y - \cos^2 y$
15. $y = e^{x \cos x}$
16. Find the second derivative of $y = \frac{x}{\ln x}$.
17. Find the third derivative of $y = \sin^2 x$.

Answers to Chapter 10 Quiz

1. $\frac{dy}{dx} = \cot x$

   This is a chain rule problem, because the input of the function is something other than a plain old x. You begin with the *ln* rule (so the derivative of $\ln(\textit{stuff})$ begins with $\frac{1}{\textit{stuff}}$), then you finish (as with all chain rule problems) by multiplying that by the derivative of the *stuff*. The *stuff* here is $\sin x$, and its derivative is $\cos x$. So, that gives you $\frac{1}{\sin x} \cdot \cos x$, which simplifies to $\cot x$. (Note that this function, $y = \ln(\sin x)$, is defined only for angles in the first and second quadrants. Its derivative has the same restrictions. For the rest of the problems, I will skip mentioning any restrictions to the domain of the function or its derivative.)

2. $\frac{dy}{dx} = 0$

   $\sin\frac{\pi}{4} = \frac{\sqrt{2}}{2}$ and $\ln\left(\frac{\sqrt{2}}{2}\right)$ is just a number. So this function is a horizontal line with a slope and derivative of zero.

3. $\frac{dy}{dx} = -\frac{\sin(\ln x)}{x}$

   Another chain rule problem. Use the cosine rule, then multiply by the derivative of the natural log.

   $$y = \cos(\ln x)$$
   $$\frac{dy}{dx} = -\sin(\ln x) \cdot \frac{1}{x} = -\frac{\sin(\ln x)}{x}$$

4. $\frac{dy}{dx} = -\sin^2 x + \cos^2 x$

   This is a straightforward product rule problem.

   $$\begin{aligned}\frac{dy}{dx} &= (\cos x)'(\sin x) + (\cos x)(\sin x)' \\ &= (-\sin x)(\sin x) + (\cos x)(\cos x) \\ &= -\sin^2 x + \cos^2 x\end{aligned}$$

5. $\frac{dy}{dx} = \cos x$

   Did you use the product rule for this one like with Problem 4? Tricked you! $(\cos x)(\tan x) = \sin x$, so the derivative is simply $\cos x$. *Don't forget to always be on the lookout for ways to simplify a problem using pre-algebra, algebra I and II, geometry, and trig.* It's easy to forget this when your mind is in calculus mode.

6. $\frac{dy}{dx} = 0$

   Did you use the derivative rules for tangent and secant for this problem? Tricked you again! One of the trig Pythagorean Identities is $1 + \tan^2 x = \sec^2 x$. You can use this to simplify the given function: $y = \tan^2 x - \sec^2 x = \tan^2 x - (1 + \tan^2 x) = -1$. This, of course, is a horizontal line, so its derivative is zero.

7) $\frac{dy}{dx} = \mathbf{2\tan(\tan x)\cdot\sec^2(\tan x)\cdot\sec^2 x}$

This is a twice-nested chain rule problem. Begin by rewriting the given function so that the trig power is on the outside (always a good idea): $y=(\tan(\tan x))^2$. You tackle this nested chain rule problem from outside to inside. So, you first use the power rule (for the power of 2), then you use the tangent rule (the derivative of tan(*stuff*) begins with $\sec^2$(*stuff*)). Then, as always, you finish by multiplying by the derivative of the *stuff*.

$$y=(\tan(\tan x))^2$$
$$\frac{dy}{dx}=2(\tan(\tan x))^1\cdot(\tan(\tan x))'$$
$$=2\tan(\tan x)\cdot\sec^2(\tan x)\cdot\sec^2 x$$

8) $\frac{dy}{dx} = \mathbf{\frac{\sin^2 x\cdot\cos x(3-2\sin x)}{(1-\sin x)^2}}$

This is a quotient rule problem where you need to use the chain rule when doing the derivative of the numerator.

$$y=\frac{\sin^3 x}{1-\sin x}$$
$$\frac{dy}{dx}=\frac{\left(\sin^3 x\right)'(1-\sin x)-\left(\sin^3 x\right)(1-\sin x)'}{(1-\sin x)^2}$$
$$=\frac{\left(3\sin^2 x\cdot\cos x\right)(1-\sin x)-\left(\sin^3 x\right)(-\cos x)}{(1-\sin x)^2}$$

That's the calculus. The rest of the solution is just simplification. One way to write your final answer is given above in bold.

9) $\frac{dy}{dx} = \boldsymbol{\pi^2 x^{\pi-1}}$

Don't be thrown by the pi symbol. Pi is just a number, so doing the derivative of $y=\pi x^\pi$ works exactly like doing the derivative of something like $y=5x^5$. That derivative is 5^2x^{5-1} (or $25x^4$).

10) $\frac{dy}{dx} = \mathbf{\frac{y\cdot e^x - e^y}{x\cdot e^y - e^x}}$ **or** $\mathbf{\frac{y\cdot e^x - e^y}{y\cdot e^x - e^x}}$

You need to use implicit differentiation for this one. You begin by using the product rule.

$$xe^y = ye^x$$
$$(x)'\left(e^y\right)+(x)\left(e^y\right)'=(y)'\left(e^x\right)+(y)\left(e^x\right)'$$

Now, do the individual derivatives, remembering that you treat a *y* just like you treat the *stuff* in a chain rule problem.

$$1\cdot e^y + x\cdot e^y\cdot y' = y'\cdot e^x + y\cdot e^x$$

Finally, collect all terms containing a y' on the left, factor out the y', and then divide.

$$x \cdot e^y \cdot y' - y' \cdot e^x = y \cdot e^x - e^y$$
$$y'(x \cdot e^y - e^x) = y \cdot e^x - e^y$$
$$y' = \frac{y \cdot e^x - e^y}{x \cdot e^y - e^x} \quad \text{or} \quad \frac{y \cdot e^x - e^y}{y \cdot e^x - e^x}$$

That's two ways to write your final answer. There are others.

11 $\frac{dy}{dx} = ke$

Don't forget that a constant behaves exactly like an ordinary number, and $e \approx 2.72$ is just a number. So, the current problem works exactly like finding the derivative of something like $y = 3\pi x$ whose derivative is the coefficient 3π. The current problem is no different.

12 $\frac{dy}{dx} = \frac{1 - x \ln x}{xe^x}$

This is a straightforward quotient rule problem.

$$\frac{dy}{dx} = \frac{(\ln x)'(e^x) - (\ln x)(e^x)'}{(e^x)^2}$$
$$= \frac{\frac{1}{x}e^x - \ln x \cdot e^x}{e^{2x}}$$
$$= \frac{e^x - x \ln x \cdot e^x}{xe^{2x}}$$
$$= \frac{1 - x \ln x}{xe^x}$$

13 $\frac{dy}{dx} = 1$

This is another problem where you've got to be on your toes looking for a possible simplification before diving into the calculus. $\ln(e^x)$ equals x. You knew that, right? The derivative of x is 1.

14 $\frac{dy}{dx} = \frac{1}{4 \sin y \cdot \cos y}$ or $\frac{1}{2 \sin(2y)}$

Be careful here. Had you been asked to find x' or $\frac{dx}{dy}$, you could have done the derivative the ordinary way. But to find $\frac{dy}{dx}$, you need to use implicit differentiation.

$$x = \sin^2 y - \cos^2 y$$
$$1 = 2 \sin y \cdot \cos y \cdot y' - 2 \cos y(-\sin y) y'$$
$$1 = y'(2 \sin y \cdot \cos y + 2 \sin y \cdot \cos y)$$
$$1 = y'(4 \sin y \cdot \cos y)$$
$$y' = \frac{1}{4 \sin y \cdot \cos y} \quad \text{or} \quad \frac{1}{2 \sin(2y)}$$

15 $\frac{dy}{dx} = e^{x\cos x}(\cos x - x\sin x)$

This is a chain rule problem (because the input of the exponential function is not just an x). The basic idea here is that the derivative of e^{stuff} is $e^{stuff} \cdot stuff'$. The *stuff* is $x\cos x$, which is a product, so to find *stuff'*, you need to use, naturally, the product rule:

$$\begin{aligned}(x\cos x)' &= x'\cos x + x(\cos x)' \\ &= \cos x - x\sin x\end{aligned}$$

Plug everything in where it belongs, and you're done:

$$\begin{aligned}\frac{dy}{dx} &= e^{stuff} \cdot stuff' \\ &= e^{x\cos x}(\cos x - x\sin x)\end{aligned}$$

16 $y'' = \frac{2-\ln x}{x\ln^3 x}$

$$\begin{aligned}y &= \frac{x}{\ln x} \\ y' &= \frac{x'\ln x - x(\ln x)'}{(\ln x)^2} \\ &= \frac{\ln x - 1}{\ln^2 x} \\ y'' &= \frac{(\ln x - 1)'\ln^2 x - (\ln x - 1)\left(\ln^2 x\right)'}{\left(\ln^2 x\right)^2} \\ &= \frac{\frac{1}{x}\ln^2 x - (\ln x - 1)(2\ln x)\frac{1}{x}}{\ln^4 x} \\ &= \frac{\ln^2 x - (\ln x - 1)(2\ln x)}{x\ln^4 x} \\ &= \frac{\ln x - 2(\ln x - 1)}{x\ln^3 x} \\ &= \frac{2 - \ln x}{x\ln^3 x}\end{aligned}$$

17 $y''' = -4\sin(2x)$ **or** $-8\sin x\cos x$

The first derivative is a chain rule problem:

$$\begin{aligned}y &= \sin^2 x \\ y' &= 2\sin x\cos x\end{aligned}$$

From here, you could get the second derivative with the product rule, and then go on and get the third derivative. But there's a shortcut: $2\sin x\cos x = \sin(2x)$, right? Thus,

$$y' = 2\sin x \cdot \cos x$$
$$= \sin(2x)$$

Now the second and third derivatives are a snap. They're both very simple chain rule derivatives.

$$y'' = 2\cos(2x)$$
$$y''' = -4\sin(2x) \quad \text{or} \quad -8\sin x\cos x$$

IN THIS CHAPTER

» Weathering the ups and downs of moody functions

» Locating extrema

» Using the first and second derivative tests

» Interpreting concavity and points of inflection

» Comparing the graphs of functions and derivatives

» Muzzling the Mean Value Theorem — *GRRRRR*

Chapter 11 Differentiation and the Shape of Curves

If you've read Chapters 9 and 10, you're probably an expert at finding derivatives. This is a good thing, because in this chapter you use derivatives to understand the shape of functions — where they rise and where they fall, where they max out and bottom out, how they curve, and so on. Then in Chapter 12, you use your knowledge about the shape of functions to solve real-world problems.

Taking a Calculus Road Trip

Consider the graph of $f(x)$ in Figure 11-1.

Imagine that you're driving along this function from left to right. Along your drive, there are several points of interest between a and l. All of them, except for the start and finish points, relate to the steepness of the road — in other words, its slope or derivative.

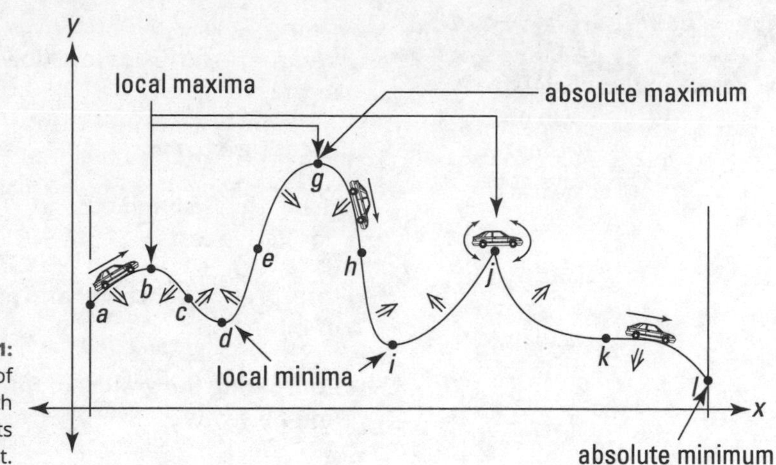

FIGURE 11-1: The graph of $f(x)$ with several points of interest.

Now, prepare yourself — I'm going to throw lots of new terms and definitions at you all at once here. You shouldn't, however, have much trouble with these ideas because they mostly involve commonsense notions like driving up or down an incline, or going over the crest of a hill.

Climb every mountain, ford every stream: Positive and negative slopes

First, notice that as you begin your trip at *a*, you're climbing up. Thus, the function is *increasing* and its slope and derivative are therefore *positive*. You climb the hill till you reach the top at *b* where the road levels out. The road is level there, so the slope and derivative equal *zero*.

Because the derivative is zero at *b*, point *b* is called a *stationary point* of the function. Point *b* is also a *local maximum* or *relative maximum* of *f* because it's the top of a hill. To be a local max, *b* just has to be the highest point in its immediate neighborhood. It doesn't matter that the nearby hill at *g* is even higher.

After reaching the crest of the hill at *b*, you start going down — duh. So, after *b*, the slope and derivative are *negative* and the function is *decreasing*. To the left of every local max, the slope is positive; to the right of a max, the slope is negative.

I can't think of a travel metaphor for this section: Concavity and inflection points

The next point of interest is *c*. Can you see that as you go down from *b* to *c*, the road gets steeper and steeper, but that after *c*, although you're still going down, the road is gradually starting to curve up again and get less steep? The little down arrow between *b* and *c* in Figure 11-1 indicates that this section of the road is curving down — the function is said to be *concave down* there. As you can see, the road is also concave down between *a* and *b*.

TIP

Concavity poetry: *Down* looks like a *frown*, *up* looks like a *cup*. A portion of a function that's concave *down* looks like a *frown*. Where it's concave *up*, like between *c* and *e*, it looks like a *cup*.

Wherever a function is concave *down*, its derivative (and slope) are *decreasing;* wherever a function is concave *up*, its derivative (and slope) are *increasing*.

Okay, so the road is concave down until *c* where it switches to concave up. Because the concavity switches at *c*, it's a *point of inflection*. The point *c* is also the steepest point on this stretch of the road. Inflection points are always at the steepest — or least steep — points in their immediate neighborhoods.

WARNING

Be careful with function sections that have a negative slope. Point *c* is the steepest point in its neighborhood because it has a bigger negative slope than any other nearby point. But remember, a big negative number is actually a *small* number, so the slope and derivative at *c* are actually the *smallest* of all the points in the neighborhood. From *b* to *c*, the derivative of the function is *decreasing* (because it's becoming a bigger negative). From *c* to *d*, the derivative is *increasing* (because it's becoming a smaller negative). Got it?

This vale of tears: A local minimum

Let's get back to your drive. After point *c*, you keep going down till you reach *d*, the bottom of a valley. Point *d* is another stationary point because the road is level there and the derivative is zero. Point *d* is also a *local* or *relative minimum* because it's the lowest point in its immediate neighborhood.

A scenic overlook: The absolute maximum

After *d*, you travel up, passing *e*, which is another inflection point. It's the steepest point between *d* and *g* and the point where the derivative is greatest. You stop at the scenic overlook at *g*, another stationary point and another local max. Point *g* is also the *absolute maximum* on the interval from *a* to *l* because it's the very highest point on the road from *a* to *l*.

Car trouble: Teetering on the corner

Going down from *g*, you pass another inflection point, *h*, another local min, *i*, then you go up to *j*, where you foolishly try to drive over the peak. Your front wheels make it over, but your car's chassis gets stuck on the precipice, leaving you teetering up and down with your wheels spinning. Your car teeters at *j* because you can't draw a tangent line there. No tangent line means no slope, and no slope means no derivative — or you can say that the derivative at *j* is *undefined*. A sharp turning point like *j* is called a *corner*. (By the way, be careful with the expressions "no slope" and "no derivative." In this context, "no" means *nonexistent*, NOT *zero*.)

It's all downhill from here

After dislodging your car, you head down, with the road getting less and less steep until it flattens out for an instant at *k*. (Again, note that because the slope and the derivative are becoming smaller and smaller *negative* numbers on the way to *k*, they are actually *increasing*.) Point *k* is another stationary point because its derivative is zero. It's also another inflection point because the concavity switches from up to down at *k*. After passing *k*, you go down to *l*, your final destination. Because *l* is the endpoint of the interval, it's not a local min — endpoints never qualify as local mins or maxes — but it is the *absolute minimum* on the interval because it's the very lowest point from *a* to *l*.

Hope you enjoyed your trip.

Your travel diary

I want to review your trip and some of the previous terms and definitions, and introduce yet a few more terms:

- The function *f* in Figure 11-1 has a derivative of zero at stationary points (level points) *b, d, g, i,* and *k*. At *j*, the derivative is *undefined*. These points where the derivative is either zero or undefined are the *critical points* of the function. The *x*-values of these critical points are called the *critical numbers* of the function. (Note that critical numbers must be within a function's domain.)
- All local maxes and mins — the peaks and valleys — must occur at critical points. However, not all critical points are necessarily local maxes or mins. Point *k,* for instance, is a critical point, but neither a max nor a min. Local maximums and minimums — or *maxima* and *minima* — are called, collectively, local *extrema* of the function. (Use a lot of these fancy plurals if you want to sound like a professor.) A single local max or min is a local *extremum.* The absolute max is the highest point on the road from *a* to *l*. The absolute min is the lowest point.
- The function is increasing whenever you're going up, where the derivative is positive; it's decreasing whenever you're going down, where the derivative is negative. The function is also decreasing at point *k,* a horizontal inflection point, even though the slope and derivative are zero there. I realize that seems a bit odd, but that's the way it works — take my word for it. At all horizontal inflection points, a function is either increasing or decreasing. At local extrema *b, d, g, i,* and *j,* the function is neither increasing nor decreasing.
- The function is concave up wherever it looks like a cup or a smile (some say where it "holds water") and concave down wherever it looks like a frown (or "spills water"). Inflection points *c, e, h,* and *k* are where the concavity switches from up to down or vice versa. Inflection points are also the steepest or least steep points in their immediate neighborhoods.

Finding Local Extrema — My Ma, She's Like, Totally Extreme

Now that you have the preceding section under your belt and know what local extrema are, you need to know how to do the math to find them. You saw in the last section that all local extrema occur at critical points of a function — that's where the derivative is zero or undefined (but don't forget that critical points aren't always local extrema). So, the first step in finding a function's local extrema is to find its critical numbers (the *x*-values of the critical points).

Cranking out the critical numbers

Find the critical numbers of $f(x)=3x^5-20x^3$. See Figure 11-2.

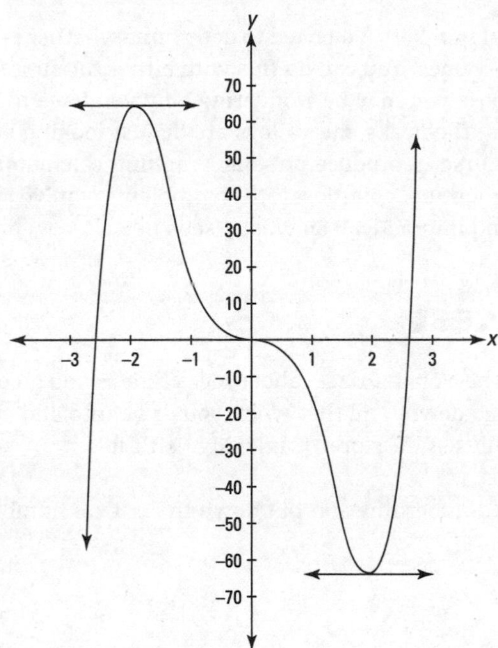

FIGURE 11-2: The graph of $f(x)=3x^5-20x^3$.

Here's what you do:

1. **Find the first derivative of *f* using the power rule.**

$$f(x)=3x^5-20x^3$$
$$f'(x)=15x^4-60x^2$$

2. **Set the derivative equal to zero and solve for *x*.**

$$15x^4-60x^2=0$$
$$15x^2\left(x^2-4\right)=0$$
$$15x^2(x+2)(x-2)=0$$

$$15x^2=0 \quad \text{or} \quad x+2=0 \quad \text{or} \quad x-2=0$$
$$x=0 \quad \text{or} \quad x=-2 \quad \text{or} \quad x=2$$

These three *x*-values are critical numbers of *f*. Additional critical numbers could exist if the first derivative were undefined at some *x*-values, but because the derivative, $15x^4-60x^2$, is defined for all input values, the above solution set, 0, –2, and 2, is the complete list of critical numbers. Because the derivative of *f* equals zero at these three critical numbers, the curve has horizontal tangents at these numbers. In Figure 11-2, you can see the little horizontal tangent lines drawn where $x=-2$ and $x=2$. The third horizontal tangent line where $x=0$ is the *x*-axis.

MATH RULES

A curve has a horizontal tangent line wherever its derivative is zero, namely, at its stationary points. A curve will have horizontal tangent lines at all of its local mins and maxes (except where the derivative is undefined like, for example, at sharp corners like point *j* in Figure 11-1) and at all of its horizontal inflection points.

Now that you've got the list of critical numbers, you need to determine whether peaks or valleys or inflection points occur at those *x*-values. You can do this with either the first derivative test or the second derivative test. I suppose you may be wondering why you have to test the critical numbers when you can see where the peaks and valleys are by just looking at the graph in Figure 11-2 — which you can, of course, reproduce on your graphing calculator. Good point. Okay, so this problem — not to mention countless other problems you've done in math courses — is somewhat contrived and impractical. So what else is new?

The first derivative test

The first derivative test is based on the Nobel-Prize-caliber ideas that as you go over the top of a hill, first you go up and then you go down, and that when you drive into and out of a valley, you go down and then up. This calculus stuff is pretty amazing, isn't it?

Here's how you use the test. Take a number line and put down the critical numbers you found above: 0, –2, and 2. See Figure 11-3.

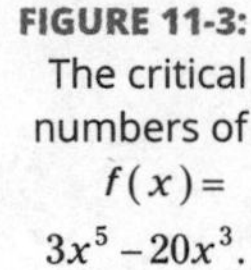

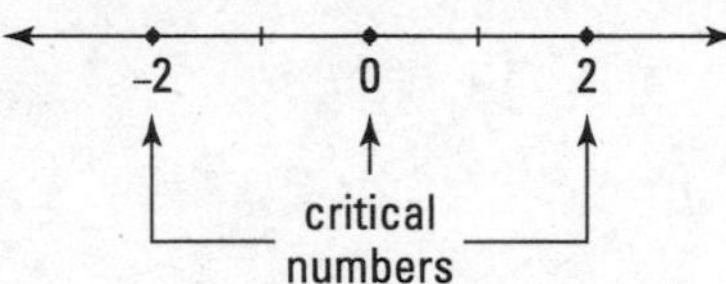

FIGURE 11-3: The critical numbers of $f(x) = 3x^5 - 20x^3$.

This number line is now divided into four regions: to the left of –2, from –2 to 0, from 0 to 2, and to the right of 2. Pick a value from each region, plug it into the first derivative, and note whether your result is positive or negative. Let's use the numbers –3, –1, 1, and 3 to test the regions:

$$f'(x) = 15x^4 - 60x^2$$

$$f'(-3) = 15(-3)^4 - 60(-3)^2 = 15 \cdot 81 - 60 \cdot 9 = 675$$
$$f'(-1) = 15(-1)^4 - 60(-1)^2 = 15 - 60 = -45$$
$$f'(1) = 15(1)^4 - 60(1)^2 = 15 - 60 = -45$$
$$f'(3) = 15(3)^4 - 60(3)^2 = 15 \cdot 81 - 60 \cdot 9 = 675$$

By the way, if you had noticed that this first derivative is an *even* function, you'd have known, without doing the computation, that $f(1) = f(-1)$ and that $f(3) = f(-3)$. (Chapter 5 discusses even functions. A polynomial function with all even powers, like $f'(x)$ shown here, is one type of even function.)

These four results are, respectively, positive, negative, negative, and positive. Now, take your number line, mark each region with the appropriate positive or negative sign, and indicate where the function is increasing (where the derivative is positive) and decreasing (where the derivative is negative). The result is a so-called *sign graph*. See Figure 11-4. (The four right-pointing arrows at the top of the figure simply indicate that "increasing" and "decreasing" tell you what's happening as you move along the function from *left to right*.)

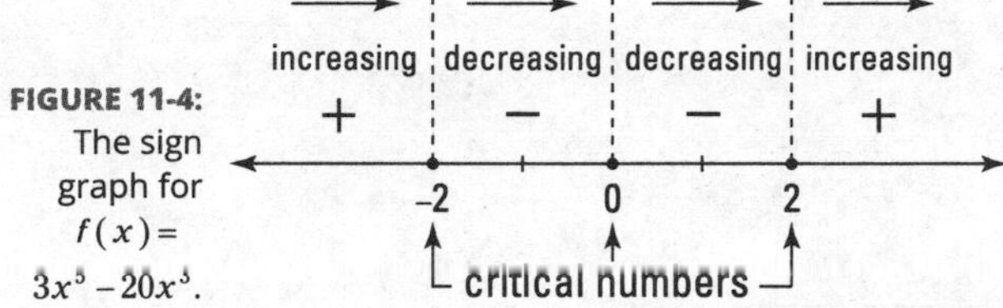

FIGURE 11-4: The sign graph for $f(x) = 3x^5 - 20x^3$.

Figure 11-4 simply tells you what you already know if you've looked at the graph of f — that the function goes up until −2, down from −2 to 0, further down from 0 to 2, and up again from 2 on.

Now here's the rocket science. The function switches from increasing to decreasing at −2; in other words, you go up to −2 and then down. So, at −2 you have the top of a hill or a local maximum. Conversely, because the function switches from decreasing to increasing at 2, you have the bottom of a valley there or a local minimum. And because the sign of the first derivative doesn't switch (from positive to negative or vice versa) at zero, there's neither a min nor a max at that x-value (you usually — like here — get a horizontal inflection point when this happens).

The last step is to obtain the function values, in other words the heights, of the two local extrema by plugging the x-values into the original function:

$$f(x) = 3x^5 - 20x^3$$

$$f(-2) = 3(-2)^5 - 20(-2)^3 = 64$$
$$f(2) = 3(2)^5 - 20(2)^3 = -64$$

Thus, the local max is located at (−2, 64) and the local min is at (2, −64). You're done.

WARNING

To use the first derivative test to check for a local extremum at a particular critical number, the function must be *continuous* at that x-value.

EXAMPLE

Q. Use the first derivative test to determine the location of the local extrema of $g(x) = 15x^3 - x^5$. Refer to the following figure.

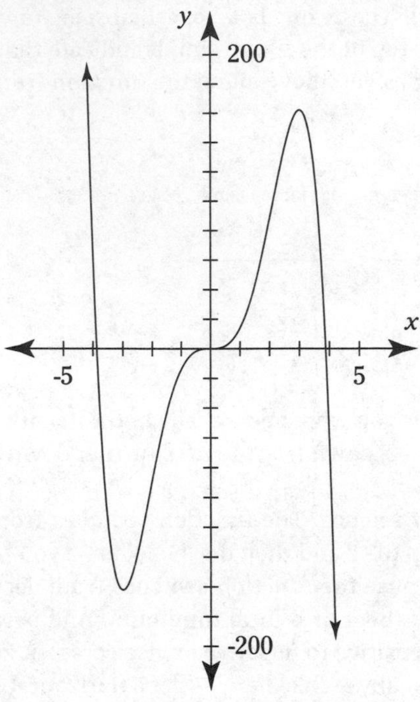

A. **The local min is at $(-3, -162)$, and the local max is at $(3, 162)$.**

1. **Find the first derivative of g using the power rule.**

$$g(x) = 15x^3 - x^5$$
$$g'(x) = 45x^2 - 5x^4$$

2. **Set the derivative equal to zero and solve for x to get the critical numbers of g.**

$$45x^2 - 5x^4 = 0$$
$$5x^2(9 - x^2) = 0$$
$$5x^2(3 - x)(3 + x) = 0$$
$$5x^2 = 0 \quad \text{or} \quad 3 - x = 0 \quad \text{or} \quad 3 + x = 0$$
$$x = 0 \qquad\qquad x = 3 \qquad\qquad x = -3$$

If the first derivative was undefined for some x-values in the domain of g, there could be more critical numbers, but because $g'(x) = 45x^2 - 5x^4$ is defined for all real numbers, 0, 3, −3 is the complete list of critical numbers of g.

3. **Plot the three critical numbers on a number line, noting that they create four regions (see the figure in Step 5).**
4. **Plug a number from each of the four regions into the derivative, noting whether the results are positive or negative.**

   If you've already factored the derivative (see Step 2), it's usually best to use the factored form of the derivative in this step. And all you need to do is note whether the results are positive or negative. There's no need to compute the exact results. To wit:

   $$g'(x) = 5x^2(3-x)(3+x)$$

   $$\begin{aligned} g'(-4) &= (Pos.)(Pos.)(Neg.) = Neg. \\ g'(-1) &= (Pos.)(Pos.)(Pos.) = Pos. \\ g'(1) &= (Pos.)(Pos.)(Pos.) = Pos. \\ g'(4) &= (Pos.)(Neg.)(Pos.) = Neg. \end{aligned}$$

   (A slight shortcut here is to notice that since $g'(x)$ is an even function, $g'(1)$ must equal $g'(-1)$, and $g'(4)$ must equal $g'(-4)$).
5. **Draw a "sign graph." Take your number line and label each region — based on your results from Step 4 — positive (increasing) or negative (decreasing). Refer to the following figure.**

   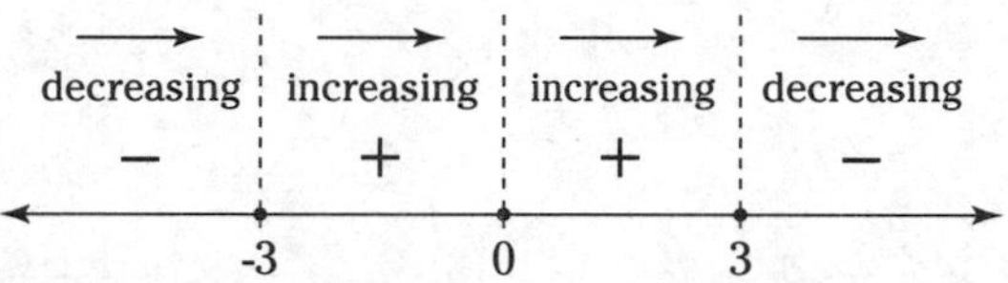

   This sign graph tells you where the function is increasing (rising as you go from left to right) and where it is decreasing (falling as you go from left to right).
6. **Use the sign graph to determine whether there's a local minimum, local maximum, or neither at each critical number.**

   Because g goes down on its way to $x = -3$ and up after $x = -3$, it must bottom out at $x = -3$, so there's a local min there. Conversely, g peaks at $x = 3$ because it rises until $x = 3$, and then falls. There is thus a local max at $x = 3$. And because g climbs on its way to $x = 0$ and then climbs further, there is neither a min nor a max at $x = 0$.
7. **Determine the y-values of the local extrema by plugging the x-values into the original function.**

   $$\begin{aligned} g(-3) &= 15(-3)^3 - (-3)^5 \\ &= -162 \\ g(3) &= 15(3)^3 - (3)^5 \\ &= 162 \end{aligned}$$

   So, the local min is at $(-3, -162)$, and the local max is at $(3, 162)$.

YOUR TURN

Use the first derivative to find the local extrema of $f(x) = 6x^{2/3} - 4x + 1$.

2 Find the local extrema of $h(x) = \frac{x}{\sqrt{2}} + \cos x - \frac{\sqrt{2}}{2}$ in the interval $(0, 2\pi)$ with the first derivative test.

Locate the local extrema of $y = \left(x^2 - 8\right)^{2/3}$ with the first derivative test.

Using the first derivative test, determine the local extrema of $s = \frac{t^4 + 4}{-2t^2}$.

The second derivative test — no, no, anything but another test!

The second derivative test is based on two more prize-winning ideas: first, that at the crest of a hill, a road has a hump shape — in other words, it's curving down or *concave down;* and second, that at the bottom of a valley, a road is cup-shaped, so it's curving up or *concave up.*

The concavity of a function at a point is given by its second derivative: a *positive* second derivative means the function is concave *up*, a *negative* second derivative means the function is concave *down*, and a second derivative of *zero* is *inconclusive* (the function could be concave up, concave down, or there could be an inflection point there).

WARNING

When the second derivative test fails. If the second derivative equals zero (or is undefined) at a particular critical number, the second derivative test fails and you learn nothing about whether there's a local extremum there. When this happens, you have to use the first derivative test to determine whether or not you have a local extremum.

Let's return to the function $f = 3x^5 - 20x^3$ that I analyzed in the previous section with the first derivative test. To use the second derivative test for *f*, all you have to do is find its second derivative and then plug in its critical numbers, which you found before (–2, 0, and 2), and note whether your results are positive, negative, or zero. To wit —

$$f(x) = 3x^5 - 20x^3$$
$$f'(x) = 15x^4 - 60x^2 \quad \text{(power rule)}$$
$$f''(x) = 60x^3 - 120x \quad \text{(power rule)}$$

$$f''(-2) = 60(-2)^3 - 120(-2) = -240$$
$$f''(0) = 60(0)^3 - 120(0) = 0$$
$$f''(2) = 60(2)^3 - 120(2) = 240$$

At $x = -2$, the second derivative is negative (–240). This tells you that *f* is concave down where *x* equals –2, and therefore that there's a local max there. The second derivative is positive (240) where *x* is 2, so *f* is concave up and thus there's a local min at $x = 2$. Because the second derivative equals zero at $x = 0$, the second derivative test fails for that critical number — it tells you nothing about the concavity at $x = 0$ or whether there's a local min or max there. When this happens, you have to use the first derivative test.

After you find a function's critical numbers, you have to decide whether to use the first or the second derivative test to find the extrema. For some functions, the second derivative test is the easier of the two because 1) the second derivative is usually easy to get, 2) you can often plug the critical numbers into the second derivative and do a quick computation, and 3) you will often get non-zero results and thus get your answers without having to do a sign graph and test regions. (Points 1, 2, and 3 all apply to what you just saw with $f''(-2)$ and $f''(2)$.) On the other hand, testing regions on a sign graph (the first derivative test) is also fairly quick and easy, and if the second derivative test fails (see the warning), you'll have to do that anyway. (You just saw this with $f''(0)$.) As you do practice problems, you'll get a feel for when to use each test.

Now go through the first and second derivative tests one more time with the following example.

EXAMPLE

Q. Find the local extrema of $g(x) = 2x - 3x^{2/3} + 4$. See the following figure.

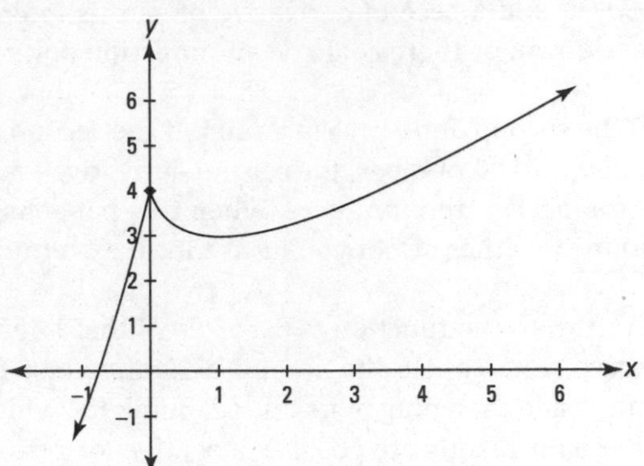

A. **There's a local min at $(1,\ 3)$, and a local max at $(0,\ 4)$.**

1. **Find the first derivative of g.**

$$g(x) = 2x - 3x^{2/3} + 4$$
$$g'(x) = 2 - 2x^{-1/3} \quad \text{(power rule)}$$

2. **Set the derivative equal to zero and solve.**

$$\begin{aligned} 2 - 2x^{-1/3} &= 0 \\ -2x^{-1/3} &= -2 \\ x^{-1/3} &= 1 \\ \left(x^{-1/3}\right)^{-3} &= 1^{-3} \\ x &= 1 \end{aligned}$$

Thus 1 is a critical number.

3. **Determine whether the first derivative is undefined for any x-values.**

The derivative, $2 - 2x^{-1/3}$, which equals $2 - \frac{2}{\sqrt[3]{x}}$, is undefined at $x = 0$. Thus, zero is another critical number. From Steps 2 and 3, you've got the complete list of critical numbers of g: 0 and 1.

4. **Plot the critical numbers on a number line, and then use the first derivative test to figure out the sign of each region.**

You can use −1, 0.5, and 2 as test numbers:

$$g'(x) = 2 - 2x^{-1/3}$$

$$\begin{aligned} &g'(-1) = 4 && \text{(pos.)} \\ &g'(0.5) \approx -0.52 && \text{(neg.)} \\ &g'(2) \approx 0.41 && \text{(pos.)} \end{aligned}$$

The following figure shows the sign graph.

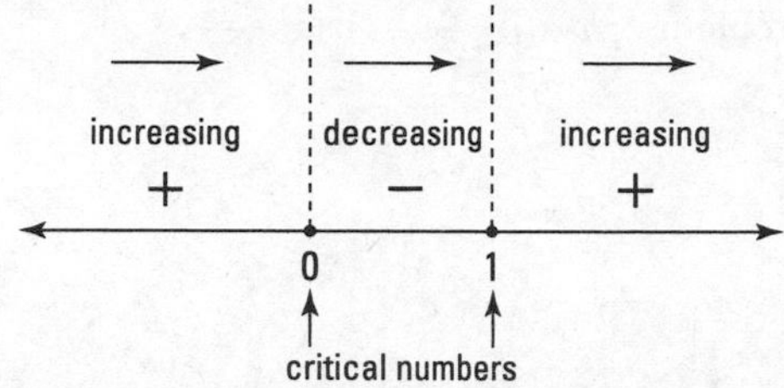

Because the first derivative of g switches from positive to negative at zero, there's a local max there. And because the first derivative switches from negative to positive at 1, there's a local min at $x = 1$.

5. **Plug the critical numbers into g to obtain the function values (the heights) of these two local extrema.**

$$g(x) = 2x - 3x^{2/3} + 4$$

$$g(0) = 4$$
$$g(1) = 3$$

So, there's a local max at $(0, 4)$ and a local min at $(1, 3)$. You're done.

You could have used the second derivative test instead of the first derivative test in Step 4. First, you need the second derivative of g, which is, as you know, the derivative of its first derivative:

$$g'(x) = 2 - 2x^{-1/3}$$
$$g''(x) = \frac{2}{3}x^{-4/3}$$

Evaluate the second derivative at $x = 1$ (the critical number from Step 2):

$$g''(1) = \frac{2}{3}$$

Because $g''(1)$ is positive, you know that g is concave up at $x = 1$ and, therefore, that there's a local min there.

At the other critical number, $x = 0$ (from Step 3), the first derivative is undefined. The second derivative test is no help where the first derivative is undefined, so you've got to use the first derivative test for that critical number.

YOUR TURN

5 Use the second derivative test to analyze the critical numbers of the function from Problem 2, $h(x)=\frac{x}{\sqrt{2}}+\cos x-\frac{\sqrt{2}}{2}$.

6 Find the local extrema of $f(x)=-2x^3+6x^2+1$ with the second derivative test.

7 Find the local extrema of $y=2x^4-\frac{1}{3}x^6$ with the second derivative test.

8 Consider the function from Problem 3, $y=\left(x^2-8\right)^{2/3}$, and the function $s=8+\frac{21t}{4}-\frac{7t^3}{4}$. Which of the two functions is easier to analyze with the second derivative test, and why? For the function you pick, use the second derivative test to find its local extrema.

Finding Absolute Extrema on a Closed Interval

The basic idea in this section is quite simple. Instead of finding all local extrema as in the previous sections (all the peaks and all the valleys), you just want to determine the single highest point and single lowest point along a *continuous* function in some *closed* interval. These *absolute extrema* can occur at a peak or valley or at an edge(s) of the interval. (*Note:* You could have, say, two peaks at the same height so there'd be a tie for the absolute max; but there would still be exactly one y-value that's the absolute maximum value on the interval.)

REMEMBER

A *closed* interval like $[2, 5]$ includes the endpoints 2 and 5. An *open* interval like $(2, 5)$ excludes the endpoints.

Before you practice with some problems, look at Figure 11-5 to see two standard absolute extrema problems (*continuous* functions on a *closed* interval), and at Figure 11-6 for four strange functions that don't have the standard single absolute max and single absolute min.

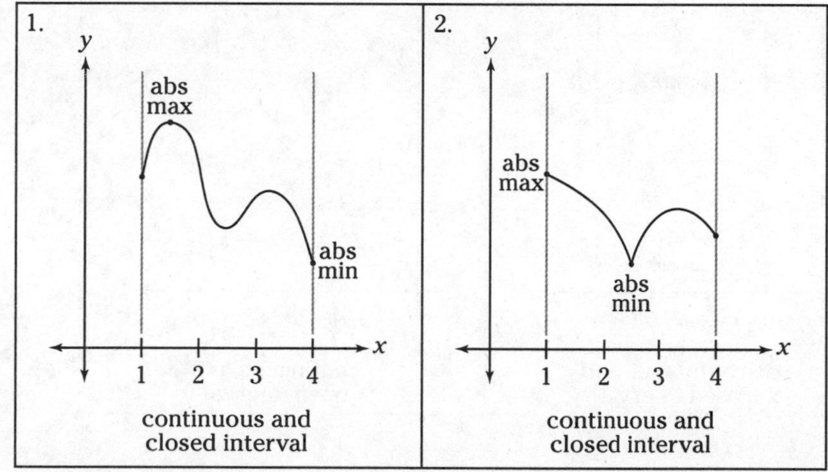

FIGURE 11-5: Two standard absolute extrema functions.

Finding the absolute max and min is a snap. All you do is compute the critical numbers of the function in the given interval, determine the height of the function at each critical number, and then figure the height of the function at the two endpoints of the interval. The greatest of this set of heights is the absolute max; and the least, of course, is the absolute min. Here's an example:

Find the absolute max and min of $h(x) = \cos(2x) - 2\sin x$ in the closed interval $\left[\frac{\pi}{2}, 2\pi\right]$.

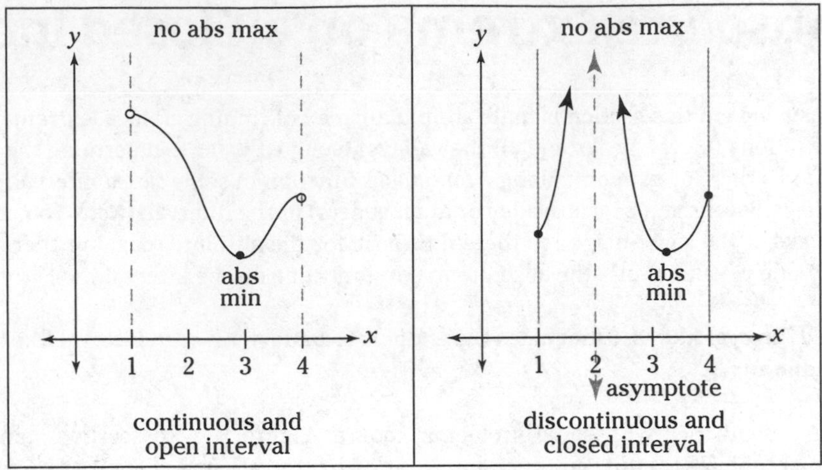

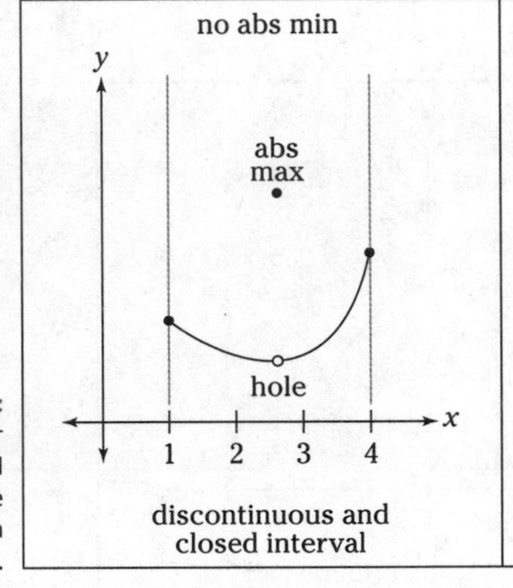

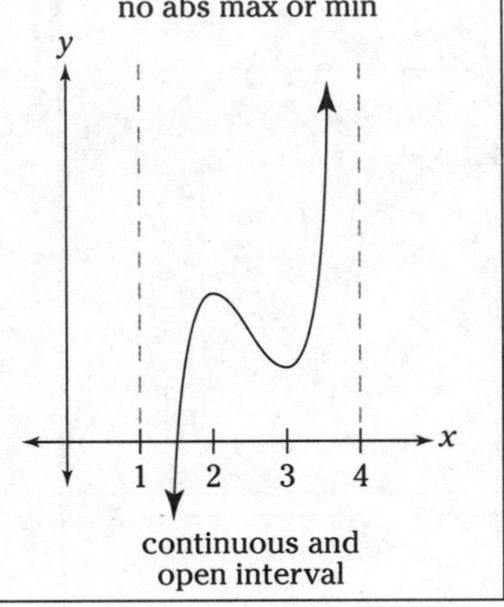

FIGURE 11-6: Four nonstandard absolute extrema functions.

1. **Find the critical numbers of h in the *open* interval $\left(\frac{\pi}{2},\ 2\pi\right)$.**

   (See Chapter 6 if you're a little rusty on trig functions.)

$$
\begin{aligned}
h(x) &= \cos(2x) - 2\sin x \\
h'(x) &= -\sin(2x)\cdot 2 - 2\cos x && \text{(by the chain rule)} \\
0 &= -2\sin(2x) - 2\cos x && \text{(now divide both sides by –2)} \\
0 &= \sin(2x) + \cos x && \text{(now use a trig identity)} \\
0 &= 2\sin x\cos x + \cos x && \text{(factor out } \cos x\text{)} \\
0 &= \cos x(2\sin x + 1)
\end{aligned}
$$

$\cos x = 0$ or $2\sin x + 1 = 0$

$x = \frac{3\pi}{2}$ $\qquad$ $\sin x = -\frac{1}{2}$

$x = \frac{7\pi}{6}, \frac{11\pi}{6}$

Thus, the zeros of h' are $\frac{7\pi}{6}$, $\frac{3\pi}{2}$, and $\frac{11\pi}{6}$, and because h' is defined for all input numbers, this is the complete list of critical numbers.

2. **Compute the function values (the heights) at each critical number.**

$$h(x) = \cos(2x) - 2\sin x$$

$$\begin{aligned} h\left(\frac{7\pi}{6}\right) &= \cos\left(2\cdot\frac{7\pi}{6}\right) - 2\sin\left(\frac{7\pi}{6}\right) \\ &= 0.5 - 2\cdot(-0.5) = 1.5 \\ h\left(\frac{3\pi}{2}\right) &= \cos\left(2\cdot\frac{3\pi}{2}\right) - 2\sin\left(\frac{3\pi}{2}\right) \\ &= -1 - 2\cdot(-1) = 1 \\ h\left(\frac{11\pi}{6}\right) &= \cos\left(2\cdot\frac{11\pi}{6}\right) - 2\sin\left(\frac{11\pi}{6}\right) \\ &= 0.5 - 2\cdot(-0.5) = 1.5 \end{aligned}$$

3. **Determine the function values at the endpoints of the interval.**

$$\begin{aligned} h\left(\frac{\pi}{2}\right) &= \cos\left(2\cdot\frac{\pi}{2}\right) - 2\sin\left(\frac{\pi}{2}\right) \\ &= -1 - 2\cdot 1 = -3 \\ h(2\pi) &= \cos(2\cdot 2\pi) - 2\sin(2\pi) \\ &= 1 - 2\cdot 0 = 1 \end{aligned}$$

So, from Steps 2 and 3, you've found five heights: 1.5, 1, 1.5, –3, and 1. The largest number in this list, 1.5, is the absolute max; the smallest, –3, is the absolute min.

The absolute max occurs at two points: $\left(\frac{7\pi}{6}, 1.5\right)$ and $\left(\frac{11\pi}{6}, 1.5\right)$. The absolute min occurs at one of the endpoints, $\left(\frac{\pi}{2}, -3\right)$, and is thus called an *endpoint extremum.*

Table 11-1 shows the values of $h(x) = \cos(2x) - 2\sin x$ at the three critical numbers in the interval from $\frac{\pi}{2}$ to 2π and at the interval's endpoints; Figure 11-7 shows the graph of h.

Table 11-1 Values of $h(x) = \cos(2x) - 2\sin x$ at the Critical Numbers and Endpoints for the Interval $\left[\frac{\pi}{2}, 2\pi\right]$

$h(x)$	–3	1.5	1	1.5	1
x	$\frac{\pi}{2}$	$\frac{7\pi}{6}$	$\frac{3\pi}{2}$	$\frac{11\pi}{6}$	2π

A couple of observations. First, as you can see in Figure 11-7, the points $\left(\frac{7\pi}{6}, 1.5\right)$ and $\left(\frac{11\pi}{6}, 1.5\right)$ are both *local* maxima of h, and the point $\left(\frac{3\pi}{2}, 1\right)$ is a local minimum of h. However, if you want only to find the *absolute* extrema on a closed interval, you don't have to pay any attention to whether critical points are local maxes, mins, or neither. And, thus, you don't have to bother to use the first or second derivative tests. All you have to do is determine the heights at the critical numbers and at the endpoints and then pick the largest and smallest numbers from this list. Second, the absolute max and min in the given interval tell you nothing about how the function behaves outside the interval. Function h, for instance, might rise higher than 1.5 outside the interval from $\frac{\pi}{2}$ to 2π (although it doesn't), and it might go lower than −3 (although it never does).

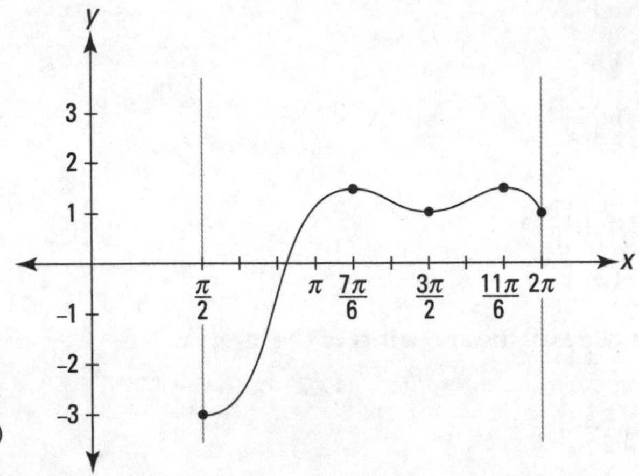

FIGURE 11-7: The graph of $h(x)=\cos(2x)-2\sin x$.

EXAMPLE

Q. Determine the absolute min and absolute max of $f(x)=\sqrt{|x|}-x$ in the interval $\left[-1, \frac{1}{2}\right]$.

A. The absolute max is 2 and the absolute min is zero.

1. Get all the critical numbers.

The first step is to determine the derivative, set it equal to zero, and solve, but before you can get the derivative, you have to split the function in two to get rid of the absolute value bars:

I. When $x \geq 0$, $\sqrt{|x|}=\sqrt{x}$ and, thus,

$$\begin{aligned} f(x) &= \sqrt{x}-x \\ f'(x) &= \frac{1}{2\sqrt{x}}-1 \\ 0 &= \frac{1}{2\sqrt{x}}-1 \\ 2\sqrt{x} &= 1 \\ x &= \frac{1}{4} \end{aligned}$$

II. When $x<0$, $\sqrt{|x|}=\sqrt{-x}$ and, thus,

$$\begin{aligned} f(x) &= \sqrt{-x}-x \\ f'(x) &= \frac{-1}{2\sqrt{-x}}-1 \\ 0 &= \frac{-1}{2\sqrt{-x}}-1 \\ 2\sqrt{-x} &= -1 \end{aligned}$$

No solution

Now, determine whether the derivative is undefined anywhere.

The derivative is undefined at $x=0$ because the denominator of the derivative can't equal zero. (If you graph this function [always a good idea], you'll also see the sharp corner at $x = 0$ and thus know immediately that the derivative is undefined there.)

The critical numbers are therefore 0 and $\frac{1}{4}$.

2. **Compute the function values (the heights) at all the critical numbers.**

$f\left(\frac{1}{4}\right)=\frac{1}{4} \qquad f(0)=0$

It's just a coincidence, by the way, that in both cases the input equals the output.

3. **Compute the function values at the two edges of the interval.**

$f(-1)=2 \qquad f\left(\frac{1}{2}\right)=\frac{\sqrt{2}}{2}-\frac{1}{2}\approx 0.207$

4. **The highest of all the function values from Steps 2 and 3 is the absolute max; the lowest of all the values from Steps 2 and 3 is the absolute min.**

Thus, 2 is the absolute max and zero is the absolute min.

(This particular problem was more involved than usual because of that extra twist in Step 1 involving the absolute value bars.)

Find the absolute extrema of $f(x)=\sin x+\cos x$ on the interval $[0, 2\pi]$.

Find the absolute extrema of $g(x)=2x^3-3x^2-5$ on the interval $[-0.5, 0.5]$.

Find the absolute extrema of $p(x)=(x+1)^{4/5}-0.5x$ on the interval $[-2, 31]$.

Find the absolute extrema of $q(x)=2\cos(2x)+4\sin x$ on the interval $\left[-\frac{\pi}{2}, \frac{5\pi}{4}\right]$.

Finding Absolute Extrema over a Function's Entire Domain

A function's *absolute max* and *absolute min* over its *entire domain* are the highest and lowest values (heights) of the function anywhere it's defined. Unlike in the previous section where you saw that a continuous function must have both an absolute max and min on a closed interval, when you consider a function's entire domain, a function can have an absolute max or min, or both, or neither. For example, the parabola $y = x^2$ has an absolute min at the point $(0, 0)$ — the bottom of its cup shape — but no absolute max because it goes up forever to the left and the right. You might think that its absolute max would be infinity, but infinity is not a number and thus it doesn't qualify as a maximum (ditto for using negative infinity as an absolute min).

On the one hand, the idea of a function's very highest point and very lowest point seems pretty simple, doesn't it? But there's a wrench in the works. The wrench is the category of things that *don't* qualify as maxes or mins.

I already mentioned that infinity and negative infinity don't qualify. Then there are empty "endpoints" like $(3, 4)$ on $f(x)$ in Figure 11-8. The function $f(x)$ doesn't have an absolute max. Its max isn't 4 because it never gets to 4, and its max can't be anything less than 4, like 3.999, because it gets higher than that, say 3.9999. Similarly, an infinitesimal hole in a function can't qualify as a max or min. For example, consider the absolute value function, $y = |x|$, you know, the V-shaped function with the sharp corner at the origin. The function $y = |x|$ has no absolute max because it goes up to infinity. Its absolute min is zero (at $(0, 0)$ of course). But now, say you alter the function slightly by plucking out the point at $(0, 0)$ and leaving an infinitesimal hole there. Now the function has no absolute minimum.

Now consider $g(x)$ in Figure 11-8. It shows another type of situation that doesn't qualify as a min (or max). The function $g(x)$ has no absolute min. Going left, g crawls along the horizontal asymptote at $y = 0$, always getting lower and lower, but never getting as low as zero. Since it never gets to zero, zero can't be the absolute min, and there can't be any other absolute min (like, say, 0.0001) because at some point way to the left, g will get below any small number you can name.

Keeping this in mind, here's an approach for locating a function's absolute maximum and minimum (if there are any):

1. **Find the height of the function at each of its critical numbers; in other words, find the function's critical points.**

   (Recall that a function's critical numbers are the x-values within the function's domain where the derivative is zero or undefined; critical points are the points on the function corresponding to the critical numbers.) *If a function has an absolute max and/or min, each must occur at a critical point, so, once you have all of a function's critical points, you have all of the candidates for an absolute max and an absolute min.*

   You just did something similar in Step 1 of both examples in the previous section, but this time you consider *all* the critical points, not just those in a given interval. The highest of these points will be the function's absolute max unless the function goes higher than that point, in which case the function won't have an absolute max. The lowest of those points will be the function's absolute min unless the function goes lower than that point, in which case it won't have an absolute min. If you apply Step 1 to $g(x)$ in Figure 11-8, you'll find that it has no critical points. When this happens, you're done. The function has neither an absolute max nor an absolute min.

2. **Graph the function to see how high and low it goes.**

   Look at the graph of the function. If you see that the function gets higher than the highest of its critical points, it has no absolute max; if it goes lower than the lowest of its critical points, it has no absolute min. Applying Steps 1 and 2 to $f(x)$ in Figure 11-8, Step 1 would reveal two critical points: the endpoint at $(3, 1)$ (where the first derivative is undefined) and the local max at roughly $(4.1, 1.3)$ (where the first derivative equals zero). Those two critical points are the only candidates for an absolute min or max. But then in Step 2, you'd see that *f* goes higher than the higher of the critical points, $(4.1, 1.3)$, and that it, therefore, has no absolute max; and you'd see that *f* goes lower than the lower of the critical points, $(3, 1)$, and, thus, that it has no absolute min.

3. **Use the first derivative test and a sign graph to analyze the shape of the function.**

   You may have your final answer after finishing Step 2 (as in this example), but if any candidates for an absolute max or min remain after finishing Step 2, you may want to use a sign graph to analyze the shape of the function — especially its end behavior (what happens to a function way to the left and way to the right) — before you can positively conclude whether there is an absolute max and/or min.

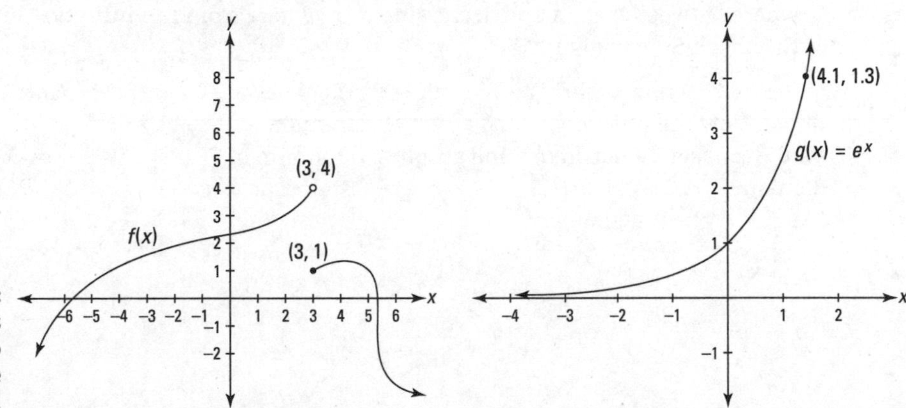

FIGURE 11-8: Two functions with no absolute extrema.

EXAMPLE

Q. Find the absolute maximum and absolute minimum of $f(x) = \ln x - x$ over its entire domain.

A. **The absolute maximum is $(1, -1)$ and *f* has no absolute minimum.**

1. Find all of the function's critical points.

To find the critical points of *f*, first set the derivative of *f* equal to zero and solve:

$$f(x) = \ln x - x$$
$$f'(x) = \frac{1}{x} - 1$$
$$0 = \frac{1}{x} - 1$$
$$x = 1$$

So, 1 is a critical number. And don't forget that you always have to also check for any x-values where the derivative is undefined. The derivative, $\frac{1}{x}-1$, is undefined at $x=0$, but 0 is not in f's domain, so 0 is not a critical number. Thus, 1 is the only critical number.

Plugging 1 into f gives you f's only critical point.

$$f(1) = \ln 1 - 1$$
$$= -1$$

Thus, f's critical point is at $(1,-1)$. This is the only candidate for an absolute max or min.

2. **Looking at the graph of f, you can see f curving down from $(1,-1)$ to the left and to the right.**

   Since $(1,-1)$ is the only candidate for an absolute max or a min, you can conclude that f has no absolute min. And it certainly appears from the graph that $(1,-1)$ is the absolute max, but you should confirm that in Step 3.

3. **The only critical number is $x = 1$, so that's the only number on your sign graph.**

   Testing any number between 0 and 1 in the derivative produces a positive result (x-values less than or equal to 0 are not in f's domain), and testing any number greater than 1 produces a negative result. That tells you that f is increasing everywhere between 0 and 1 and decreasing everywhere from 1 to infinity. Thus, $(1,-1)$ is the absolute maximum.

YOUR TURN

13 Find the absolute maximum and absolute minimum of $f(x)=x^4-2x^2$ over its entire domain.

---

14 **Find the absolute maximum and absolute minimum of functions A, B, C, and D shown in the following figure.**

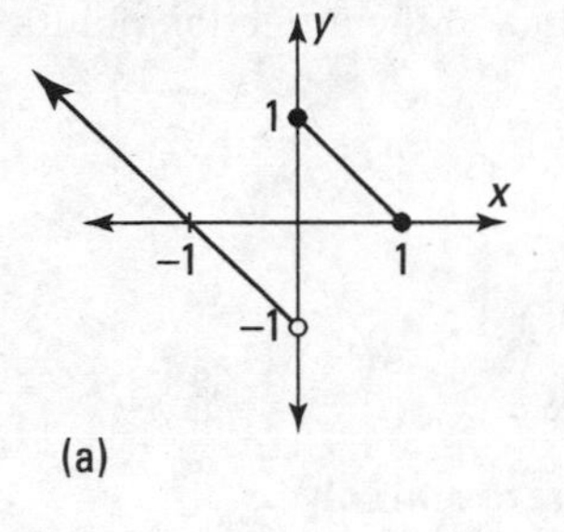

(a)

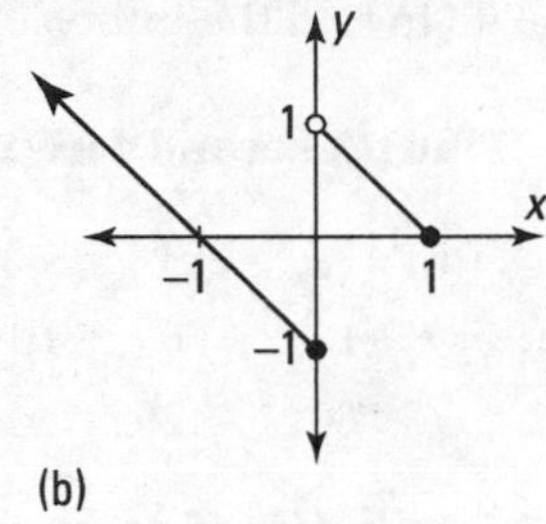

(b)

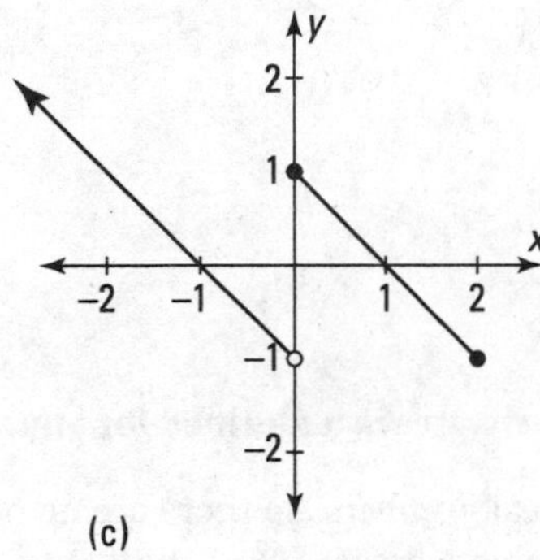

(c)

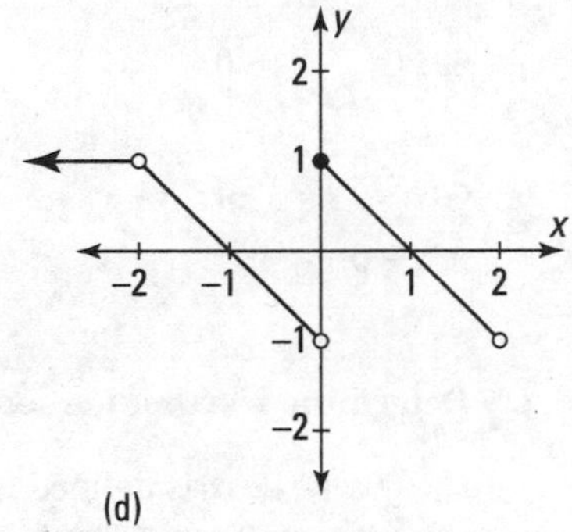

(d)

Locating Concavity and Inflection Points

Look back at the function $f(x)=3x^5-20x^3$ in Figure 11-2. You used the three critical numbers of f, −2, 0, and 2, to find the function's local extrema: (−2, 64) and (2, −64). This section investigates what happens elsewhere on this function — specifically, where it's concave up or down and where the concavity switches (the inflection points).

The process for finding concavity and inflection points is analogous to using the first derivative test and the sign graph to find local extrema, except that now you use the second derivative. (See the section "Finding Local Extrema — My Ma, She's Like, Totally Extreme.") Here's what you do to find the intervals of concavity and the inflection points of $f(x) = 3x^5 - 20x^3$:

1. **Find the second derivative of *f*.**

$$f(x) = 3x^5 - 20x^3$$
$$f' = 15x^4 - 60x^2 \quad \text{(the power rule)}$$
$$f'' = 60x^3 - 120x \quad \text{(the power rule)}$$

2. **Set the second derivative equal to zero and solve.**

$$60x^3 - 120x = 0$$
$$60x(x^2 - 2) = 0$$

$$60x = 0 \quad \text{or} \quad x^2 - 2 = 0$$
$$x = 0 \qquad\qquad x^2 = 2$$
$$x = \pm\sqrt{2}$$

3. **Determine whether the second derivative is undefined for any *x*-values.**

$f'' = 60x^3 - 120x$ is defined for all real numbers, so there are no other *x*-values to add to the list from Step 2. Thus, the complete list is $-\sqrt{2}$, 0, and $\sqrt{2}$.

Steps 2 and 3 give you what you could call "second derivative critical numbers" of *f* because they're analogous to the critical numbers of *f* that you find using the first derivative. But, as far as I'm aware, this set of numbers has no special name. The important thing to know is that this list is made up of the zeros of f'' plus any *x*-values where f'' is undefined.

4. **Plot these numbers on a number line and test the regions with the *second* derivative.**

Use $-2, -1, 1$, and 2 as test numbers.

$$f''(x) = 60x^3 - 120x$$

$$f''(-2) = -240 \quad \text{(neg.)}$$
$$f''(-1) = 60 \quad \text{(pos.)}$$
$$f''(1) = -60 \quad \text{(neg.)}$$
$$f''(2) = 240 \quad \text{(pos.)}$$

Figure 11-9 shows the sign graph.

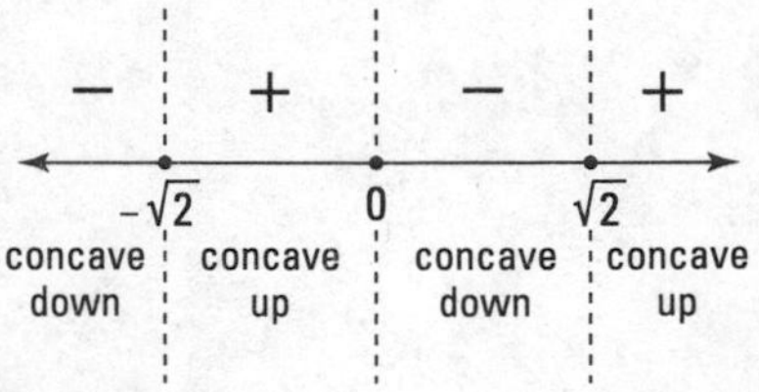

FIGURE 11-9: A second derivative sign graph for $f(x) = 3x^5 - 20x^3$.

A positive sign on this sign graph tells you that the function is concave up in that interval; negative means concave down. The function has an inflection point (usually) at any x-value where the signs switch from positive to negative or vice versa.

Because the signs switch at $-\sqrt{2}$, 0, and $\sqrt{2}$, and because these three numbers are zeros of f'', inflection points occur at these x-values. If, however, you have a problem where the signs switch at a number where f'' is undefined, you have to check one additional thing before concluding that there's an inflection point there. An inflection point exists at a given x-value only if you can draw a tangent line to the function at that number. This is the case if the first derivative exists at that number or if the tangent line is vertical there; both of these situations are covered by a simple rule: *If the concavity switches at a point where the curve is smooth, you have an inflection point there.*

WARNING

All inflection points have a second derivative of zero (if the second derivative exists), but **not all points with a second derivative of zero are inflection points.** This is no different from "all ships are boats but not all boats are ships." (For example, $y = x^4$, which resembles a parabola, has a second derivative equal to zero at the point $(0, 0)$, but that point is *not* an inflection point — it's a local minimum.)

5. **Plug these three x-values into f to obtain the function values of the three inflection points.**

$$f(x) = 3x^5 - 20x^3$$

$$f(-\sqrt{2}) \approx 39.6$$
$$f(0) = 0$$
$$f(\sqrt{2}) \approx -39.6$$

The square root of 2 equals about 1.4, so there are inflection points at about $(-1.4, 39.6)$, $(0, 0)$, and about $(1.4, -39.6)$. You're done.

Figure 11-10 shows f's inflection points as well as its local extrema and its intervals of concavity.

YOUR TURN

15 Find the intervals of concavity and the inflection points of $f(x) = -2x^3 + 6x^2 - 10x + 5$.

16 Find the intervals of concavity and the inflection points of $g(x) = x^4 - 12x^2$.

17 Find the intervals of concavity and the inflection points of $p(x) = \dfrac{x}{x^2 + 9}$.

18 Find the intervals of concavity and the inflection points of $q(x) = \sqrt[5]{x} - \sqrt[3]{x}$. You'll want to use your calculator for this one.

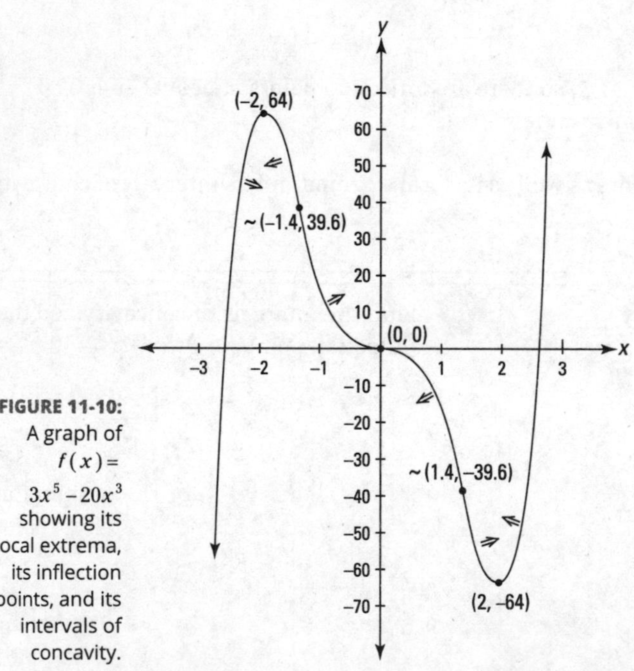

FIGURE 11-10: A graph of $f(x) = 3x^5 - 20x^3$ showing its local extrema, its inflection points, and its intervals of concavity.

Looking at Graphs of Derivatives Till They Derive You Crazy

You can learn a lot about functions and their derivatives by looking at their graphs side by side and comparing their important features. Let's keep going with the same function, $f(x) = 3x^5 - 20x^3$; you're going to travel along f from left to right (see Figure 11-11), pausing to note its points of interest and also observing what's happening to the graph of $f' = 15x^4 - 60x^2$ at the same points. But first, check out the following (long) warning.

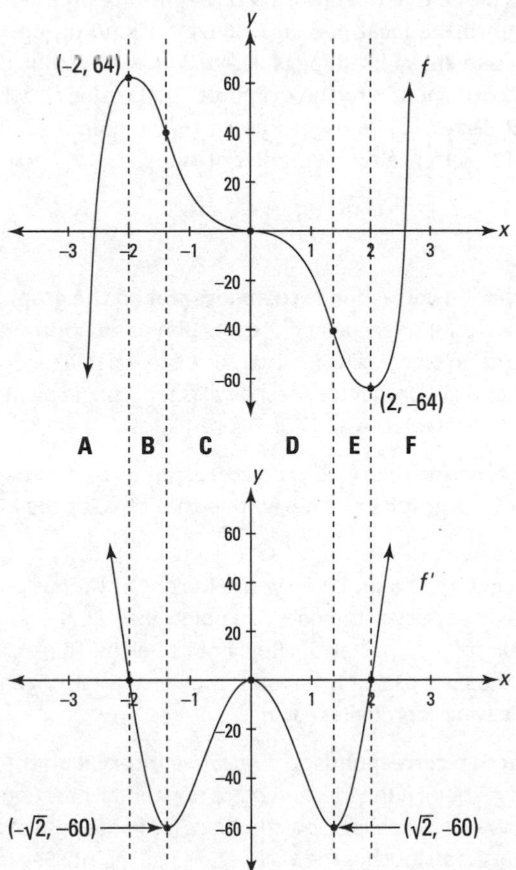

FIGURE 11-11: $f = 3x^5 - 20x^3$ and its first derivative, $f' = 15x^4 - 60x^2$.

WARNING

This is NOT the function! As you look at the graph of f' in Figure 11-11, or the graph of any other derivative, you may need to slap yourself in the face every minute or so to remind yourself that "This is the *derivative* I'm looking at, *not* the function!" You've looked at hundreds and hundreds of graphs of functions over the years, so when you start looking at graphs of derivatives, you can easily lapse into thinking of them as regular functions. You might, for instance, look at an interval that's going up on the graph of a derivative and mistakenly conclude that the original function must also be going up in the same interval — an understandable mistake. You know the first derivative is the same thing as slope. So, when you see the graph of the first derivative going up, you may think, "Oh, the first derivative (the slope) is going up, and when the slope goes up that's like going up a hill, so the original function must be rising." This sounds reasonable because, *loosely* speaking, you can describe the front side of a hill as a slope that's going up, increasing. But *mathematically* speaking, the front side of a hill has a *positive* slope, not necessarily an *increasing* slope. So, where a function is increasing, the graph of its derivative will be *positive*, but that derivative graph might be going up or down. Say you're going up a hill. As you approach the top of the hill, you're still going *up*, but, in general, the *slope* (the steepness) is going *down*. It might be 3, then 2, then 1, and then, at the top of the hill, the slope is zero. So, the slope is getting smaller or *decreasing*, even as you're climbing the hill or *increasing*. In such an interval, the graph of the function is *increasing*, but the graph of its derivative is *decreasing*. Got that?

Okay, let's get back to the f and its derivative in Figure 11-11. Beginning on the left and traveling toward the right, f increases until the local max at $(-2, 64)$. It's going up, so its slope is *positive*, but f is getting less and less steep, so its slope is *decreasing* — the slope decreases until it becomes zero at the peak. This corresponds to the graph of f' (the slope), which is *positive* (because it's above the x-axis) but *decreasing* as it goes down to the point $(-2, 0)$. Let's summarize your entire trip along f and f' with the following list of rules.

MATH RULES

Rules are rules:

- **An *increasing* interval on a function corresponds to an interval on the graph of its derivative that's *positive*** (or *zero* for a single point if the function has a horizontal inflection point). In other words, a function's increasing interval corresponds to a part of the derivative graph that's above the x-axis (or that touches the axis for a single point in the case of a horizontal inflection point). See intervals A and F in Figure 11-11.
- **A local *max* on the graph of a function (like $(-2, 64)$) corresponds to a *zero* — an x-intercept — on an interval of the graph of its derivative that crosses the x-axis going *down* (like at $(-2, 0)$).**
- **On a derivative graph, you've got an m-axis.** When you're looking at various points on the derivative graph, don't forget that the y-coordinate of a point, like $(-2, 0)$, on a graph of a first derivative tells you the *slope* of the original function, not its height. Think of the y-axis on the first derivative graph as the *slope*-axis or the m-axis; you could think of general points on the first derivative graph as having coordinates (x, m).
- **A *decreasing* interval on a function corresponds to a *negative* interval on the graph of the derivative** (or *zero* for a single point if the function has a horizontal inflection point). The negative interval on the derivative graph is below the x-axis (or in the case of a horizontal inflection point, the derivative graph touches the x-axis at a single point). See intervals B, C, D, and E in Figure 11-11 (but consider them as a single section), where f goes down all the way from the local max at $(-2, 64)$ to the local min at $(2, -64)$ and where f' is negative between $(-2, 0)$ and $(2, 0)$ except for at the point $(0, 0)$ on f' which corresponds to the horizontal inflection point on f.
- **A local *min* on the graph of a function corresponds to a zero (an x-intercept) on an interval of the graph of its derivative that crosses the x-axis going up (like at $(2, 0)$).**

Now let's take a second trip along f to consider its intervals of concavity and its inflection points. First, consider intervals A and B in Figure 11-11. The graph of f is concave down — which means the same thing as a *decreasing* slope — until it gets to the inflection point at about $(-1.4, 39.6)$.

So, the graph of f' *decreases* until it bottoms out at about $(-1.4, -60)$. These coordinates tell you that the inflection point at -1.4 on f has a slope of -60. Note that the inflection point on f at $(-1.4, 39.6)$ is the steepest point on that stretch of the function, but it has the *smallest* slope because its slope is a larger *negative* than the slope at any other nearby point.

Between $(-1.4, 39.6)$ and the next inflection point at $(0, 0)$, f is concave up, which means the same thing as an *increasing* slope. So, the graph of f' *increases* from about -1.4 to where it hits a local max at $(0, 0)$. See interval C in Figure 11-11. Let's take a break from our trip for some more rules.

MATH RULES

More rules:

- **A concave *down* interval on the graph of a function corresponds to a *decreasing* interval on the graph of its derivative** (intervals A, B, and D in Figure 11-11). And a concave *up* interval on the function corresponds to an *increasing* interval on the derivative (intervals C, E, and F).
- **An *inflection point* on a function (except for a vertical inflection point where the derivative is undefined) corresponds to a *local extremum* on the graph of its derivative.** An inflection point of *minimum* slope (in its neighborhood) corresponds to a local *min* on the derivative graph; an inflection point of *maximum* slope (in its neighborhood) corresponds to a local *max* on the derivative graph.

Resuming our trip, after $(0, 0)$, f is concave down till the inflection point at about $(1.4, -39.6)$ — this corresponds to the decreasing section of f' from $(0, 0)$ to its min at $(1.4, -60)$ (interval D in Figure 11-11). Finally, f is concave up the rest of the way, which corresponds to the increasing section of f' beginning at $(1.4, -60)$ (intervals E and F in the figure).

Well, that pretty much brings you to the end of the road. Going back and forth between the graphs of a function and its derivative can be *very* trying at first. If your head starts to spin, take a break and come back to this stuff later.

If I haven't already succeeded in *deriving* you crazy — aren't these calculus puns fantastic? — perhaps this final point will do the trick. Look again at the graph of the derivative, f', in Figure 11-11 and also at the sign graph for f' in Figure 11-9. That sign graph, because it's a second derivative sign graph, bears exactly (well, almost exactly) the same relationship to the graph of f' as a first derivative sign graph bears to the graph of a regular function. In other words, *negative* intervals on the sign graph in Figure 11-9 (to the left of $-\sqrt{2}$ and between zero and $\sqrt{2}$) show you where the graph of f' is *decreasing*; *positive* intervals on the sign graph (between $-\sqrt{2}$ and zero and to the right of $\sqrt{2}$) show you where f' is *increasing*. And points where the signs switch from positive to negative or vice versa (at $-\sqrt{2}$, zero, and $\sqrt{2}$) show you where f' has local extrema. Clear as mud, right?

19 Given the graph of the derivative of *f* shown in the following figure, and the fact that *f* contains the points $(-2, 0)$, $(0, 2)$, and $(2, 0)$, sketch a graph of *f* showing its local extrema, its inflection points, and where it's concave up and concave down.

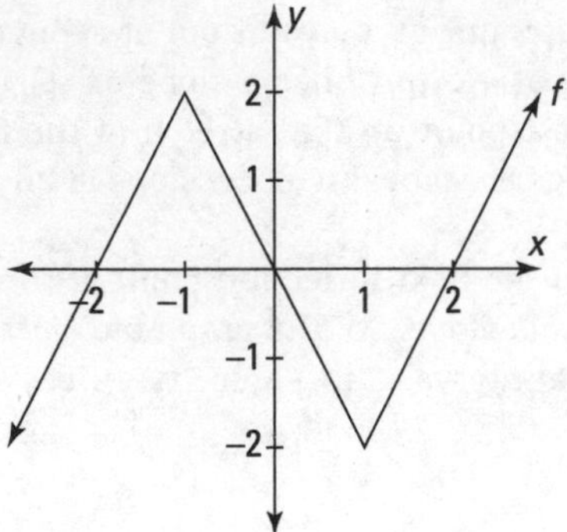

20 Given the graph of the derivative of *g* shown in the following figure, and the fact that *g* contains the point $(0, 1)$, sketch a graph of *g* and identify the location of its *x*-intercepts.

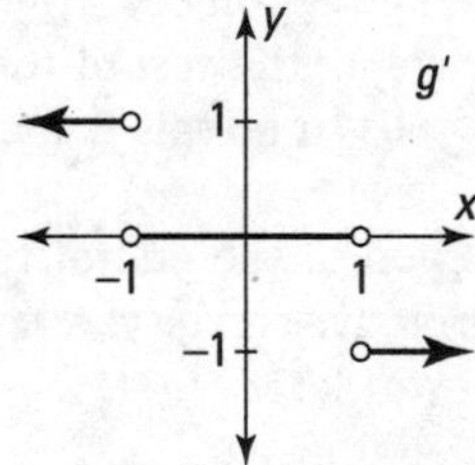

The Mean Value Theorem — Go Ahead, Make My Day

You won't use the Mean Value Theorem a lot, but it's a famous theorem — one of the two or three most important in all of calculus — so you really should learn it. It's very simple and has a nice connection to the Mean Value Theorem for integrals, which I show you in Chapter 17. Look at Figure 11-12.

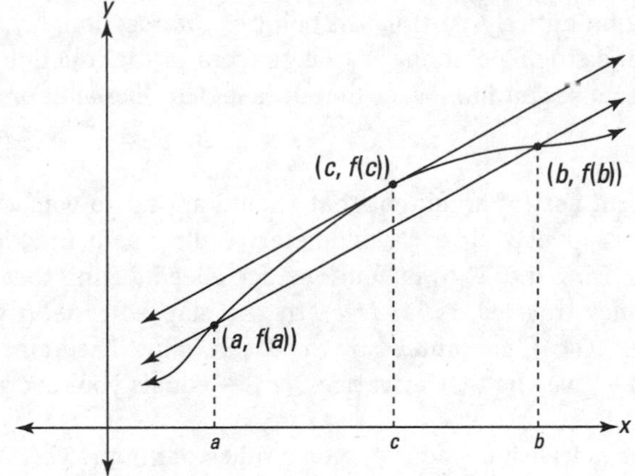

FIGURE 11-12: An illustration of the Mean Value Theorem.

Here's the formal definition of the theorem.

MATH RULES

The Mean Value Theorem: If f is continuous on the closed interval $[a, b]$ and differentiable on the open interval (a, b), then there exists at least one number c in (a, b) such that

$$f'(c) = \frac{f(b) - f(a)}{b - a}$$

Now for the plain-English version. First, you need to take care of the fine print. The requirements in the theorem that the function be continuous and differentiable guarantee that the function is a regular, smooth function without things like gaps or corners. Most (but not all) ordinary functions will satisfy these requirements, but don't neglect to check that they are in fact satisfied. (Keep your eyes peeled on quizzes and tests for functions that don't satisfy these requirements.)

Here's what the theorem means. The secant line connecting points $(a, f(a))$ and $(b, f(b))$ in Figure 11-12 has a slope given by the slope formula:

$$Slope = \frac{y_2 - y_1}{x_2 - x_1} = \frac{f(b) - f(a)}{b - a}$$

Note that this is the same as the right side of the equation in the Mean Value Theorem. The derivative at a point is the same thing as the slope of the tangent line at that point, so the theorem just says that there must be at least one point between a and b where the slope of the tangent is the same as the slope of the secant line from a to b. The result is parallel lines, like in Figure 11-12.

Why must this be so? Here's a visual argument. Imagine that you grab the secant line connecting $(a, f(a))$ and $(b, f(b))$, and then you slide it up, keeping it parallel to the original secant line. Can you see that the two points of intersection between this sliding line and the function — the two points that begin at $(a, f(a))$ and $(b, f(b))$ — will gradually get closer and closer to each other until they come together at $(c, f(c))$? If you raise the line any further, you break away from the function entirely. At this last point of intersection, $(c, f(c))$, the sliding line touches the function at a single point and is thus tangent to the function there, and it has the same slope as the original secant line. Well, that does it. This explanation is a bit oversimplified, but it'll do.

Here's a completely different sort of argument that should appeal to your common sense. If the function in Figure 11-12 gives your car's odometer reading as a function of time, then the slope of the secant line from a to b gives your average speed during that interval of time, because dividing the distance traveled, $f(b)-f(a)$, by the elapsed time, $b-a$, gives you the average speed. The point $(c, f(c))$, guaranteed by the Mean Value Theorem, is a point where your instantaneous speed — given by the derivative $f'(c)$ — equals your average speed.

Now, imagine that you take a drive and you average 50 miles per hour. The Mean Value Theorem says that there must be at least one point during your trip when your speed was exactly 50 mph. But you don't need a fancy-pants calculus theorem to tell you that. It's just common sense. Think about it. Your average speed can't be 50 mph if you go slower than 50 the whole way or if you go faster than 50 the whole way. To average 50 mph, either you go exactly 50 for the whole drive, or you have to go slower than 50 for part of the drive and faster than 50 at other times. In the former case, the theorem is obviously satisfied because you're driving at exactly 50 at every point in time. And in the latter case, the theorem is also satisfied because when you speed up or slow down from going slower than 50 to going faster than 50 (or vice versa), you have to hit exactly 50 mph at some point. You can't jump over 50 — like going 49 one moment then 51 the next — because speeds go up by *sliding* up the scale, not jumping. At some point, your speedometer slides past 50, and for at least one instant, you're going exactly 50 mph. That's all the Mean Value Theorem says.

With the Mean Value Theorem, you figure an average rate or slope over an interval and then use the first derivative to find one or more points in the interval where the instantaneous rate or slope equals the average rate or slope. Here's an example:

EXAMPLE

Q. Given $f(x) = x^3 - 4x^2 - 5x$, find all numbers c in the open interval $(2, 4)$ where the instantaneous rate equals the average rate over the interval.

A. **The only answer is $\dfrac{4+2\sqrt{7}}{3}$.**

Basically, you're finding the points along the curve in the interval where the slope is the same as the slope from $(2, f(2))$ to $(4, f(4))$. Mathematically speaking, you find all numbers c where $f'(c) = \dfrac{f(4)-f(2)}{4-2}$.

1. **Get the first derivative.**

$$f(x) = x^3 - 4x^2 - 5x$$
$$f'(x) = 3x^2 - 8x - 5$$

2. **Using the slope formula, $m = \frac{y_2 - y_1}{x_2 - x_1}$, figure the slope from $(2,\ f(2))$ to $(4,\ f(4))$.**

$$\begin{aligned} f(4) &= 4^3 - 4\cdot 4^2 - 5\cdot 4 \\ &= -20 \\ f(2) &= 2^3 - 4\cdot 2^2 - 5\cdot 2 \\ &= -18 \end{aligned} \qquad \begin{aligned} m &= \frac{f(4) - f(2)}{4 - 2} \\ &= \frac{-20 - (-18)}{2} \\ &= -1 \end{aligned}$$

3. **Set the derivative equal to this slope and solve.**

$$3x^2 - 8x - 5 = -1$$
$$3x^2 - 8x - 4 = 0$$
$$\begin{aligned} x &= \frac{8 \pm \sqrt{(-8)^2 - 4(3)(-4)}}{6} \\ &= \frac{8 \pm 4\sqrt{7}}{6} \\ &= \frac{4 + 2\sqrt{7}}{3} \quad \text{or} \quad \frac{4 - 2\sqrt{7}}{3} \\ &\approx 3.10 \quad \text{or} \quad \approx -0.43 \end{aligned}$$

Because -0.43 is outside the interval $(2,\ 4)$, your only answer is $\frac{4 + 2\sqrt{7}}{3}$.

YOUR TURN

For $g(x) = x^3 + x^2 - x$, find all the values c in the interval $(-2,\ 1)$ that satisfy the Mean Value Theorem.

For $s(t) = t^{4/3} - 3t^{1/3}$, find all the values of c in the interval $(0,\ 3)$ that satisfy the Mean Value Theorem.

Practice Questions Answers and Explanations

1 **Local min at $(0,\ 1)$; local max at $(1,\ 3)$.**

1. **Find the first derivative using the power rule.**

$$f(x)=6x^{2/3}-4x+1$$
$$f'(x)=4x^{-1/3}-4$$

2. **Find the critical numbers of *f*.**

a. Set the derivative equal to zero and solve.

$$4x^{-1/3}-4=0$$
$$x^{-1/3}=1$$
$$x=1$$

b. Determine the *x*-values where the derivative is undefined.

$$f'(x)=4x^{-1/3}-4=\frac{4}{\sqrt[3]{x}}-4$$

Because the denominator is not allowed to equal zero, $f'(x)$ is undefined at $x=0$. Thus the critical numbers of *f* are 0 and 1.

3. **Plot the critical numbers on a number line.**

I'm skipping the figure this time because I assume you can imagine a number line with dots at 0 and 1. Don't disappoint me!

4. **Plug a number from each of the three regions into the derivative.**

$$f'(-1)=4(-1)^{-1/3}-4=-4-4=-8$$
$$f'\left(\frac{1}{2}\right)=4\left(\frac{1}{2}\right)^{-1/3}-4=4(2)^{1/3}-4=\text{positive}$$
$$f'(8)=4(8)^{-1/3}-4=2-4=-2$$

Make your life easy. Note how the numbers I picked for the first and third computations made the math easy. With the second computation, you can save a little time and skip the final calculation because all you care about is whether the result is positive or negative (this assumes that you know that the cube root of 2 is more than 1 — you'd better!).

5. **Draw your sign graph.**

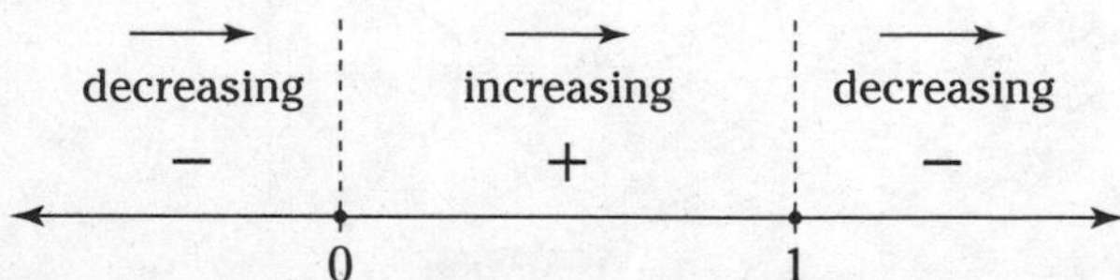

6. **Determine whether there's a local min or max or neither at each critical number.**

f goes down to where $x=0$ and then up, so there's a local min at $x=0$, and *f* goes up to where $x=1$ and then down, so there's a local max at $x=1$.

7. **Figure the y-value of the two local extrema.**

$$f(0)=6(0)^{2/3}-4(0)+1=1$$
$$f(1)=6(1)^{2/3}-4(1)+1=3$$

Thus, there's a local min at $(0, 1)$ and a local max at $(1, 3)$. Check this answer by looking at a graph of f on your graphing calculator.

(2) **Local max at $\left(\frac{\pi}{4}, \frac{\pi\sqrt{2}}{8}\right)$; local min at $\left(\frac{3\pi}{4}, \frac{3\pi\sqrt{2}}{8}-\sqrt{2}\right)$.**

1. **Find the first derivative.**

$$h(x)=\frac{x}{\sqrt{2}}+\cos x-\frac{\sqrt{2}}{2}$$
$$h'(x)=\frac{1}{\sqrt{2}}-\sin x$$

2. **Find the critical numbers of h.**

   a. Set the derivative equal to zero and solve:

$$\frac{1}{\sqrt{2}}-\sin x=0$$
$$\sin x=\frac{\sqrt{2}}{2}$$
$$x=\frac{\pi}{4} \text{ or } \frac{3\pi}{4} \quad \text{(These are the solutions in the given interval.)}$$

   b. Determine the x-values where the derivative is undefined.

   The derivative isn't undefined anywhere, so the critical numbers of h are $\frac{\pi}{4}$ and $\frac{3\pi}{4}$.

3. **Test numbers from each region on your number line.**

$$h'\left(\frac{\pi}{6}\right)=\frac{1}{\sqrt{2}}-\sin\frac{\pi}{6} = \frac{\sqrt{2}}{2}-\frac{1}{2} = \text{positive}$$
$$h'\left(\frac{\pi}{2}\right)=\frac{1}{\sqrt{2}}-\sin\frac{\pi}{2} = \frac{\sqrt{2}}{2}-1 = \text{negative}$$
$$h'(\pi)=\frac{1}{\sqrt{2}}-\sin\pi = \frac{\sqrt{2}}{2}-0 = \text{positive}$$

4. **Draw a sign graph.**

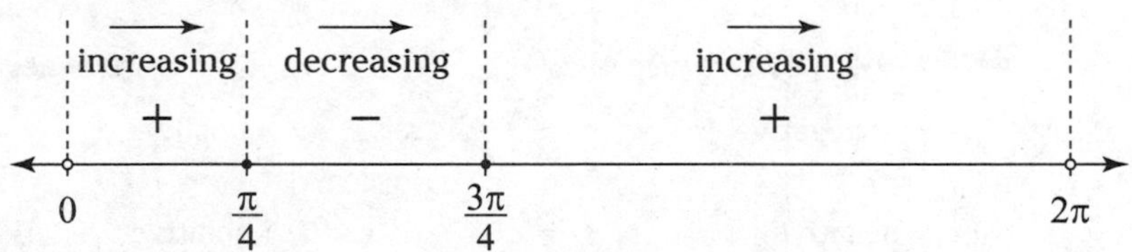

5. **Decide whether there's a local min, max, or neither at each of the two critical numbers.**

Going from left to right along the function, you go up until $x=\frac{\pi}{4}$ and then down, so there's a local max at $x=\frac{\pi}{4}$. It's vice versa for $x=\frac{3\pi}{4}$, so there's a local min there.

6. **Compute the y-values of these two extrema.**

$$h\left(\frac{\pi}{4}\right)=\frac{\frac{\pi}{4}}{\sqrt{2}}+\cos\frac{\pi}{4}-\frac{\sqrt{2}}{2} \qquad h\left(\frac{3\pi}{4}\right)=\frac{\frac{3\pi}{4}}{\sqrt{2}}+\cos\frac{3\pi}{4}-\frac{\sqrt{2}}{2}$$

$$=\frac{\pi}{4\sqrt{2}}+\frac{\sqrt{2}}{2}-\frac{\sqrt{2}}{2} \qquad =\frac{3\pi\sqrt{2}}{8}-\frac{\sqrt{2}}{2}-\frac{\sqrt{2}}{2}$$

$$=\frac{\pi\sqrt{2}}{8} \qquad =\frac{3\pi\sqrt{2}}{8}-\sqrt{2}$$

So, you have a max at $\left(\frac{\pi}{4},\ \frac{\pi\sqrt{2}}{8}\right)$ and a min at $\left(\frac{3\pi}{4},\ \frac{3\pi\sqrt{2}}{8}-\sqrt{2}\right)$.

(3) **Local mins at $(-2\sqrt{2},\ 0)$ and $(2\sqrt{2},\ 0)$; a local max at $(0,\ 4)$.**

Same basic steps as in Problems 1 and 2, but abbreviated a bit.

1. **Find the derivative.**

$$y=\left(x^2-8\right)^{2/3}$$

$$y'=\frac{2}{3}\left(x^2-8\right)^{-1/3}(2x)=\frac{4x}{3\sqrt[3]{x^2-8}}$$

2. **Find the critical numbers.**

a. $$\frac{4x}{3\sqrt[3]{x^2-8}}=0$$

$$x=0$$

b. The first derivative will be undefined when the denominator is zero, so

$$3\sqrt[3]{x^2-8}=0$$

$$\sqrt[3]{x^2-8}=0$$

$$x^2-8=0$$

$$x=\pm 2\sqrt{2}$$

Thus, the critical numbers are $-2\sqrt{2}$, 0, and $2\sqrt{2}$.

3. **Test a number from each of the four regions.**

$$y'(-10)=\frac{2}{3}\left((-10)^2-8\right)^{-1/3}(2\cdot(-10)) \qquad y'(-1)=\frac{2}{3}\left((-1)^2-8\right)^{-1/3}(2\cdot(-1))$$

$$=\frac{2}{3}(\text{positive})^{-1/3}\cdot\text{negative} \qquad =\frac{2}{3}(\text{negative})^{-1/3}\cdot\text{negative}$$

$$=\frac{2}{3}\cdot\text{positive}\cdot\text{negative} \qquad =\frac{2}{3}\cdot\text{negative}\cdot\text{negative}$$

$$=\text{negative} \qquad =\text{positive}$$

$$y'(1)=\text{negative} \qquad y'(10)=\text{positive}$$

4. **Make a sign graph.**

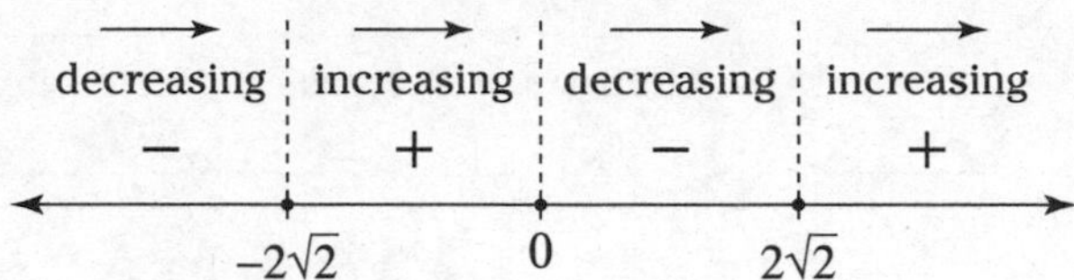

5. **Find the *y*-values at the critical numbers.**

$y = \left(\left(-2\sqrt{2} \right)^2 - 8 \right)^{2/3} = 0$. There's a local min at $\left(-2\sqrt{2},\ 0 \right)$.

$y = \left(0^2 - 8 \right)^{2/3} = (-8)^{2/3} = 4$. There's a local max at $(0,\ 4)$.

$y = \left(\left(2\sqrt{2} \right)^2 - 8 \right)^{2/3} = 0$. There's another local min at $\left(2\sqrt{2},\ 0 \right)$.

Check out this interesting curve on your graphing calculator.

(4) **Local maxes at $\left(-\sqrt{2},\ -2 \right)$ and $\left(\sqrt{2},\ -2 \right)$; no local minima.**

1. **Do the differentiation thing.**

$$s = \frac{t^4 + 4}{-2t^2}$$

$$s' = \frac{\left(t^4 + 4 \right)' \left(-2t^2 \right) - \left(t^4 + 4 \right) \left(-2t^2 \right)'}{\left(-2t^2 \right)^2} = \frac{\left(4t^3 \right) \left(-2t^2 \right) - \left(t^4 + 4 \right) (-4t)}{4t^4} = \frac{-t^4 + 4}{t^3}$$

2. **Find the critical numbers.**

$$\frac{-t^4 + 4}{t^3} = 0$$

$$4 - t^4 = 0$$

$$\left(2 - t^2 \right) \left(2 + t^2 \right) = 0$$

$$\left(\sqrt{2} - t \right) \left(\sqrt{2} + t \right) \left(2 + t^2 \right) = 0$$

$t = \sqrt{2}$ or $-\sqrt{2}$

So $-\sqrt{2}$ and $\sqrt{2}$ are two critical numbers of *s*.

REMEMBER

Numbers to include on your sign graph. Remember that $t = 0$ is a third important number because $t = 0$ makes the derivative's denominator equal zero, so you need to include zero on your sign graph in order to define test regions. Note, however, that $t = 0$ is *not* a critical number of *s* because *s* is undefined at $t = 0$. And because there is no point on *s* at $t = 0$, there can't be a local extremum at $t = 0$.

3. **Test a number from each of the four regions: You're on your own.**

4. **Make a sign graph.**

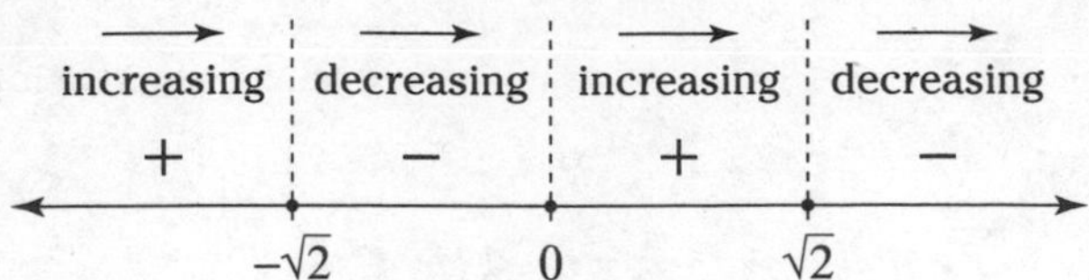

She loves me; she loves me not; she loves me; she loves me not.

5. **Find the y-values.**

$s(-\sqrt{2})=\frac{(-\sqrt{2})^4+4}{-2(-\sqrt{2})^2}=\frac{4+4}{-4}=-2$. You climb up the hill to $(-\sqrt{2},\ -2)$, then down, so there's a local max there.

$s(0)=\frac{0^4+4}{-2(0)^2}=$ undefined (which you already knew). Therefore, there's no local extremum at $t=0$. Remember that if a problem asks you to identify only the x-values and not the y-values of the local extrema, and you only consider the sign graph, you will incorrectly conclude — using the current problem as an example — that there's a local min at $t=0$. So you should always check where your function is undefined.

$s(\sqrt{2})=\frac{\sqrt{2}^4+4}{-2\sqrt{2}^2}=-2$. Up, then down again, so there's another local max at $(\sqrt{2},\ -2)$.

As always, you should check out this function on your graphing calculator.

(5) **Local max at $x=\frac{\pi}{4}$; local min at $x=\frac{3\pi}{4}$.**

1. **Find the second derivative.**

$$h(x)=\frac{x}{\sqrt{2}}+\cos x-\frac{\sqrt{2}}{2}$$
$$h'(x)=\frac{1}{\sqrt{2}}-\sin x$$
$$h''(x)=-\cos x$$

2. **Plug in the critical numbers (from Problem 2).**

$$h''\left(\frac{\pi}{4}\right)=-\cos\frac{\pi}{4}=-\frac{\sqrt{2}}{2} \qquad h''\left(\frac{3\pi}{4}\right)=-\cos\frac{3\pi}{4}=\frac{\sqrt{2}}{2}$$

You're done. You determine that h is concave down at $x=\frac{\pi}{4}$, so there's a local max there, and h is concave up at $x=\frac{3\pi}{4}$, so there's a local min at that x-value. (In Problem 2, you already determined the y-values for these extrema.)

Note that h is an example of a function where the second derivative test is quick and easy.

6 **Local min at $(0, 1)$; local max at $(2, 9)$.**

1. **Find the critical numbers.**

$$f(x) = -2x^3 + 6x^2 + 1$$
$$f'(x) = -6x^2 + 12x$$
$$0 = -6x^2 + 12x$$
$$0 = -6x(x-2)$$
$$x = 0,\ 2$$

2. **Find the second derivative.**

$$f'(x) = -6x^2 + 12x$$
$$f''(x) = -12x + 12$$

3. **Plug in the critical numbers.**

$$f''(0) = -12\ (0) + 12 = 12 \quad \text{(concave up: min)}$$
$$f''(2) = -12\ (2) + 12 = -12 \quad \text{(concave down: max)}$$

4. **Determine the *y*-coordinates for the extrema.**

$$f(0) = -2(0)^3 + 6(0)^2 + 1 = 1$$
$$f(2) = -2(2)^3 + 6(2)^2 + 1 = 9$$

So, there's a min at $(0,\ 1)$ and a max at $(2,\ 9)$.

Note that *f* is another function where the second derivative test works like a charm.

7 **You find local maxes at $x = -2$ and $x = 2$ with the second derivative test; you find a local min at $x = 0$ with street smarts.**

1. **Find the critical numbers.**

$$y = 2x^4 - \frac{1}{3}x^6$$
$$y' = 8x^3 - 2x^5$$
$$8x^3 - 2x^5 = 0$$
$$2x^3(4 - x^2) = 0$$
$$2x^3(2-x)(2+x) = 0$$

Thus, $x = 0,\ 2,\ -2$.

2. **Get the second derivative.**

$$y' = 8x^3 - 2x^5$$
$$y'' = 24x^2 - 10x^4$$

3. **Plug in.**

$$y''(-2) = 24(-2)^2 - 10(-2)^4 = 96 - 160 = \text{negative, thus a max.}$$

$$y''(0) = 24(0)^2 - 10(0)^4 = 0\text{, thus inconclusive.}$$

$$y''(2) = 24(2)^2 - 10(2)^4 = \text{same as } y''(-2) = \text{negative, thus a max.}$$

The second derivative test fails at $x = 0$, so you have to use the first derivative test for that critical number.

TIP

When the first derivative test may be preferable. If — as in the function for this problem — one of the critical numbers is $x = 0$, and you can quickly see that the second derivative will equal zero at $x = 0$ (because, for example, all the terms of the second derivative will be simple powers of x), then the second derivative test will fail for $x = 0$, and you might want to use the first derivative test instead.

However, because this problem involves a continuous function and because there's only one critical number between the two maxes you found, the only possibility is that there's a min at $x = 0$. (Try this streetwise logic out on your teacher and let me know if it works.)

Your pick should be $s = 8 + \frac{21t}{4} - \frac{7t^3}{4}$; local min at $(-1, 4.5)$ and local max at $(1, 11.5)$.

TIP

The second derivative test fails where the second derivative is undefined (in addition to failing where the second derivative equals zero).

To pick, look at the first derivative of each function:

$$y = \left(x^2 - 8\right)^{2/3}$$

$$y' = \frac{2}{3}\left(x^2 - 8\right)^{-1/3}(2x) = \frac{4x}{3\left(x^2 - 8\right)^{1/3}}$$

$$s = 8 + \frac{21t}{4} - \frac{7t^3}{4}$$

$$s' = \frac{21}{4} - \frac{21}{4}t^2$$

Do you see the trouble you're going to run into with $y(x)$? The first derivative is undefined at $x = \pm 2\sqrt{2}$. And the second derivative will also be undefined at those x-values, because when you take the second derivative with the quotient rule, squaring the bottom, the denominator will contain that same factor, $\left(x^2 - 8\right)$. The second derivative test will thus fail at $\pm 2\sqrt{2}$, and you'll have to use the first derivative test. In contrast to $y(x)$, the second derivative test works great with $s(t)$:

1. **Get the critical numbers.**

$$s' = \frac{21}{4} - \frac{21}{4}t^2$$

$$0 = \frac{21}{4} - \frac{21}{4}t^2$$

$$\frac{21}{4}t^2 = \frac{21}{4}$$

$$t = \pm 1$$

s' is not undefined anywhere, so -1 and 1 are the only critical numbers.

2. **Do the second derivative.**

$$s' = \frac{21}{4} - \frac{21}{4}t^2$$

$$s'' = -\frac{21}{2}t$$

3. **Plug in the critical numbers.**

$$s''(-1) = \frac{21}{2} \quad \text{(concave up: min)}$$

$$s''(1) = -\frac{21}{2} \quad \text{(concave down: max)}$$

4. **Get the heights of the extrema.**

$$s(-1) = 8 + \frac{21(-1)}{4} - \frac{7(-1)^3}{4} = 4.5$$

$$s(1) = 8 + \frac{21(1)}{4} - \frac{7(1)^3}{4} = 11.5$$

You're done; s has a local min at $(-1,\ 4.5)$ and a local max at $(1,\ 11.5)$.

(9) **Absolute max at $\left(\frac{\pi}{4}, \sqrt{2}\right)$; absolute min at $\left(\frac{5\pi}{4}, -\sqrt{2}\right)$.**

1. **Find critical numbers.**

$$f(x) = \sin x + \cos x$$

$$f'(x) = \cos x - \sin x$$

$$0 = \cos x - \sin x$$

$$\sin x = \cos x \quad \text{(divide both sides by } \cos x\text{)}$$

$$\tan x = 1$$

$$x = \frac{\pi}{4}, \frac{5\pi}{4} \quad \text{(the solutions in the given interval)}$$

The derivative is never undefined, so these are the only critical numbers.

WARNING

If you divide both sides of an equation by something that can equal zero at one or more x-values (like you did here when dividing both sides by $\cos x$), you may miss one or more solutions. You have to check whether any of those x-values is a solution. In this problem, $\cos x = 0$ at $\frac{\pi}{2}$ and $\frac{3\pi}{2}$, and it's easy to check (in Line 4 of Step 1) that $\sin x$ does not equal $\cos x$ at either of those values, so there's no problem here. But if $\sin x$ did equal $\cos x$ at either of those values, you'd have one or two more solutions and one or two more critical numbers. (Note that you have to check any such values in the line of the solution immediately above the line where you do the dividing — the way you just used Line 4; you couldn't use Line 5 for the check.)

2. **Evaluate the function at the critical numbers.**

$$f\left(\frac{\pi}{4}\right) = \sin\frac{\pi}{4} + \cos\frac{\pi}{4} = \frac{\sqrt{2}}{2} + \frac{\sqrt{2}}{2} = \sqrt{2}$$

$$f\left(\frac{5\pi}{4}\right) = \sin\frac{5\pi}{4} + \cos\frac{5\pi}{4} = -\frac{\sqrt{2}}{2} - \frac{\sqrt{2}}{2} = -\sqrt{2}$$

3. **Evaluate the function at the endpoints of the interval.**

$$f(0) = \sin 0 + \cos 0 = 1$$
$$f(2\pi) = \sin(2\pi) + \cos(2\pi) = 1$$

4. **The largest of the four answers from Steps 2 and 3 is the absolute max; the smallest is the absolute min.**

The absolute max is at $\left(\frac{\pi}{4},\ \sqrt{2}\right)$. The absolute min is at $\left(\frac{5\pi}{4},\ -\sqrt{2}\right)$.

10 **Absolute min at $(-0.5, -6)$; absolute max at $(0, -5)$.**

1. **Find critical numbers.**

$$\begin{aligned} g(x) &= 2x^3 - 3x^2 - 5 \\ g'(x) &= 6x^2 - 6x \\ 0 &= 6x^2 - 6x \\ 0 &= 6x(x-1) \\ x &= 0,\ 1 \end{aligned}$$

The solution $x = 1$ is rejected because it's outside the given interval; $x = 0$ is your only critical number.

2. **Evaluate the function at $x = 0$.**

$$g(0) = 2(0)^3 - 3(0)^2 - 5 = -5$$

3. **Do the endpoint thing.**

$$\begin{aligned} g(-0.5) &= 2(-0.5)^3 - 3(-0.5)^2 - 5 \\ &= 2(-0.125) - 3(0.25) - 5 \\ &= -6 \end{aligned}$$

$$\begin{aligned} g(0.5) &= 2(0.5)^3 - 3(0.5)^2 - 5 \\ &= 2(0.125) - 3(0.25) - 5 \\ &= -5.5 \end{aligned}$$

4. **Pick the smallest and largest answers from Steps 2 and 3.**

The absolute min is at the left endpoint, $(-0.5,\ -6)$. The absolute max is smack dab in the middle, $(0,\ -5)$.

(11) Absolute max at $(-2, 2)$; absolute mins at $(-1, 0.5)$ and $(31, 0.5)$.

I think you know the steps by now.

$$p(x)=(x+1)^{4/5}-0.5x$$
$$p'(x)=\frac{4}{5}(x+1)^{-1/5}-0.5$$
$$=\frac{4}{5(x+1)^{1/5}}-0.5$$
$$0=\frac{4}{5(x+1)^{1/5}}-0.5$$
$$0.5=\frac{4}{5(x+1)^{1/5}}$$

$$2.5(x+1)^{1/5}=4$$
$$(x+1)^{1/5}=\frac{8}{5}$$
$$(x+1)=\left(\frac{8}{5}\right)^5$$
$$x=9.48576$$

That's one critical number, but $x=-1$ is also one because it produces an undefined derivative.

$$p(-1)=(-1+1)^{4/5}-0.5(-1)$$
$$=0.5$$

$$p(9.48576)=(9.48576+1)^{4/5}-0.5(9.48576)$$
$$=1.81072$$

Left endpoint: $p(-2)=(-2+1)^{4/5}-0.5(-2)=2$

Right endpoint: $p(31)=(31+1)^{4/5}-0.5(31)=16-15.5=0.5$

Your absolute max is at the left endpoint: $(-2, 2)$. There's a tie for the absolute min: at the cusp $(-1, 0.5)$ and at the right endpoint $(31, 0.5)$.

(12) Absolute min at $\left(-\frac{\pi}{2}, -6\right)$; absolute maxes at $\left(\frac{\pi}{6}, 3\right)$ and $\left(\frac{5\pi}{6}, 3\right)$.

$$q(x)=2\cos(2x)+4\sin x$$
$$q'(x)=-2\sin(2x)\cdot 2+4\cos x$$
$$0=-4\sin(2x)+4\cos x$$
$$0=\sin(2x)-\cos x \qquad \text{(dividing by } -4\text{)}$$
$$0=2\sin x\cos x-\cos x \qquad \text{(trig identity)}$$
$$0=\cos x(2\sin x-1)$$

$$0=\cos x \qquad\qquad 2\sin x-1=0$$
$$x=-\frac{\pi}{2}, \frac{\pi}{2} \qquad \text{or} \qquad \sin x=\frac{1}{2}$$
$$x=\frac{\pi}{6}, \frac{5\pi}{6}$$

MATH RULES

Technically, $x = -\frac{\pi}{2}$ is not one of the critical numbers; being at an endpoint of an interval, it is refused membership in the critical number club. It's a moot point, though, because you have to evaluate the endpoints anyway.

$$\begin{aligned} q\left(\frac{\pi}{6}\right) &= 2\cos\left(2\cdot\frac{\pi}{6}\right) + 4\sin\frac{\pi}{6} \\ &= 2\cdot\frac{1}{2} + 4\cdot\frac{1}{2} = 3 \\ q\left(\frac{\pi}{2}\right) &= 2\cos\left(2\cdot\frac{\pi}{2}\right) + 4\sin\frac{\pi}{2} \\ &= -2 + 4 = 2 \\ q\left(\frac{5\pi}{6}\right) &= 2\cos\left(2\cdot\frac{5\pi}{6}\right) + 4\sin\frac{5\pi}{6} \\ &= 2\cdot\frac{1}{2} + 4\cdot\frac{1}{2} = 3 \end{aligned}$$

Left endpoint: $q\left(-\frac{\pi}{2}\right) = 2\cos\left(2\cdot\left(-\frac{\pi}{2}\right)\right) + 4\sin\left(-\frac{\pi}{2}\right) = -2 + 4(-1) = -6$

Right endpoint: $q\left(\frac{5\pi}{4}\right) = 2\cos\left(2\cdot\frac{5\pi}{4}\right) + 4\sin\frac{5\pi}{4} = 2\cdot 0 + 4\left(-\frac{\sqrt{2}}{2}\right) \approx -2.828$

Pick your winners: absolute min at left endpoint $\left(-\frac{\pi}{2},\ -6\right)$, and a tie for absolute max at $\left(\frac{\pi}{6},\ 3\right)$ and $\left(\frac{5\pi}{6},\ 3\right)$.

13 **The function's absolute min of –1 occurs at two points, $(-1, -1)$ and $(1, -1)$; *f* has no absolute max.**

Find the height of the function at each of its critical numbers; in other words, find the function's critical points.

$$\begin{aligned} f(x) &= x^4 - 2x^2 \\ f'(x) &= 4x^3 - 4x \\ &= 4x(x^2 - 1) \\ &= 4x(x-1)(x+1) \end{aligned}$$

So, $x = 0$, $x = 1$, and $x = -1$ are the critical numbers, and, since $f(0) = 0^4 - 2\cdot 0^2 = 0$, $f(1) = 1^4 - 2\cdot 1^2 = -1$, and $f(-1) = (-1)^4 - 2\cdot(-1)^2 = -1$, the critical points of *f* are $(-1,\ -1)$, $(0,\ 0)$, and $(1,\ -1)$. That's the complete list of candidates for the absolute extrema of *f*.

When you look at a graph of *f*, you'll see that its absolute min of –1 occurs at two points, $(-1,\ -1)$ and $(1,\ -1)$. And, because *f* obviously rises higher than the highest of the critical points, $(0,\ 0)$, *f* has no absolute max.

A graph of *f* makes this solution obvious, but, if you want to be thorough, you should confirm that the absolute min is –1 with the first derivative test. That will show that *f* decreases from negative infinity to –1, then increases from –1 to 0, then decreases from 0 to 1, and, finally, increases from 1 to infinity. That confirms that fact that *f* never gets below –1.

14 **(a) Function A has neither an absolute max nor an absolute min.**

As I'm sure you know, when you see an arrow like the one on the left on function A, you have to assume that that line (actually a ray with an open endpoint) continues in the direction shown indefinitely. Duh. Okay, so function A goes up forever to the left. And, thus — since infinity does not qualify as a max — function A has no absolute max.

Nor does it have an absolute min. You can see that the function goes down to the hollow dot at −1. If that dot were solid, the absolute min would be −1. But because it's hollow, there is no absolute min. Is the min −0.99999? No, function A goes below that: say, to −0.999999999. No matter what number you pick above −1 as a candidate for the absolute minimum, I can always give you another number below your candidate number and above −1. So, no number can qualify as the absolute minimum.

(b) Function B has no absolute max. Its absolute min is −1

(c) Function C has no absolute max. Its absolute min is −1.

(d) The absolute maximum of Function D is 1. This maximum height occurs at $(0,\ 1)$ and at all x-values in the interval $(-\infty,\ -2)$ along the line $y=1$. **Function D has no absolute minimum.**

15 **The function f is concave up from negative infinity to the inflection point at $(1, -1)$, then concave down from there to positive infinity.**

1. **Get the second derivative.**

$$f(x)=-2x^3+6x^2-10x+5$$
$$f'(x)=-6x^2+12x-10$$
$$f''(x)=-12x+12$$

2. **Set equal to 0 and solve.**

$$-12x+12=0$$
$$x=1$$

3. **Check for x-values where the second derivative is undefined. There are none.**

4. **Test your two regions — to the left and to the right of $x=1$ — and make your sign graph.**

$$f''(x)=-12x+12$$

$$f''(0)=12$$
$$f''(2)=-12$$

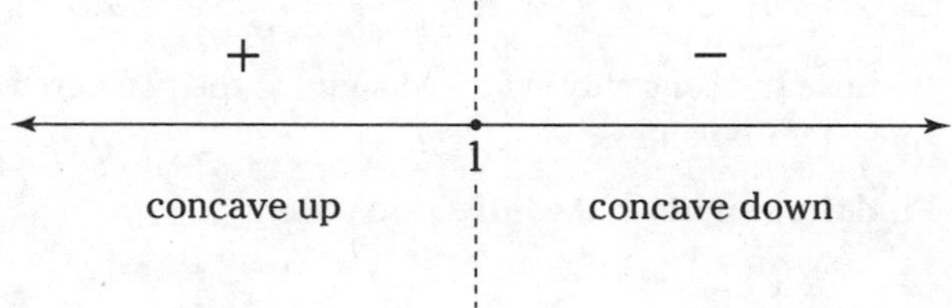

Because the concavity switches at $x=1$ and because f'' equals zero there, there's an inflection point at $x=1$.

5. **Find the height of the inflection point.**

$$f(x) = -2x^3 + 6x^2 - 10x + 5$$
$$f(1) = -1$$

Thus, *f* is concave up from negative infinity to the inflection point at $(1, -1)$, and then concave down from there to infinity. As always, you should check your result on your graphing calculator. *Hint:* To get a good feel for the look of this function, you need a fairly odd graphing window — try something like Xmin = −2, Xmax = 4, Ymin = −20, Ymax = 20.

(16) **The function *g* is concave up from negative infinity to the inflection point at $\left(-\sqrt{2}, -20\right)$, then concave down to an inflection point at $\left(\sqrt{2}, -20\right)$, then concave up again to infinity.**

1. **Find the second derivative.**

$$g(x) = x^4 - 12x^2$$
$$g'(x) = 4x^3 - 24x$$
$$g''(x) = 12x^2 - 24$$

2. **Set to 0 and solve.**

$$12x^2 - 24 = 0$$
$$x^2 = 2$$
$$x = \pm\sqrt{2}$$

3. **Is the second derivative undefined anywhere? No.**

4. **Test the three regions and make a sign graph. See the following figure.**

$$g''(x) = 12x^2 - 24$$

$$g''(-2) = 24$$
$$g''(0) = -24$$
$$g''(2) = 24$$

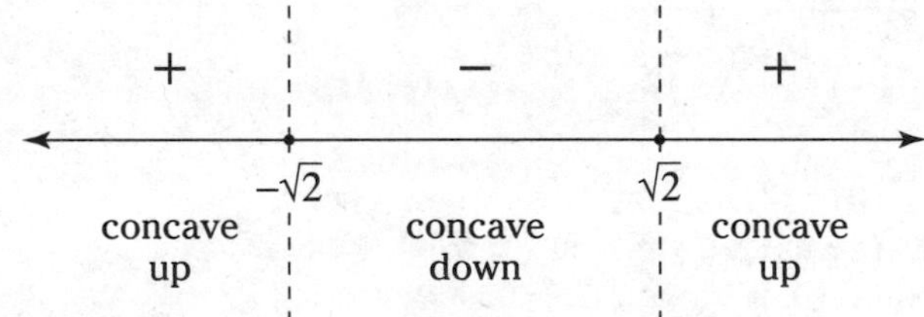

Because the concavity switched signs at the two zeros of g'', there are inflection points at these two *x*-values.

5. **Find the heights of the inflection points.**

$$g(x) = x^4 - 12x^2$$

$$g\left(-\sqrt{2}\right) = -20$$
$$g\left(\sqrt{2}\right) = -20$$

The function g is concave up from negative infinity to the inflection point at $\left(-\sqrt{2},\ -20\right)$, concave down from there to another inflection point at $\left(\sqrt{2},\ -20\right)$, and then concave up again from there to infinity.

(17) **Concave down from negative infinity to an inflection point at $\left(-3\sqrt{3}, -\frac{\sqrt{3}}{12}\right)$, then concave up till the inflection point at $(0, 0)$, then concave down again till the third inflection point at $\left(3\sqrt{3}, \frac{\sqrt{3}}{12}\right)$, and, finally, concave up to infinity.**

1. **Get the second derivative.**

$$p'(x)=\frac{x'\left(x^2+9\right)-x\left(x^2+9\right)'}{\left(x^2+9\right)^2}$$
$$=\frac{x^2+9-2x^2}{\left(x^2+9\right)^2}$$
$$=\frac{9-x^2}{\left(x^2+9\right)^2}$$

$$p''=\frac{\left(9-x^2\right)'\left(x^2+9\right)^2-\left(9-x^2\right)\left(\left(x^2+9\right)^2\right)'}{\left(x^2+9\right)^4}$$
$$=\frac{-2x\left(x^2+9\right)^2-\left(9-x^2\right)2\left(x^2+9\right)2x}{\left(x^2+9\right)^4}$$
$$=\frac{\left(x^2+9\right)\left[-2x\left(x^2+9\right)-4x\left(9-x^2\right)\right]}{\left(x^2+9\right)^4}$$
$$=\frac{-2x^3-18x-36x+4x^3}{\left(x^2+9\right)^3}$$
$$=\frac{2x\left(x^2-27\right)}{\left(x^2+9\right)^3}$$

2. **Set equal to zero and solve.**

$$\frac{2x\left(x^2-27\right)}{\left(x^2+9\right)^3}=0$$
$$2x(x^2-27)=0$$

$$2x=0 \qquad \text{or} \qquad x^2-27=0$$
$$x=0 \qquad\qquad x=\pm 3\sqrt{3}$$

3. **Check for undefined points of the second derivative; there are none.**

4. **Test four regions with the second derivative. You can skip the sign graph.**

TIP

Doing mental math builds math muscles. You might be able to do all of this in your head (try it!) because all that matters is whether the answers are positive or negative.

$$p''=\frac{2x\left(x^2-27\right)}{\left(x^2+9\right)^3}$$

$$p''(-10)=\frac{2(-10)\left((-10)^2-27\right)}{\left((-10)^2+9\right)^3} \qquad p''(-1)=\frac{2(-1)\left((-1)^2-27\right)}{\left((-1)^2+9\right)^3}$$

$$=\frac{2(N)(P)}{P^3} \qquad =\frac{2(N)(N)}{P^3}$$

$$=\frac{N}{P} \qquad =\frac{P}{P}$$

$$=N \qquad =P$$

$$p''(1)=\frac{2(1)\left(1^2-27\right)}{\left(1^2+9\right)^3} \qquad p''(10)=\frac{2(10)\left(10^2-27\right)}{\left(10^2+9\right)^3}$$

$$=\frac{2(P)(N)}{P^3} \qquad =\frac{2(P)(P)}{P^3}$$

$$=\frac{N}{P} \qquad =\frac{P}{P}$$

$$=N \qquad =P$$

The concavity goes *negative, positive, negative, positive,* so there's an inflection point at each of the three zeros of p''.

5. **Find the heights of the inflection points.**

$$p(x)=\frac{x}{x^2+9}$$

$$p\left(-3\sqrt{3}\right)=\frac{-3\sqrt{3}}{\left(-3\sqrt{3}\right)^2+9} \qquad p(0)=0 \qquad p\left(3\sqrt{3}\right)=\frac{3\sqrt{3}}{\left(3\sqrt{3}\right)^2+9}$$

$$=\frac{-3\sqrt{3}}{27+9}=\frac{-\sqrt{3}}{12} \qquad =\frac{\sqrt{3}}{12}$$

Taking a drive on highway p, you'll be turning right from negative infinity to $\left(-3\sqrt{3},\ -\frac{\sqrt{3}}{12}\right)$, then you'll be turning left till $(0,\ 0)$, then right again till $\left(3\sqrt{3},\ \frac{\sqrt{3}}{12}\right)$, and on your final leg to infinity, you round a *very long* bend to the left. (At each of the three inflection points, you'd be going straight for an infinitesimal moment.)

(18) **Concave down from negative infinity till an inflection point at about $(-0.085,\ -0.171)$, then concave up till a vertical inflection point at $(0,\ 0)$, then concave down till a third inflection point at about $(0.085,\ 0.171)$, then concave up out to infinity.**

You know the routine.

$$q(x)=\sqrt[5]{x}-\sqrt[3]{x}$$

$$q'(x)=\frac{1}{5}x^{-4/5}-\frac{1}{3}x^{-2/3}$$

$$q''(x)=\frac{-4}{25}x^{-9/5}+\frac{2}{9}x^{-5/3}$$

$$0=\frac{-4}{25x^{9/5}}+\frac{2}{9x^{5/3}}$$

Whoops, I guess this algebra's kind of messy. Better get the zeros on your calculator: Just graph and find the x-intercepts. There are two: $x \approx -0.085$ and $x \approx 0.085$. So you have two "critical numbers," right? Wrong! Don't forget to check for undefined points of the second derivative. Because $q''(x) = \frac{-4}{25x^{9/5}} + \frac{2}{9x^{5/3}}$, q'' is undefined at $x = 0$. Since $q(x)$ is defined at $x = 0$, zero is another "critical number." So, you have three "critical numbers" and four regions. You can test them with −1, −0.01, 0.01, and 1:

$$q''(x) = \frac{-4}{25}x^{-9/5} + \frac{2}{9}x^{-5/3}$$

$$q''(-1) = -\frac{14}{225} \qquad q''(-0.01) \approx 158 \qquad q''(0.01) \approx -158 \qquad q''(1) = \frac{14}{225}$$

Thus the concavity goes *down, up, down, up*. Because the second derivative is zero at −0.085 and 0.085, and because the concavity switches there, you can conclude that there are inflection points at those two x-values. But because both the first and second derivatives are undefined at $x = 0$, you have to check whether there's a vertical tangent there. You can see that there is by just looking at the graph, but if you want to be rigorous about it, you figure the limit of the first derivative as x approaches zero. Since that equals infinity, you have a vertical tangent at $x = 0$, and thus there's an inflection point there.

Now plug −0.085, 0, and 0.085 into q to get the y-values, and you're done.

19 **Absolute precision is not the point of this exercise, but your graph should look close to the graph of *f* shown here.**

Note the following features of the graph of *f*. Your graph should show local mins at $(-2, 0)$ and $(2, 0)$ and a local max at $(0, 2)$. And your graph should show inflection points at roughly $(-1, 1)$ and $(1, 1)$. Finally, your graph should show that *f* is concave up everywhere to the left of $(-1, 1)$ and everywhere to the right of $(1, 1)$. In between those points, *f* should be concave down.

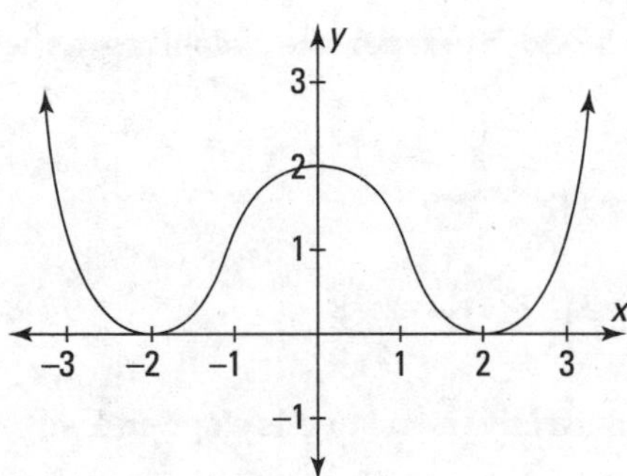

(20) Unlike with Problem 19, your graph of g for this problem should be precisely what is shown here.

You were given that g contains the point $(0,\ 1)$, and the graph of g' shows that the derivative of g (its slope) is zero between -1 and 1. These facts tell you that there's a section of g that's a horizontal line between $(-1,\ 1)$ and $(1,\ 1)$. To the left of -1, the graph of g' shows that the derivative (the slope) equals 1. Thus, going left from $(-1,\ 1)$, g must go down one, left one, crossing the x-axis at -2. The logic is very similar going right from $(1,\ 1)$ to the x-intercept at 2.

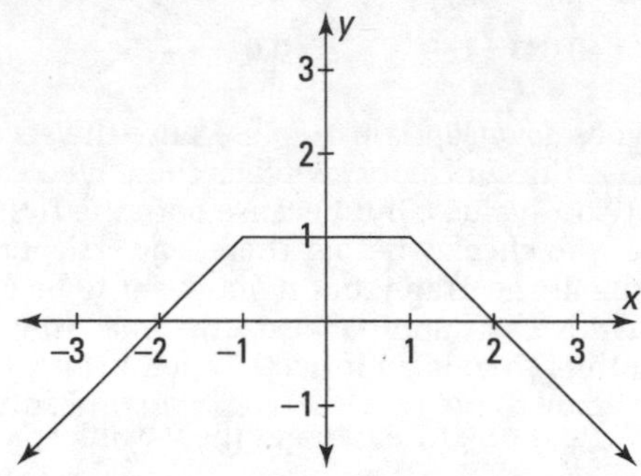

(21) The values of c are $\frac{-1-\sqrt{7}}{3}$ and $\frac{-1+\sqrt{7}}{3}$.

1. **Find the first derivative.**

$$g(x) = x^3 + x^2 - x$$
$$g'(x) = 3x^2 + 2x - 1$$

2. **Figure the slope between the endpoints of the interval.**

$$g(-2) = (-2)^3 + (-2)^2 - (-2) \qquad\qquad g(1) = 1$$
$$= -2$$

$$m = \frac{g(-2) - g(1)}{-2-1} = \frac{-2-1}{-2-1} = 1$$

3. **Set the derivative equal to this slope and solve.**

$$3x^2 + 2x - 1 = 1$$
$$3x^2 + 2x - 2 = 0$$

$$x = \frac{-2 \pm \sqrt{4-(-24)}}{6}$$
$$= \frac{-2 \pm 2\sqrt{7}}{6}$$
$$= \frac{-1-\sqrt{7}}{3} \text{ or } \frac{-1+\sqrt{7}}{3}$$

Both are inside the given interval, so you have two answers.

(22) **The value of c is $\frac{3}{4}$.**

1. **Find the first derivative.**

$$s(t) = t^{4/3} - 3t^{1/3}$$

$$s'(t) = \frac{4}{3}t^{1/3} - t^{-2/3}$$

2. **Figure the slope between the endpoints of the interval.**

$$s(0) = 0$$

$$s(3) = 3^{4/3} - 3 \cdot 3^{1/3} = 0$$

$$m = \frac{s(3) - s(0)}{3 - 0} = \frac{0 - 0}{3} = 0$$

3. **Set the derivative equal to the result from Step 2 and solve.**

$$\frac{4}{3}t^{1/3} - t^{-2/3} = 0$$

$$t^{-2/3}\left(\frac{4}{3}t^{1} - 1\right) = 0$$

$$t^{-2/3} = 0 \quad \text{or} \quad \frac{4}{3}t^{1} - 1 = 0$$

$$\varnothing \qquad\qquad t = \frac{3}{4}$$

Graph s to confirm that its slope at $t = \frac{3}{4}$ is zero.

If you're ready to test your skills a bit more, take the following chapter quiz that incorporates all the chapter topics.

Whaddya Know? Chapter 11 Quiz

Quiz time! Complete each problem to test your knowledge on the various topics covered in this chapter. You can then find the solutions and explanations in the next section.

1 Use the first derivative test to find any local extrema of $f(x)=e^x-x$.

2 Use the first derivative test to find any local extrema of $g(x)=x^2-\ln x$.

3 Use the second derivative test to find any local extrema of f from Problem 1.

4 Use the second derivative test to find any local extrema of g from Problem 2.

5 Find the absolute max and min of $f(x)=e^x+e^{-x}$ on the interval $[-\ln 2,\ 2]$.

6 Find the absolute max and min of $g(x)=\ln\left(e^{x^2}\right)$ on the interval $[-1,\ 1]$.

7 Find the absolute max and min of $f(x)=\sqrt[3]{x}-x^2$ over its entire domain.

8 Find the absolute max and min of $g(x)=\sqrt[4]{x^2}$ over its entire domain.

9 Find the intervals of concavity and the inflection points of $f(x)=\cos x+\dfrac{1}{2}x^2$.

10 Find the intervals of concavity and the inflection points of $g(x)=2x^6-5x^4$.

11 **a.** For $f(x)=\dfrac{1}{x-1}$, find all values c in the interval $(0,\ 2)$ that satisfy the Mean Value Theorem.

b. For $g(x)=\sqrt[3]{x}$, find all values c in the interval $(-1,\ 1)$ that satisfy the Mean Value Theorem.

12 For $g(x)=\sqrt[3]{x}$, find all values c in the interval $(0,\ 8)$ that satisfy the Mean Value Theorem.

Answers to Chapter 11 Quiz

(1) **There's a local min at $(0, 1)$.** That's the only local extremum.

1. **Find the first derivative of *f* using the e^x rule and the power rule.**

   $$f(x) = e^x - x$$
   $$f'(x) = e^x - 1$$

2. **Set the derivative equal to zero and solve for *x* to get the critical numbers of *f*.**

   $$e^x - 1 = 0$$
   $$e^x - 1$$
   $$x = 0$$

   If the first derivative were undefined for some *x*-value(s) in the domain of *f*, there could be more critical numbers, but because $f'(x) = e^x - 1$ is defined for all real numbers, 0 is the only critical number of *f*.

3. **Plot the single critical number on a number line, noting that it creates two regions.**

   I'm going to skip the figure for this problem, because it's so simple. You should have an ordinary number line with a single point marked at $x = 0$.

4. **Plug a number from each of the two regions into the derivative, noting whether the results are positive or negative.**

   $$f'(x) = e^x - 1$$

   $$f'(\text{any negative number}) = \text{negative}$$
   $$f'(\text{any positive number}) = \text{positive}$$

5. **Draw a sign graph. Take your number line and label each region, based on your results from Step 4 — positive (increasing) or negative (decreasing).**

   Let's skip this simple figure as well. Your sign graph should be a number line with a minus sign over the negative side of the line and a plus sign over the positive side of the line.

   This sign graph tells you that *f* is decreasing from negative infinity to zero and increasing from zero to positive infinity.

6. **Use the sign graph to determine whether there's a local minimum, local maximum, or neither at each critical number.**

   Because *f* goes down on its way from negative infinity to $x = 0$ and up after $x = 0$, it must bottom out at $x = 0$, so there's a local min there.

7. **Determine the *y*-value of the local min by plugging the *x*-value into the original function.**

   $$f(0) = e^0 - 0 = 1$$

   So, the local min is at $(0, 1)$. That's the only local extremum.

② **There's a local min at $\left(\frac{\sqrt{2}}{2},\ \frac{1}{2}+\frac{1}{2}\ln 2\right)$.** That's the only local extremum.

1. **Find the first derivative of g using the power rule and the natural log rule.**

$$g(x)=x^2-\ln x$$
$$g'(x)=2x-\frac{1}{x}$$

2. **Set the derivative equal to zero and solve for x to get the critical numbers of g.**

$$2x-\frac{1}{x}=0$$
$$2x=\frac{1}{x}$$
$$x^2=\frac{1}{2}$$
$$x=\pm\frac{\sqrt{2}}{2}$$

You can disregard $-\frac{\sqrt{2}}{2}$ because that's outside g's domain. You can also disregard the fact that g' is undefined at $x=0$ because zero is also outside the domain. Thus, $\frac{\sqrt{2}}{2}$ is the only critical number of g.

3. **Plot the critical number, $\frac{\sqrt{2}}{2}$, on a number line, and also put a hollow dot at $x=0$ to show that that's the end of g's domain, namely, the interval $(0,\ \infty)$.**

4. **Plug a number from each of the two regions into the derivative, noting whether the results are positive or negative.**

$$g'(x)=2x-\frac{1}{x}$$

$$g'\left(\frac{1}{2}\right)=2\cdot\frac{1}{2}-\frac{1}{\frac{1}{2}}=-1$$
$$g'(1)=2\cdot 1-\frac{1}{1}=1$$

5. **Draw a sign graph. Take your number line and label each region, based on your results from Step 4 — positive (increasing) or negative (decreasing).**

See the following sign graph.

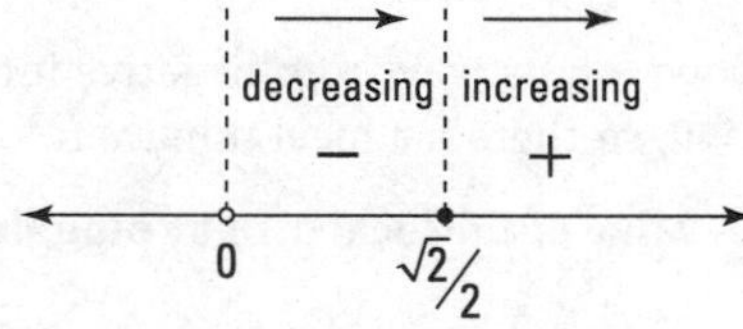

6. **Use the sign graph to determine whether there's a local minimum, local maximum, or neither at each critical number.**

   Because g goes down from zero to $\frac{\sqrt{2}}{2}$ and up from there, it must bottom out at $x = \frac{\sqrt{2}}{2}$, so there's a local min there.

7. **Determine the y-value of the local min by plugging the x-value into the original function.**

$$g(x) = x^2 - \ln x$$
$$g\left(\frac{\sqrt{2}}{2}\right) = \left(\frac{\sqrt{2}}{2}\right)^2 - \ln\frac{\sqrt{2}}{2}$$
$$= \frac{1}{2} + \frac{1}{2}\ln 2$$

   (Did you follow that simplification I did with the $-\ln\frac{\sqrt{2}}{2}$? See if you can work it out.) So, the local min is at $\left(\frac{\sqrt{2}}{2}, \frac{1}{2} + \frac{1}{2}\ln 2\right)$, or roughly $(0.707, 0.847)$. That's the only local extremum.

(3) There's a local min at $x = 0$.

1. **Find the critical numbers.**

   You have the single critical number from Problem 1: $x = 0$.

2. **Get the second derivative.**

$$f(x) = e^x - x$$
$$f'(x) = e^x - 1$$
$$f''(x) = e^x$$

3. **Plug the critical number into the second derivative.**

   $f''(0) = e^0 = 1$

   This *positive* result tells you that f is concave *up* at $x = 0$, and, therefore, that there's a local *min* there. This confirms your result from Problem 1.

(4) There's a local min at $x = \frac{\sqrt{2}}{2}$.

1. **Find the critical numbers.**

   You have the critical number from Problem 2. The only critical number of g is $x = \frac{\sqrt{2}}{2}$.

2. **Get the second derivative.**

$$g(x) = x^2 - \ln x$$
$$g'(x) = 2x - \frac{1}{x}$$
$$g''(x) = 2 + \frac{1}{x^2}$$

3. **Plug the critical number into the second derivative.**

$$g''\left(\frac{\sqrt{2}}{2}\right)=2+\frac{1}{\left(\frac{\sqrt{2}}{2}\right)^2}=2+2=4$$

This positive result tells you that g is concave *up* at $x=\frac{\sqrt{2}}{2}$, and, therefore, that there's a local min there. This confirms your result from Problem 2.

(5) **The absolute max occurs at $\left(2,\ e^2+\frac{1}{e^2}\right)$. The absolute min occurs at $(0,\ 2)$.**

1. **Find the critical numbers of f in the *open* interval $(-\ln 2,\ 2)$.**

$$\begin{aligned} f(x)&=e^x+e^{-x}\\ f'(x)&=e^x-e^{-x}\\ 0&=e^x-e^{-x}\\ e^{-x}&=e^x\\ x&=0 \end{aligned}$$

Thus, $x=0$ is the only zero of f' (in the given interval or anywhere else, for that matter), and because f' is defined for all input numbers, zero is the only critical number.

2. **Compute the function value (the height) at the critical number.**

$$\begin{aligned} f(x)&=e^x+e^{-x}\\ f(0)&=e^0+e^{-0}=2 \end{aligned}$$

3. **Determine the function values at the endpoints of the interval.**

$$f(x)=e^x+e^{-x}$$

$$\begin{aligned} f(-\ln 2)&=e^{-\ln 2}+e^{\ln 2}\\ &=\frac{1}{2}+2=2.5\\ f(2)&=e^2+e^{-2}\\ &=e^2+\frac{1}{e^2} \end{aligned}$$

So, from Steps 2 and 3, you've found three heights: 2, 2.5, and $e^2+\frac{1}{e^2}$. The largest number in this list, $e^2+\frac{1}{e^2}$, is the absolute max; the smallest, 2, is the absolute min.

The absolute max occurs at $\left(2,\ e^2+\frac{1}{e^2}\right)$, an *endpoint extremum*. The absolute min occurs at $(0,\ 2)$.

6 **The absolute max occurs at both $(-1, 1)$ and $(1, 1)$. The absolute min occurs at $(0, 0)$.**

1. **Find the critical numbers of *f* in the *open* interval $(-1, 1)$.**

$$g(x) = \ln\left(e^{x^2}\right)$$
$$g'(x) = \frac{1}{e^{x^2}} \cdot e^{x^2} \cdot 2x$$
$$= 2x$$
$$0 = 2x$$
$$x = 0$$

Just like in Problem 5, $x = 0$ is the only zero of g' (in the given interval or anywhere else), and because g' is defined for all input numbers, zero is the only critical number.

2. **Compute the function value (the height) at the critical number.**

$$g(x) = \ln\left(e^{x^2}\right)$$
$$g(0) = \ln\left(e^{0^2}\right)$$
$$= \ln 1 = 0$$

3. **Determine the function values at the endpoints of the interval.**

$$g(x) = \ln\left(e^{x^2}\right)$$

$$g(-1) = \ln\left(e^{(-1)^2}\right)$$
$$= \ln e$$
$$= 1$$
$$g(1) = 1$$

So, from Steps 2 and 3, you've found three heights (two different heights): 0, 1, and 1 again. The largest number in this list, 1, is the absolute max; the smallest, 0, is the absolute min.

The absolute max occurs at both $(-1, 1)$ and $(1, 1)$. The absolute min occurs at $(0, 0)$.

This was a trick problem of sorts. I went through the hard-way solution here, thinking that many might not have seen the following easy way. Congrats if you saw that $\ln\left(e^{x^2}\right) = x^2$, and, thus, that $g(x) = x^2$, the simple parabola you know well. And the absolute max and min of $g(x) = x^2$ on the interval $[-1, 1]$ should be obvious without doing any calculus.

(7) $f(x)=\sqrt[3]{x}-x^2$ **has an absolute max at (~ 0.341, ~ 0.582). It has no absolute min.**

1. **Find the height of the function at each of its critical numbers; in other words, find the function's critical points.**

$$f(x)=\sqrt[3]{x}-x^2$$
$$f'(x)=\frac{1}{3}x^{-2/3}-2x$$
$$2x=\frac{1}{3}x^{-2/3}$$
$$6x=x^{-2/3}$$
$$x^{5/3}=\frac{1}{6}$$
$$x=\left(\frac{1}{6}\right)^{3/5}$$

So, $\left(\frac{1}{6}\right)^{3/5}\approx 0.341$ is a critical number. Zero is another critical number, because f' is undefined at $x=0$. The heights at those critical numbers are $\frac{5}{6}\left(\frac{1}{6}\right)^{1/5}\approx 0.582$ and zero, respectively. Those are your only candidates for an absolute max and an absolute min.

2. **Graph the function to see how high and low it goes.**

You can see from the graph of f that it never goes above (~ 0.341, ~ 0.582), so that's the absolute max. And f obviously goes below the other critical point at $(0,\ 0)$, so f has no absolute min.

3. **Use the first derivative test and a sign graph to analyze the shape of the function.**

You should confirm your absolute max result from Step 2, by using the first derivative test. That will show that f is rising from negative infinity until (~ 0.341, ~ 0.582) and falling from there to infinity. That clinches it.

(8) $g(x)=\sqrt[4]{x^2}$ **has no absolute max. Its absolute min is at $(0,\ 0)$.**

1. **Find the height of the function at each of its critical numbers; in other words, find the function's critical points.**

$$g(x)=\sqrt[4]{x^2}$$
$$=\left(x^2\right)^{1/4}$$
$$g'(x)=\frac{1}{4}\left(x^2\right)^{-3/4}\cdot 2x$$
$$=\frac{x}{2\sqrt[4]{x^6}}\quad\text{or}\quad\frac{x}{2|x|\sqrt[4]{x^2}}$$

The derivative g' is undefined at $x=0$, so that's a critical number of g; g' has no zeros, so $x=0$ is the only critical number. Plugging zero into g gives you $(0,\ 0)$ for the only critical point on g and your only candidate for an absolute max or min.

(By the way, as you may have noticed, $g(x)=\sqrt[4]{x^2}$ is a peculiar function. When you graph it, you'll see that it's the same as the ordinary square root function, $f(x)=\sqrt{x}$, in the first quadrant with another branch symmetric to that in the second quadrant. You've got to be very careful when doing the derivative of g. There are a few ways to go awry. For example, you'll get the wrong derivative if you simplify the second line to $x^{1/2}$.)

2. **Graph the function to see how high and low it goes.**

   You can see from the graph of g that it never goes below $(0, 0)$, so that's the absolute min. And g obviously goes above $(0, 0)$, so g has no absolute max.

9. **$f(x) = \cos x + \frac{1}{2}x^2$ is concave up everywhere. It has no inflection points.**

Find the second derivative, set it equal to zero, and solve:

$$f(x) = \cos x + \frac{1}{2}x^2$$
$$f'(x) = -\sin x + x$$
$$f''(x) = -\cos x + 1$$
$$0 = -\cos x + 1$$
$$\cos x = 1$$
$$x = 2k\pi \quad (k \text{ is any integer})$$

Don't forget to also check for any x-values where the second derivative is undefined. There are none. So, the zeros of f'' are the only numbers you need to consider for this concavity-and-inflection-point inquiry.

Notice that $f''(x) = -\cos x + 1$ will be positive for all input values, except when x equals a multiple of 2π. At those multiples, the second derivative equals zero.

Consider one of those zeros, say, $x = 2\pi$. To the left of $x = 2\pi$, f is concave up (because f'' is positive). And to the right of $x = 2\pi$, f is also concave up. And that means that f is concave up at $x = 2\pi$ as well. This applies to all the zeros of f'', and, thus, f is concave up everywhere.

I find this a fascinating function. Graph it and take a look at it. Close in, it looks sort of like a parabola, but with a very flat bottom. Then, as you zoom out, it more closely resembles the simple parabola $y = \frac{1}{2}x^2$. But at every multiple of 2π along the function, where f'' equals zero, the curve sort of straightens out for an infinitesimal moment (these straighter stretches are very difficult to see, but if you look very carefully using the right scale, you can make them out). Interesting!

10. **$g(x) = 2x^6 - 5x^4$ is concave up on two intervals: $(-\infty, -1)$ and $(1, \infty)$. The function g is concave down on the interval $(-1, 1)$; g has inflection points at $(-1, -3)$ and $(1, -3)$.**

Find the second derivative, set it equal to zero, and solve:

$$g(x) = 2x^6 - 5x^4$$
$$g'(x) = 12x^5 - 20x^3$$
$$g''(x) = 60x^4 - 60x^2$$
$$0 = 60x^4 - 60x^2$$
$$0 = 60x^2(x^2 - 1)$$
$$0 = 60x^2(x - 1)(x + 1)$$
$$x = 0,\ 1, \text{ and } -1$$

Check for any x-values where the second derivative is undefined. There are none, so −1, 0, and 1 are your "second derivative critical numbers." They divide the number line into four intervals: $(-\infty, -1), (-1, 0), (0, 1)$, and $(1, \infty)$. Test any number from each interval in the second derivative:

$$g''(x) = 60x^2(x-1)(x+1)$$

$$g''(-2) = (\text{pos})(\text{neg})(\text{neg}) = \text{pos}$$
$$g''(-0.5) = (\text{pos})(\text{neg})(\text{pos}) = \text{neg}$$
$$g''(0.5) = (\text{pos})(\text{neg})(\text{pos}) = \text{neg}$$
$$g''(2) = (\text{pos})(\text{pos})(\text{pos}) = \text{pos}$$

These results tell you that g is concave up from negative infinity to –1, concave down from –1 to 1, and concave up again from 1 to infinity. Because the concavity switches signs at –1 and 1, there are inflection points at those x-values.

11 **(a) The MVT does not apply.**

Hope you didn't fall for this trap. The MVT does not apply here, because $f(x) = \frac{1}{x-1}$ is not continuous over the interval $(0, 2)$.

(b) The MVT does not apply.

Another trap problem (with a twist). The MVT does not apply here, because $g(x) = \sqrt[3]{x}$ is not everywhere differentiable on the open interval $(-1, 1)$. The first derivative, $g'(x) = \frac{1}{3\sqrt[3]{x^2}}$, is undefined at $x = 0$.

The twist here is that if you go ahead and do the MVT math, you'll correctly find two values for c, namely, $\pm\frac{\sqrt{3}}{9}$, where the derivative (the slope of the tangent line) does in fact equal the slope of the secant line from $(-1, -1)$ to $(1, 1)$ (the points on the function at the endpoints of the given interval). So, the MVT did work correctly. My hunch is that the MVT will work for a larger class of functions than are technically allowed by the theorem — namely, all continuous and smooth functions without any corners or cusps. Bottom line: Never mind this! Go by the book. That's what your professor will expect.

12 **For $g(x) = \sqrt[3]{x}$ in the interval $(0, 8)$, $c = \frac{8\sqrt{3}}{9}$ satisfies the MVT.**

Unlike in Problem 11b, $g(x) = \sqrt[3]{x}$ *does* satisfy the MVT fine print over the interval $[0, 8]$. Note that the MVT requires that the function be continuous on the closed interval $[0, 8]$, which $g(x) = \sqrt[3]{x}$ is. But $g(x) = \sqrt[3]{x}$ needs to be differentiable only over the *open* interval $(0, 8)$ — which it is. The fact that $g(x) = \sqrt[3]{x}$ is not differentiable at $x = 0$ — which was the hitch in Problem 11b — is irrelevant here.

By the MVT, $g'(c) = \frac{g(b) - g(a)}{b - a}$. You've got $a = 0$ and $b = 8$; determine the three other things you need here, so you can plug them in.

$$g'(x) = \frac{1}{3\sqrt[3]{x^2}} \text{ so } g'(c) = \frac{1}{3\sqrt[3]{c^2}}.$$

$$g(0) = \sqrt[3]{0} = 0$$

$$g(8) = \sqrt[3]{8} = 2$$

Plug everything into the MVT formula and finish:

$$\frac{1}{3\sqrt[3]{c^2}} = \frac{2-0}{8-0}$$

$$\frac{1}{3\sqrt[3]{c^2}} = \frac{1}{4}, \text{ etc.}$$

$$c = \pm\frac{8\sqrt{3}}{9}$$

You can reject the negative answer, because it's outside the given interval.

IN THIS CHAPTER

- » Getting the most bang for your buck — optimization problems
- » Getting up to speed with position, velocity, and acceleration
- » Related rates — brace yourself

Chapter 12

Your Problems Are Solved: Differentiation to the Rescue!

In Chapter 1, I argue that calculus has changed the world in countless ways, that its impact is not limited to Ivory Tower mathematics, but is all around us in down-to-earth things like microwave ovens, cellphones, and cars. Well, it's now Chapter 12, and I'm *finally* ready to show you how to use calculus to solve some practical problems.

Getting the Most (or Least) Out of Life: Optimization Problems

One of the most practical uses of differentiation is finding the maximum or minimum value of a real-world function: the maximum output of a factory, the maximum strength of a beam, the minimum cost of running some business, the maximum range of a missile, and so on. Let's see how this works by walking through some problems.

The maximum volume of a box

A box with no top is to be manufactured from a 30-inch-by-30-inch piece of cardboard by cutting and folding it, as shown in Figure 12-1.

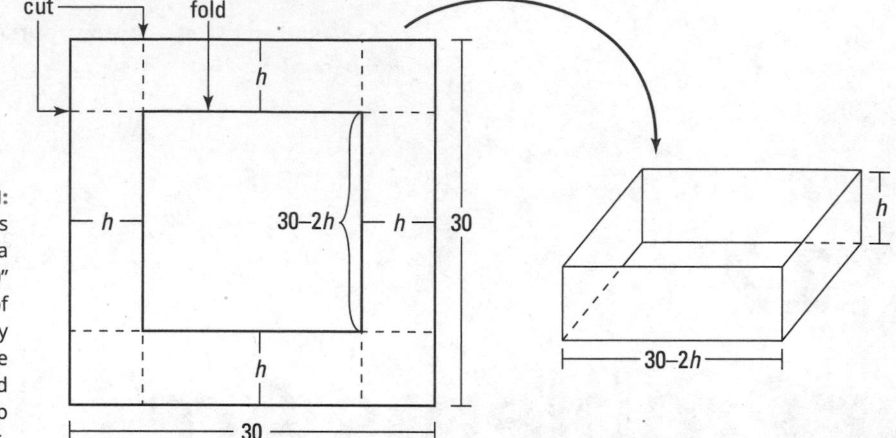

FIGURE 12-1: The box is made from a 30"-by-30" piece of cardboard by cutting off the corners and folding up the sides.

What dimensions will produce a box with the maximum volume? Mathematics often seems abstract and impractical, but here's an honest-to-goodness practical problem (well . . . almost). If a manufacturer can sell bigger boxes for more and is making 100,000 boxes, you'd better believe they want the exact answer to this question. Here's how you do it:

1. **Express the thing you want maximized, the volume, as a function of the unknown, the height of the box (which is the same as the length of the cut).**

   $$V = l \cdot w \cdot h$$

   $$V(h) = (30-2h)(30-2h) \cdot h$$ (You can see in Figure 12-1 that both the *length* and *width* equal $30-2h$.)

   $$= \left(900 - 120h + 4h^2\right) \cdot h$$

   $$= 4h^3 - 120h^2 + 900h$$

2. **Determine the domain of your function.**

   The height can't be negative, and because the length (and width) of the box equals $30-2h$, which can't be negative, h can't be greater than 15. Thus, sensible values for h are $0 \le h \le 15$. You now want to find the maximum value of $V(h)$ in this interval. You use the method from the section, "Finding Absolute Extrema on a Closed Interval," in Chapter 11.

3. **Find the critical numbers of $V(h)$ in the open interval $(0,\ 15)$ by setting its derivative equal to zero and solving. And don't forget to check for numbers where the derivative is undefined.**

$$
\begin{aligned}
V(h) &= 4h^3 - 120h^2 + 900h \\
V'(h) &= 12h^2 - 240h + 900 \quad \text{(power rule)} \\
0 &= 12h^2 - 240h + 900 \\
0 &= h^2 - 20h + 75 \quad \text{(dividing both sides by 12)} \\
h &= (h-15)(h-5) \quad \text{(ordinary trinomial factoring)} \\
h &= 15 \text{ or } 5
\end{aligned}
$$

Because 15 is not in the open interval $(0, 15)$, it doesn't qualify as a critical number (though this is a moot point because you end up testing it in Step 4 anyway). And because this derivative is defined for all input values and is, thus, never undefined, there are no additional critical numbers. So, 5 is the only critical number.

4. **Evaluate the function at the critical number, 5, and at the endpoints of the interval, 0 and 15, to locate the function's max.**

$$V(h) = 4h^3 - 120h^2 + 900h$$

$$
\begin{aligned}
V(0) &= 0 \\
V(5) &= 2{,}000 \\
V(15) &= 0
\end{aligned}
$$

WARNING

Test the endpoints. The *extremum* (dig that fancy word for *maximum* or *minimum*) you're looking for doesn't often occur at an endpoint, but it can — so don't fail to evaluate the function at the interval's two endpoints.

So, a height of 5 inches produces the box with maximum volume (2000 cubic inches). Because the length and width equal $30 - 2h$, a height of 5 gives a length and width of $30 - 2 \cdot 5$, or 20, and thus the dimensions of the desired box are $5'' \times 20'' \times 20''$. That's it.

The maximum area of a corral — yeehaw!

A rancher can afford 300 feet of fencing to build a corral that's divided into two equal rectangles. See Figure 12-2.

What dimensions will maximize the corral's area? This is another practical problem. The rancher wants to give his animals as much room as possible given the length of fencing he can afford. Like all businesspeople, he wants the most bang for his buck.

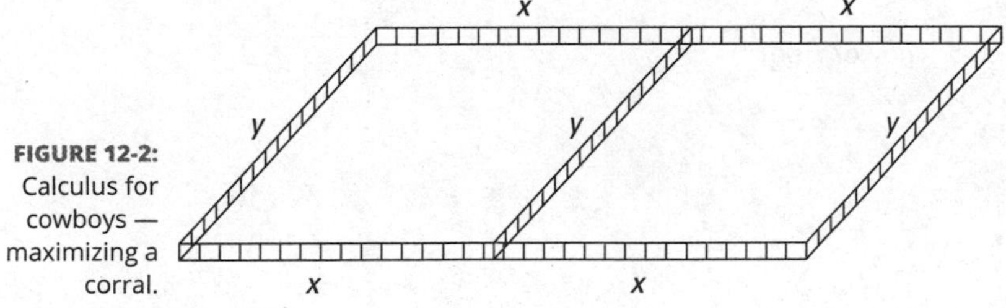

FIGURE 12-2: Calculus for cowboys — maximizing a corral.

1. a. **Express the thing you want maximized (area) as a function of the two unknowns (*x* and *y*).**

$$A = l \cdot w$$
$$= 2x \cdot y$$

In the above cardboard box example, you can easily write the volume as a function of *one* variable — which is always what you need. But here, the area is a function of two variables (*x* and *y*), so Step 1 has the following two extra sub-steps that will eliminate one of the variables.

1. b. **Use the given information to relate the two variables to each other.**

The 300 feet of fencing is used for seven sections, thus

$$300 = x + x + x + x + y + y + y$$
$$300 = 4x + 3y$$

1. c. **Solve this equation for *y* and plug the result in for *y* in the equation from Step 1a. This gives you what you need — a function of one variable.**

$$4x + 3y = 300$$
$$3y = 300 - 4x$$
$$y = \frac{300 - 4x}{3}$$
$$y = 100 - \frac{4}{3}x \quad \text{(Now do the substitution.)}$$

$$A = (2x)(y)$$
$$A(x) = (2x)\left(100 - \frac{4}{3}x\right)$$
$$A(x) = 200x - \frac{8}{3}x^2$$

2. **Determine the domain of the function.**

You can't have a negative length of fence, so *x* can't be negative, and the most *x* can be is 300 divided by 4, or 75. Thus, $0 \le x \le 75$.

3. **Find the critical numbers of $A(x)$ in the open interval $(0, 75)$ by setting its derivative equal to zero and solving (and check whether the derivative is undefined anywhere in the interval).**

$$A(x) = 200x - \frac{8}{3}x^2$$
$$A'(x) = 200 - \frac{16}{3}x \quad \text{(power rule)}$$
$$0 = 200 - \frac{16}{3}x$$
$$\frac{16}{3}x = 200$$
$$= 37.5$$

Because A' is defined for all *x*-values, 37.5 is the only critical number.

4. Evaluate the function at the critical number, 37.5, and at the endpoints of the interval, 0 and 75.

$$A(x) = 200x - \frac{8}{3}x^2$$

$$A(0) = 0$$
$$A(37.5) = 3750$$
$$A(75) = 0$$

Note: Evaluating a function at the endpoints of a closed interval is a standard step in finding an absolute extremum on the interval. However, you could have skipped this step here had you noticed that $A(x)$ is an upside-down parabola and that, therefore, its peak at $(37.5,\ 3750)$ must be higher than either endpoint.

The maximum value in the interval is 3750, and thus, an x-value of 37.5 feet maximizes the corral's area. The length is 2x, or 75 feet. The width is y, which equals $100 - \frac{4}{3}x$. Plugging in 37.5 gives you $100 - \frac{4}{3}(37.5)$, or 50 feet. So, the rancher will build a 75'-by-50' corral with an area of 3750 square feet.

YOUR TURN

1 What are the dimensions of the soup can of greatest volume that can be made with 50 square inches of tin? (The entire can, including the top and bottom, is made of tin.) And what's its volume?

2 A Norman window is in the shape of a semicircle above a rectangle. If the straight edges of the frame cost $20 per linear foot and the circular frame costs $25 per linear foot, and you want a window with an area of 20 square feet, what dimensions will minimize the cost of the frame?

---

3 A right triangle is placed in the first quadrant with its legs on the x- and y-axes. Given that its hypotenuse must pass through the point $(2, 5)$, what are the dimensions and area of the smallest such triangle?

4 You're designing an open-top cardboard box for a purveyor of nuts. The top will be made of clear plastic, but the plastic-box-top designer is handling that. The box must have a square base and two cardboard pieces that divide the box into four sections for the almonds, cashews, pecans, and walnuts. See the following figure. Given that you want a box with a volume of 72 cubic inches, what dimensions will minimize the total cardboard area and thus minimize the cost of the cardboard? What's the total area of cardboard?

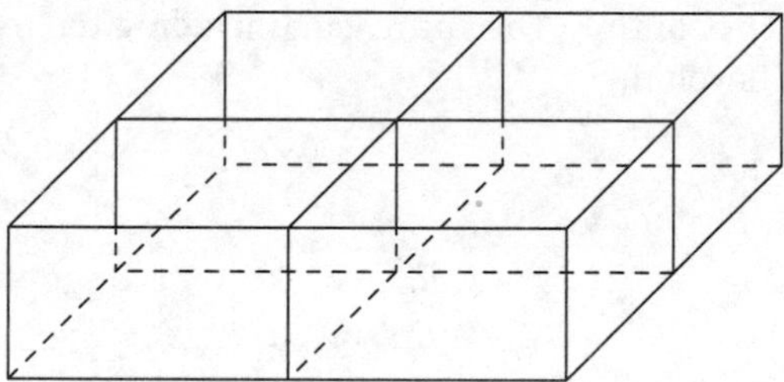

Yo-Yo a Go-Go: Position, Velocity, and Acceleration

Every time you get in your car, you witness differentiation. Your speed is the first derivative of your position. And when you step on the accelerator or the brake — accelerating or decelerating — you experience a second derivative.

MATH RULES

The derivative of position is velocity, and the derivative of velocity is acceleration. If a function gives the position of something as a function of time, you differentiate the position function to get the velocity function, and you differentiate the velocity function to get the acceleration function. Stated a different but equivalent way, the first derivative of position is velocity, and the second derivative of position is acceleration.

Here's a problem. A yo-yo moves straight up and down. Its height above the ground, as a function of time, is given by the function $H(t)=t^3-6t^2+5t+30$, where t is in seconds and $H(t)$ is in inches. At $t=0$, it's 30 inches above the ground, and after 4 seconds, it's at a height of 18 inches. See Figure 12-3.

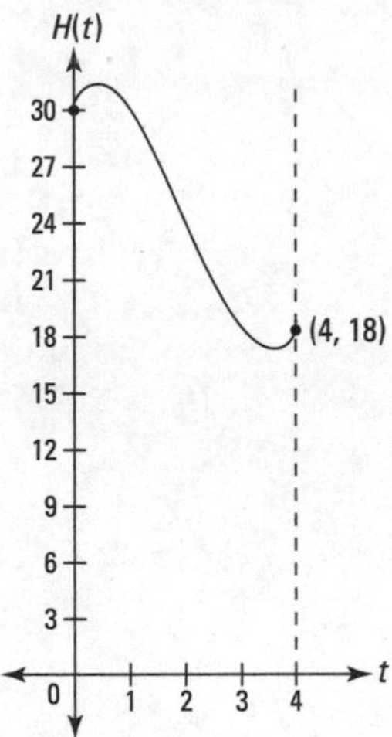

FIGURE 12-3: The yo-yo's height, from 0 to 4 seconds.

Velocity, $V(t)$, is the derivative of position (height, in this problem), and acceleration, $A(t)$, is the derivative of velocity. Thus:

$$H(t)=t^3-6t^2+5t+30$$
$$V(t)=H'(t)=3t^2-12t+5 \quad \text{(power rule)}$$
$$A(t)=V'(t)=H''(t)=6t-12 \quad \text{(power rule)}$$

Take a look at the graphs of these three functions in Figure 12-4.

Using the three functions and their graphs, I want to discuss several things about the yo-yo's motion:

- Maximum and minimum height
- Maximum, minimum, and average velocity
- Total displacement
- Maximum, minimum, and average speed
- Total distance traveled
- Positive and negative acceleration
- Speeding up and slowing down

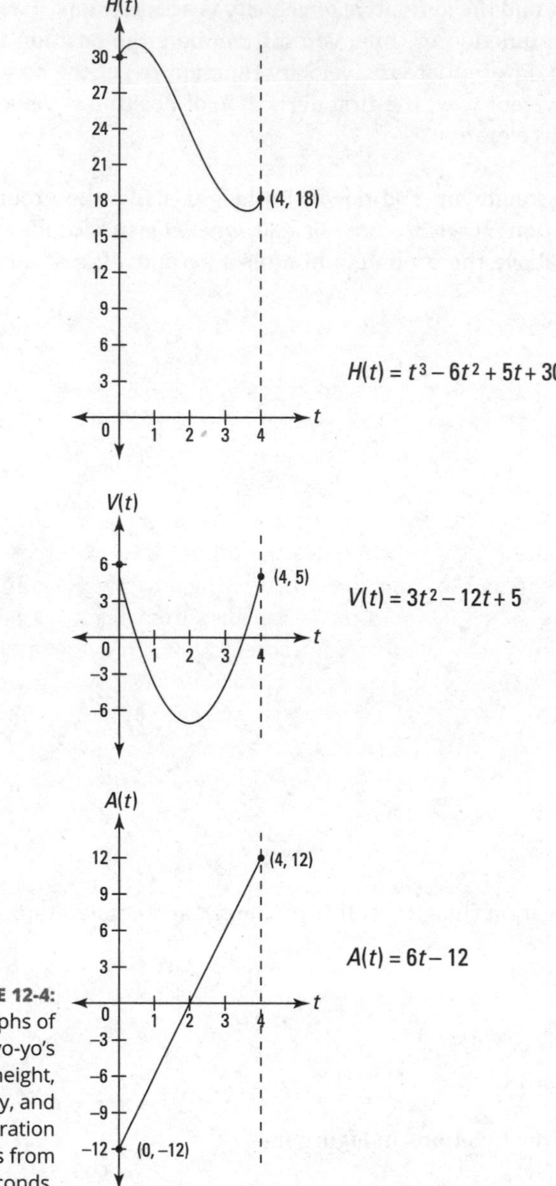

FIGURE 12-4: The graphs of the yo-yo's height, velocity, and acceleration functions from 0 to 4 seconds.

Because this is a lot to cover, I'll cut some corners — like not always checking endpoints when looking for extrema if it's obvious that they don't occur at the endpoints. Do you mind? I didn't think so. (Position, velocity, and acceleration problems make use of several ideas from Chapter 11 — local extrema, concavity, inflection points — so you may want to take a look back at those definitions if you're a little hazy.) But before tackling the bulleted topics, let's go over a few things about velocity, speed, and, acceleration.

Velocity, speed, and acceleration

None of your friends will complain — or even notice — if you use the words "velocity" and "speed" interchangeably, but your friendly mathematician *will* complain. Here's the difference.

For the velocity function in Figure 12-4, *upward* motion by the yo-yo is defined as a *positive* velocity, and *downward* motion is a *negative* velocity. This is the standard way velocity is treated in most calculus and physics problems. (Or, if the motion is horizontal, going *right* is a *positive* velocity and going *left* is a *negative* velocity.)

Speed, on the other hand, is always positive (or zero). If a car goes by at 50 mph, for instance, you say its speed is 50, and you mean *positive* 50, regardless of whether it's going to the right or the left. For velocity, the direction matters; for speed, it does not. In everyday life, speed is a simpler idea than velocity because it's always positive and because it agrees with our commonsense notion about how fast something is moving. But in calculus, speed is actually the trickier idea because it doesn't fit nicely into the three-function scheme shown in Figure 12-4.

WARNING

You've got to keep the velocity-speed distinction in mind when analyzing velocity and acceleration. The way we talk about velocity, speed, and acceleration — in calculus class, as opposed to in everyday life — can get pretty weird. For example, if an object is going down (or to the left) faster and faster, its speed is increasing, but its velocity is *decreasing* because its velocity is becoming a bigger and bigger *negative* (and bigger negatives are smaller numbers). This seems weird, but that's the way it works. And here's another strange thing: Acceleration is defined as the rate of change of velocity, not speed. So, if an object is slowing down while going in the downward direction and thus has an *increasing* velocity — because the velocity is becoming a smaller and smaller negative — the object has a *positive* acceleration. In everyday English, you'd say that the object is decelerating (slowing down), but in calculus class, though you could still say the object is slowing down, you'd say that the object has a negative velocity and a positive acceleration. (By the way, *deceleration* isn't exactly a technical term, so you should probably avoid it in calculus class. It's best to use the following vocabulary: *positive acceleration, negative acceleration, positive velocity, negative velocity, speeding up,* and *slowing down.*) I could go on with this stuff, but I bet you've had enough.

Maximum and minimum height

The maximum and minimum height of the yo-yo, in other words, the max and min of $H(t)$, occur at the local extrema you can see in Figure 12-4. To locate them, set the derivative of $H(t)$ (that's $V(t)$) equal to zero and solve:

$$\begin{aligned}
H'(t) = V(t) &= 3t^2 - 12t + 5 \\
0 &= 3t^2 - 12t + 5 \\
t &= \frac{-(-12) \pm \sqrt{(-12)^2 - 4(3)(5)}}{2 \cdot 3} \quad \text{(quadratic formula)} \\
&= \frac{12 \pm \sqrt{84}}{6} \\
&= \frac{12 \pm 2\sqrt{21}}{6} \\
&= \frac{6 \pm \sqrt{21}}{3} \\
&= \sim 0.47 \text{ or } \sim 3.53
\end{aligned}$$

These two numbers are the zeros of $H'(t)$ (which is $V(t)$) and the t-coordinates, that's *time*-coordinates, of the max and min of $H(t)$, which you can see in Figure 12-4. In other words, these are the *times* when the yo-yo reaches its maximum and minimum heights. Plug these numbers into $H(t)$ to obtain the heights:

$$H(0.47) \approx 31.1$$
$$H(3.53) \approx 16.9$$

So, the yo-yo gets as high as about 31.1 inches above the ground at $t \approx 0.47$ seconds and as low as about 16.9 inches at $t \approx 3.53$ seconds. (By the way, do you see why the max and min of the yo-yo's height would occur when the yo-yo's velocity is zero?)

Velocity and displacement

As I explain in the section "Velocity, speed, and acceleration," *velocity* is basically like *speed*, except that while speed is always positive (or zero), velocity can be positive (when going up or to the right) or negative (when going down or to the left). The connection between *displacement* and *distance traveled* is similar: Distance traveled is always positive (or zero), but going down (or left) counts as *negative* displacement. In everyday speech, speed and distance traveled are the more user-friendly ideas, but when it comes to calculus and physics, velocity and displacement are the more fundamental ideas.

Total displacement

Let's get back to my yo-yo analysis. Total displacement is defined as final position minus initial position.

$$\textit{Total displacement} = \textit{final position} - \textit{initial position}$$

So, because the yo-yo starts at a height of 30 and ends at a height of 18,

$$\textit{Total displacement} = 18 - 30 = -12 \text{ inches}$$

This is negative because the net movement is *downward*.

Average velocity

Average velocity is given by total displacement divided by elapsed time:

$$\textit{Average velocity} = \frac{\textit{total displacement}}{\textit{elapsed time}}$$

I just calculated the total displacement (-12 inches), and time runs from 0 seconds to 4 seconds, so the elapsed time is 4 seconds. Thus,

$$\textit{Average velocity} = \frac{-12 \text{ inches}}{4 \text{ seconds}} = -3 \text{ inches per second}$$

This answer of *negative* 3 tells you that the yo-yo is, on average, going *down* 3 inches per second.

Maximum and minimum velocity

To determine the yo-yo's maximum and minimum velocity during the interval from 0 to 4 seconds, set the derivative of $V(t)$ — that's $A(t)$ — equal to zero and solve:

$$V'(t) = A(t) = 6t - 12$$
$$6t - 12 = 0$$
$$t = 2$$

(Look again at Figure 12-4. At $t = 2$, you get the zero of $A(t)$, the local min of $V(t)$, and the inflection point of $H(t)$. But you already knew that, right? If not, check out Chapter 11.)

Now, evaluate $V(t)$ at the critical number, 2, and at the interval's endpoints, 0 and 4:

$$V(t) = 3t^2 - 12t + 5$$

$$V(0) = 5$$
$$V(2) = -7$$
$$V(4) = 5$$

So, the yo-yo has a maximum velocity of 5 inches per second twice — at both the beginning and the end of the interval. It reaches a minimum velocity of −7 inches per second at $t = 2$ seconds.

Speed and distance traveled

As mentioned in the previous section, *velocity* and *displacement* are the more technical concepts, while *speed* and *distance traveled* are the more commonsense ideas. *Speed,* of course, is the thing you read on your speedometer, and you can read *distance traveled* on your odometer or your "tripometer" after setting it to zero.

Total distance traveled

To determine total distance, add up the distances traveled on each leg of the yo-yo's trip: the up leg, the down leg, and the second up leg.

First, the yo-yo goes up from a height of 30 inches to about 31.1 inches (where the first turn-around point is). That's a distance of about 1.1 inches. Next, it goes down from about 31.1 to about 16.9 (the height of the second turn-around point). That's a distance of 31.1 minus 16.9, or about 14.2 inches. Finally, the yo-yo goes up again from about 16.9 inches to its final height of 18 inches. That's another 1.1 inches. Add these three distances to obtain the total distance traveled: $\sim 1.1 + \sim 14.2 + \sim 1.1 \approx 16.4$ inches. (*Note:* Compare this answer to the total displacement of −12. The displacement is negative because the net movement is downward. And the positive amount of the displacement, namely 12, is less than the distance traveled of 16.4 because with displacement, the up legs of the yo-yo's trip cancel out part of the down leg distance. Check out the math: $1.1 - 14.2 + 1.1 = -12$. Get it?) In a nutshell, the total distance traveled is how far you go along the path you're on — like the way your car's odometer works or, if you're walking, a pedometer. Total displacement, on the other hand, is a *net* distance, or you could say the *crow-fly* distance from start to finish (remembering, though, that this can be a negative number for going down or to the left).

Average speed

The yo-yo's average speed is given by the total distance traveled divided by the elapsed time.

$$\textit{Average speed} = \frac{\textit{total distance}}{\textit{elapsed time}}$$

Thus,

$$\textit{Average speed} \approx \frac{16.4}{4} \approx 4.1 \text{ inches per second}$$

Maximum and minimum speed

You previously determined the yo-yo's maximum velocity (5 inches per second) and its minimum velocity (–7 inches per second). A velocity of –7 is a speed of 7, so that's the yo-yo's maximum speed. Its minimum speed of zero occurs at the two turnaround points.

REMEMBER

When you switch directions, your velocity is zero (for an infinitesimal moment).

A good way to analyze maximum and minimum speed is to consider the speed function and its graph. (Or, if you're a glutton for punishment, check out the following mumbo jumbo.)

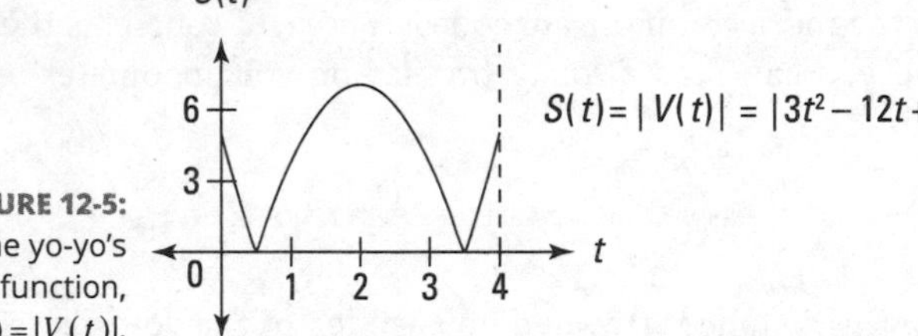

FIGURE 12-5: The yo-yo's speed function, $S(t) = |V(t)|$.

Speed equals the absolute value of velocity. So, for the yo-yo problem, the speed function, $S(t)$, equals $|V(t)| = |3t^2 - 12t + 5|$. Check out the graph of $S(t)$ in Figure 12-5. Looking at this graph, it's easy to see that the yo-yo's maximum speed occurs at $t = 2$ (the maximum speed is $S(2) = |3(2)^2 - 12(2) + 5| = 7$), and that the minimum speed is zero at the two x-intercepts.

MATH RULES

Minimum and maximum speed. For a continuous velocity function, the *minimum speed* is zero whenever the maximum and minimum velocities are of opposite signs or when one of them is zero. When the maximum and minimum velocities are both positive or both negative, the *minimum* speed is the *lesser* of the absolute values of the maximum and minimum velocities. In all cases, the *maximum* speed is the *greater* of the absolute values of the maximum and minimum velocities. Is that a mouthful or what?

Burning some rubber with acceleration

Let's go over acceleration: Put your pedal to the metal.

Positive and negative acceleration

The graph of the acceleration function at the bottom of Figure 12-4 is a simple line, $A(t) = 6t - 12$. It's easy to see that the acceleration of the yo-yo goes from a minimum of $-12 \frac{\text{inches per second}}{\text{second}}$ at $t = 0$ seconds to a maximum of $12 \frac{\text{inches per second}}{\text{second}}$ at $t = 4$ seconds, and that the acceleration is zero at $t = 2$ when the yo-yo reaches its minimum velocity (and maximum speed). When the acceleration is *negative* — on the interval $[0, 2)$ — the velocity is *decreasing*. When the acceleration is *positive* — on the interval $(2, 4]$ — the velocity is *increasing*.

Speeding up and slowing down

Figuring out when the yo-yo is speeding up and slowing down is probably more interesting and descriptive of its motion than the info in the preceding section. An object is speeding up (what we call "acceleration" in everyday speech) whenever the velocity and the calculus acceleration are both positive or both negative. And an object is slowing down ("deceleration" in everyday speech) when the velocity and the calculus acceleration are of opposite signs.

Look at all three graphs in Figure 12-4 again. From $t = 0$ to about $t = 0.47$, the velocity is positive and the acceleration is negative, so the yo-yo is slowing down while moving upward (till its velocity becomes zero and it reaches its maximum height). In plain English, the yo-yo is decelerating from 0 to about 0.47 seconds. The greatest deceleration occurs at $t = 0$, when the deceleration is $12 \frac{\text{inches per second}}{\text{second}}$ (the graph shows *negative* 12, but I'm calling it positive 12 because I'm calling it a deceleration, get it?).

From about $t = 0.47$ to $t = 2$, both velocity and acceleration are negative, so the yo-yo is speeding up while moving downward. From $t = 2$ to about $t = 3.53$, velocity is negative and acceleration is positive, so the yo-yo is slowing down again as it continues downward (till it bottoms out at its lowest height). Finally, from about $t = 3.53$ to $t = 4$, both velocity and acceleration are positive, so the yo-yo is speeding up again. The yo-yo reaches its greatest acceleration of $12 \frac{\text{inches per second}}{\text{second}}$ at $t = 4$ seconds.

Tying it all together

Note the following connections among the three graphs in Figure 12-4. The *negative* section on the graph of $A(t)$ — from $t = 0$ to $t = 2$ — corresponds to a *decreasing* section of the graph of $V(t)$ and a *concave down* section of the graph of $H(t)$. The *positive* interval on the graph of $A(t)$ — from $t = 2$ to $t = 4$ — corresponds to an *increasing* interval on the graph of $V(t)$ and a *concave up* interval on the graph of $H(t)$. When $t = 2$ seconds, $A(t)$ has a zero, $V(t)$ has a *local minimum*, and $H(t)$ has an *inflection point*.

WHAT THE HECK IS A SECOND SQUARED?

Note that I use the unit $\frac{\text{inches per second}}{\text{second}}$ for acceleration instead of the equivalent but weird-looking unit, $\frac{\text{inches}}{\text{second}^2}$. You often see acceleration given in terms of a unit of distance over second^2, or you might see something like $\text{inches per second}^2$. But what the heck is a second^2? It's meaningless, and something like inches/second^2 is a bad way to think about acceleration. The best way to understand acceleration is as a change in speed per unit of time. If a car can go from 0 to 60 mph in 6 seconds, that's an increase in speed of 60 mph in 6 seconds, or, on average, 10 mph each second — that's an acceleration of $10\frac{\text{mph}}{\text{second}}$. It's slightly more confusing when the speed has a unit like feet/second and the unit of time for the acceleration is also a second, because then the word *second* appears twice. But it still works like the car example. Say an object starts at rest and speeds up to 10 feet/second after 1 second, then up to 20 feet/second after 2 seconds, to 30 feet/second after 3 seconds, and so on. Its speed is increasing 10 feet/second each second, and that's an acceleration of $10\frac{\text{feet per second}}{\text{second}}$ or $10\frac{\text{feet/second}}{\text{second}}$. (By the way, it's helpful to write the acceleration unit in either of these ways (using a vertical fraction) as a speed over the unit of time — instead of horizontally like 10 feet per second per second or 10 feet/second/second — to emphasize that acceleration is a change in speed per unit of time.) Think of acceleration this way, not in terms of that weird second^2 thing.

YOUR TURN

For Problems 5, 6, and 7, a duck-billed platypus is swimming back and forth along the side of your boat, blithely unaware that he's the subject for calculus problems in rectilinear motion. The back of your boat is at the zero position, and the front of your boat is in the positive direction (see the following figure). Note that $s(t)$ gives the platypus's position (in feet) as a function of time (seconds). Find his a) position, b) velocity, c) speed, and d) acceleration, at $t = 2$ seconds.

 $s(t) = 5t^2 + 4$

6 $s(t) = 3t^4 - 5t^3 + t - 6$

7 $s(t) = \frac{1}{t} + \frac{8}{t^3} - 3$

For Problems 8, 9, and 10, a three-toed sloth is hanging onto a tree branch and moving right and left along the branch. (The tree trunk is at zero and the positive direction goes out from the trunk.) Note that $s(t)$ gives its position (in feet) as a function of time (seconds). Between $t = 0$ and $t = 5$, for each problem, find a) the intervals when the sloth is moving away from the trunk, the intervals when it is moving toward the trunk, and when and where it turns around; b) its total distance moved and its average speed; and c) its total displacement and average velocity.

8 $s(t) = 2t^3 - t^2 + 8t - 5$

9. $s(t) = t^4 + t^2 - t$

10. $s(t) = \frac{t+1}{t^2+4}$

Related Rates — They Rate, Relatively

Say you're filling up your swimming pool and you know how fast water is coming out of your hose, and you want to calculate how fast the water level in the pool is rising. You know one rate (how fast the water is being poured in), and you want to determine another rate (how fast the water level is rising). These rates are called *related rates* because one depends on the other — the rate that the water level is rising will depend on the rate that the water is being poured in. In a related rates problem, you're given one or more rates that are either constant or changing, and you have to figure out a related rate that is usually changing. You have to determine this related rate at one particular point in time. (If this isn't crystal clear, you'll see what I mean in a minute when you work through the following problems.)

Related rates problems are the Waterloo for many a calculus student. But they're not that bad after you get the basic technique down. The best way to get the hang of them is by working through lots of examples, so let's get started.

Blowing up a balloon

You're blowing up a balloon at a rate of 300 cubic inches per minute. When the balloon's radius is 3 inches, how fast is the radius increasing?

1. **Draw a diagram, labeling it with any *unchanging* measurements (there aren't any in this unusually simple problem) and making sure to assign a variable to anything in the problem that's *changing* (unless it's irrelevant to the problem). See Figure 12-6.**

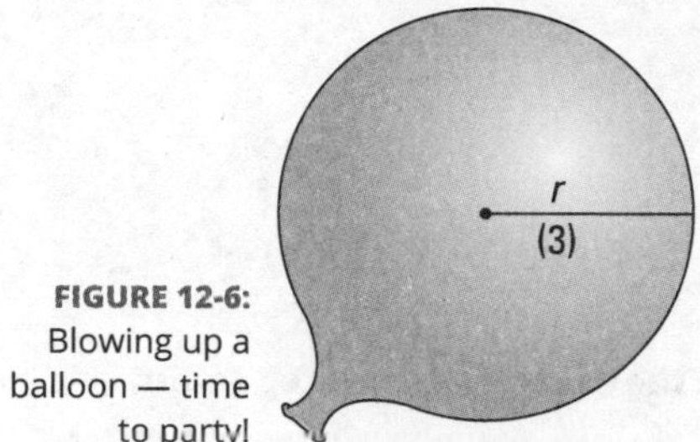

FIGURE 12-6: Blowing up a balloon — time to party!

Notice that the radius in Figure 12-6 is labeled with the variable *r*. The radius needs a variable because as the balloon is being blown up, the radius is *changing*. I put the 3 in parentheses to emphasize that the number 3 is *not* an unchanging measurement. The problem asks you to determine something *when* the radius is 3 inches, but remember, the radius is constantly changing.

TIP

Changing or unchanging? In related rates problems, it's important to distinguish between what is changing and what is *not* changing.

The volume of the balloon is also changing, so you need a variable for volume, *V*. You could put a *V* on your diagram to indicate the changing volume, but there's really no easy way to mark part of the balloon with a *V* like you can show the radius with an *r*.

2. **List the given rate and the rate you're asked to determine as derivatives with respect to time.**

You're pumping up the balloon at 300 cubic inches per minute. That's a rate — a change in volume (cubic inches) per change in time (minutes). So,

$$\frac{dV}{dt} = 300 \text{ cubic inches per minute}$$

You have to figure out a related rate, namely, how fast the radius is changing, so

$$\frac{dr}{dt} = ?$$

3. **Write down the formula that connects the variables in the problem, *V* and *r*.**

Here's the formula for the volume of a sphere:

$$V = \frac{4}{3}\pi r^3$$

4. **Differentiate your formula with respect to time, *t*.**

When you differentiate in a related rates problem, all variables are treated like the *y*'s are treated in a typical implicit differentiation problem.

$$\frac{dV}{dt} = \frac{4}{3}\pi \cdot 3r^2 \frac{dr}{dt}$$
$$= 4\pi r^2 \frac{dr}{dt}$$

You need to add the $\frac{dr}{dt}$ just like you would add on a y' or a $\frac{dy}{dx}$ with implicit differentiation.

5. **Substitute known values for the rate and variables in the equation from Step 4, and then solve for the thing you're asked to determine.**

It's given that $\frac{dV}{dt} = 300$, and you're asked to figure out $\frac{dr}{dt}$ when $r = 3$, so plug in these numbers and solve for $\frac{dr}{dt}$.

WARNING

Differentiate before you plug in. Be sure to differentiate (Step 4) *before* you plug the given information into the unknowns (Step 5).

$$\frac{dV}{dt} = 4\pi r^2 \frac{dr}{dt}$$
$$300 = 4\pi \cdot 3^2 \frac{dr}{dt}$$
$$300 = 36\pi \frac{dr}{dt}$$
$$\frac{300}{36\pi} = \frac{dr}{dt}$$
$$\frac{dr}{dt} = \frac{25}{3\pi} \approx 2.65 \text{ inches per minute}$$

So, the radius is increasing at a rate of about 2.65 inches per minute when the radius measures 3 inches. Think of all the balloons you've blown up since your childhood. Now you finally have the answer to the question that's been bugging you all these years.

By the way, if you plug 5 into *r* instead of 3, you get an answer of about 0.95 inches per minute. This should agree with your balloon-blowing-up experience: The bigger the balloon gets, the slower it grows. It's a good idea to check things like this every so often to see that the math agrees with common sense.

TIP

After finishing a math problem, ask yourself whether your answer makes sense. Asking this question can help you increase your success in math (and science). I'm not referring here to the things you can do to confirm an answer — things like plugging an answer into a variable, or solving backwards — confirming your answers like that is another great strategy for success. What I'm talking about here is asking yourself whether your answer to a problem is within the ballpark of what's reasonable (as opposed to something that's ridiculous or impossible). And while it's not always possible to decide whether a math answer is reasonable, you should always keep this important question in mind.

Filling up a trough

Here's another related rates problem. A trough is being filled up with swill. The trough is 10 feet long, and its cross-section is an isosceles triangle with a base of 2 feet and a height of 2 feet 6 inches (with the vertex at the bottom, of course). Swill's being poured in at a rate of 5 cubic feet per minute. When the depth of the swill is 1 foot 3 inches, how fast is the swill level rising?

1. **Draw a diagram, labeling it with any *unchanging* measurements and assigning variables to any *changing* things. See Figure 12-7.**

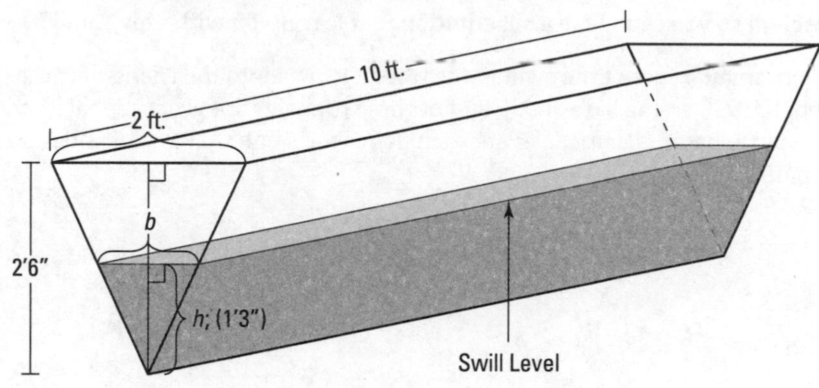

FIGURE 12-7: Filling a trough with swill — lunch time.

Note that Figure 12-7 shows the *unchanging* dimensions of the trough (2 feet, 2 feet 6 inches, and 10 feet), and that these dimensions do *not* have variable names like l for length or h for height. And note that the *changing* things — the height (or depth) of the swill and the width of the surface of the swill (which gets wider as the swill gets deeper) — have variable names, h for height and b for base (I say *base* instead of *width* because it's the base of the upside-down triangle shape made by the swill). The volume of the swill is also changing, so you can call that V.

2. **List the given rate and the rate you're asked to figure out as derivatives with respect to *time*.**

$$\frac{dV}{dt} = 5 \text{ cubic feet per minute}$$

$$\frac{dh}{dt} = ?$$

3. a. **Write down the formula that connects the variables in the problem: V, h, and b.**

I'm *absolutely positive* that you remember the formula for the volume of a right prism (the shape of the swill in the trough):

$$V = (\textit{area of base})(\textit{height})$$

Note that this "base" is the base of the prism (the whole triangle at the end of the trough), not the base of the triangle that is labeled b in Figure 12-7. Also, this "height" is the height of the prism (the length of the trough), not the height labeled h in Figure 12-7. Sorry about the confusion. Deal with it.

The area of the triangular base equals $\frac{1}{2}bh$, and the "height" of the prism is 10 feet, so the formula becomes

$$V = \frac{1}{2}bh \cdot 10$$
$$V = 5bh$$

Now, unlike the formula in the balloon example, this formula contains a variable, b, that you don't see in your list of derivatives in Step 2. So, Step 3 has a second part — getting rid of this extra variable.

3. b. Find an equation that relates the unwanted variable, *b*, to some other variable in the problem so you can make a substitution that leaves you with only *V* and *h*.

The triangular face of the swill in the trough is *similar* to the triangular face of the trough itself, so the base and height of these triangles are proportional. (Recall from geometry that *similar triangles* are triangles of the same shape; their sides are proportional.) Thus,

$$\frac{b}{2} = \frac{h}{2.5} \qquad \text{(Be careful: 2'6" is } \textit{not} \text{ 2.6 feet.)}$$
$$b = \frac{2h}{2.5}$$
$$b = 0.8h$$

TIP

Be on the lookout for similar triangles. Similar triangles come up a lot in related rates problems. Look for them whenever the problem involves a triangle, a triangular prism, or a cone shape.

Now substitute $0.8h$ for b in your formula from Step 3a.

$$V = 5bh$$
$$V = (5)(0.8h)(h)$$
$$V = 4h^2$$

4. Differentiate this equation with respect to *t*.

$$\frac{dV}{dt} = 8h\frac{dh}{dt} \qquad \left(\text{the power rule with the implicit differentiation } \frac{dh}{dt}\right)$$

5. Substitute known values for the rate and variable in the equation from Step 4 and then solve.

You know that $\frac{dV}{dt} = 5$ cubic feet per minute, and you want to determine $\frac{dh}{dt}$ when h equals 1 foot 3 inches, or 1.25 feet, so plug in 5 and 1.25 and solve for $\frac{dh}{dt}$:

$$\frac{dV}{dt} = 8h\frac{dh}{dt}$$
$$5 = 8 \cdot 1.25 \cdot \frac{dh}{dt}$$
$$\frac{dh}{dt} = \frac{1}{2}$$

That's it. The swill's level is rising at a rate of $\frac{1}{2}$ foot per minute when the swill is 1 foot 3 inches deep. Dig in.

Fasten your seat belt: You're approaching a calculus crossroads

Ready for another common related rates problem? One car leaves an intersection traveling north at 50 mph, and another is driving west toward the intersection at 40 mph. At one point, the north-bound car is three-tenths of a mile north of the intersection, and the west-bound car is four-tenths of a mile east of it. At this point, how fast is the distance between the cars changing?

1. **Do the diagram thing. See Figure 12-8.**

TIP

Variable or fixed? Before going on with this problem, I want to mention the problem you'll do next. It involves a ladder leaning against and sliding down a wall. Check out the figure that goes with the next problem, but note that when you draw your own diagram for a ladder-against-a-wall problem like this, you won't draw a house like I did. Your diagram would be very similar to Figure 12-8, except that the y-axis would represent the wall, the x-axis would be the ground, and the diagonal line would be the ladder. The ladder problem and the car problem are quite similar, but there's an important difference. The distance between the cars is *changing,* so the diagonal line in Figure 12-8 is labeled with a variable, s. A ladder, on the other hand, has a *fixed* length, so the diagonal line in your diagram for a ladder problem would be labeled with a number, not a variable.

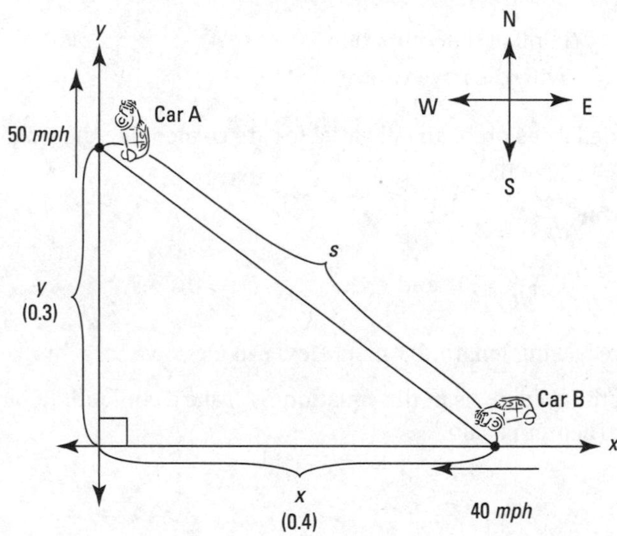

FIGURE 12-8: Calculus — it's a drive in the country.

2. **List all given rates and the unknown rate.**

As Car A travels north, the distance y is growing at 50 miles per hour. That's a rate, a change in distance per change in time. So,

$$\frac{\textit{change in distance in y direction}}{\textit{change in time}} = \frac{dy}{dt} = 50 \text{ mph}$$

As Car B travels west, the distance x is *shrinking* at 40 miles per hour. That's a *negative* rate:

$$\frac{\text{change in distance in } x \text{ direction}}{\text{change in time}} = \frac{dx}{dt} = -40 \text{ mph}$$

You have to figure out how fast s is changing, so,

$$\frac{\text{change in distance in } s \text{ direction}}{\text{change in time}} = \frac{ds}{dt} = ?$$

3. **Write the formula that relates the variables in the problem: x, y, and s.**

   The Pythagorean Theorem, $a^2 + b^2 = c^2$, will do the trick for this right triangle problem. In this problem, x and y are the legs of the right triangle, and s is the hypotenuse, so $x^2 + y^2 = s^2$.

TIP

The Pythagorean Theorem is used a lot in related rates problems. If there's a right triangle in your problem, it's quite likely that $a^2 + b^2 = c^2$ is the formula you'll need.

Because this formula contains the variables x, y, and s, which all appear in your list of derivatives in Step 2, you don't have to tweak this formula like you did in the trough problem.

4. **Differentiate with respect to t.**

$$s^2 = x^2 + y^2$$

$$2s\frac{ds}{dt} = 2x\frac{dx}{dt} + 2y\frac{dy}{dt} \quad \text{(implicit differentiation with the power rule)}$$

(Remember, in a related rates problem, all variables are treated like the y's in an implicit differentiation problem.)

5. **Substitute and solve for $\frac{ds}{dt}$.**

$$x = 0.4, \ y = 0.3, \ \frac{dx}{dt} = -40, \ \frac{dy}{dt} = 50, \ \text{and } s = \ \ldots$$

"Holy devoid distance lacking length, Batman. How can we solve for $\frac{ds}{dt}$ unless we have values for the rest of the unknowns in the equation?" "Take a chill pill, Robin — just use the Pythagorean Theorem again."

$$s^2 = x^2 + y^2$$
$$s^2 = 0.4^2 + 0.3^2$$
$$s^2 = 0.16 + 0.09$$
$$s^2 = 0.25$$
$$s = \pm 0.5 \quad \text{(square rooting both sides)}$$

You can reject the negative answer because s obviously has a positive length. So $s = 0.5$.

Now plug everything into your equation:

$$2s\frac{ds}{dt}=2x\frac{dx}{dt}+2y\frac{dy}{dt}$$
$$2\cdot 0.5\cdot\frac{ds}{dt}=2\cdot 0.4\cdot(-40)+2\cdot 0.3\cdot 50$$
$$1\cdot\frac{ds}{dt}=-32+30$$
$$\frac{ds}{dt}=-2$$

This negative answer means that the distance, s, is *decreasing*.

Thus, when Car A is 3 blocks north of the intersection and Car B is 4 blocks east of it, the distance between them is decreasing at a rate of 2 mph.

Try this at your own risk

One final related rates problem. A homeowner decides to paint his home. He picks up a home improvement book, which recommends that a ladder should be placed against a wall such that the distance from the foot of the ladder to the bottom of the wall is one-third the length of the ladder. Not being the sharpest tool in the shed, the homeowner gets mixed up and thinks that it's the distance from the *top* of the ladder to the base of the wall that should be a third of the ladder's length. He sets up his 18-foot ladder accordingly, and — despite this unstable ladder placement — he manages to climb the ladder and start painting. (Perhaps the foot of the ladder is caught on a tree root or something.) His luck doesn't last long, and the ladder begins to slide rapidly down the wall. One foot before the top of the ladder hits the ground, it's falling at a rate of 20 feet/second. At this moment, how fast is the foot of the ladder moving away from the wall?

1. **Draw a diagram, labeling it with any *unchanging* measurements and assigning variables to any *changing* things.**

   See the following figure.

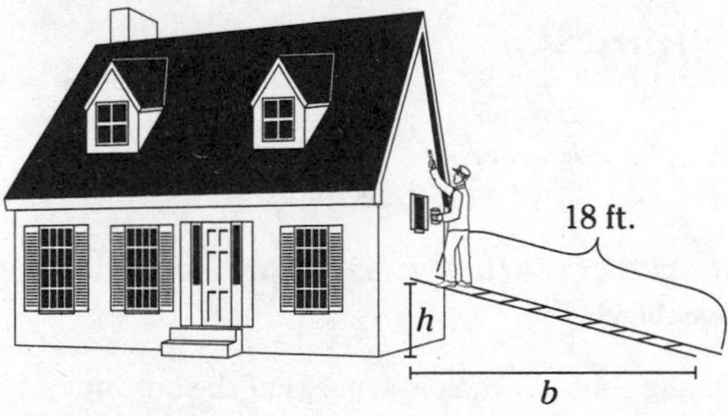

   You don't have to draw the house — the basic triangle is enough. But I've sketched a fuller picture of this scenario to make clear what a knucklehead this guy is.

2. **List all given rates and the rate you're asked to figure out. Write these rates as derivatives with respect to time.**

   Note that h is the distance from the top of the ladder to the bottom of the wall; b is the distance from the base of the ladder to the wall.

   You're told that the ladder is *falling* at a rate of 20 feet/second. Going down is *negative*, so

   $$\frac{dh}{dt}=-20 \quad (\text{when } h=1) \qquad \frac{db}{dt}=?$$

3. **Write down the formula that connects the variables in the problem, h and b.**

   That's the Pythagorean Theorem, of course: $a^2+b^2=c^2$, thus $h^2+b^2=18^2$.

4. **Differentiate with respect to time.**

   $$\begin{aligned} h^2+b^2&=18^2\\ 2h\frac{dh}{dt}+2b\frac{db}{dt}&=0 \end{aligned}$$

5. **Substitute known values for the rates and variables in the equation from Step 4, and then solve for the thing you're asked to determine.**

   You're trying to determine $\frac{db}{dt}$, so you have to plug numbers into everything else. But, as often happens, you don't have a number for b, so you need to use a formula to get the number you need. This will often be the same formula you already used.

   $$\begin{aligned} h^2+b^2&=18^2\\ 1^2+b^2&=18^2\\ b&=\pm\sqrt{323}\\ &\approx\pm 17.97 \text{ feet} \end{aligned}$$

   (Obviously, you can reject the negative answer.)

   Now you have what you need to finish the problem.

   $$\begin{aligned} 2h\frac{dh}{dt}+2b\frac{db}{dt}&=0\\ 2(1)(-20)+2(17.97)\frac{db}{dt}&=0\\ \frac{db}{dt}&=\frac{40}{35.94}\\ &\approx 1.11 \text{ feet/second} \end{aligned}$$

6. **Ask yourself whether your answer is reasonable. (This optional step is always a good idea when feasible.)**

   Yes, it does make sense. It makes sense that the bottom of the ladder is moving out much more slowly than the top of the ladder is moving down. Hold a yardstick against a wall so the bottom of it is on the floor and the top of it is on the wall about 4 or 5 inches from the floor. Then push the top of the yardstick 4 or 5 inches down to the floor. You'll see that the bottom barely moves farther out from the wall. Right triangles with a fixed hypotenuse always work like that. If one leg is much shorter than the other, the short leg can change a lot while the long leg barely changes. It's a byproduct of the Pythagorean Theorem.

YOUR TURN

11 This problem involves the same trough as the one in the example problem (it's 10 feet long and its cross section is an isosceles triangle with a base of 2 feet and a height of 2 feet 6 inches). But this time, the farmer is pouring swill into the trough at a rate of 1 cubic foot per minute, and just as the swill reaches the brim, three hogs start violently sucking down the swill at a rate of $\frac{1}{2}$ cubic foot per minute for each hog. They're going at it so vigorously that another $\frac{1}{2}$ cubic foot of swill is being splashed out of the trough each minute. The farmer keeps pouring in swill, but she's no match for her hogs. When the depth of the swill falls to 1 foot 8 inches, how fast is the swill level falling?

12 A pitcher delivers a fastball, which the batter pops up — it goes straight up above home plate. When it reaches a height of 60 feet, it's moving up at a rate of 50 feet per second. At this point, how fast is the distance from the ball to second base growing? *Note:* The distance between the bases of a baseball diamond is 90 feet (home to first, first to second, and so on).

13 A 6-foot-tall man looking over his shoulder sees his shadow that's cast by a 15-foot-tall lamppost in front of him. The shadow frightens him, so he starts running away from it — toward the lamppost. Unfortunately, this only makes matters worse, as it causes the frightening head of the shadow to gain on him. He starts to panic and run even faster. Five feet before he crashes into the lamppost, he's running at a speed of 15 miles per hour. At this point, how fast is the tip of the shadow moving?

14 Salt is being unloaded onto a conical pile at a rate of 200 cubic feet per minute. If the height of the cone-shaped pile is always equal to the radius of the cone's base, how fast is the height of the pile increasing when it's 18 feet tall?

Practice Questions Answers and Explanations

1. **The dimensions are 3.25 inches wide and 3.25 inches tall. The volume is about 27.14 cubic inches.**

   1. **Draw your diagram (see the following figure).**

2. a. **Write a formula for the thing you want to maximize, the volume:**

$V = \pi r^2 h$

2. b. **Use the given information to relate *r* and *h*.**

$$Surface\ Area = \overbrace{2\pi r^2}^{\text{top and bottom}} + \overbrace{2\pi rh}^{\text{lateral area}}$$

$$50 = 2\pi r^2 + 2\pi rh$$

$$25 = \pi r^2 + \pi rh$$

2. c. **Solve for *h* and substitute to create a function of one variable.**

$$25 = \pi r^2 + \pi rh, \text{ so } h = \frac{25 - \pi r^2}{\pi r} = \frac{25}{\pi r} - r$$

$$V = \pi r^2 h$$

$$V(r) = \pi r^2 \left(\frac{25}{\pi r} - r\right)$$

$$= 25r - \pi r^3$$

3. **Figure the domain.**

$r > 0$ is obvious

$h > 0$ is also obvious

And because $25 = \pi r^2 + \pi rh$ (from Step 2b), when $h = 0$, $r = \sqrt{\frac{25}{\pi}}$; so to make $h > 0$, r must be less than $\sqrt{\frac{25}{\pi}}$, or about 2.82 inches.

4. **Find the critical numbers of $V(r)$.**

$$V(r) = 25r - \pi r^3$$
$$V'(r) = 25 - 3\pi r^2$$
$$0 = 25 - 3\pi r^2$$
$$r^2 = \frac{25}{3\pi}$$
$$r = \pm\sqrt{\frac{25}{3\pi}}$$
$$\approx 1.63 \text{ inches}$$ (You can reject the negative answer because it's outside the domain.)

5. **Evaluate the volume at the critical number.**

$$V(1.63) = 25 \cdot 1.63 - \pi(1.63)^3$$
$$\approx 27.14 \text{ cubic inches}$$

The can will be $2 \cdot 1.63$ or about 3.25 inches wide and $\frac{25}{\pi \cdot 1.63} - 1.63$ or about 3.25 inches tall. Isn't that nice? The largest can has the same width and height and would thus fit perfectly into a cube. Geometric optimization problems frequently have results where the dimensions have some nice, simple mathematical relationship to each other.

By the way, did you notice that I skipped evaluating the volume at the endpoints of the domain? Can you guess why I did that? *Hint:* What's the volume for the smallest and largest value of the radius?

(2) **The dimensions of the window need to be about 4′3″ wide and about 5′1″ high. The minimum cost is roughly $373.**

1. **Draw a diagram with variables (see the following figure).**

2. a. **Express the thing you want to minimize, the cost.**

$$\text{Cost} = (\textit{length of curved frame}) \cdot (\textit{cost per linear foot}) + (\textit{length of straight frame}) \cdot (\textit{cost per linear foot})$$
$$= (\pi x)(25) + (2x + 2y)(20)$$
$$= 25\pi x + 40x + 40y$$

2. b. Relate the two variables to each other.

$$Area = Semicircle + Rectangle$$

$$20 = \frac{\pi x^2}{2} + 2xy$$

2. c. Solve for y and substitute.

$$2xy = 20 - \frac{\pi x^2}{2}$$

$$y = \frac{20}{2x} - \frac{\pi x^2}{4x}$$

$$= \frac{10}{x} - \frac{\pi x}{4}$$

$$Cost = 25\pi x + 40x + 40y$$

$$C(x) = 25\pi x + 40x + 40\left(\frac{10}{x} - \frac{\pi x}{4}\right)$$

$$= 25\pi x + 40x + \frac{400}{x} - 10\pi x$$

$$= 15\pi x + 40x + \frac{400}{x}$$

3. Find the domain.

$x > 0$ is obvious. And when x gets large enough, the entire window of 20 square feet in area will be one big semicircle, so

$$20 = \frac{\pi x^2}{2}$$

$$40 = \pi x^2$$

$$x^2 = \frac{40}{\pi}$$

$$x = \sqrt{\frac{40}{\pi}} \approx 3.57$$

Thus, x must be less than or equal to 3.57.

4. Find the critical numbers of $C(x)$.

$$C(x) = 15\pi x + 40x + \frac{400}{x}$$

$$C'(x) = 15\pi + 40 + (-400)x^{-2}$$

$$0 = 15\pi + 40 - 400x^{-2}$$

$$400x^{-2} = 15\pi + 40$$

$$x^2 = \frac{400}{15\pi + 40}$$

$$x = \pm\sqrt{\frac{400}{15\pi + 40}}$$

$$x \approx \pm 2.143$$

Omit −2.143 because it's outside the domain. So, 2.143 is the only critical number.

5. Evaluate the cost at the critical number and at the endpoints.

$$C(x) = 15\pi x + 40x + \frac{400}{x}$$
$$C(0) = undefined$$
$$C(2.143) \approx \$373$$
$$C(3.57) \approx \$423$$

You know $C(2.143) \approx \$373$ is a min (not a max) because the cost goes up to \$423 as x increases from 2.143, and as x decreases to zero, the cost also goes up (imagine plugging some tiny number like $x = 0.001$ into $C(x)$; you would get an enormous cost).

So, the least expensive frame for a 20-square-foot window will cost about \$373 and will be 2×2.143, or about 4.286 feet or 4′3″ wide at the base. Because $y = \frac{10}{x} - \frac{\pi x}{4}$, the height of the rectangular lower part of the window will be 2.98, or about 3′ tall. The total height will thus be 2.98 plus 2.14, or about 5′11″.

3 **The hypotenuse meets the y-axis at $(0, 10)$ and the x-axis at $(4, 0)$; the triangle's area is 20.**

1. Draw a diagram (see the following figure).

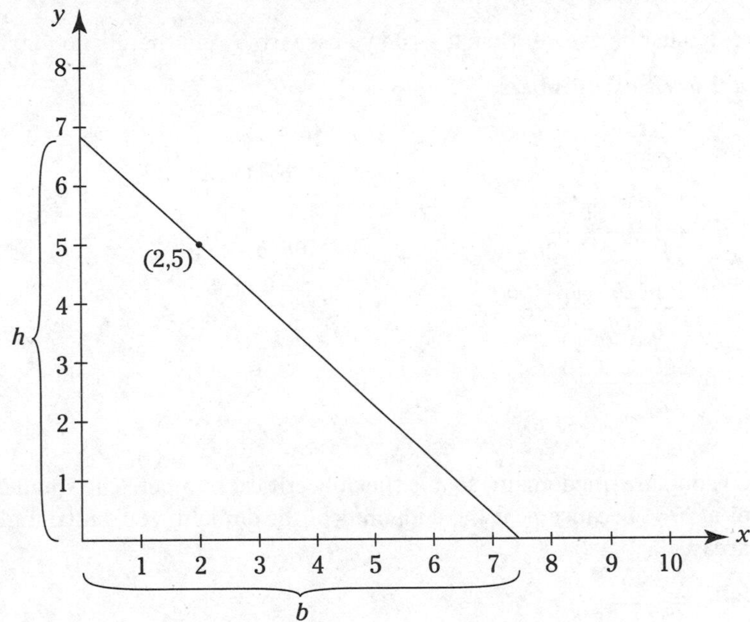

2. a. Write a formula for the thing you want to minimize, the area:

$$A = \frac{1}{2}bh$$

2. b. Use the given constraints to relate *b* and *h*.

This is a bit tricky — *Hint:* Consider similar triangles. If you draw a horizontal line from $(0, 5)$ to $(2, 5)$, you create a little triangle in the upper-left corner that's similar to the whole triangle. (You can prove their similarity with AA — remember your geometry? — as both triangles have a right angle and both share the top angle.)

Because the triangles are similar, their sides are proportional:

$$\frac{height_{\text{big triangle}}}{base_{\text{big triangle}}} = \frac{height_{\text{small triangle}}}{base_{\text{small triangle}}}$$

$$\frac{h}{b} = \frac{h-5}{2}$$

2. c. **Solve for one variable in terms of the other — take your pick — and substitute into your formula to create a function of a single variable.** (I realized after doing the math below that it would have been a bit quicker to solve for b instead of h. My bad. But the rest of the problem works the same and is just as easy either way.)

$$\begin{aligned} 2h &= b(h-5) \\ 2h &= bh - 5b \\ h(2-b) &= -5b \\ h &= \frac{5b}{b-2} \end{aligned}$$

$$\begin{aligned} A &= \frac{1}{2}bh \\ A(b) &= \frac{1}{2}b \cdot \left(\frac{5b}{b-2}\right) \\ &= \frac{5b^2}{2b-4} \end{aligned}$$

3. **Find the domain.**

Here, b must be greater than 2 — do you see why? And there's no maximum value for b.

4. **Find the critical numbers.**

$$\begin{aligned} A(b) &= \frac{5b^2}{2b-4} \\ A'(b) &= \frac{(5b^2)'(2b-4) - (5b^2)(2b-4)'}{(2b-4)^2} \\ &= \frac{10b(2b-4) - 10b^2}{(2b-4)^2} \\ &= \frac{10b^2 - 40b}{(2b-4)^2} \end{aligned}$$

$$\begin{aligned} \frac{10b^2 - 40b}{(2b-4)^2} &= 0 \\ 10b^2 - 40b &= 0 \\ 10b(b-4) &= 0 \\ b &= 0 \text{ or } 4 \end{aligned}$$

Zero is outside the domain, so 4 is the only critical number. The smallest triangle must occur at $b = 4$ because near the endpoints of the domain, you get triangles with astronomical areas.

5. **Finish.**

$$b = 4 \quad \text{and} \quad h = \frac{5b}{b-2}, \text{ so, } \quad h = \frac{5 \cdot 4}{4-2} = 10$$

And the triangle's area is thus 20.

(4) The minimizing dimensions are 6-by-6-by-2, made with 108 square inches of cardboard.

1. **Draw a diagram and label with variables (see the following figure).**

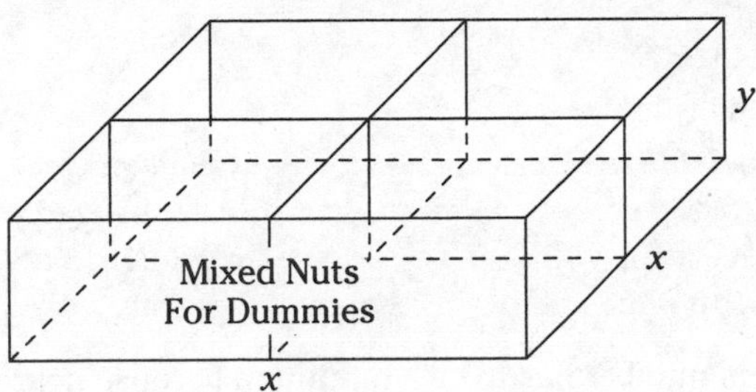

2. a. **Express the thing you want to minimize, the cardboard area, as a function of the variables.**

$$\textit{Cardboard area} = \overbrace{x^2}^{\textit{square base}} + \overbrace{4xy}^{\textit{four sides}} + \overbrace{2xy}^{\textit{two dividers}}$$

$$A = x^2 + 6xy$$

2. b. **Use the given constraint to relate x to y.**

$$Vol = l \cdot w \cdot h$$

$$72 = x \cdot x \cdot y$$

2. c. **Solve for y and substitute in the equation from Step 2a to create a function of one variable.**

$$y = \frac{72}{x^2}$$

$$A = x^2 + 6xy$$

$$A(x) = x^2 + 6x\left(\frac{72}{x^2}\right)$$

$$= x^2 + \frac{432}{x}$$

3. **Find the domain.**

$x > 0$ is obvious

$y > 0$ is also obvious

And if you made y small enough, say the height of a proton — great box, eh? — x would have to be astronomically big to make the volume 72 cubic inches. So, technically, there is no maximum value for x.

4. **Find the critical numbers.**

$$A(x) = x^2 + \frac{432}{x}$$
$$A'(x) = 2x - 432x^{-2}$$
$$0 = 2x - \frac{432}{x^2}$$
$$\frac{432}{x^2} = 2x$$
$$x = \sqrt[3]{216} = 6$$

You know this number has to be a minimum because near the endpoints, say when $x = 0.0001$ or $y = 0.0001$, you get absurd boxes — either thin and tall like a mile-high toothpick or short and flat like a square piece of cardboard as big as a city block with a microscopic lip. Both of these would have an *enormous* area and would be of interest only to calculus professors. (Whoops; another slight math omission. Do you see it?)

5. **Finish.**

$x = 6$, so the total area is

$$A(6) = 6^2 + \frac{432}{6} = 36 + 72 = 108 \qquad \text{and} \qquad y = \frac{72}{x^2} = \frac{72}{6^2} = 2$$

That's it — a 6-by-6-by-2 box made with 108 square inches of cardboard.

5 **(a)** At $t = 2$, the platypus's position is $\mathbf{s(2) = 24}$ **feet from the back of your boat.**

(b) $v(t) = s'(t) = 10t$, so at $t = 2$, the platypus's velocity is $\mathbf{s'(2) = 20}$ **feet/second** (20 is positive so that's toward the front of the boat).

(c) Speed is the absolute value of velocity, so the speed is also **20 feet/second.**

(d) Acceleration, $a(t)$, equals $v'(t) = s''(t) = 10$. That's a constant, so the platypus's acceleration is $\mathbf{10 \frac{feet/second}{second}}$ **at all times.**

6 **(a)** $s(2)$ gives the platypus's position at $t = 2$; that's $3 \cdot 2^4 - 5 \cdot 2^3 + 2 - 6$, or **4 feet, from the back of the boat.**

(b) $v(t) = s'(t) = 12t^3 - 15t^2 + 1$. At $t = 2$, the velocity is thus **37 feet per second.**

(c) Speed is also **37 feet per second.**

(d) $a(t) = v'(t) = s''(t) = 36t^2 - 30t$. $a(2)$ equals $\mathbf{84 \frac{feet/second}{second}}$.

7 **(a)** At $t = 2$, $s(t)$ equals $\frac{1}{2} + 1 - 3$, or -1.5 feet. This means that the platypus is **1.5 feet behind the back of the boat.**

(b) $\mathbf{-1\frac{3}{4}}$ **feet/second**

$$v(t) = s'(t) = -t^{-2} - 24t^{-4}$$

$$\begin{aligned} v(2) = s'(2) &= -2^{-2} - 24(2)^{-4} \\ &= -\frac{1}{4} - \frac{24}{16} \\ &= -1\frac{3}{4} \text{ feet/second} \end{aligned}$$

A negative velocity means that the platypus is swimming "backward"; in other words, he's swimming toward the left, moving away from the back of the boat.

(c) *Speed* = |*velocity*|, so the platypus's speed is $\mathbf{1\frac{3}{4}}$ **feet/second.**

(d) $a(t) = v'(t) = s''(t) = 2t^{-3} + 96t^{-5}$, or $\frac{2}{t^3} + \frac{96}{t^5}$. Therefore, $a(2)$ is $\frac{2}{8} + \frac{96}{32}$, or $\mathbf{3\frac{1}{4}\ \frac{feet/second}{second}}$.

Give yourself a pat on the back if you figured out that this positive acceleration with a negative velocity means the platypus is actually slowing down.

8 **(a) The sloth never turns around; it moves away from the trunk for the entire interval from 0 to 5 seconds.**

Find the zeros of the velocity:

$$\begin{aligned} v(t) = s'(t) &= 6t^2 - 2t + 8 \\ 0 &= 6t^2 - 2t + 8 \\ 0 &= 3t^2 - t + 4 \end{aligned}$$

No solutions.

TIP

A quadratic has no solutions when the discriminant ($b^2 - 4ac$) is negative.

The fact that the velocity is never zero means that the sloth never turns around. At $t = 0$, $v(t) = 8\frac{\text{feet}}{\text{second}}$ which is positive, so the sloth moves away from the trunk for the entire interval.

(b and c) Both total distance and total displacement are 265 feet; both average speed and average velocity are 53 feet/second.

Displacement $= s(5) - s(0) = 260 - (-5) = 265$

Because there are no turnaround points and because the motion is in the positive direction, the total distance and total displacement are the same: 265 feet.

$$\textit{Average velocity} = \frac{\textit{total displacement}}{\textit{total time}} = \frac{s(5) - s(0)}{5 - 0} = \frac{265}{5} = 53\frac{\text{feet}}{\text{second}}$$

Whenever the total distance equals the total displacement, average speed also equals average velocity: 53 feet/second.

9 (a) **The sloth is going left from 0 seconds to 0.385 seconds, and right from 0.385 to 5 seconds. The sloth turns around at 0.385 seconds when it is at $s(0.385) = 0.385^4 + 0.385^2 - 0.385$, or -0.215 feet. That's 0.215 feet to the left of the trunk.**

Find the zeros of $v(t)$: $v(t) = s'(t) = 4t^3 + 2t - 1$
You'll need your calculator for this:
Graph $y = 4t^3 + 2t - 1$ and locate the x-intercepts. There's just one: $x \approx 0.385$. That's the only zero of $s'(t) = v(t)$.

WARNING

Don't forget that a zero of a derivative can be a horizontal inflection as well as a local extremum. You get turnaround points only at the local extrema of the position function. Because $v(0) = -1$ (a leftward velocity) and $v(1) = 5$ (a rightward velocity), $s(0.385)$ must be a turnaround point (and it's also a local min on the position graph). Does the first derivative test ring a bell?

Thus, the sloth is going left from 0 to 0.385 seconds and right from 0.385 to 5 seconds. The sloth turns around, obviously, at 0.385 seconds when it is at $s(0.385) = 0.385^4 + 0.385^2 - 0.385$, or -0.215 feet. (I presume you figured out that there must be another branch on the tree on the other side of the trunk to allow the sloth to go left to a negative position.)

(b) **The sloth moved a total distance of 645.43 feet at an average speed of 129.1 feet/second.**

There are two legs of the sloth's trip. It goes left from $t = 0$ till $t = 0.385$, then right from $t = 0.385$ till $t = 5$. Just add up the *positive* lengths of the two legs.

$$\begin{aligned} length_{\text{leg }1} &= |s(0.385) - s(0)| \\ &= |-0.215 - 0| \\ &= 0.215 \text{ feet} \end{aligned}$$

$$\begin{aligned} length_{\text{leg }2} &= |s(5) - s(0.385)| \\ &= \left|5^4 + 5^2 - 5 - (-0.215)\right| \\ &= 645.215 \text{ feet} \end{aligned}$$

The total distance is thus $0.215 + 645.215$, or 645.43 feet. That's one big tree! The branch is longer than two football fields.

The sloth's average speed is $645.43 / 5$, or about 129.1 feet/second. That's one fast sloth! About 88 per hour!

(c) **The sloth's total displacement is 645 feet and its average velocity is 129 feet/second.**

Total displacement is $s(5) - s(0)$; that's $645 - 0 = 645$ feet. Lastly, the sloth's average velocity is simply total displacement divided by total time — that's $645 / 5$, or 129 feet per second.

10 (a) **The sloth goes right from $t = 0$ till $t = 1.236$ seconds; then turns around at $s(1.236)$, or about 0.405 feet to the right of the trunk; and then goes left till $t = 5$.**

Find the zeros of $v(t)$:

$$v(t)=s'(t)=\frac{(t+1)'(t^2+4)-(t+1)(t^2+4)'}{(t^2+4)^2}$$

$$=\frac{t^2+4-(2t^2+2t)}{(t^2+4)^2}$$

$$=\frac{-t^2-2t+4}{(t^2+4)^2}$$

Set this equal to zero and solve:

$$\frac{-t^2-2t+4}{(t^2+4)^2}=0$$

$$-t^2-2t+4=0$$

$$t=\frac{-2\pm\sqrt{4-(-16)}}{2}$$

$$\approx -3.236 \text{ or } 1.236$$

Reject the negative solution because it's outside the interval of interest: $t=0$ to $t=5$. So, the only zero velocity occurs at $t=1.236$ seconds.

Because $v(0)=0.25$ feet per second and $v(5)\approx -0.037$, the first derivative test tells you that $s(1.236)$ must be a local max and therefore a turnaround point.

The sloth thus goes right from 0 to 1.236 seconds; then turns around at $s(1.236)$, which is about 0.405 feet to the right of the trunk; and then it goes left till $t=5$.

(b) The total distance moved equals 0.353 feet; the average speed is 0.071 feet per second.

The sloth's total distance is the sum of the lengths of the two legs:

$$\text{going right}=|s(1.236)-s(0)|$$

$$=|0.405-0.25|$$

$$\approx 0.155$$

$$\text{going left}=|s(5)-s(1.236)|$$

$$\approx 0.198$$

Total distance is therefore about $0.155+0.198=0.353$ feet. The sloth's average speed is thus about 0.353/5, or 0.071 feet per second. That's roughly 4.26 feet/minute — much more like it for a sloth.

(c) The total displacement is about −0.043 feet; the average velocity is −0.0086 feet per second.

Total displacement is defined as final position minus initial position, so that's

$$s(5)-s(0)=\frac{6}{29}-\frac{1}{4}\approx -0.043 \text{ feet}$$

And thus the sloth's average velocity is about –0.043/5, or –0.0086 feet per second.

11 It's falling at a rate of 0.9 inches per minute.

1. **Draw a diagram, labeling the diagram with any *unchanging* measurements and assigning variables to any *changing* things.**

See the following figure.

Note that the height of 1 foot 8 inches — which is the height only at one particular point in time — is in parentheses to distinguish it from the other *unchanging* dimensions.

2. **List all given rates and the rate you're asked to figure out.**

   Express these rates as derivatives with respect to time. Give yourself a high-five if you realized that the thing that matters about the changing volume of swill is the *net* rate of change of volume.

   Swill is coming in at 1 cubic foot per minute. It's going out at $3 \cdot \frac{1}{2}$ cubic foot per minute (for the three hogs) plus another $\frac{1}{2}$ cubic foot per minute (the splashing) — that's a total of 2 cubic feet per minute going out. So, the net is 1 cubic foot per minute going out — that's a *negative* rate of change. In calculus language, you write:

   $\frac{dV}{dt} = -1$ cubic foot per minute.

   You're asked to determine how fast the height is changing, so write: $\frac{dh}{dt} = ?$

3. **Steps 3a and 3b here are the same as in the example trough problem. So, just like in that problem, you've got the following formula:**

   $V = 4h^2$

4. **Differentiate with respect to *t*.**

   $$\frac{dV}{dt} = 8h\frac{dh}{dt}$$

5. **Substitute all known quantities into this equation and solve for $\frac{dh}{dt}$.**

   You were given that $h = 1'8''$ (you must convert this to feet — that's $\frac{5}{3}$ feet), and you figured out in Step 2 that $\frac{dV}{dt} = -1$, so

$$-1 = 8 \cdot \frac{5}{3} \cdot \frac{dh}{dt}$$

$$\frac{dh}{dt} = \frac{-1}{\frac{40}{3}}$$

$$= \frac{-3}{40} \text{ feet / minute}$$

$$= \frac{-9}{10} \text{ inches / minute}$$

   Thus, when the swill level drops to a depth of 1 foot 8 inches, it's falling at a rate of 0.9 inches per minute. Mmm, mmm, good!

6. **Ask whether this answer makes sense.**

   It's not easy to come up with a common-sense explanation of why this answer is or is not reasonable. But there's another type of check that works here and in many other related rates problems.

   Take a very small increment of time — something much less than the time unit of the rates used in the problem. This problem involves rates per *minute*, so use 1 second for your time increment. Now ask yourself what happens in this problem in 1 second. The swill is leaving the trough at 1 cubic foot/minute; so, in 1 second, $\frac{1}{60}$ cubic feet will leave the trough. What does that do to the swill height? Because of similar triangles, when the swill falls to a depth of 1 foot 8 inches, which is $\frac{2}{3}$ of the height of the trough, the width of the surface of the swill must be $\frac{2}{3}$ of the width of the trough — and that comes to $1\frac{1}{3}$ feet. So, the surface area of the swill is $1\frac{1}{3} \times 10$ feet.

   Assuming the trough walls are straight (this type of simplification always works in this type of checking process), the swill that leaves the trough will form the shape of a *very, very* short box ("box" sounds funny because this shape is so thin; maybe "thin piece of plywood" is a better image).

   The volume of a box equals $length \cdot width \cdot height$, thus

$$\frac{1}{60} = 10 \cdot 1\frac{1}{3} \cdot height$$

$$height = 0.00125$$

This tells you that in 1 second, the height should fall 0.00125 feet or something very close to it. (This process sometimes produces an exact answer and sometimes an answer with a very small error.) Now, finally, see whether this number agrees with the answer. Your answer was –0.9 inches/minute. Convert this to feet/second:

$$-\frac{9}{10} \div 12 \div 60 = -0.00125$$

It checks.

(12) **The distance is growing at about 21.3 feet per second.**

1. **Draw your diagram and label it. See the following figure.**

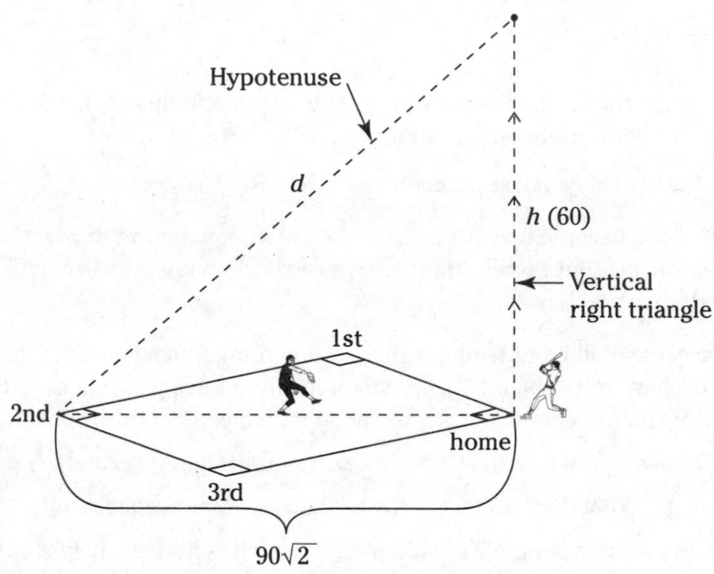

2. **List all given rates and the rate you're asked to figure out.**

$$\frac{dh}{dt} = 50\frac{\text{ft}}{\text{sec}} \quad (\text{when } h = 60) \qquad \frac{dd}{dt} = ?$$

3. **Write a formula that involves the variables:**

$$h^2 + \left(90\sqrt{2}\right)^2 = d^2$$

4. **Differentiate with respect to time:**

$$2h\frac{dh}{dt} = 2d\frac{dd}{dt}$$

5. **Substitute known values and solve for $\frac{dd}{dt}$:**

   You're missing a needed value, *d*. So, use the Pythagorean Theorem to get it:

$$h^2 + \left(90\sqrt{2}\right)^2 = d^2$$
$$60^2 + \left(90\sqrt{2}\right)^2 = d^2$$
$$d \approx \pm 140.7 \text{ feet} \quad \text{(You can reject the negative answer.)}$$

   Now do the substitutions:

$$2h\frac{dh}{dt} = 2d\frac{dd}{dt}$$
$$2 \cdot 60 \cdot 50 = 2 \cdot 140.7\frac{dd}{dt}$$
$$\frac{dd}{dt} = \frac{2 \cdot 60 \cdot 50}{2 \cdot 140.7}$$
$$\approx 21.3 \text{ feet/second}$$

6. **Check whether this answer makes sense.**

   For this one, you're on your own. *Hint:* Use the Pythagorean Theorem to calculate $d\ \frac{1}{50}$ of a second after the critical moment. Do you see why I picked this time increment?

13 **It's moving at 25 miles per hour.**

1. **The diagram thing: See the figure.**

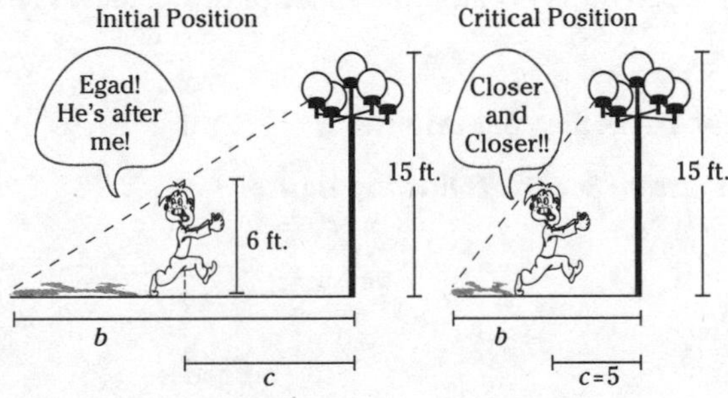

2. **List the known and unknown rates.**

$$\frac{dc}{dt} = -15\frac{\text{miles}}{\text{hour}} \quad (\text{when } c = 5). \quad \text{This is negative because } c \text{ is shrinking.}$$
$$\frac{db}{dt} = ?$$

3. **Write a formula that connects the variables.**

This is another similar triangle situation. For your two similar triangles, use the triangles in the initial position diagram.

$$\frac{height_{big\ triangle}}{height_{little\ triangle}} = \frac{base_{big\ triangle}}{base_{little\ triangle}}$$

$$\frac{15}{6} = \frac{b}{b-c}$$

$$15b - 15c = 6b$$

$$9b = 15c$$

$$3b = 5c$$

4. **Differentiate with respect to *t*.**

$$3\frac{db}{dt} = 5\frac{dc}{dt}$$

5. **Substitute known values.**

$$3\frac{db}{dt} = 5(-15)$$

$$\frac{db}{dt} = -25\text{miles/hour}$$

Thus, the top of the shadow is moving toward the lamppost at 25 miles per hour (and is thus gaining on the man at a rate of 25 – 15 = 10 miles/hour).

A somewhat unusual twist in this problem is that you never had to plug in the given distance of 5 feet. This is because the speed of the shadow is independent of the man's position.

(14) **It's increasing at $2\frac{1}{3}$ inches per minute.**

1. **Draw your diagram: See the following figure.**

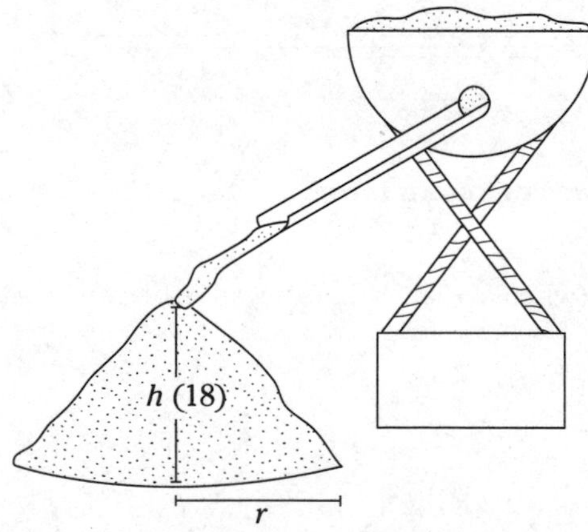

2. **List the rates: $\frac{dV}{dt}$ = 200 cubic feet per minute, $\frac{dh}{dt}$ = ?**

3. a. The formula thing:

$$V_{cone} = \frac{1}{3}\pi r^2 h$$

3. b. Write an equation relating *r* and *h* so that you can get rid of *r*:

$$r = h$$

What could be simpler? Now get rid of *r*:

$$V = \frac{1}{3}\pi h^2 h = \frac{1}{3}\pi h^3$$

4. Differentiate:

$$\frac{dV}{dt} = \pi h^2 \frac{dh}{dt}$$

5. Substitute and solve for $\frac{dh}{dt}$:

$$200 = \pi \cdot 18^2 \frac{dh}{dt}$$

$$\frac{dh}{dt} \approx 0.196 \text{ feet/minute}$$

$$\approx 2\frac{1}{3} \text{ inches/minute}$$

6. Check whether this answer makes sense.

Calculate the increase in the height of the cone from the critical moment ($h = 18$) to $\frac{1}{200}$ minute after the critical moment. When $h = 18$, $V = \frac{1}{3}\pi(18)^3$, or about 6107.256 cubic feet. $\frac{1}{200}$ minute later, the volume (which grows at a rate of 200 cubic feet per minute) will increase by 1 cubic foot to about 6108.256 cubic feet. Now solve for *h*:

$$6108.256 = \frac{1}{3}\pi h^3$$

$$h = \sqrt[3]{\frac{6108.256}{\frac{1}{3}\pi}}$$

$$\approx 18.000982$$

Thus, in $\frac{1}{200}$ minute, the height would grow from 18 feet to 18.000982 feet. That's a change of 0.000982 feet. Multiply that by 200 to get the change in 1 minute:

$0.000982 \cdot 200 \approx 0.196.$

It checks.

If you're ready to test your skills a bit more, take the following chapter quiz that incorporates all the chapter topics.

Whaddya Know? Chapter 12 Quiz

For Problems 1 through 4, you've got a corral problem similar to the one in the section, "The maximum area of a corral — yeehaw!" in this chapter. Your corral for Problems 1 through 4 is similar to the one in Figure 12-2, except that this time the rancher wants to have three equal rectangles side by side instead of two. And this time, the rancher wants to build a corral with 7200 square feet of area. He wants to construct the corral using as little fencing as possible. (You should sketch a diagram similar to Figure 12-2 with 4 y's instead of 3 and 6 x's instead of 4.)

1. Express the thing you want to minimize or maximize as a function of x.

2. Find the derivative of your function, set it equal to zero, and solve.

3. What's the domain of your function (mathematically, not practically, speaking), and what are its critical numbers?

4. What dimensions of the corral minimize the total length of fencing? And how much fencing is needed?

For Problems 5, 6, and 7, you've got a particle moving right and/or left along an ordinary number line. The function $s(t) = t^3 - 4t^2 + 4t - 8$ gives the particle's position on the line (in meters) as a function of time (in seconds).

5. **(a)** What's the velocity function, and what's the acceleration function?

   (b) What is $s'''(t)$, and explain what it means, what it tells you.

6. Does this particle ever reverse direction? If so, when? And, if so, does the particle switch from going left to going right or vice versa?

7. **(a)** What's the particle's total displacement from $t = 0$ to $t = 4$?

   (b) What's the particle's total distance traveled from $t = 0$ to $t = 4$?

For Problems 8 through 10, you've got a right triangle with legs of lengths x and y. The x leg is growing at a rate of 3 inches/minute and the y leg is shrinking at a rate of 5 inches/minute.

8. Express the two given rates with calculus notation.

9. At a certain point in time, the area of the triangle is 50 square inches and the two legs are equal. At this point, how fast is the area changing?

10. One minute later, how fast is the area changing?

Answers to Chapter 12 Quiz

1. $\boldsymbol{L(x) = 6x + \frac{9600}{x}}$

   There are 6 x-lengths of fencing and 4 y-lengths, so the total length of fencing is $L(x) = 6x + 4y$.

   You need a function of x, so express the given area in terms of x and y, solve for y, then make a substitution.

$$7200 = (3x)(y) \qquad L(x) = 6x + 4\left(\frac{2400}{x}\right)$$
$$y = \frac{2400}{x} \qquad = 6x + \frac{9600}{x}$$

2. $\boldsymbol{x = 40}$

$$\begin{aligned} L(x) &= 6x + 9600x^{-1} \\ L'(x) &= 6 - 9600x^{-2} \\ &= 6 - \frac{9600}{x^2} \\ 0 &= 6 - \frac{9600}{x^2}, \text{ etc.} \\ x &= \pm 40 \end{aligned}$$

   You can reject -40, of course.

3. **The domain of the length function is $(0, \infty)$. It has one critical number: 40.**

   x can't be zero or negative; it can be any positive number. In interval notation, the domain of the length function is $(0, \infty)$.

   A critical number of a function is a number in the function's domain where the derivative equals zero or is undefined. $L'(x)$ is defined for all x-values in the domain. And you found its zeros in Problem 2, namely, ± 40; only positive 40 is in the domain. That's the only critical number.

4. **A 120-foot-by-60-foot corral will minimize the total fencing needed; 480 feet of fencing is needed to build the corral.**

   $L'(x)$ is negative for x-values less than 40 and positive for x-values greater than 40. Thus, $L(x)$ is decreasing until 40 and increasing after 40, and, therefore $L(x)$ hits a minimum at $x = 40$. An x-value of 40 gives you a 120-by-60-foot corral with 480 total feet of fencing.

5. **(a)** $\boldsymbol{v(t) = 3t^2 - 8t + 4}$

   $\boldsymbol{a(t) = 6t - 8}$

$$\begin{aligned} s(t) &= t^3 - 4t^2 + 4t - 8 \\ v(t) &= s'(t) = 3t^2 - 8t + 4 \\ a(t) &= v'(t) = s''(t) = 6t - 8 \end{aligned}$$

   (b) $\boldsymbol{s'''(t) = 6}$. This tells you the rate of change of the acceleration.

$s'''(t) = 6$. The third derivative of s is the first derivative of a, the acceleration. Thus $s'''(t)$ tells you the rate of change of the acceleration. Acceleration for the particle in this problem would be given in terms of $\frac{\text{meters per second}}{\text{second}}$. The constant rate of change of the acceleration, namely 6, means that the acceleration of the particle is increasing 6 $\frac{\text{meters per second}}{\text{second}}$ each second. If it's a bit difficult to wrap your mind around that, no worries, you're not alone.

6 **Yes it does. From $t = 0$ seconds till $t = \frac{2}{3}$ seconds, the particle is moving to the right. At $\frac{2}{3}$ seconds, it reverses direction, and moves to the left till $t = 2$ seconds. It then reverses direction again and moves to the right indefinitely.**

Set the velocity equal to zero and solve:

$$v(t) = 3t^2 - 8t + 4$$
$$0 = 3t^2 - 8t + 4$$
$$0 = (3t - 2)(t - 2) \qquad t = \frac{2}{3} \text{ or } t = 2$$

If there are any turnaround points, they must occur when the velocity is zero. But a zero velocity does not guarantee that you've got a turnaround point. You must also check that the particle changes direction. You can easily check (in your head) that if you plug a positive number less than $\frac{2}{3}$, say $\frac{1}{2}$, into t in the factored form of $v(t)$, you'll get a negative times a negative. That's a positive, so from zero seconds till $\frac{2}{3}$ second, the particle is moving to the right. Plugging 1 into t gives you positive times negative, or a negative answer, so from $\frac{2}{3}$ second till 2 seconds, the particle is moving to the left. Finally, plugging 3 into t gives you positive times positive, or a positive answer, so the particle is moving right after 2 seconds.

7 **(a) Total displacement is 16.**

Total displacement equals final position minus initial position, that's $s(4)$ minus $s(0)$:

$$s(t) = t^3 - 4t^2 + 4t - 8$$

$$s(4) = 4^3 - 4 \cdot 4^2 + 4 \cdot 4 - 8$$
$$= 64 - 64 + 16 - 8$$
$$= 8$$
$$s(0) = -8$$

$$s(4) - s(0) = 8 - (-8) = 16$$

(b) Total distance traveled is $18\frac{10}{27}$.

To get the total distance traveled, you need to add the lengths of the three legs of the trip. The moving-right leg from zero seconds till $\frac{2}{3}$ seconds, the moving-left leg from $\frac{2}{3}$ seconds till 2 seconds, and the moving-right leg from 2 seconds till 4 seconds.

$$s(t) = t^3 - 4t^2 + 4t - 8$$

$$s(0) = -8$$
$$s\left(\frac{2}{3}\right) = -\frac{184}{27}$$
$$s(2) = -8$$
$$s(4) = 8$$

The particle moves $\frac{32}{27}$ to the right, then $\frac{32}{27}$ to the left, then 16 to the right for a total distance of $18\frac{10}{27}$.

8 $\boldsymbol{\frac{dx}{dt} = 3}$ **and** $\boldsymbol{\frac{dy}{dt} = -5}$**.**

Note that since you were told that the *y* leg is *shrinking*, $\frac{dy}{dt}$ must be negative.

9 **−10 square inches per minute.**

The area of a right triangle equals one-half the product of the legs, and it's given that the legs are equal when the area is 50:

$$A = \frac{1}{2}xy$$
$$50 = \frac{1}{2}xy$$
$$100 = xy$$

So, *x* and *y* are, obviously, both equal to 10. But, don't forget, do not plug in these values till after you differentiate the area function.

$$A = \frac{1}{2}xy$$
$$\frac{dA}{dt} = \frac{1}{2}\left(\frac{dx}{dt}y + \frac{dy}{dt}x\right)$$
$$= \frac{1}{2}(3 \cdot 10 - 5 \cdot 10) = -10$$

10 **−25 square inches per minute.**

At the critical moment in Problem 9, both *x* and *y* equal 10; *x* is growing at 3 inches/minute, and *y* is shrinking at 5 inches/minute. Thus, one minute after that critical moment, *x* will equal 13 and *y* will equal 5. Just plug in and finish.

$$\frac{dA}{dt} = \frac{1}{2}\left(\frac{dx}{dt}y + \frac{dy}{dt}x\right)$$
$$= \frac{1}{2}(3 \cdot 5 - 5 \cdot 13) = -25$$

IN THIS CHAPTER

» Tangling with tangents

» Negotiating normals

» Lining up for linear approximations

» Profiting from business and economics problems

» Doing $\sqrt{37}$ in your head

Chapter 13
More Differentiation Problems: Going Off on a Tangent

In this chapter, you see three more applications of differentiation: tangent and normal line problems, linear approximation problems, and economics problems. The common thread tying these problems together is the idea of a line tangent to a curve — which should come as no surprise since the meaning of the derivative of a curve is the slope of the tangent line. The problems in this chapter are all "practical" applications of differentiation in a sense, but some of them are — to be honest — much more likely to be found in a math book than in the real world. But at the other end of the spectrum, you encounter problems here like the economics problem of finding maximum profit. What could be more practical than that?

Tangents and Normals: Joined at the Hip

In everyday life, it's perfectly normal to go off on a tangent now and then. In calculus, on the other hand, there is nothing at all normal about a tangent. You need only note a couple of points before you're ready to try some problems:

- At its point of *tangency,* a tangent line has the same slope as the curve it's tangent to. In calculus, whenever a problem involves slope, you should immediately think derivative. The derivative is the key to all tangent line problems.
- At its point of intersection to a curve, a *normal* line is *perpendicular* to the tangent line drawn at that same point. Perpendicular lines have slopes that are opposite reciprocals. So, for a normal line problem you just use the derivative to get the slope of the tangent line, and then the opposite reciprocal of that gives you the slope of the normal line.

Ready to try a few problems? Say, that reminds me. I once had this problem with my carburetor. I took my car into the shop, and the mechanic told me the problem would be easy to fix, but when I went back to pick up my car . . . Hey, wait a minute. Where was I? I guess I sort of went off on a *tangent. Ha Ha Ha Ha.* I really crack myself up.

The tangent line problem

I bet there have been several times, just in the last month, when you've wanted to determine the location of a line through a given point that's tangent to a given curve. Let's walk through a problem to see how you do it.

Determine the points of tangency of the lines through the point $(1, -1)$ that are tangent to the parabola $y = x^2$.

Here's what you do. If you graph the parabola and plot the point, you can see that there are two ways to draw a tangent line from $(1, -1)$: up to the right and up to the left. See Figure 13-1.

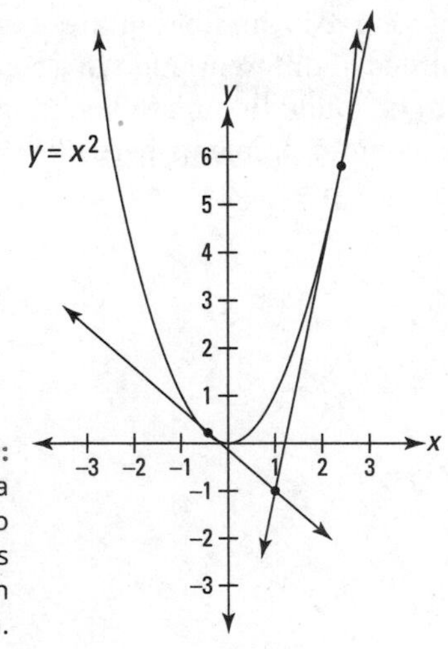

FIGURE 13-1: The parabola $y = x^2$ and two tangent lines through $(1, -1)$.

The key to this problem is in the meaning of the derivative. Don't forget: *The derivative of a function at a given point is the slope of the tangent line at that point.* So, all you have to do is set the derivative of the parabola equal to the slope of the tangent line and solve:

1. **Because the equation of the parabola is $y = x^2$, you can take a general point on the parabola, (x, y), and substitute x^2 for y.**

   So, label the two points of tangency (x, x^2).

2. **Take the derivative of the parabola.**

   $$y = x^2$$
   $$y' = 2x$$

3. **Using the slope formula, $\frac{y_2 - y_1}{x_2 - x_1}$, set the slope of each tangent line from $(1, -1)$ to (x, x^2) equal to the derivative at (x, x^2), which is $2x$, and solve for x.**

   (By the way, the math you do in this step may make more sense to you if you think of it as applying to just one of the tangent lines — say, the one going up to the right — but the math actually applies to both tangent lines simultaneously.)

   $$\frac{y_2 - y_1}{x_2 - x_1} \text{ (the slope of the tangent line)} = y' \text{ (the derivative)}$$

   $$\frac{x^2 - (-1)}{x - 1} = 2x$$
   $$x^2 - (-1) = 2x(x - 1)$$
   $$x^2 + 1 = 2x^2 - 2x$$
   $$0 = x^2 - 2x - 1$$

   $$x = \frac{2 \pm \sqrt{(-2)^2 - 4(1)(-1)}}{2 \cdot 1} \quad \text{(quadratic formula)}$$
   $$= \frac{2 \pm \sqrt{8}}{2}$$
   $$= \frac{2 \pm 2\sqrt{2}}{2}$$
   $$= 1 \pm \sqrt{2}$$

   So, the x-coordinates of the points of tangency are $1 + \sqrt{2}$ and $1 - \sqrt{2}$.

4. **Plug each of these x-coordinates into $y = x^2$ to obtain the y-coordinates.**

   $$y = (1 + \sqrt{2})^2 \qquad y = (1 - \sqrt{2})^2$$
   $$= 1 + 2\sqrt{2} + 2 \qquad = 1 - 2\sqrt{2} + 2$$
   $$= 3 + 2\sqrt{2} \qquad = 3 - 2\sqrt{2}$$

   Thus, the two points of tangency are $(1 + \sqrt{2}, 3 + 2\sqrt{2})$ and $(1 - \sqrt{2}, 3 - 2\sqrt{2})$, or about $(2.4, 5.8)$ and $(-0.4, 0.2)$.

The normal line problem

Here's the companion problem to the tangent line problem in the previous section. Find the points of perpendicularity for all normal lines to the parabola, $y = \frac{1}{16}x^2$, that pass through the point $(3, 15)$.

Graph the parabola and plot the point $(3, 15)$. Now, before you do the math, try to estimate the locations of all normal lines. How many can you see? It's fairly easy to see that, starting at $(3, 15)$, one normal line goes down and to the right and another goes down to the left. But did you see that there's actually a second normal line that goes down to the left? No worries if you didn't see it, because when you do the math, you get all three solutions.

TIP

Making common-sense estimates enhances mathematical understanding. When doing calculus, or any math for that matter, come up with a common-sense, ballpark estimate of the solution to a problem before doing the math (when possible and time permitting). This deepens your understanding of the concepts involved and provides a check to the mathematical solution. (This is a powerful math strategy — take my word for it — despite the fact that in this particular problem, most people will find, at most, two of the three normal lines using an eyeball estimate.)

Figure 13-2 shows the parabola and the three normal lines.

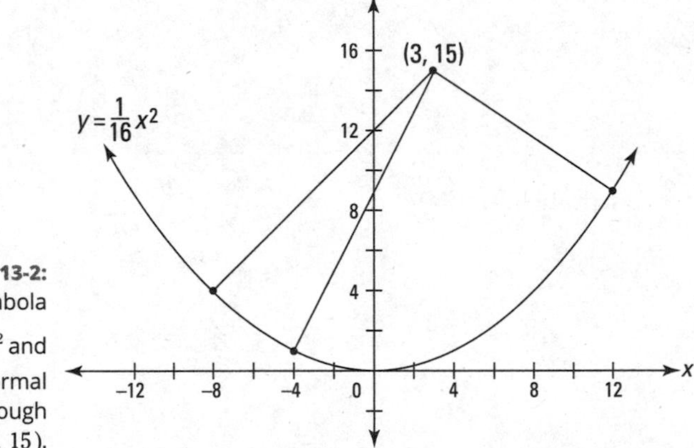

FIGURE 13-2: The parabola $y = \frac{1}{16}x^2$ and three normal lines through $(3, 15)$.

Looking at the figure, you can appreciate how practical this problem is. It'll really come in handy if you happen to find yourself standing inside the curve of a parabolic wall, and you want to know the precise location of the three points on the wall where you could throw a ball and have it bounce straight back to you.

The solution is very similar to the solution of the tangent line problem, except that in this problem you use the rule for the slopes of perpendicular lines.

Each normal line in Figure 13-2 is perpendicular to the tangent line drawn at the point where the normal meets the curve. So, the slope of each normal line is the opposite reciprocal of the slope of the corresponding tangent line — which, of course, is given by the derivative. So here goes:

1. **Take a general point, (x, y), on the parabola $y = \frac{1}{16}x^2$, and substitute $\frac{1}{16}x^2$ for *y*.**

   So, label each point of perpendicularity $\left(x, \frac{1}{16}x^2\right)$.

2. **Take the derivative of the parabola.**

$$y = \frac{1}{16}x^2$$

$$y' = \frac{1}{8}x$$

3. **Using the slope formula, $\frac{y_2 - y_1}{x_2 - x_1}$, set the slope of each normal line from $(3, 15)$ to $\left(x, \frac{1}{16}x^2\right)$ equal to the opposite reciprocal of the derivative at $\left(x, \frac{1}{16}x^2\right)$, and solve for *x*.**

$$\frac{\frac{1}{16}x^2 - 15}{x - 3} = -\frac{8}{x}$$

(The derivative is $\frac{1}{8}x$ or $\frac{x}{8}$ and its opposite reciprocal is thus $-\frac{8}{x}$.)

$$\frac{1}{16}x^3 - 15x = -8x + 24$$

(by cross-multiplication and distribution)

$$x^3 - 112x - 384 = 0$$

(after bringing all terms to one side and multiplying both sides by 16)

Now, there's no easy way to get exact solutions to this cubic (3rd-degree) equation like the way the quadratic formula gives you the exact solutions to a 2nd-degree equation. Instead, you can graph $y = x^3 - 112x - 384$, and the *x*-intercepts give you the solutions, but with this method, there's no guarantee you'll get exact solutions. (If you don't, the decimal approximations you get will be good enough.) Here, however, you luck out — actually, I had something to do with it — and get the exact solutions of –8, –4, and 12. (You should graph the cubic function so you see how this works.)

4. **Plug each of these *x*-coordinates into $y = \frac{1}{16}x^2$ to obtain the *y*-coordinates.**

$$y = \frac{1}{16}(-8)^2 = 4 \qquad y = \frac{1}{16}(-4)^2 = 1 \qquad y = \frac{1}{16}(12)^2 = 9$$

Thus, the three points of normalcy are $(-8, 4)$, $(-4, 1)$, and $(12, 9)$ — play ball!

YOUR TURN

1 Two lines through the point $(1,\ -3)$ are tangent to the parabola $y = x^2$. Determine the points of tangency.

2 The Earth has a radius of 4000 miles. Say you're standing on the shore and your eyes are 5′ 3.36″ above the surface of the water. How far out can you see to the horizon before the Earth's curvature makes the water dip below the horizon? See the following figure.

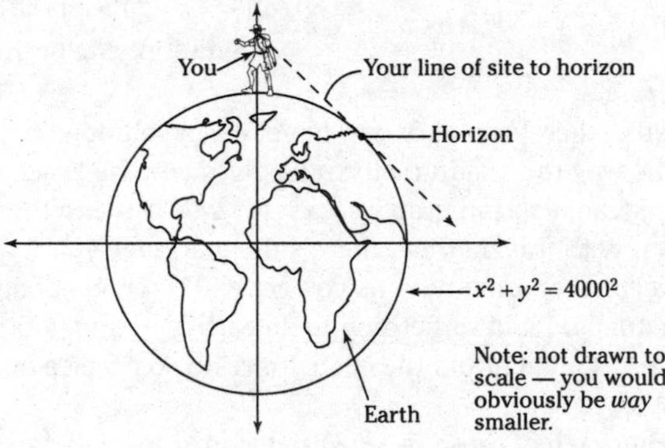

3 Find the points of normalcy for all lines through $(0, 1)$ normal to the curve $y = x^4$. The results might surprise you. Before you begin solving this problem, graph $y = x^4$ and put the cursor at $(0, 1)$. Now guess where normal lines will be and whether they represent the shortest paths or longest paths from $(0, 1)$ to $y = x^4$. *Note:* Do *ZoomSqr* to get the distances on the graph to appear properly proportional to each other.

---

4 An ill-prepared adventurer has run out of water on a hot, sunny day in the desert. He's 30 miles due north and 7 miles due east of his camp. His map shows a winding river — that by some odd coincidence happens to flow according to the function $y = 10\sin\frac{x}{10} + 10\cos\frac{x}{5} + x$ (where the camp lies at the origin; see the following figure). What point along the river is closest to him? He figures that he and his camel can just barely make it another 15 miles or so. (*Hint:* The closest point must occur at a point of normalcy.)

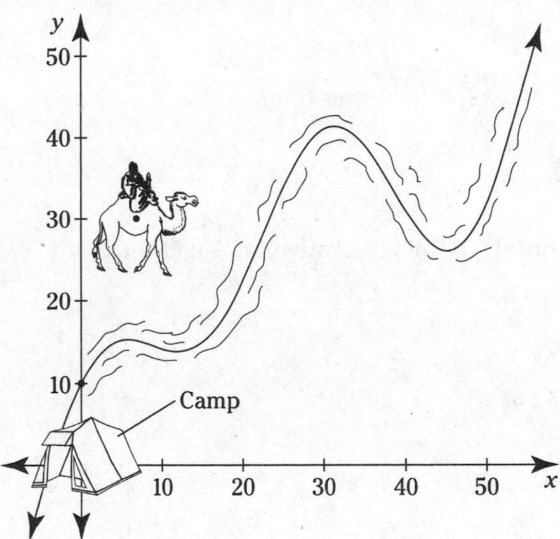

Straight Shooting with Linear Approximations

Because ordinary functions are locally *linear* (that's straight) — and the further you zoom in on them, the straighter they look — a line tangent to a function is a good approximation of the function near the point of tangency. Figure 13-3 shows the graph of $f(x)=\sqrt{x}$ and a line tangent to the function at the point $(9, 3)$. You can see that near $(9, 3)$, the curve and the tangent line are virtually indistinguishable.

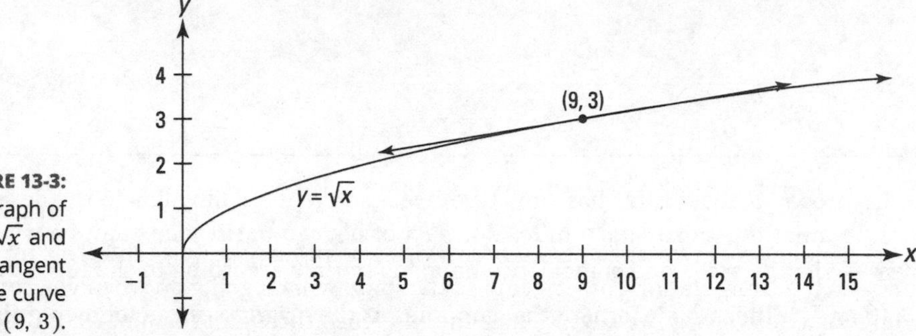

FIGURE 13-3: The graph of $f(x)=\sqrt{x}$ and a line tangent to the curve at $(9, 3)$.

Determining the equation of this tangent line is a breeze. You've got a point, $(9, 3)$, and the slope is given by the derivative of f at 9:

$$f(x)=\sqrt{x}=x^{1/2}$$

$$f'(x)=\frac{1}{2}x^{-1/2}=\frac{1}{2\sqrt{x}} \qquad \text{(power rule)}$$

$$f'(9)=\frac{1}{2\sqrt{9}}=\frac{1}{6}$$

Now just take this slope (this derivative) of $\frac{1}{6}$ and the point $(9, 3)$, and plug them into point-slope form:

$$y-y_1=m(x-x_1)$$

$$y-3=\frac{1}{6}(x-9)$$

$$y=3+\frac{1}{6}(x-9)$$

That's the equation of the line tangent to $f(x)=\sqrt{x}$ at $(9, 3)$. I suppose you might be wondering why I wrote the equation as $y=3+\frac{1}{6}(x-9)$. It might seem more natural to put the 3 to the right of $\frac{1}{6}(x-9)$, which, of course, would also be correct. And I could have simplified the equation further, writing it in $y=mx+b$ form. I explain later in this section why I wrote it the way I did.

(If you have your graphing calculator handy, graph $f(x)=\sqrt{x}$ and the tangent line. Zoom in on the point $(9, 3)$ a couple times. You'll see that — as you zoom in — the curve gets straighter and straighter and the curve and tangent line get closer and closer to each other.)

Now, say you want to approximate the square root of 10. Because 10 is pretty close to 9, and because you can see from Figure 13-3 that $f(x)$ and its tangent line are close to each other at $x = 10$, the y-coordinate of the line at $x = 10$ is a good approximation of the function value at $x = 10$, namely $\sqrt{10}$.

Just plug 10 into the line equation for your approximation:

$$\begin{aligned} y &= 3 + \frac{1}{6}(x - 9) \\ &= 3 + \frac{1}{6}(10 - 9) \\ &= 3 + \frac{1}{6} = 3\frac{1}{6} \end{aligned}$$

Thus, the square root of 10 is about $3\frac{1}{6}$. This is only about 0.004 more than the exact answer of 3.1623. . . . The error is roughly a tenth of one percent.

Now I can explain why I wrote the equation for the tangent line the way I did. This form makes it easier to do the computation and easier to understand what's going on when you compute an approximation. Here's why. You know that the line goes through the point $(9, 3)$, right? And you know the slope of the line is $\frac{1}{6}$. So, you can start at $(9, 3)$ and go to the right (or left) along the line in a stair-step fashion, as shown in Figure 13-4: over 1, up $\frac{1}{6}$; over 1, up $\frac{1}{6}$; and so on. (Note that since $slope = \frac{rise}{run}$, when the run is 1 [as shown in Figure 13-4], the rise equals the slope.)

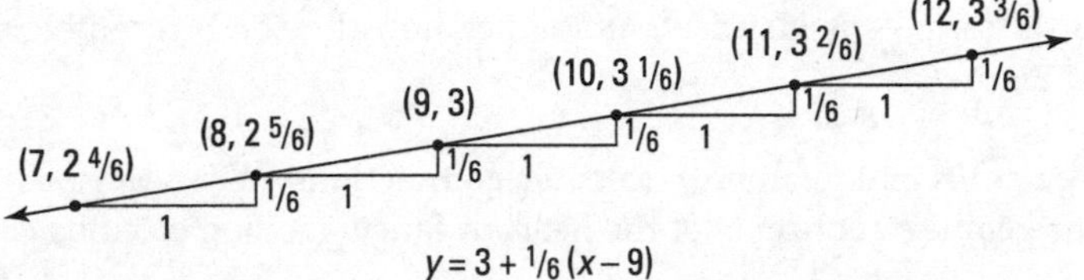

FIGURE 13-4: The linear approximation line and several of its points.

So, when you're doing an approximation, you start at a y-value of 3 and go up $\frac{1}{6}$ for each 1 you go to the right. Or if you go to the left, you go down $\frac{1}{6}$ for each 1 you go to the left. When the line equation is written like $y = 3 + \frac{1}{6}(x - 9)$, the computation of an approximation parallels this stair-step scheme.

Figure 13-4 shows the approximate values for the square roots of 7, 8, 10, 11, and 12. Here's how you come up with these values. To get to 8, for example, from $(9, 3)$, you go 1 to the left, so you go down $\frac{1}{6}$ to $2\frac{5}{6}$; or to get to 11 from $(9, 3)$, you go *two* to the right, so you go up *two*-sixths to $3\frac{2}{6}$ or $3\frac{1}{3}$. (If you go to the right *one-half* to $9\frac{1}{2}$, you go up *half* of a sixth, that's a twelfth, to $3\frac{1}{12}$, the approximate square root of $9\frac{1}{2}$.)

The following list shows the size of the errors for the approximations shown in Figure 13-4. Note that the errors grow as you get further from the point of tangency (9, 3). Also, the errors grow faster when going down from 9 to 8 then 7, and so on, than going up from 9 to 10 then 11, and so on; errors often grow faster in one direction than the other with linear approximations because of the shape of the curve.

$\sqrt{7}$: 0.8% error

$\sqrt{8}$: 0.2% error

$\sqrt{10}$: 0.1% error

$\sqrt{11}$: 0.5% error

$\sqrt{12}$: 1.0% error

MATH RULES

Linear approximation equation: Here's the general form for the equation of the tangent line that you use for a linear approximation. The values of a function $f(x)$ can be approximated by the values of the tangent line $l(x)$ near the point of tangency, $(x_0, f(x_0))$, where

$$l(x) = f(x_0) + f'(x_0)(x - x_0)$$

This is less complicated than it looks. It's just the gussied-up calculus version of the point-slope equation of a line you've known since Algebra I, $y - y_1 = m(x - x_1)$, with the y_1 moved to the right side:

$$y = y_1 + m(x - x_1)$$

This equation and the equation for $l(x)$ differ only in the symbols used; the *meaning* of both equations — term for term — is identical. And notice how they both resemble the equation of the tangent line in Figure 13-4.

TIP

Look for algebra-calculus and geometry-calculus connections. Whenever possible, try to see the basic algebra or geometry concepts at the heart of fancy-looking calculus concepts.

Once you get the hang of linear approximation, you can impress your friends by approximating things like $\sqrt[3]{70}$ in your head — like this: Bingo! $4\frac{1}{8}$. How did I do it? Here you go.

EXAMPLE

Q. Use linear approximation to estimate $\sqrt[3]{70}$.

A. $4\frac{1}{8}$

1. **Find a perfect cube root near $\sqrt[3]{70}$.**

   You notice that $\sqrt[3]{70}$ is near a no-brainer, $\sqrt[3]{64}$, which, of course, is 4. That gives you the point $(64, 4)$ on the graph of $y = \sqrt[3]{x}$.

2. **Find the slope of $y = \sqrt[3]{x}$ (which is the slope of the tangent line) at $x = 64$.**

   $y' = \frac{1}{3}x^{-2/3}$, so the slope at 64 is $\frac{1}{48}$.

This tells you that — to approximate cube roots near 64 — you add (or subtract) $\frac{1}{48}$ to 4 for each increase (or decrease) of one from 64. For example, the cube root of 65 is about $4\frac{1}{48}$; the cube root of 66 is about $4\frac{2}{48}$, or $4\frac{1}{24}$; the cube root of 67 is about $4\frac{3}{48}$, or $4\frac{1}{16}$; and the cube root of 63 is about $3\frac{47}{48}$. Since 70 is 6 more than 64, the answer to the question is that $\sqrt[3]{70}$ is about $4\frac{6}{48}$, or $4\frac{1}{8}$. You're done, but let's keep going and obtain the linear approximation equation. You can use this equation to approximate cube roots of numbers near 64 (integers and non-integers).

3. **Use the point-slope form to write the equation of the tangent line at (64, 4).**

$$y - y_1 = m(x - x_1)$$
$$y - 4 = \frac{1}{48}(x - 64)$$
$$y = 4 + \frac{1}{48}(x - 64)$$

4. **Plug 70 into *x* for your approximation:**

$$y = 4 + \frac{1}{48}(70 - 64)$$
$$= 4\frac{6}{48} = 4\frac{1}{8}$$

You didn't need this equation to answer the current question, but you should know how to come up with a linear approximation equation and how to use it.

YOUR TURN

For Problems 5 through 8, make the indicated estimates, and, whether you need it or not, determine the applicable linear approximation equation. (Some students may prefer to always use the linear approximation equation; others might like to use the shortcut that I just explained.)

5 Estimate the 4th root of 17.

6 Approximate 3.01^5.

7 Estimate $\sin\frac{\pi}{180}$ ($\frac{\pi}{180}$ is one degree, of course).

8 Approximate $\ln\left(e^{10}+5\right)$.

Business and Economics Problems

Believe it or not, calculus is actually used in the real world of business and economics — learn calculus and increase your profits! Tell me: When you're driving around an upscale part of town and you pass by a *huge* home, what's the first thing that comes to your mind? I bet it's "Just look at that home! That guy (or gal) must know calculus."

Managing marginals in economics

Look again at Figures 13-3 and 13-4 in the previous section. Recall that the derivative and thus the slope of $y=\sqrt{x}$ at $(9, 3)$ is $\frac{1}{6}$, and that the tangent line at this point can be used to approximate the function near the point of tangency. So, as you go over 1 from 9 to 10 along the function itself, you go up *about* $\frac{1}{6}$. And, thus, $\sqrt{10}$ is about $\frac{1}{6}$ more than $\sqrt{9}$. In economics, that little bit (like the $\frac{1}{6}$) is called a *marginal*.

MATH RULES

Marginal cost, marginal revenue, and marginal profit work a lot like linear approximation. *Marginal cost*, *marginal revenue*, and *marginal profit* all involve how much a function goes up (or down) as you go over 1 to the right — just like with linear approximation.

Say you've got a cost function that gives you the total cost, $C(x)$, of producing x items. See Figure 13-5.

Look at the blown-up square on the right in the figure. The derivative of $C(x)$ at the point of tangency gives you the slope of the tangent line and thus the amount you go up as you go 1 to the right along the tangent line. (This amount is labeled in the figure as *marginal cost.*) Going 1 to the right along the cost function itself shows you the increase in cost of producing one more item. (This is labeled as the *extra cost.*) Because the tangent line is a good approximation of the cost function, the derivative of C — called the *marginal cost* — is the *approximate* increase in cost of producing one more item. Marginal revenue and marginal profit work the same way.

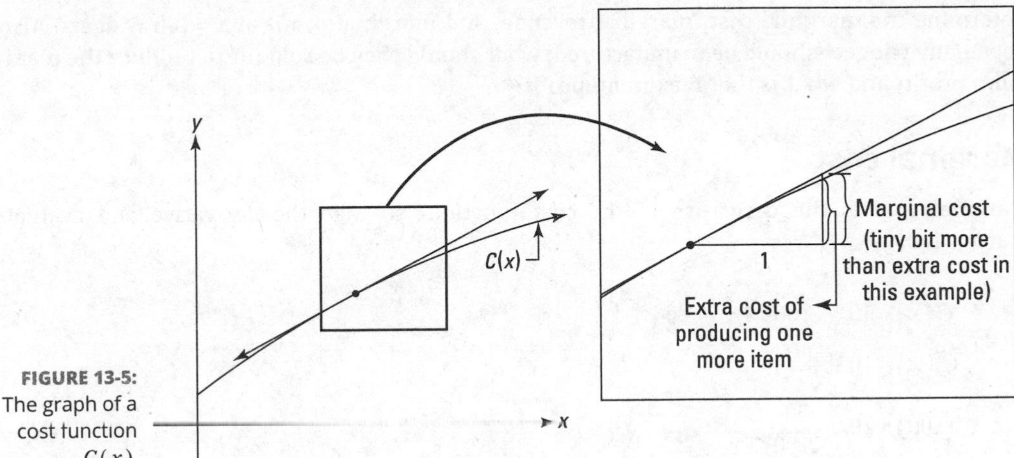

FIGURE 13-5: The graph of a cost function $C(x)$.

MATH RULES

Definitions of *marginal cost, marginal revenue,* and *marginal profit:*

Marginal cost equals the derivative of the cost function.

Marginal revenue equals the derivative of the revenue function.

Marginal profit equals the derivative of the profit function.

(Marginal cost and marginal revenue are almost always positive; marginal profit can be positive or negative.)

Before doing a problem involving marginals, there's one more piece of business to take care of. A *demand function* tells you how many items will be purchased (what the demand will be) given the price. The lower the price, of course, the higher the demand; and the higher the price, the lower the demand. You'd think that the number purchased should be a function of the price — input a price and find out how many items people will buy at that price — but traditionally, a demand function is written the other way around. The price is expressed as a function of the number demanded. I know that seems a bit odd, but don't sweat it — the function works either way. Think of it like this: If a retailer wants to sell a given number of items, the demand function tells the retailer what they should set the selling price at.

Okay, so here's the problem. A widget manufacturer determines that the demand function for their widgets is

$$p = \frac{1000}{\sqrt{x}}$$

where p is the price of a widget and x is the number of widgets demanded. (Note that a demand function like this can also be called a *price function*.) The cost of producing x widgets is given by the following cost function:

$$C(x) = 10x + 100\sqrt{x} + 10{,}000$$

Determine the marginal cost, marginal revenue, and marginal profit at $x=100$ widgets. Also, how many widgets should be manufactured, what should they be sold for to produce the maximum profit, and what is that maximum profit?

Marginal cost

Marginal cost is the derivative of the cost function, so take the derivative and evaluate it at $x=100$:

$$\begin{aligned} C(x) &= 10x+100\sqrt{x}+10{,}000 \\ C'(x) &= 10+\frac{50}{\sqrt{x}} \quad \text{(power rule)} \\ C'(100) &= 10+\frac{50}{\sqrt{100}} \\ &= 10+\frac{50}{10}=15 \end{aligned}$$

Thus, the marginal cost at $x=100$ is $15 — this is the approximate cost of producing the 101st widget.

Marginal revenue

Revenue, $R(x)$, equals the number of items sold, x, times the price, p:

$$\begin{aligned} R(x) &= x\cdot p \\ &= x\cdot\frac{1000}{\sqrt{x}} \quad \text{(using the earlier demand function)} \\ &= \frac{1000x}{\sqrt{x}}\cdot\frac{\sqrt{x}}{\sqrt{x}} \quad \text{(rationalizing the denominator)} \\ &= \frac{1000x\sqrt{x}}{x} \\ &= 1000\sqrt{x} \end{aligned}$$

Marginal revenue is the derivative of the revenue function, so take the derivative of $R(x)$ and evaluate it at $x=100$:

$$\begin{aligned} R(x) &= 1000\sqrt{x} \\ R'(x) &= \frac{500}{\sqrt{x}} \quad \text{(power rule)} \\ R'(100) &= \frac{500}{\sqrt{100}}=50 \end{aligned}$$

Thus, the approximate revenue from selling the 101st widget is $50.

Marginal profit

Profit, $P(x)$, equals revenue minus cost. So,

$$\begin{aligned}P(x) &= R(x) - C(x)\\ &= 1000\sqrt{x} - \left(10x + 100\sqrt{x} + 10{,}000\right)\\ &= -10x + 900\sqrt{x} - 10{,}000\end{aligned}$$

Marginal profit is the derivative of the profit function, so take the derivative of $P(x)$, and evaluate it at $x = 100$:

$$\begin{aligned}P(x) &= -10x + 900\sqrt{x} - 10{,}000\\ P'(x) &= -10 + \frac{450}{\sqrt{x}}\\ P'(100) &= -10 + \frac{450}{\sqrt{100}}\\ &= -10 + 45 = 35\end{aligned}$$

Selling the 101st widget brings in an approximate profit of \$35.

TIP

Marginal profit shortcuts: Did you notice either of the two shortcuts you could have taken here? First, you can use the fact that

$$P'(x) = R'(x) - C'(x)$$

to determine $P'(x)$ directly, without first determining $P(x)$. Then, after getting $P'(x)$, you just plug 100 into x for your answer.

And, if all you want to know is $P'(100)$, you can use the following really short shortcut:

$$\begin{aligned}P'(100) &= R'(100) - C'(100)\\ &= 50 - 15\\ &= 35\end{aligned}$$

This is common sense. If it costs you about \$15 to produce the 101st widget and you sell it for about \$50, then your profit is about \$35.

I did it the long way because you need both the profit function, $P(x)$, and the marginal profit function, $P'(x)$, for the following problems. (You will often need to do it the long way.)

Maximum profit

To determine maximum profit, set the derivative of profit — that's marginal profit — equal to zero, solve for x, and then plug the result into the profit function:

$$\begin{aligned} P'(x) &= -10 + \frac{450}{\sqrt{x}} \\ 0 &= -10 + \frac{450}{\sqrt{x}} \\ 10 &= \frac{450}{\sqrt{x}} \\ 10\sqrt{x} &= 450 \\ \sqrt{x} &= 45 \\ x &= 2025 \end{aligned}$$

So, maximum profit occurs when 2025 widgets are sold. Plug this into $P(x)$:

$$\begin{aligned} P(x) &= -10x + 900\sqrt{x} - 10{,}000 \\ P(2025) &= -10 \cdot 2025 + 900\sqrt{2025} - 10{,}000 \\ &= -20{,}250 + 900 \cdot 45 - 10{,}000 \\ &= 10{,}250 \end{aligned}$$

Thus, the maximum profit is $10,250. (Extra credit: Did you see where I got a bit lazy here? The derivative of the profit function is zero at $x = 2025$, but that doesn't guarantee that there's a max at that x-value. There could instead be a min or an inflection point there. You could use either the first or second derivative test [see Chapter 11] to show that it's actually a max. But I just peeked at a graph of the profit function and saw that it's sort of an upside-down cup shape, so I knew that there was a max at the top of the cup at $x = 2025$.)

Finally, plug the number sold into the demand function to determine the profit-maximizing price:

$$\begin{aligned} p &= \frac{1000}{\sqrt{x}} \\ p &= \frac{1000}{\sqrt{2025}} \approx 22.22 \end{aligned}$$

So, the maximum profit of $10,250 occurs when the price is set at $22.22. At this price, 2025 widgets will be sold. Figure 13-6 sums up these results. Note that because profit equals revenue minus cost, the vertical distance or gap between the revenue and cost functions at a given x-value gives the profit at that x-value. Maximum profit occurs where the gap is greatest.

And here's another thing. Because maximum profit occurs where $P'(x)=0$, and because $P'(x)=R'(x)-C'(x)$, it follows that the profit will be greatest where $0=R'(x)-C'(x)$ — in other words, where $R'(x)=C'(x)$. And when $R'(x)=C'(x)$, the slopes of the functions' tangent lines are equal. So, if you were to draw tangent lines to $R(x)$ and $C(x)$ where the gap between the two is greatest, these tangents would be parallel. Right about now, you're probably thinking something like, "Such symmetry, such simple elegance, such beauty! Verily, the mathematics muse seduces the heart as much as the mind." Yeah, it's nice all right, but let's not get carried away.

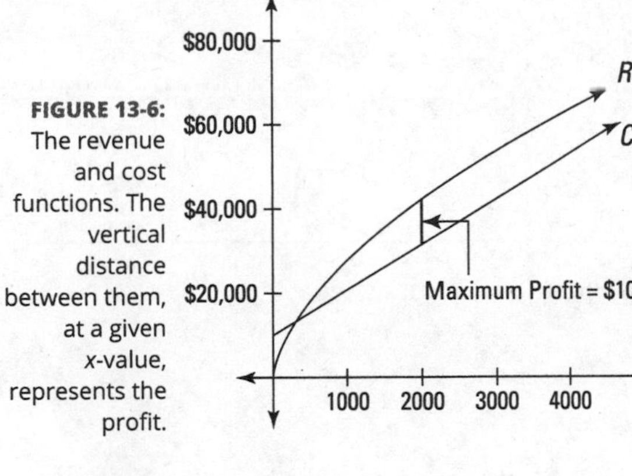

FIGURE 13-6: The revenue and cost functions. The vertical distance between them, at a given x-value, represents the profit.

(Note that although the scale of this graph makes $C(x)=10x+100\sqrt{x}+10{,}000$ look like a straight line, its middle term of $100\sqrt{x}$ means that it is not exactly straight.)

YOUR TURN

For Problems 9 through 12, use the following demand (or price) and cost functions for the production and sale of some widgets.

$$p(x)=400-0.0002x^{1.5}$$
$$C(x)=50{,}000+100x+0.0001x^3$$

9 **(a) What's the marginal cost at $x=100$?**

(b) What's the cost of producing the 201st widget?

10 **(a)** What's the marginal revenue function?

(b) What additional revenue is generated for the firm by the 101st, 401st, and 901st widgets?

---

11 What's the profit generated by the 401st, 901st, and 1601st widgets?

---

12 **(a)** How many widgets should be manufactured and sold to maximize the firm's profit?

(b) What is that maximum profit?

(c) What price should the widgets be sold for to achieve this maximum profit?

Practice Questions Answers and Explanations

1 **The points of tangency are $(-1,\ 1)$ and $(3,\ 9)$.**

1. **Express a point on the parabola in terms of *x*.**

   The equation of the parabola is $y = x^2$, so you can take a general point on the parabola $(x,\ y)$ and substitute x^2 for *y*. So your point is $\left(x,\ x^2\right)$.

2. **Take the derivative of the parabola.**

   $y = x^2$
   $y' = 2x$

3. **Using the slope formula, $m = \dfrac{y_2 - y_1}{x_2 - x_1}$, set the slope of the tangent line from $(1, -3)$ to $\left(x,\ x^2\right)$ equal to the derivative. Then solve for *x*.**

   $$\frac{x^2 - (-3)}{x - 1} = 2x$$
   $$x^2 + 3 = 2x^2 - 2x$$
   $$x^2 - 2x - 3 = 0$$
   $$(x+1)(x-3) = 0$$

   $x = -1$ or 3

4. **Plug these *x*-coordinates into $y = x^2$ to get the *y*-coordinates.**

   $y = (-1)^2 = 1$ and $y = 3^2 = 9$

   So, there's one line through $(1, -3)$ that's tangent to the parabola at $(-1,\ 1)$, and another through $(1, -3)$ that's tangent at $(3,\ 9)$. You may want to confirm these answers by graphing the parabola and your two tangent lines:

   $y = -2(x+1) + 1$ and $y = 6(x-3) + 9$

2 **The horizon is about 2.83 miles away.**

1. **Write the equation of the Earth's circumference as a function of *y* (see the figure in the problem).**

   $$x^2 + y^2 = 4000^2$$
   $$y = \pm\sqrt{4000^2 - x^2}$$

   You can disregard the negative half of this circle because your line of sight will obviously be tangent to the upper half of the Earth.

2. **Express a point on the circle in terms of *x*: $\left(x,\ \sqrt{4000^2 - x^2}\right)$**

3. **Take the derivative of the circle.**

$$y=\sqrt{4000^2-x^2}$$

$$y'=\frac{1}{2}\left(4000^2-x^2\right)^{-1/2}(-2x) \quad \text{(chain rule)}$$

$$=\frac{-x}{\sqrt{4000^2-x^2}}$$

4. **Using the slope formula, set the slope of the tangent line from your eyes to $\left(x, \sqrt{4000^2-x^2}\right)$ equal to the derivative, and then solve for *x*.**

Your eyes are 5′ 3.36″ above the top of the Earth at the point $(0, 4000)$ on the circle. Convert your height to miles; that's exactly 0.001 miles (what an amazing coincidence!). So, the coordinates of your eyes are $(0, 4000.001)$.

$$\frac{y_2-y_1}{x_2-x_1}=m$$

$$\frac{\sqrt{4000^2-x^2}-4000.001}{x-0}=\frac{-x}{\sqrt{4000^2-x^2}}$$

$$-x^2=\left(4000^2-x^2\right)-4000.001\sqrt{4000^2-x^2} \quad \text{(cross-multiplication)}$$

$$-4000^2=-4000.001\sqrt{4000^2-x^2} \quad \text{(Use your calculator.)}$$

$$3999.999=\sqrt{4000^2-x^2} \quad \text{(Now square both sides.)}$$

$$15{,}999{,}992=4000^2-x^2$$

$$x^2=8$$

$$x=2\sqrt{2}\approx 2.83 \text{ miles}$$

Many people are surprised that the horizon is so close. What do you think?

By the way, you can solve this problem much more quickly with some basic high school geometry; no calculus is needed. Can you do it?

(3) **The points of normalcy are $(-0.915, 0.702)$, $(-0.519, 0.073)$, $(0, 0)$, $(0.519, 0.073)$, and $(0.915, 0.702)$.**

1. **Express a point on the curve in terms of *x*: A general point is $\left(x, x^4\right)$.**
2. **Take the derivative.**

$$y=x^4$$

$$y'=4x^3$$

3. **Set the slope from $(0, 1)$ to $\left(x, x^4\right)$ equal to the opposite reciprocal of the derivative and solve.**

$$\frac{x^4-1}{x-0}=\frac{-1}{4x^3}$$

$$4x^7-4x^3+x=0$$

$$x\left(4x^6-4x^2+1\right)=0$$

$$x=0 \quad \text{or} \quad 4x^6-4x^2+1=0$$

Unless you have a special gift for solving 6th-degree equations, you better use your calculator — just graph $y = 4x^6 - 4x^2 + 1$ and find all the x-intercepts. There are x-intercepts at about −0.915, −0.519, 0.519, and 0.915. Dig those palindromic numbers!

4. **Plug these four solutions into $y = x^4$ to get the y-coordinates (there's also the $x = 0$ no-brainer).**

$$(-0.519)^4 = (0.519)^4 = \ \sim 0.073$$
$$(-0.915)^4 = (0.915)^4 = \ \sim 0.702$$

You're done. Five normal lines can be drawn. The points of normalcy are $(-0.915,\ 0.702)$, $(\ 0.519,\ 0.073)$, $(0,\ 0)$, $(0.519,\ 0.073)$, and $(0.915,\ 0.702)$.

I find this result interesting, first, because there are so many normal lines, and second, because the normal lines from $(0,\ 1)$ to $(-0.915,\ 0.702)$, $(0,\ 0)$, and $(0.915,\ 0.702)$ are all the shortest paths (compared to other points in their respective vicinities). The other two normal lines are the longest paths. This is curious: When a curve is concave away from a point, a normal to the curve can only be a local shortest path, so you might think that in the current problem, where $y = x^4$ is everywhere concave *toward* $(0,\ 1)$, you could get only locally longest paths. But it turns out that when a curve is concave toward a point, you can get either a local shortest or a local longest path.

WARNING

Don't report me! I played slightly fast and loose with the math for the $x = 0$ solution. Did you notice that $x = 0$ doesn't work if you plug it back into the equation $\frac{x^4 - 1}{x - 0} = \frac{-1}{4x^3}$ because both denominators become zero? However — promise not to leak this to your calculus teacher — this is okay here because both sides of the equation become $\frac{\text{non-zero number}}{\text{zero}}$. (Actually, they're both $\frac{-1}{0}$, but something like $\frac{5}{0} = \frac{2}{0}$ would also work.) Non-zero over zero means a vertical line with undefined slope. So, the $\frac{-1}{0} = \frac{-1}{0}$ tells you that you have a vertical normal line at $x = 0$.

(4) **The closest point is $(6.11, 15.26)$, which is 14.77 miles away.**

1. **Express a point on the curve in terms of x:**

$$\left(x,\ 10\sin\frac{x}{10} + 10\cos\frac{x}{5} + x\right)$$

2. **Take the derivative.**

$$y = 10\sin\frac{x}{10} + 10\cos\frac{x}{5} + x$$
$$y' = 10\cos\left(\frac{x}{10}\right)\cdot\frac{1}{10} - 10\sin\left(\frac{x}{5}\right)\cdot\frac{1}{5} + 1$$
$$= \cos\frac{x}{10} - 2\sin\frac{x}{5} + 1$$

3. **Set the slope from $(7, 30)$ to the general point equal to the opposite reciprocal of the derivative, and solve.**

$$\frac{30 - \left(10\sin\frac{x}{10} + 10\cos\frac{x}{5} + x\right)}{7 - x} = \frac{-1}{\cos\frac{x}{10} - 2\sin\frac{x}{5} + 1}$$

WARNING

Unless you wear a pocket protector, don't even think about solving this equation without a calculator.

Solve on your calculator by graphing the following equation and finding the

x intercepts: $y = \dfrac{30 - \left(10\sin\frac{x}{10} + 10\cos\frac{x}{5} + x\right)}{7 - x} - \dfrac{-1}{\cos\frac{x}{10} - 2\sin\frac{x}{5} + 1}$

TIP

Your calculator's window settings. It's a bit tricky to find the x-intercepts for this gnarly function. You have to play around with your calculator's window settings a bit. And don't forget that your calculator will draw vertical asymptotes that look like zeros of the function, but are not. Now, it turns out that this function has an infinite number of x-intercepts (I think). There's one between $x = -18$ and -19 and there are more at bigger negatives. And there's one between $x = 97$ and 98 and there are more at bigger positives. But these zeros represent points on the river so far away that they need not be considered. Only three zeros are plausible candidates for the closest trip to the river. To see the first candidate zero, set Xmin $= -1$, Xmax $= 10$, Xscl $= 1$, Ymin $= -5$, Ymax $= 25$, and Yscl $= 5$. To see the other two, set Xmin $= 10$, Xmax $= 30$, Xscl $= 1$, Ymin $= -2$, Ymax $= 10$, and Yscl $= 1$. These zeros are at roughly 6.11, 13.75, and 20.58.

4. **Plug the zeros into the original function to obtain the y-coordinates.**

   You get the following points of normalcy: $(6.11, 15.26)$, $(13.75, 14.32)$, and $(20.58, 23.80)$.

5. **Use the distance formula, $D = \sqrt{(x_2 - x_1)^2 + (y_2 - y_1)^2}$, to find the distance from our parched adventurer to the three points of normalcy.**

   The distances are 14.77 miles to $(6.11, 15.26)$, 17.07 miles to $(13.75, 14.32)$, and 14.93 miles to $(20.58, 23.80)$. Using his trusty compass, he heads mostly south and a little west to $(6.11, 15.26)$. An added benefit of this route is that it's in the direction of his camp.

(5) **The approximation is $2\frac{1}{32}$ or 2.03125.**

1. **Find a "round" number near 17 where the 4th root is very easy to get: that's 16.**

   $\sqrt[4]{16} = 2$. So, the point $(16, 2)$ is on $f(x) = \sqrt[4]{x}$.

2. **Determine the slope of the tangent line to f at $(16, 2)$.**

$$f(x) = \sqrt[4]{x}$$
$$f'(x) = \frac{1}{4}x^{-3/4}$$
$$= \frac{1}{4\sqrt[4]{x}^3}$$
$$f'(16) = \frac{1}{32}$$

   Because 17 is 1 more than 16, you should have your answer. Do you see it? Whether you do or not, keep going to finish with the standard classroom approach.

3. **Use the point-slope form of a line to write the equation of the tangent line at $(16, 2)$.**

$$y - 2 = \frac{1}{32}(x - 16)$$

4. **Plug your number into the tangent line to get your approximation (which you probably already know).**

$$y = 2 + \frac{1}{32}(17 - 16)$$
$$= 2\frac{1}{32} \text{ or } 2.03125$$

The exact answer is about 2.03054. Your estimate is only $\frac{3}{100}$ of 1 percent too big! Not too shabby. Extra-credit question: No matter what 4th root you estimate with the linear approximation technique, your answer will be too big. Do you see why?

6 **The approximation is 247.05.**

1. **Find your round number.**

   That's 3, duh. Since $3^5 = 243$, the point $(3, 243)$ is on $g(x) = x^5$.

2. **Find the slope at your point.**

$$g(x) = x^5$$
$$g'(x) = 5x^4$$
$$g'(3) = 405$$

3. **Write the tangent line equation.**

$$y - y_1 = m(x - x_1)$$
$$y - 243 = 405(x - 3)$$

4. **Get your approximation.**

$$y = 243 + 405(3.01 - 3) = 247.05$$

   Only $\frac{1}{100}$ of 1 percent off.

7 **The approximation is $\frac{\pi}{180}$.**

You know the routine (the angle size near 1 degree whose sine you can easily compute is zero degrees):

$$f(x) = \sin x$$
$$f(0) = 0, \quad \text{so } (0, 0) \text{ is your point}$$

$$f'(x) = \cos x$$
$$f'(0) = 1, \quad \text{so 1 is the slope at } (0, 0)$$

$$y - y_1 = m(x - x_1)$$
$$y - 0 = 1(x - 0)$$
$$y = x \quad \text{is the tangent line}$$

Your number is $x = \frac{\pi}{180}$, so, since $y = x$, you get $y = \frac{\pi}{180}$.

This shows that for very small angles, the sine of the angle and the angle itself (in radians) are approximately equal. (The same is true of the tangent of an angle, by the way.) The approximation of $\frac{\pi}{180}$ is only $\frac{1}{200}$ of 1 percent too big.

8 **The approximation is $10+\frac{5}{e^{10}}$.**

Just imagine all the situations where such an approximation will come in handy!

$$q(x)=\ln(x)$$
$$q\left(e^{10}\right)=10, \text{ so } \left(e^{10}, 10\right) \text{ is your point}$$

$$q'(x)=\frac{1}{x}$$
$$q'\left(e^{10}\right)=\frac{1}{e^{10}}, \text{ so } \frac{1}{e^{10}} \text{ is the slope at } \left(e^{10}, 10\right)$$

$$y-y_1=m(x-x_1)$$
$$y-10=\frac{1}{e^{10}}\left(x-e^{10}\right)$$
$$y=10+\frac{1}{e^{10}}\left(x-e^{10}\right) \text{ is the tangent line}$$

Now you can plug in your number, $x=e^{10}+5$:

$$y=10+\frac{1}{e^{10}}\left(\left(e^{10}+5\right)-e^{10}\right)$$
$$y=10+\frac{5}{e^{10}}$$

Hold on to your hat. This approximation is a mere 0.00000026% too big!

9 **(a) $103.00**

$$C(x)=50{,}000+100x+0.0001x^3$$
$$C'(x)=100+0.0003x^2$$
$$C'(100)=100+0.0003(100)^2$$
$$=100+3=103$$

(b) $112.00

$$C'(200)=100+0.0003(200)^2$$
$$=100+12=112$$

10 **(a) $R'(x)=400-0.0005x^{1.5}$**

1. **Find the revenue function.**

$$\textit{Revenue}=(\textit{\# of items sold})(\textit{price per item})$$
$$R(x)=x\left(400-0.0002x^{1.5}\right) \quad \text{(using the price function)}$$
$$=400x-0.0002x^{2.5}$$

2. **Take its derivative.**

$$R'(x) = 400 - 0.0005x^{1.5}$$

(b) **$399.50, $396.00, and $386.50, respectively.**

$$R'(100) = 400 - 0.0005(100)^{1.5} = 399.50$$
$$R'(400) = 400 - 0.0005(400)^{1.5} = 396.00$$
$$R'(900) = 400 - 0.0005(900)^{1.5} = 386.50$$

11 **$248.00, $43.50, and –$500.00, respectively.**

$$\begin{aligned} \textit{Marginal profit} &= \textit{marginal revenue} - \textit{marginal cost} \\ P'(x) &= R'(x) - C'(x) \\ &= \left(400 - 0.0005x^{1.5}\right) - \left(100 + 0.0003x^2\right) \\ &= 300 - 0.0005x^{1.5} - 0.0003x^2 \end{aligned}$$

$$P'(400) = 300 - 0.0005(400)^{1.5} - 0.0003(400)^2 = 248$$
$$P'(900) = 300 - 0.0005(900)^{1.5} - 0.0003(900)^2 = 43.5$$
$$P'(1600) = 300 - 0.0005(1600)^{1.5} - 0.0003(1600)^2 = -500$$

This negative profit for the 1601st widget tells you that the firm would lose money if it were to produce and sell that widget. Therefore, it will obviously want to produce and sell fewer widgets than that. See the solution to the next problem.

12 (a) **974 widgets**

Like with any maximization problem, to find the maximum profit, you set the first derivative equal to zero and solve for x.

$$\begin{aligned} P'(x) &= 300 - 0.0005x^{1.5} - 0.0003x^2 \\ 0 &= 300 - 0.0005x^{1.5} - 0.0003x^2 \\ x &\approx 974.33 \end{aligned}$$

(You have to use your calculator to find that solution.)

Thus, the firm should produce and sell 974 widgets to maximize profits. (It's kind of obvious in this problem that the profit function hits a maximum at this x-value; but, if you want to be more rigorous, you should show that this x-value is indeed where a maximum occurs, as opposed to a minimum or a horizontal inflection point.) I did this problem like any maximization problem, without mentioning marginals. But, as you know, the first derivative of the profit is the marginal profit. So, the preceding math shows that the marginal profit is zero when 974 widgets are sold. Do you see why the maximum profit should occur where the marginal profit equals zero?

(b) $143,877.52

Determine the profit function and evaluate it at $x = 974$. (This is a very unusual calculus problem, by the way, where you determined the derivative, $P'(x)$, before you had the function itself, $P(x)$.)

$$\begin{aligned} P(x) &= R(x) - C(x) \\ &= 400x - 0.0002x^{2.5} - \left(50{,}000 + 100x + 0.0001x^3\right) \\ &= -50{,}000 + 300x - 0.0002x^{2.5} - 0.0001x^3 \\ P(974) &\approx 143{,}877.52 \end{aligned}$$

(c) $393.92

Just plug 974 into the price function.

$$\begin{aligned} p(x) &= 400 - 0.0002x^{1.5} \\ p(974) &= 400 - 0.0002(974)^{1.5} \\ &\approx 400 - 6.08 \\ &\approx 393.92 \end{aligned}$$

If you're ready to test your skills a bit more, take the following chapter quiz that incorporates all the chapter topics.

Whaddya Know? Chapter 13 Quiz

Quiz time! Complete each problem to test your knowledge on the various topics covered in this chapter. You can then find the solutions and explanations in the next section.

1. Determine the equation of the line(s) tangent to $f(x)=e^x$ that pass through $(0, 0)$, and give the point(s) of tangency.

2. Determine the equation of the line(s) normal to $f(x)=\ln x$ that pass through $(0, 1)$, and give the point(s) of normalcy.

3. Determine the equation of the line(s) normal to $f(x)=\sin x$ that pass through $\left(\frac{\pi}{2}, -\frac{\pi}{2}\right)$, and give the point(s) of normalcy.

4. Use the linear approximation method to estimate 4.99^4.

5. Use the linear approximation method to estimate $\ln 10.1$.

6. Use the linear approximation method to estimate $\sqrt[5]{33}$, $\sqrt[5]{34}$, and $\sqrt[5]{35}$.

7. Use the linear approximation method to estimate $e^{0.99}$.

For Problems 8 to 10, use the following demand (or price) and cost functions for the production and sale of some thingamajobs.

$$p=\frac{900}{\sqrt[3]{x}}$$

$$C(x)=20x+150\sqrt[3]{x^2}+1000$$

8. **(a)** What's the marginal cost function?

   (b) What's the approximate cost of producing the 1001st thingamajob?

9. **(a)** What's the revenue function and what's the marginal revenue function?

   (b) What's the approximate revenue from the sale of the 1001st thingamajob?

10. **(a)** What's the marginal profit function?

    (b) What's the approximate profit earned from the sale of the 1001st thingamajob?

Answers to Chapter 13 Quiz

1. **$y = ex$, which passes through $(0,\ 0)$, is tangent to $f(x) = e^x$ at $(1,\ e)$.**

2. **$y = -x + 1$, which passes through $(0,\ 1)$, is normal to $f(x) = \ln x$ at $(1,\ 0)$.**

3. **There are three normal lines to $f(x) = \sin x$ that pass through $\left(\frac{\pi}{2}, -\frac{\pi}{2}\right)$:**

   The vertical line $x = \frac{\pi}{2}$ is normal to f at $\left(\frac{\pi}{2},\ 1\right)$,

   $y = -x$ is normal to f at the origin, and

   $y = x - \pi$ is normal to f at $(\pi,\ 0)$.

4. **620**

5. **$\ln 10 + 0.01$**

6. **$2\frac{1}{80}$, $2\frac{2}{80}$, and $2\frac{3}{80}$**

7. **$0.99e$**

8. **(a) $C'(x) = 20 + \frac{100}{\sqrt[3]{x}}$**

$$C(x) = 20x + 150\sqrt[3]{x^2} + 1000$$
$$= 20x + 150x^{2/3} + 1000$$
$$C'(x) = 20 + 150 \cdot \frac{2}{3}x^{-1/3}$$
$$= 20 + \frac{100}{\sqrt[3]{x}}$$

   (b) The approximate cost of producing the 1001st thingamajob is \$30.

$$C'(x) = 20 + \frac{100}{\sqrt[3]{x}}$$
$$C'(1000) = 20 + \frac{100}{\sqrt[3]{1000}} = 30$$

9. **(a) $R(x) = 900\sqrt[3]{x^2}$ and $R'(x) = \frac{600}{\sqrt[3]{x}}$**

$$R(x) = x \cdot p$$
$$= x \cdot \frac{900}{\sqrt[3]{x}}$$
$$= 900\sqrt[3]{x^2}$$
$$R'(x) = 900 \cdot \frac{2}{3}x^{-1/3}$$
$$= \frac{600}{\sqrt[3]{x}}$$

(b) The approximate revenue from the sale of the 1001st thingamajob is $60.

$$R'(x)=\frac{600}{\sqrt[3]{x}}$$

$$R'(1000)=\frac{600}{\sqrt[3]{1000}}=60$$

10 **(a)** $\boldsymbol{P'(x)=\frac{500}{\sqrt[3]{x}}-20}$

$$\begin{aligned}P'(x)&=R'(x)-C'(x)\\&=\frac{600}{\sqrt[3]{x}}-\left(20+\frac{100}{\sqrt[3]{x}}\right)\\&=\frac{500}{\sqrt[3]{x}}-20\end{aligned}$$

(b) The approximate profit earned from the sale of the 1001st thingamajob is $30.

$$P'(x)=\frac{500}{\sqrt[3]{x}}-20$$

$$P'(1000)=\frac{500}{\sqrt[3]{1000}}-20=30$$

Or, even better, you could simply subtract $C'(1000)$ from $R'(1000)$ to obtain $P'(1000)$.

5

Integration and Infinite Series

In This Unit . . .

CHAPTER 14: Intro to Integration and Approximating Area

Integration: Just Fancy Addition
Finding the Area Under a Curve
Approximating Area
Getting Fancy with Summation Notation
Finding Exact Area with the Definite Integral
Approximating Area with the Trapezoid Rule and Simpson's Rule
Practice Questions Answers and Explanations
Whaddya Know? Chapter 14 Quiz
Answers to Chapter 14 Quiz

CHAPTER 15: Integration: It's Backwards Differentiation

Antidifferentiation
Vocabulary, Voshmabulary: What Difference Does It Make?
The Annoying Area Function
The Power and the Glory of the Fundamental Theorem of Calculus
The Fundamental Theorem of Calculus: Take Two
Finding Antiderivatives: Three Basic Techniques
Finding Area with Substitution Problems
Practice Questions Answers and Explanations
Whaddya Know? Chapter 15 Quiz
Answers to Chapter 15 Quiz

CHAPTER 16: Integration Techniques for Experts

Integration by Parts: Divide and Conquer
Tricky Trig Integrals
Your Worst Nightmare: Trigonometric Substitution
The *A*'s, *B*'s, and *Cx*'s of Partial Fractions
Practice Questions Answers and Explanations
Whaddya Know? Chapter 16 Quiz
Answers to Chapter 16 Quiz

CHAPTER 17: Who Needs Freud? Using the Integral to Solve Your Problems

The Mean Value Theorem for Integrals and Average Value
The Area between Two Curves — Double the Fun
Volumes of Weird Solids: No, You're Never Going to Need This
Analyzing Arc Length
Surfaces of Revolution — Pass the Bottle 'Round
Practice Questions Answers and Explanations
Whaddya Know? Chapter 17 Quiz
Answers to Chapter 17 Quiz

CHAPTER 18: Taming the Infinite with Improper Integrals

L'Hôpital's Rule: Calculus for the Sick
Improper Integrals: Just Look at the Way That Integral Is Holding Its Fork!
Practice Questions Answers and Explanations
Whaddya Know? Chapter 18 Quiz
Answers to Chapter 18 Quiz

CHAPTER 19: Infinite Series: Welcome to the Outer Limits

Sequences and Series: What They're All About
Convergence or Divergence? That Is the Question
Alternating Series
Keeping All the Tests Straight
Practice Questions Answers and Explanations
Whaddya Know? Chapter 19 Quiz
Answers to Chapter 19 Quiz

IN THIS CHAPTER

- » Integrating — adding it all up
- » Approximating areas and sizing up sigma sums
- » Using the definite integral to get exact areas
- » Reconnoitering rectangles
- » Totaling up trapezoids
- » Applying Simpson's rule: Calculus for Bart and Homer

Chapter 14

Intro to Integration and Approximating Area

Since you're still reading this book, I presume that means you survived differentiation (Chapters 9 through 13). Now you begin the second major topic in calculus: integration. Just as two simple ideas lie at the heart of differentiation — *rate* (like *miles per hour*) and the steepness or *slope* of a curve — integration can also be understood in terms of two simple ideas: *adding up* small pieces of something and the *area* under a curve. In this chapter, I introduce you to these two fundamental concepts.

By the way, much of the material in this chapter and the first section of Chapter 15 is both more difficult and less useful than what follows it. If ever there was a time for the perennial complaint — "What is the point of learning this stuff?" — this is it. Now, some calculus teachers would give you all sorts of fancy arguments and pedagogical justifications for why this material is taught, but, let's be honest, the sole purpose of teaching these topics is to inflict maximum pain on calculus students. Well, you're stuck with it, so deal with it. The good news is that this material will make much of what comes later seem easy by comparison.

Integration: Just Fancy Addition

Consider the lamp on the left in Figure 14-1. Say you want to determine the volume of the lamp's base. Why would you want to do that? Beats me. Anyway, a formula for the volume of such a weird shape doesn't exist, so you can't calculate the volume directly. (Archimedes had his "Eureka!" moment when he calculated the volume of another weird shape. Remember his method? *Hint:* If you try his method on a lamp, make sure you unplug it first!)

Okay, so there's no formula for the volume of the lamp base. However, with integration, you can calculate the volume. Imagine that the base is cut up into thin, horizontal slices as shown on the right in Figure 14-1.

FIGURE 14-1: A lamp with a curvy base and the base cut into thin horizontal slices.

Can you see that each slice is shaped like a thin pancake? Now, because there *is* a formula for the volume of a pancake (a pancake is just a very short cylinder), you can determine the total volume of the base by simply calculating the volume of each pancake-shaped slice and then adding up the volumes. That's integration in a nutshell.

But, of course, if that's all there was to integration, there wouldn't be such a fuss about it — certainly not enough to vault Newton, Leibnitz, and other calculus all-stars into the mathematics hall of fame. What makes integration one of the great achievements in the history of mathematics is that — to continue with the lamp example — it gives you the *exact* volume of the lamp's base by sort of cutting it into an *infinite* number of *infinitely* thin slices. Now *that is* something. If you cut the lamp into fewer than an infinite number of slices, you can get only a very good approximation of the total volume — not the exact answer — because each pancake-shaped slice would have a weird, curved edge, which would cause a small error when computing the volume of the slice with the cylinder formula.

Integration has an elegant symbol: $\int$. You've probably seen it before — maybe in one of those cartoons with some Einstein guy in front of a blackboard filled with indecipherable gobbledygook. Soon, this will be *you*. That's right: You'll be filling up pages in your notebook with equations containing the integration symbol. Onlookers will be amazed and envious.

You can think of the integration symbol as just an elongated *S* for "sum up." So, for our lamp problem, you can write

$$\int_{bottom}^{top} dB = B$$

where dB means a little bit of the base — actually an infinitely small piece. So the equation just means that if you sum up all the little pieces of the base from the *bottom* to the *top*, the result is B, the volume of the whole base.

This is a bit oversimplified — I can hear the siren of the math police now — but it's a good way to think about integration. By the way, thinking of dB as a little or infinitesimal piece of B is an idea you saw before with differentiation (see Chapter 9), where the derivative or slope, $\frac{dy}{dx}$, is equal to the ratio of a little bit of y to a little bit of x, as you shrink the $\frac{rise}{run}$ stair step down to an infinitesimal size (see Figure 9-13). Thus, both differentiation and integration involve infinitesimals.

So, whenever you see something like

$$\int_a^b \textit{little piece of mumbo jumbo}$$

it just means that you add up all the little (infinitesimal) pieces of the mumbo jumbo from a to b to get the total of all of the mumbo jumbo from a to b. Or you might see something like

$$\int_{t=0\text{ sec.}}^{t=20\text{ sec.}} \textit{little piece of distance}$$

which means to add up the little pieces of distance traveled between 0 and 20 seconds to get the total distance traveled during that time span.

To sum up — that's a pun! — the mathematical expression to the right of the integration symbol stands for a little bit of something, and integrating such an expression means to add up all the little pieces between some starting point and some ending point to determine the total between the two points.

Finding the Area Under a Curve

As I discuss in Chapter 9, the most fundamental meaning of a derivative is that it's a rate, a *this per that* like *miles per hour*, and that when you graph the *this* as a function of the *that* (like *miles* as a function of *hours*), the derivative becomes the slope of the function. In other words, the derivative is a rate, which on a graph appears as a slope.

It works in a similar way with integration. The most fundamental meaning of integration is to add up (you might be adding up distances or volumes, for example). And when you depict integration on a graph, you can see the adding-up process as a summing up of little bits of area to arrive at the total area under a curve. Consider Figure 14-2.

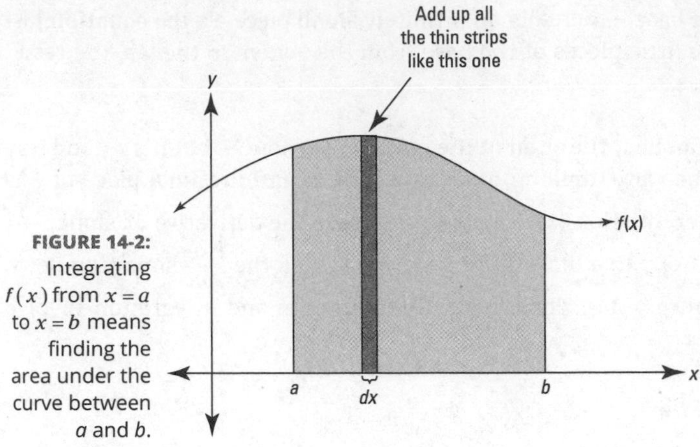

FIGURE 14-2: Integrating $f(x)$ from $x = a$ to $x = b$ means finding the area under the curve between a and b.

The shaded area in Figure 14-2 can be calculated with the following integral:

$$\int_a^b f(x)dx$$

Look at the thin rectangle in Figure 14-2. It has a height of $f(x)$ and a width of dx (a little bit of x), so its area (*length* times *width*, of course) is given by $f(x)\cdot dx$. The above integral tells you to add up the areas of all the narrow rectangular strips between a and b under the curve $f(x)$. As the strips get narrower and narrower, you get a better and better estimate of the area. The power of integration lies in the fact that it gives you the *exact* area by sort of adding up an infinite number of infinitely thin rectangles.

If you're doing a problem where both the x and y axes are labeled in a unit of length, say, *feet*, then each thin rectangle measures so many feet by so many feet, and its area — *length* times *width* — is some number of *square feet*. In this case, when you integrate to get the total area under the curve between a and b, your final answer will be an amount of — what else? — area. But you can use this adding-up-areas-of-rectangles scheme to add up tiny bits of anything — distance, volume, or energy, for example. In other words, the area under the curve doesn't have to stand for an actual area.

If, for example, the units on the x-axis are *hours* and the y-axis is labeled in *miles per hour*, then, because *rate* times *time* equals *distance* (and because $\frac{miles}{hour}\cdot hours = miles$), the area of each rectangle represents an amount of distance (in miles), and the total area gives you the total distance traveled during the given time interval. Or if the x-axis is labeled in *hours* and the y-axis in *kilowatts* of electrical power — in which case the curve gives power usage as a function of time — then the area of each rectangular strip (*kilowatts* times *hours*) represents a number of *kilowatt-hours* of energy. In that case, the total area under the curve gives you the total number of kilowatt-hours of energy consumption between two points in time.

Figure 14-3 shows how you would do the lamp volume problem, from earlier in this chapter, by adding up areas. In this graph, the function $A(h)$ gives the cross-sectional *area* of a thin pancake slice of the lamp as a function of its height measured from the bottom of the lamp. So, this time, the h-axis is labeled in *inches* (that's h as in *height* from the bottom of the lamp), and the y-axis is labeled in *square inches*, and thus each thin rectangle has a width measured in inches and a height measured in square inches. The area of each rectangle, therefore, represents *inches* times *square inches*, or *cubic inches* of volume.

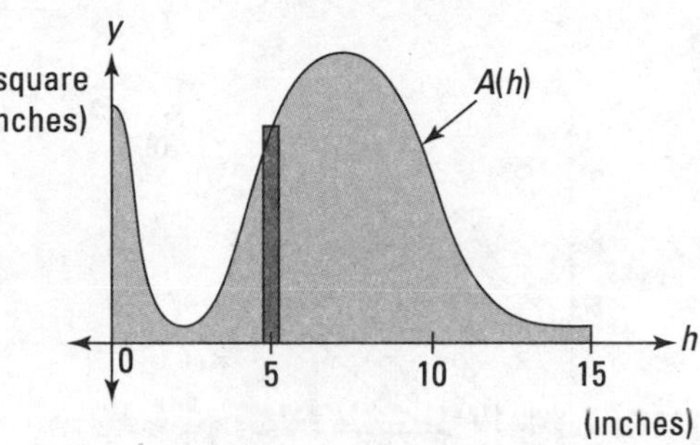

FIGURE 14-3: This shaded *area* gives you the *volume* of the base of the lamp in Figure 14-1. (*Note:* The shape of $A(h)$ is close to what it should be, but it's not precise.)

The area of the thin rectangle in Figure 14-3 represents the *volume* of the thin pancake slice of the lamp 5 inches up from the bottom of the base. The total shaded area and thus the volume of the lamp's base is given by the following integral:

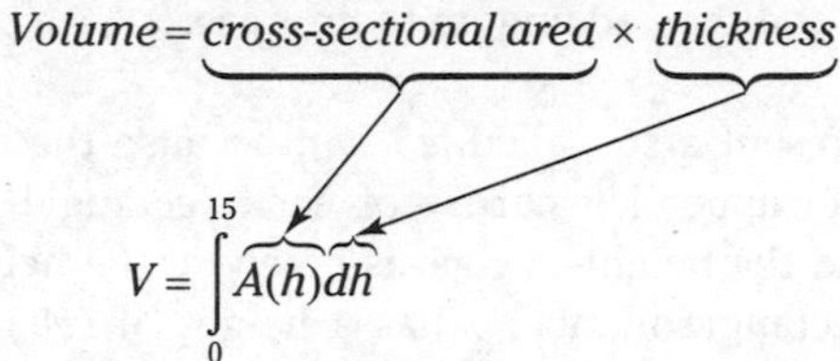

Volume = cross-sectional area × thickness
V = ∫₀¹⁵ A(h)dh

This integral tells you to add up the volumes of all the thin pancake slices from 0 to 15 inches (that is, from the bottom to the top of the lamp's base), each slice having a volume given by $A(h)$ (its cross-sectional area) times dh (its height or thickness). (By the way, Figure 14-3 resembles the left half of the lamp's base [tilted on its side], but it's not that. It has a similar shape because where the lamp base is wide, the corresponding circular slice has a large cross-sectional area.)

Okay, enough of this introductory stuff. In the next section, you actually calculate some areas.

Approximating Area

Before explaining how to calculate exact areas, I want to show you how to approximate areas. The approximation method is useful not only because it lays the groundwork for the exact method — integration — but because for some curves, integration is impossible, and an approximation of area is the best you can do.

The material in this section — using rectangles to approximate the area of strange shapes — is part of every calculus course because integration rests on this foundation. But, in a sense, this material doesn't involve calculus at all. You could do everything in this section without calculus, and if calculus had never been invented, you could still approximate area with the methods described here.

Approximating area with left sums

Say you want the exact area under the curve $f(x) = x^2 + 1$ between $x = 0$ and $x = 3$. See the shaded area on the graph on the left in Figure 14-4.

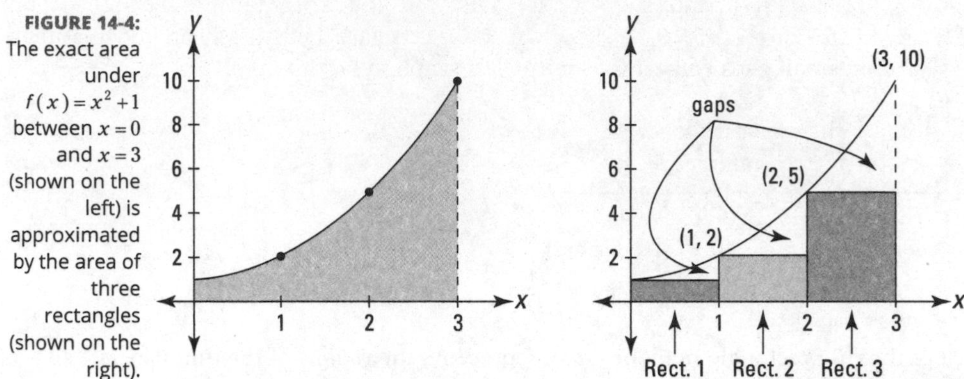

FIGURE 14-4: The exact area under $f(x) = x^2 + 1$ between $x = 0$ and $x = 3$ (shown on the left) is approximated by the area of three rectangles (shown on the right).

You can get a rough estimate of the total area by drawing three rectangles under the curve, as shown on the right in Figure 14-4, and then adding up their areas.

The rectangles in Figure 14-4 represent a so-called *left sum* because the height of each rectangle is determined by where the upper *left* corner of each rectangle touches the curve. Each rectangle has a width of 1 and the height of each is given by the height of the function at the rectangle's left edge. So, rectangle number 1 has a height of $f(0) = 0^2 + 1 = 1$; its area ($length \cdot width$ or $height \cdot width$) is thus $1 \cdot 1$ or 1. Rectangle 2 has a height of $f(1) = 1^2 + 1 = 2$, so its area is $2 \cdot 1$, or 2. And rectangle 3 has a height of $f(2) = 2^2 + 1 = 5$, so its area is $5 \cdot 1$, or 5. Adding these three areas gives you a total of $1 + 2 + 5$, or 8. You can see that this is an underestimate of the total area under the curve because of the three gaps between the rectangles and the curve shown in Figure 14-4.

For a better estimate, double the number of rectangles to six. Figure 14-5 shows six "left" rectangles under the curve and also how the six rectangles begin to fill up the three gaps you see in Figure 14-4.

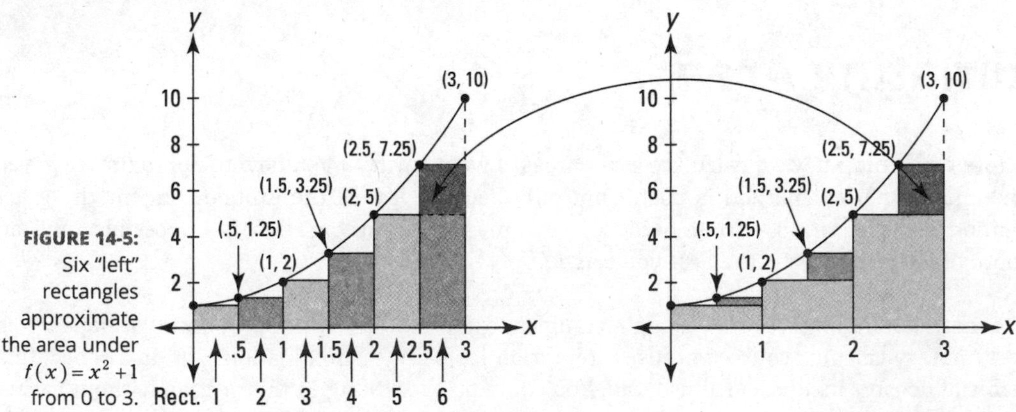

FIGURE 14-5: Six "left" rectangles approximate the area under $f(x) = x^2 + 1$ from 0 to 3.

See the three small shaded rectangles in the graph on the right in Figure 14-5? They sit on top of the three rectangles from Figure 14-4 and represent how much the area estimate has improved by using six rectangles instead of three.

Now total up the areas of the six rectangles. Each has a width of 0.5 and the heights are $f(0), f(0.5), f(1), f(1.5)$, and so on. I'll spare you the arithmetic. Here's the total: $0.5+0.625+1+1.625+2.5+3.625=9.875$. This is a better estimate, but it's still an underestimate because of the six small gaps you can see on the left graph in Figure 14-5.

Table 14-1 shows the area estimates given by 3, 6, 12, 24, 48, 96, 192, and 384 rectangles. You don't have to double the number of rectangles each time like I've done here. You can use any number of rectangles of equal width that you want. I just like the doubling scheme because, with each doubling, the gaps are plugged up more and more in the way shown in Figure 14-5. Any guesses as to where the estimates in Table 14-1 are headed? Looks like 12 to me.

Table 14-1 Estimates of the Area Under $f(x)=x^2+1$ from 0 to 3 Given by Increasing Numbers of "Left" Rectangles

Number of Rectangles	Area Estimate
3	8
6	9.875
12	~10.906
24	~11.445
48	~11.721
96	~11.860
192	~11.930
384	~11.965

Here's the fancy-pants formula for a left-rectangle sum.

MATH RULES

The left rectangle rule: You can approximate the exact area under a curve between a and b, $\int_a^b f(x)dx$, with a sum of *left* rectangles of equal width given by the following formula. In general, the more rectangles, the better the estimate.

$$L_n=\frac{b-a}{n}\left[f(x_0)+f(x_1)+f(x_2)+\ldots+f(x_{n-1})\right]$$

where n is the number of rectangles, $\frac{b-a}{n}$ is the width of each rectangle, x_0 through x_{n-1} are the x-coordinates of the left edges of the n rectangles, and the function values are the heights of the rectangles.

I'd better explain this formula a bit. Look back to the six rectangles shown in Figure 14-5. The width of each rectangle equals the length of the total span from 0 to 3 (which, of course, is $3-0$, or 3) divided by the number of rectangles, 6. That's what the $\frac{b-a}{n}$ does in the formula.

Now, what about those x's with the subscripts? The x-coordinate of the *left* edge of rectangle 1 in Figure 14-5 is called x_0, the *right* edge of rectangle 1 (which is the same as the left edge of rectangle 2) is at x_1, the right edge of rectangle 2 is at x_2, the right edge of rectangle 3 is at x_3, and so on all the way up to the right edge of rectangle 6, which is at x_6. For the six rectangles in Figure 14-5, x_0 is 0, x_1 is 0.5, x_2 is 1, x_3 is 1.5, x_4 is 2, x_5 is 2.5, and x_6 is 3. The heights of the six left rectangles in Figure 14-5 occur at their left edges, which are at x_0 through x_5. You don't use the right edge of the last rectangle, x_6, in a left sum. That's why the list of function values in the formula stops at x_{n-1}. This all becomes clearer — cross your fingers — when you look at the formula for *right* rectangles in the next section.

Here's how to use the formula for the six rectangles in Figure 14-5:

$$\begin{aligned}L_6 &= \frac{3-0}{6}[f(x_0)+f(x_1)+f(x_2)+f(x_3)+f(x_4)+f(x_5)]\\&=\frac{1}{2}[f(0)+f(0.5)+f(1)+f(1.5)+f(2)+f(2.5)]\\&=\frac{1}{2}(1+1.25+2+3.25+5+7.25)\\&=\frac{1}{2}(19.75)=9.875\end{aligned}$$

Note that had I distributed the width of $\frac{1}{2}$ to each of the heights after the third line in the solution, you'd have seen the sum of the areas of the six rectangles — which you saw in the paragraph where I totaled up the area of the six rectangles. The formula just uses the shortcut of first adding up the heights and then multiplying by the width.

Approximating area with right sums

Now estimate the same area under $f(x)=x^2+1$ from 0 to 3 with *right* rectangles. This method works like the left sum method, except each rectangle is drawn so that its *right* upper corner touches the curve. See Figure 14-6.

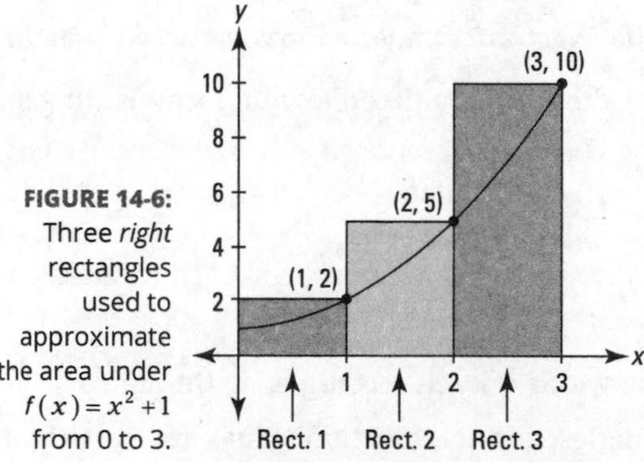

FIGURE 14-6: Three *right* rectangles used to approximate the area under $f(x)=x^2+1$ from 0 to 3.

The heights of the three rectangles in Figure 14-6 are given by the function values at their *right* edges: $f(1)=2$, $f(2)=5$, and $f(3)=10$. Each rectangle has a width of 1, so the areas are 2, 5, and 10, which total 17. You don't have to be a rocket scientist to see that this time you get an *over*estimate of the actual area under the curve, as opposed to the *under*estimate that you get with the left-rectangle method I detail in the previous section (more on that in a minute). Table 14-2 shows the improving estimates you get with more and more right rectangles.

Table 14-2 Estimates of the Area Under $f(x)=x^2+1$ from 0 to 3 Given by Increasing Numbers of "Right" Rectangles

Number of Rectangles	Area Estimate
3	17
6	14.375
12	~13.156
24	~12.570
48	~12.283
96	~12.141
192	~12.070
384	~12.035

Looks like these estimates are also headed toward 12. Here's the formula for a right rectangle sum.

MATH RULES

The right rectangle rule: You can approximate the exact area under a curve between a and b, $\int_a^b f(x)dx$, with a sum of *right* rectangles given by the following formula. In general, the more rectangles, the better the estimate.

$$R_n=\frac{b-a}{n}[f(x_1)+f(x_2)+f(x_3)+.....+f(x_n)],$$

where n is the number of rectangles, $\frac{b-a}{n}$ is the width of each rectangle, x_1 through x_n are the x-coordinates of the right edges of the n rectangles, and the function values are the heights of the rectangles.

If you compare this formula to the one for a left rectangle sum, you get the complete picture about those subscripts. The two formulas are the same except for one thing. Look at the sums of the function values in both formulas. The right sum formula has one value, $f(x_n)$, that the left sum formula doesn't have, and the left sum formula has one value, $f(x_0)$, that the right sum formula doesn't have. All the function values between those two appear in both formulas. You can get a better handle on this by comparing the three left rectangles from Figure 14-4 to the three right rectangles from Figure 14-6. Their areas and totals, which I earlier calculated, are

$$\text{Three left rectangles: } 1+2+5=8$$
$$\text{Three right rectangles: } 2+5+10=17$$

The values used in the sums of the areas are the same except for the left-most left rectangle value and the right-most right rectangle value. Both sums include rectangles with areas 2 and 5. If you look at how the rectangles are constructed, you can see that the second and third rectangles in Figure 14-4 are the same as the first and second rectangles in Figure 14-6.

Approximating area with midpoint sums

A third way to approximate areas with rectangles is to make each rectangle cross the curve at the midpoint of its top side. A midpoint sum is usually a *much* better estimate of area than either a left or a right sum. Figure 14-7 shows why.

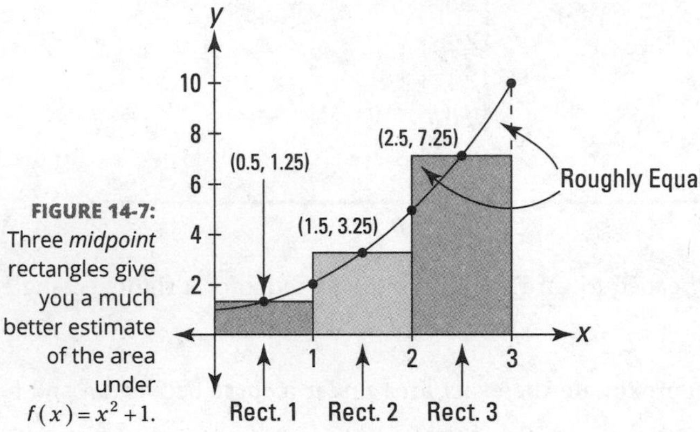

FIGURE 14-7: Three *midpoint* rectangles give you a much better estimate of the area under $f(x) = x^2 + 1$.

You can see in Figure 14-7 that the part of each rectangle that's above the curve looks about the same size as the gap between the rectangle and the curve. A midpoint sum produces such a good estimate because these two errors roughly cancel out each other.

For the three rectangles in Figure 14-7, the widths are 1 and the heights are $f(0.5) = 1.25$, $f(1.5) = 3.25$, and $f(2.5) = 7.25$. The total area comes to 11.75. Table 14-3 lists the midpoint sums for the same number of rectangles used in Tables 14-1 and 14-2.

Table 14-3 Estimates of the Area Under $f(x) = x^2 + 1$ from 0 to 3 Given by Increasing Numbers of "Midpoint" Rectangles

Number of Rectangles	Area Estimate
3	11.75
6	11.9375
12	~11.9844
24	~11.9961
48	~11.9990
96	~11.9998
192	~11.9999
384	~11.99998

If you had any doubts that the left and right sums in Tables 14-1 and 14-2 were heading to 12, Table 14-3 should dispel them. Spoiler alert: Yes, in fact, the exact area is 12. (I show you how to calculate that in several pages in the section "Finding Exact Area with the Definite Integral.") And to see how much faster the midpoint approximations approach the exact answer of 12 than the left or right approximations, compare the three tables. The error with 6 midpoint rectangles is about the same as the error with 192 left or right rectangles! Here's the mumbo jumbo.

MATH RULES

The midpoint rule: You can approximate the exact area under a curve between a and b, $\int_a^b f(x)dx$, with a sum of *midpoint* rectangles given by the following formula. In general, the more rectangles, the better the estimate.

$$M_n = \frac{b-a}{n}\left[f\left(\frac{x_0+x_1}{2}\right)+f\left(\frac{x_1+x_2}{2}\right)+f\left(\frac{x_2+x_3}{2}\right)+\ldots\ldots+f\left(\frac{x_{n-1}+x_n}{2}\right)\right],$$

where n is the number of rectangles, $\frac{b-a}{n}$ is the width of each rectangle, x_0 through x_n are the $n+1$ evenly spaced points from a to b, and the function values are the heights of the rectangles.

MATH RULES

Definition of *Riemann sum*: All three sums — left, right, and midpoint — are called Riemann sums, after the great German mathematician Bernhard Riemann (1826–1866). Basically, any approximating sum made up of rectangles is a Riemann sum, including weird sums consisting of rectangles of unequal width. Luckily, you won't have to deal with those in this book or your calculus course.

The left, right, and midpoint sums in Tables 14-1, 14-2, and 14-3 are all heading toward 12, and if you could slice up the area into an infinite number of rectangles, you'd get the exact area of 12. But I'm getting ahead of myself.

EXAMPLE

Q. Using 10 right rectangles, estimate the area under $f(x)=\ln x$ from 1 to 6.

A. **The approximate area is 6.181.**

1. **Sketch $f(x)=\ln x$ and divide the interval from 1 to 6 into ten equal increments.**

   Each increment has a length of $\frac{1}{2}$, of course. See the figure in Step 2.

2. **Draw a *right* rectangle for each of the ten increments.**

   You're doing *right* rectangles, so put your pen on the *right* end of the base of the first rectangle (that's at $x=1.5$), draw straight up till you hit the curve, and then straight left till you're directly above the left end of the base ($x=1$). Finally, going straight down, draw the left side of the first rectangle. See the following figure. I've indicated with arrows how you draw the first rectangle. Draw the rest the same way.

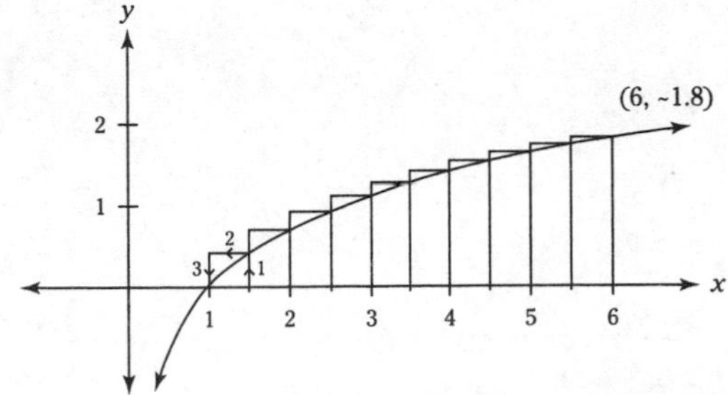

3. **Use your calculator to calculate the height of each rectangle.**

   The heights are given by $f(1.5)$, $f(2)$, $f(2.5)$, and so on, which are $\ln 1.5$, $\ln 2$, and so on.

4. **Here's the final computation:**

$$\begin{aligned}
&\frac{1}{2}(\ln 1.5+\ln 2+\ln 2.5+\ln 3+\ln 3.5+\ln 4+\ln 4.5+\ln 5+\ln 5.5+\ln 6)\\
&\approx \frac{1}{2}(0.405+0.693+0.916+1.099+1.253+1.386+1.504+1.609+1.705+1.792)\\
&\approx \frac{1}{2}(12.362)\approx 6.181
\end{aligned}$$

YOUR TURN

(a) Estimate the area under $f(x)=\ln x$ from 1 to 6 (as in the example), but this time with 10 left rectangles.

(b) How is this approximation related to the area obtained with 10 right rectangles? (*Hint:* Compare individual rectangles from both estimates.)

Approximate the same area again with 10 midpoint rectangles.

3 Rank the approximations from the example and Problems 1 and 2 from best to worst and defend your ranking. Obviously, you're not allowed to cheat by first finding the exact area with your calculator.

4 Use 8 left, right, and midpoint rectangles to approximate the area under $\sin x$ from 0 to π.

Getting Fancy with Summation Notation

Before I get to the formal definition of the *definite integral* — that's the incredible calculus tool that sort of cuts up an area into an infinite number of rectangles and thereby gives you the *exact* area — there's one more thing to take care of: summation notation.

Summing up the basics

For adding up long series of numbers like the rectangle areas in a left, right, or midpoint sum, summation or *sigma* notation comes in handy. Sigma notation may look fancy and difficult, but it's really just a shorthand way of writing a long sum. Here's how it works. Say you wanted to add up the first 100 multiples of 5 — that's from 5 to 500. You could write out the sum like this:

$$5+10+15+20+25+.....+490+495+500$$

But with sigma notation (sigma, $\sum$, is the 18th letter of the Greek alphabet), the sum is much more condensed and efficient, and, let's be honest, it looks pretty cool:

$$\sum_{i=1}^{100} 5i$$

This notation just tells you to plug 1 in for the i in $5i$, then plug 2 into the i in $5i$, then 3, then 4, and so on, up to 100. Then you add up the results. So that's $5 \cdot 1$ plus $5 \cdot 2$ plus $5 \cdot 3$, and so on, up to $5 \cdot 100$. This produces the same thing as writing out the sum the long way.

The letter *i* in this example is called the *index of summation*. The particular letter you use has no significance, though *i* and *k* are customary.

Here's one more. If you want to add up $10^2+11^2+12^2+.....+29^2+30^2$, you can write the sum with sigma notation as follows:

$$\sum_{k=10}^{30} k^2$$

There's really nothing to it.

Let's walk through a couple examples step by step. But, first, a piece of business.

REMEMBER

Pulling stuff out. In a sigma sum problem, you can pull anything through the sigma symbol to the outside except for a function of the *index of summation*.

EXAMPLE

Q. Evaluate $\sum_{i=4}^{12} 5i^2$.

A. **The sum is 3180.**

1. **Pull the 5 through the sigma symbol:** $5\sum_{i=4}^{12} i^2$
2. **Plug 4 into *i*, then 5, then 6, and so on up to 12, adding up all the terms.**

   $=5\left(4^2+5^2+6^2+7^2+8^2+9^2+10^2+11^2+12^2\right)$
3. **Finish on your calculator.**

   $=5(636)=3180$

Q. Express $50^3+60^3+70^3+80^3+\ldots+150^3$ with sigma notation.

A. $\mathbf{1000\sum_{i=1}^{11}(i+4)^3}$

1. **Create the argument (that's the input) of the sigma function.**

   The jump amount between terms in a long sum will become the coefficient of the index of summation in a sigma sum, so you know that 10*i* is the basic term of your argument. You want to cube each term, so that gives you the following:

   $\sum(10i)^3$
2. **Set the range of the sum.**

   Ask yourself what *i* must be to make the first term equal 50^3: That's 5, of course. And ask the same question about the last term of 150^3: *i* must be 15. Put the 5 and the 15 on the sigma symbol, like this:

   $=\sum_{i=5}^{15}(10i)^3$

   Check that plugging 5, then 6, then 7, and so on up to 15 into *i* produces the original sum. It works. (It's not a bad idea to do a check like this if you're new to sigma notation.)

3. **Simplify.**

$$= \sum_{i=5}^{15} 10^3 i^3$$

$$= \sum_{i=5}^{15} 1000 i^3$$

$$= 1000 \sum_{i=5}^{15} i^3$$

4. **(Optional) Set the *i* to begin at zero or one.**

   It's often desirable to have i begin at 0 or 1. To turn the 5 into a 1, you subtract 4. Then subtract 4 from the 15 as well. To compensate for this subtraction, you *add* 4 to the i in the argument:

   $$1000 \sum_{i=1}^{11} (i+4)^3$$

   If you want i to start at zero, you have

   $$1000 \sum_{i=0}^{10} (i+5)^3$$

YOUR TURN

 5 Evaluate $\sum_{i=1}^{10} 4$.

 6 Evaluate $\sum_{i=0}^{9} (-1)^i (i+1)^2$.

7 Evaluate $\sum_{i=1}^{50} \left(3i^2 + 2i\right)$.

 8 Express the following sum with sigma notation:
$30 + 35 + 40 + 45 + 50 + 55 + 60$

9 Express the following sum with sigma notation: $8+27+64+125+216$

10 Use sigma notation to express the following: $-2+4-8+16-32+64-128+256-512+1024$

Writing Riemann sums with sigma notation

Let's use sigma notation to write out the right-rectangle sum for the curve $y=x^2+1$ that you looked at in the section "Approximating Area." This can get pretty gnarly. Brace yourself.

Recall the formula for a right sum from the earlier section "Approximating area with right sums":

$$R_n=\frac{b-a}{n}[f(x_1)+f(x_2)+f(x_3)+...+f(x_n)]$$

Here's the same formula written with sigma notation:

$$R_n=\sum_{i=1}^{n}\left[f(x_i)\cdot\frac{b-a}{n}\right]$$

(Note that I could have written this instead as $R_n=\frac{b-a}{n}\sum_{i=1}^{n}f(x_i)$, which would have more nicely mirrored the formula where the $\frac{b-a}{n}$ is on the outside. Either way is fine — they're equivalent — but I chose to keep the $\frac{b-a}{n}$ on the inside so that the $\sum$ sum is actually a sum of rectangles. In other words, with the $\frac{b-a}{n}$ on the inside, the expression after the $\sum$ symbol, $f(x_i)\cdot\frac{b-a}{n}$, which the $\sum$ symbol tells you to add up, is the area of each rectangle, namely *height* times *base*.)

Now work this out for the six right rectangles in Figure 14-8.

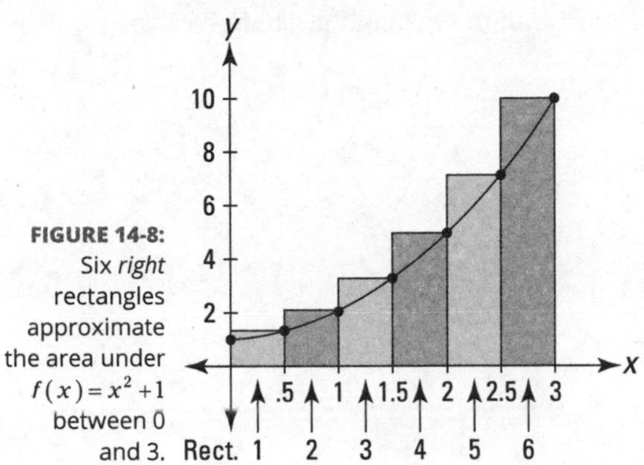

FIGURE 14-8: Six *right* rectangles approximate the area under $f(x)=x^2+1$ between 0 and 3.

You're figuring the area under x^2+1 between $x=0$ and $x=3$ with six rectangles, so the width of each, $\frac{b-a}{n}$, is $\frac{3-0}{6}$ or $\frac{3}{6}$ or $\frac{1}{2}$. So now you've got

$$R_6 = \sum_{i=1}^{6}\left[f(x_i)\cdot\frac{1}{2}\right]$$

Next, because the width of each rectangle is $\frac{1}{2}$, the right edges of the six rectangles fall on the first six multiples of $\frac{1}{2}$: 0.5, 1, 1.5, 2, 2.5, and 3. These numbers are the x-coordinates of the six points x_1 through x_6; they can be generated by the expression $\frac{1}{2}i$, where i equals 1 through 6. You can check that this works by plugging 1 in for i in $\frac{1}{2}i$, then 2, then 3, up to 6. So now you can replace the x_i in the formula with $\frac{1}{2}i$, giving you

$$R_6 = \sum_{i=1}^{6}\left[f\left(\frac{1}{2}i\right)\cdot\frac{1}{2}\right]$$

Your function, $f(x)$, is x^2+1 so $f\left(\frac{1}{2}i\right)=\left(\frac{1}{2}i\right)^2+1$, and so now you can write

$$R_6 = \sum_{i=1}^{6}\left[\left(\left(\frac{1}{2}i\right)^2+1\right)\cdot\frac{1}{2}\right]$$

If you plug 1 into i, then 2, then 3, and so on up to 6 and do the math, you get the sum of the areas of the rectangles in Figure 14-8. This sigma notation is just a fancy way of writing the sum of the six rectangles.

Are we having fun? Hold on, it gets worse — sorry. Now you're going to write out the general sum for an unknown number, n, of right rectangles. The total span of the area in question is 3, right? You divide this span by the number of rectangles to get the width of each rectangle. With

6 rectangles, the width of each is $\frac{3}{6}$; with n rectangles, the width of each is $\frac{3}{n}$. And the right edges of the n rectangles are generated by $\frac{3}{n}i$, for i equals 1 through n. That gives you

$$R_n = \sum_{i=1}^{n}\left[f\left(\frac{3}{n}i\right)\cdot\frac{3}{n}\right]$$

Or, because $f(x) = x^2 + 1$,

$$\begin{aligned}
R_n &= \sum_{i=1}^{n}\left[\left(\left(\frac{3}{n}i\right)^2 + 1\right)\cdot\frac{3}{n}\right]\\
&= \sum_{i=1}^{n}\left[\left(\frac{9i^2}{n^2} + 1\right)\cdot\frac{3}{n}\right]\\
&= \sum_{i=1}^{n}\left[\frac{27i^2}{n^3} + \frac{3}{n}\right]\\
&= \sum_{i=1}^{n}\frac{27i^2}{n^3} + \sum_{i=1}^{n}\frac{3}{n} \qquad \text{(Take my word for it.)}\\
&= \frac{27}{n^3}\sum_{i=1}^{n}i^2 + \frac{3}{n}\sum_{i=1}^{n}1
\end{aligned}$$

For this last step, you pull the $\frac{27}{n^3}$ and the $\frac{3}{n}$ through the summation symbols — recall that you're allowed to pull out anything except for a function of i.

You've now arrived at a critical step. With a sleight of hand, you're going to turn the Riemann sum here into a formula in terms of n. (This formula is what you use in the next section to obtain the exact area under the curve.)

Now, as almost no one knows, the sum of the first n square numbers, $1^2 + 2^2 + 3^2 + \ldots + n^2$, equals $\frac{n(n+1)(2n+1)}{6}$. (By the way, this 6 has nothing to do with the fact that you used 6 rectangles a couple pages back.) So, you can substitute that expression for the $\sum_{i=1}^{n} i^2$ in the last line of the sigma notation solution, and at the same time substitute n for $\sum_{i=1}^{n} 1$ (because that just tells you to add up n 1's):

$$\begin{aligned}
R_n &= \frac{27}{n^3}\sum_{i=1}^{n}i^2 + \frac{3}{n}\sum_{i=1}^{n}1\\
&= \frac{27}{n^3}\cdot\frac{n(n+1)(2n+1)}{6} + \frac{3}{n}\cdot n\\
&= \frac{27}{n^3}\cdot\frac{2n^3 + 3n^2 + n}{6} + 3\\
&= \frac{27}{n^3}\cdot\left(\frac{n^3}{3} + \frac{n^2}{2} + \frac{n}{6}\right) + 3\\
&= 9 + \frac{27}{2n} + \frac{9}{2n^2} + 3\\
&= 12 + \frac{27}{2n} + \frac{9}{2n^2}
\end{aligned}$$

The end. Finally! This is the formula for the area of n right rectangles between $x=0$ and $x=3$ under the function $f(x)=x^2+1$. You can use this formula to produce the approximate areas given in Table 14-2. But once you've got such a formula, it'd be kind of pointless to produce a table of approximate areas, because you can use the formula to determine the *exact* area. And it's a snap. I get to that in a minute in the next section.

But first, here are the formulas for n left rectangles and n midpoint rectangles between $x=0$ and $x=3$ under the same function, x^2+1. These formulas generate the area approximations in Tables 14-1 and 14-3. The algebra for deriving these formulas is even worse than what we just did for the right rectangle formula, so I decided to skip it. Do you mind? I didn't think so.

$$L_n = 12 - \frac{27}{2n} + \frac{9}{2n^2}$$

$$M_n = 12 - \frac{9}{4n^2}$$

YOUR TURN

11 Use sigma notation to express a 20-right-rectangle approximation of the area under $g(x)=x^2+3x$ from 0 to 5. Then compute the approximation.

12 Using your result from Problem 11, write a formula for approximating the area under g from 0 to 5 with n rectangles.

And now, what you've all been waiting for . . .

Finding Exact Area with the Definite Integral

Having laid all the necessary groundwork, we're finally ready to move on to determining exact areas — which is the whole point of integration. You don't need calculus to do all the approximation stuff we just did.

As you saw with the left, right, and midpoint rectangles in the section "Approximating Area," the more rectangles you use, the better the approximation. So, "all" you'd have to do to get the exact area under a curve is to use an infinite number of rectangles. Now, you can't really do that, but with the fantastic invention of limits, this is sort of what happens. Here's the definition of the definite integral that's used to compute exact areas.

MATH RULES

The *definite integral* ("simple" definition): The exact area under a curve between $x=a$ and $x=b$ is given by the definite integral, which is defined as the limit of a Riemann sum:

$$\int_a^b f(x)dx = \lim_{n\to\infty} \sum_{i=1}^{n} \left[f(x_i) \cdot \frac{b-a}{n} \right]$$

Is that a thing of beauty or what? This summation (everything to the right of "lim") is identical to the formula for n right rectangles, R_n, that I give a few pages back. The only difference here is that you take the limit of that formula as the number of rectangles approaches infinity (∞).

This definition of the definite integral is the simple version based on the right rectangle formula. I give you the real-McCoy definition later, but because all Riemann sums for a specific problem have the same limit — in other words, it doesn't matter what type of rectangles you use — you might as well use the right-rectangle definition. It's the least complicated and it'll always suffice.

Let's have a drum roll. Here, finally, is the exact area under our old friend $f(x)=x^2+1$ between $x=0$ and $x=3$:

$$\int_0^3 (x^2+1)dx = \lim_{n\to\infty} \sum_{i=1}^{n} \left[f(x_i) \cdot \frac{b-a}{n} \right]$$

$$= \lim_{n\to\infty} \left(12 + \frac{27}{2n} + \frac{9}{2n^2} \right)$$ (This is what we got in the section "Writing Reimann sums with sigma notation," after all those steps.)

$$= 12 + \frac{27}{2\cdot\infty} + \frac{9}{2\cdot\infty^2}$$

$$= 12 + \frac{27}{\infty} + \frac{9}{\infty}$$

$$= 12 + 0 + 0$$ (Remember, in a limit problem, any number divided by infinity equals zero.)

$$= 12$$

Big surprise.

This result is pretty amazing if you think about it. Using the limit process, you get an *exact* answer of 12 — sort of like 12.00000000 . . . to an infinite number of decimal places — for the area under the smooth, curving function $f(x)=x^2+1$, based on the areas of flat-topped rectangles that run along the curve in a jagged, sawtooth fashion. Remarkable!

Finding the exact area of 12 by using the limit of a Riemann sum is a lot of work (remember, we first had to determine the formula for n right rectangles). This complicated method of integration is comparable to determining a derivative the hard way by using the formal definition that's based on the difference quotient (see Chapter 9). And just as you stopped using the formal definition of the derivative after you learned the differentiation shortcuts, you won't have to use the formal definition of the definite integral based on a Riemann sum after you learn the shortcut methods in Chapters 15 and 16 — except, that is, on your final exam.

Because the limit of all Riemann sums is the same, the limits at infinity of n left rectangles and n midpoint rectangles (for $f(x) = x^2 + 1$ between $x = 0$ and $x = 3$) should give us the same result as the limit at infinity of n right rectangles. The expressions after the following limit symbols are the formulas for n left rectangles and n midpoint rectangles that appear at the end of the section "Writing Riemann sums with sigma notation," earlier in the chapter. Here's the left rectangle limit:

$$\begin{aligned}\int_0^3 (x^2+1)\,dx = L_\infty &= \lim_{n\to\infty}\left(12 - \frac{27}{2n} + \frac{9}{2n^2}\right)\\ &= 12 - \frac{27}{2\cdot\infty} + \frac{9}{2\cdot\infty^2}\\ &= 12 - \frac{27}{\infty} + \frac{9}{\infty}\\ &= 12 - 0 + 0 = 12\end{aligned}$$

And here's the midpoint rectangle limit:

$$\begin{aligned}\int_0^3 (x^2+1)\,dx = M_\infty &= \lim_{n\to\infty}\left(12 - \frac{9}{4n^2}\right)\\ &= 12 - \frac{9}{4\cdot\infty^2}\\ &= 12 - \frac{9}{\infty}\\ &= 12 - 0 = 12\end{aligned}$$

If you're somewhat incredulous that these limits actually give you the *exact* area under $f(x) = x^2 + 1$ between 0 and 3, you're not alone. After all, in these limits, as in all limit problems, the arrow-number (∞ in this example) is only *approached*; it's never actually reached. And on top of that, what would it mean to reach infinity? You can't do it. And regardless of how many rectangles you have, you always have that jagged, sawtooth edge. So how can such a method give you the exact area?

Look at it this way. You can tell from Figures 14-4 and 14-5 that the sum of the areas of left rectangles, regardless of their number, will always be an *under*estimate (this is the case for functions that are increasing over the span in question). And from Figure 14-6, you can see that the sum of the areas of right rectangles, regardless of how many you have, will always be an *over*estimate (for increasing functions). So, because the limits at infinity of the underestimate and the overestimate are both equal to 12, that must be the exact area. (A similar argument works for decreasing functions.)

MATH RULES

All Riemann sums for a given problem have the same limit. Not only are the limits at infinity of left, right, and midpoint rectangles the same for a given problem, but the limit of any Riemann sum also gives you the same answer. You can have a series of rectangles with unequal widths; you can have a mix of left, right, and midpoint rectangles; or you can construct the rectangles so they touch the curve somewhere other than at their left or right upper corners or at the midpoints of their top sides. The only thing that matters is that, in the limit, the width of all the rectangles tends to zero (and from this, it follows that the number of rectangles approaches infinity). This brings us to the following totally extreme, down-and-dirty integration mumbo jumbo that takes all these possibilities into account.

MATH RULES

The *definite integral* (real-McCoy definition): The definite integral from $x = a$ to $x = b$, $\int_a^b f(x)dx$, is the number to which all Riemann sums tend as the width of all rectangles tends to zero and as the number of rectangles approaches infinity:

$$\int_a^b f(x)dx = \lim_{\max \Delta x_i \to 0} \sum_{i=1}^{n} f(c_i)\Delta x_i,$$

where Δx_i is the width of the ith rectangle and c_i is the x-coordinate of the point where the ith rectangle touches $f(x)$. (That "$\max \Delta x_i \to 0$" simply guarantees that the width of all the rectangles approaches zero and that the number of rectangles approaches infinity.)

YOUR TURN

13 In Problem 11, you estimate the area under $g(x) = x^2 + 3x$ from 0 to 5 with 20 right rectangles. The result is about 84.2 square units. Then in Problem 12, you write a formula for the area under g for n rectangles. Use your result from Problem 12 to approximate the area under g from 0 to 5 with 50, 100, 1000, and 10,000 right rectangles.

14 Use your result from Problem 12 and the definition of the definite integral to determine the *exact* area under $g(x) = x^2 + 3x$ from 0 to 5.

Approximating Area with the Trapezoid Rule and Simpson's Rule

This section covers two more ways to estimate the area under a function. You can use them if for some reason you only want an estimate and not an exact answer — maybe because you're asked for that on an exam. But these approximation methods and the others you've gone over are useful for another reason. There are certain types of functions for which the exact area method doesn't work. (It's beyond the scope of this book to explain why this is the case or exactly what these functions are like, so just take my word for it.) So, using an approximation method may be your only choice if you happen to get one of these uncooperative functions.

The trapezoid rule

With the trapezoid rule, instead of approximating area with rectangles, you do it with — can you guess? — trapezoids. See Figure 14-9.

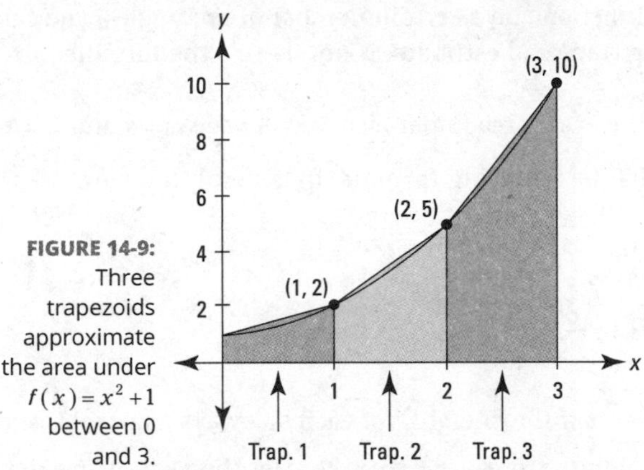

FIGURE 14-9: Three trapezoids approximate the area under $f(x) = x^2 + 1$ between 0 and 3.

Because of the way trapezoids hug the curve, they give you a much better area estimate than either left or right rectangles. And it turns out that a trapezoid approximation is the average of the left rectangle and right rectangle approximations. Can you see why? (*Hint:* The area of a trapezoid — say trapezoid 2 in Figure 14-9 — is the average of the areas of the two corresponding rectangles in the left and right sums, namely, rectangle 2 in Figure 14-4 and rectangle 2 in Figure 14-6.)

Table 14-4 lists the trapezoid approximations for the area under $f(x) = x^2 + 1$ between $x = 0$ and $x = 3$.

From the look of Figure 14-9, you might expect a trapezoid approximation to be better than a midpoint estimate, but in fact, as a general rule, midpoint estimates are about twice as good as trapezoid estimates. You can confirm this by comparing Tables 14-3 and 14-4. For instance, Table 14-3 lists an area estimate of 11.9990 for 48 midpoint rectangles. This differs from the exact area of 12 by 0.001. The area estimate with 48 trapezoids given in Table 14-4, namely 12.002, differs from 12 by twice as much.

Table 14-4 Estimates of the Area Under $f(x) = x^2 + 1$ between $x = 0$ and $x = 3$ Given by Increasing Numbers of Trapezoids

Number of Trapezoids	Area Estimate
3	12.5
6	12.125
12	~12.031
24	~12.008
48	~12.002
96	~12.0005
192	~12.0001
384	~12.00003

TIP

A trapezoid approximation is the average of the corresponding left-rectangle approximation and the right-rectangle approximation. If you've already worked out the left- and right-rectangle approximations for a particular function and a certain number of rectangles, you can just average them to get the corresponding trapezoid estimate. If not, here's the formula.

MATH RULES

The trapezoid rule: You can approximate the exact area under a curve between $x = a$ and $x = b$, $\int_a^b f(x)dx$, with a sum of trapezoids given by the following formula. In general, the more trapezoids, the better the estimate.

$$T_n = \frac{b-a}{2n}[f(x_0) + 2f(x_1) + 2f(x_2) + 2f(x_3) + \ldots.. + 2f(x_{n-1}) + f(x_n)],$$

where n is the number of trapezoids, $\frac{b-a}{2n}$ is half the "height" of each sideways trapezoid, and x_0 through x_n are the $n+1$ evenly spaced points from $x = a$ to $x = b$. (By the way, using that half-the-height expression is completely unintuitive considering that the formula for the area of a trapezoid uses its height, not half its height. For extra credit, see if you can figure out why that $b - a$ is divided by $2n$ instead of just n.)

Even though the formal definition of the definite integral is based on the sum of an infinite number of *rectangles*, I prefer to think of integration as the limit of the trapezoid rule at infinity. The further you zoom in on a curve, the straighter it gets. When you use a greater and greater number of trapezoids and then zoom in on where the trapezoids touch the curve, the tops of the trapezoids get closer and closer to the curve. If you zoom in "infinitely," the tops of the "infinitely many" trapezoids *become* the curve and, thus, the sum of their areas gives you the exact area under the curve. This is a good way to think about why integration produces the exact area — and it makes sense conceptually — but it's not actually done this way.

Simpson's rule — that's Thomas (1710–1761), not Homer (1987–)

Now I really get fancy and draw shapes that are sort of like trapezoids except that instead of having slanting tops, they have curved, parabolic tops. See Figure 14-10.

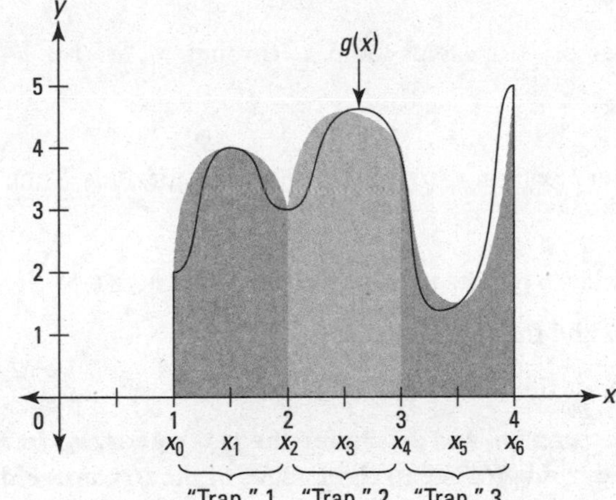

FIGURE 14-10: Three curvy-topped "trapezoids" approximate the area under $g(x)$ between 1 and 4.

Note that with Simpson's rule, each "trapezoid" spans two intervals instead of one; in other words, "trapezoid" 1 goes from x_0 to x_2, "trapezoid" 2 goes from x_2 to x_4, and so on. Because of this, the total span must always be divided into an even number of intervals.

Simpson's rule is by far the most accurate approximation method discussed in this chapter. In fact, it gives the *exact* area for any polynomial function of degree three or less. In general, Simpson's rule gives a much better estimate than either the midpoint rule or the trapezoid rule.

TIP

You can use a midpoint sum with a trapezoid sum to calculate a Simpson sum. A Simpson's rule sum is sort of an average of a midpoint sum and a trapezoid sum, except that you use the midpoint sum twice in the average. So, if you already have the midpoint sum and the trapezoid sum for some number of rectangles/trapezoids, you can obtain the Simpson's rule approximation with the following simple average:

$$S_{2n} = \frac{M_n + M_n + T_n}{3}$$

Note the subscript of $2n$. This means that if you use, say, M_3 and T_3, you get a result for S_6. But S_6, which has six intervals, has only three curvy "trapezoids" because each of them spans two intervals. Thus, this formula always involves the same number of rectangles, trapezoids, and Simpson's rule "trapezoids."

If you don't have the midpoint and trapezoid sums for this shortcut, you can use the following formula for Simpson's rule.

MATH RULES

Simpson's rule: You can approximate the exact area under a curve between $x = a$ and $x = b$, $\int_a^b f(x)dx$, with a sum of parabola-topped "trapezoids" given by the following formula. In general, the more "trapezoids," the better the estimate.

$$S_n = \frac{b-a}{3n}[f(x_0) + 4f(x_1) + 2f(x_2) + 4f(x_3) + 2f(x_4) + \ldots\ldots + 4f(x_{n-1}) + f(x_n)],$$

where n is twice the number of "trapezoids" and x_0 through x_n are the $n+1$ evenly spaced points from $x = a$ to $x = b$.

EXAMPLE

Q. Estimate the area under $f(x) = \ln x$ from 1 to 6 with 10 trapezoids. Then compute the percent error.

A. **The approximate area is 5.733. The percent error is about 0.31%.**

1. **Sketch the function and the 10 trapezoids.**

   You're on your own for this sketch.

2. **List the values for *a*, *b*, and *n*, and determine the 11 *x*-values, x_0 through x_{10} (the left edge of the first trapezoid plus the 10 right edges of the 10 trapezoids).**

   Note that in this and all similar problems, a equals x_0 and b equals x_n (x_{10} here).

   $a = 1, \quad b = 6, \quad n = 10$

   $x_0 = 1, \; x_1 = 1.5, \; x_2 = 2, \; x_3 = 2.5, \; \ldots\ldots, \; x_{10} = 6$

3. **Plug these values into the trapezoid rule formula and solve.**

$$T_{10} = \frac{6-1}{2 \cdot 10}(\ln 1 + 2\ln 1.5 + 2\ln 2 + 2\ln 2.5 + 2\ln 3 + 2\ln 3.5 + 2\ln 4 + 2\ln 4.5 + 2\ln 5 + 2\ln 5.5 + \ln 6)$$

$$\approx \frac{5}{20}(0 + 0.811 + 1.386 + 1.833 + 2.197 + 2.506 + 2.773 + 3.008 + 3.219 + 3.409 + 1.792)$$

$$\approx 5.733$$

4. **Compute the percent error.**

   My calculator tells me that the exact area is 5.7505568153635. For this problem, round that off to 5.751. The *relative error* is given by the error divided by the exact area. Multiplying that by 100% gives you the percent error. So that gives you:

$$\text{relative error} \approx \frac{5.751 - 5.733}{5.751} \approx 0.0031 = 0.31\%$$

   Compare this to the 10-midpoint-rectangle error you compute in the solution to Problem 2: a 0.14% error. As mentioned earlier, the error with a trapezoid estimate is roughly twice the corresponding midpoint-rectangle error.

Q. Estimate the area under $f(x) = \ln x$ from 1 to 6 with 10 Simpson's rule "trapezoids." Then compute the percent error.

A. **The approximate area is 5.751. The percent error is a mere 0.00069%.**

1. **List the values for *a*, *b*, and *n*, and determine the 21 x-values, x_0 through x_{20} (the 11 edges and the 10 base midpoints of the 10 curvy-topped "trapezoids").**

   $a = 1,\ \ b = 6,\ \ n = 20$

   $x_0 = 1,\ \ x_1 = 1.25,\ \ x_2 = 1.5,\ \ x_3 = 1.75,\ \ldots\ldots,\ \ x_{20} = 6$

2. **Plug these values into the formula.**

$$\begin{aligned} S_{20} &= \frac{6-1}{3\cdot 20}(\ln 1 + 4\ln 1.25 + 2\ln 1.5 + 4\ln 1.75 + 2\ln 2 + \ldots + 4\ln 5.75 + \ln 6) \\ &\approx \frac{5}{60}(69.006202893232) \\ &\approx 5.7505169 \end{aligned}$$

3. **Figure the percent error.**

   The exact answer, again, is 5.7505568153635. Round that off to 5.7505568.

$$\begin{aligned} \text{relative error} &\approx \frac{5.7505568 - 5.7505169}{5.7505568} \\ &\approx 0.0000069 \\ &\approx 0.00069\% \end{aligned}$$

   This is way better than either the midpoint or trapezoid estimate. Impressed?

YOUR TURN

15 Continuing with Problem 4, estimate the area under $y = \sin x$ from 0 to π with 8 trapezoids, and compute the percent error.

16 Estimate the same area as Problem 15 with 16 and 24 trapezoids, and compute the percent errors.

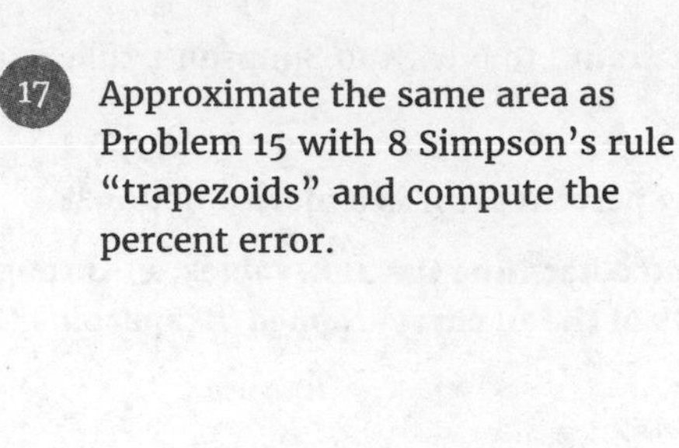

17 Approximate the same area as Problem 15 with 8 Simpson's rule "trapezoids" and compute the percent error.

18 Use the Simpson's rule shortcut to figure S_{20} for the area under $\ln x$ from 1 to 6. (Use the results from Problem 2 and the first example in this section.)

To close this chapter, here's a warning about functions that go below the x-axis. I didn't include any such functions in this chapter, because I thought you already had enough to deal with. You see the full explanation and an example in Chapter 17.

Areas *below* the x-axis count as *negative* areas. Whether approximating areas with right-, left-, or midpoint rectangles or with the trapezoid rule or Simpson's rule, or computing exact areas with the definite integral, areas below the x-axis and above the curve count as *negative* areas.

Practice Questions Answers and Explanations

1 (a) **The area is 5.285.**

1. **Sketch a graph and divide the intervals into 10 subintervals.**

2. a. **Draw the "first" left rectangle by putting your pen at the *left* end of the first base (that's at $x = 1$) and going straight up till you hit the function.**

   Whoops. You're already *on* the function at $x = 1$, right? So, guess what? For this particular problem, there is no first rectangle — or you could say it's a rectangle with a height of zero and an area of zero.

2. b. **Draw the "second" rectangle by putting your pen at $x = 1.5$ and going straight up till you hit $f(x) = \ln x$; then go *right* till you're directly above $x = 2$; and then go down to the x-axis.**

   See the following figure.

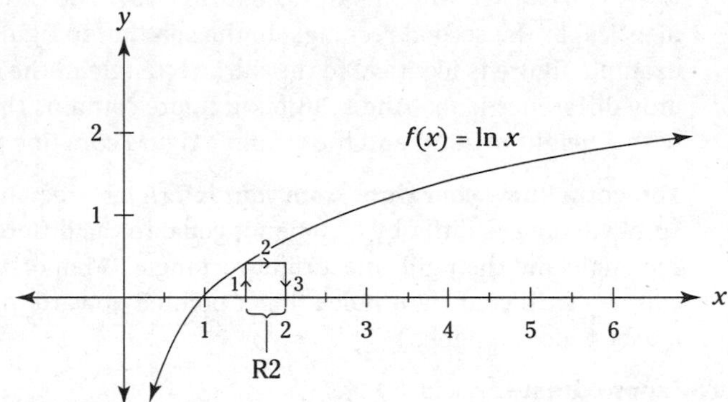

3. **Draw the rest of the rectangles.**

   See the following figure.

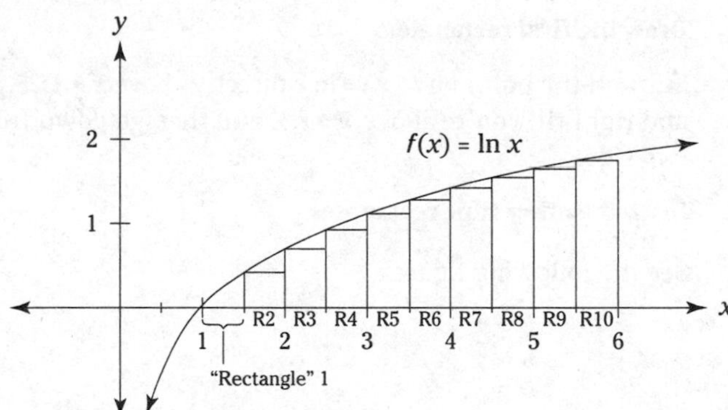

4. **Compute your approximation.**

$$\begin{aligned} Area_{10\,\text{LRs}} &= \frac{1}{2}(\ln 1 + \ln 1.5 + \ln 2 + \ln 2.5 + \ln 3 + \ln 3.5 + \ln 4 + \ln 4.5 + \ln 5 + \ln 5.5) \\ &= \frac{1}{2}(0 + 0.405 + 0.693 + 0.916 + 1.099 + 1.253 + 1.386 + 1.504 + 1.609 + 1.705) \\ &= \frac{1}{2}(10.57) = 5.285 \end{aligned}$$

(b) **The only difference is that the sum for left rectangles has a 0 at the left end and the sum for right rectangles has a 1.792 at the right end.** The other 9 numbers in both sums are the same. Look at the second line in the computation in Step 4. Note that the sum of the 10 numbers inside the parentheses includes the first 9 numbers in the computation for right rectangles, which you see in Step 4 of the answer to the first example in this chapter. The only difference in the two sums is the left-most number in the left-rectangle sum and the right-most number in the right-rectangle sum.

If you look at the figure in Step 2 of the example and at the figure in Step 3 of the solution to 1a, you can see why this works out this way. The first rectangle in the example figure is identical to the second rectangle in the solution 1a figure; the second rectangle in the example figure is identical to the third rectangle in the solution 1a figure, and so on. The only difference is that the solution 1a figure contains the left-most "rectangle" (the one with a height of zero) and the example figure contains the tall, right-most rectangle.

You gotta know your right from your left. A left-rectangle sum and a right-rectangle sum will always differ by an amount equal to the difference in area of the left-most left rectangle and the right-most right rectangle. (Memorize this sentence and recite it in class — with your right index finger pointed upward for effect. You'll instantly become a babe [dude] magnet.)

(2) **The approximate area is 5.759.**

1. **Sketch your curve and the 10 subintervals again.**
2. **Compute the midpoints of the bases of all rectangles.**

   This should be a no-brainer: 1.25, 1.75, 2.25, , 5.75.
3. **Draw the first rectangle.**

   Start on the point on $f(x) = \ln x$ directly above $x = 1.25$, then go left till you're above $x = 1$ and right till you're above $x = 1.5$, and then go down from both these points to make the two sides.
4. **Draw the other nine rectangles.**

   See the following figure.

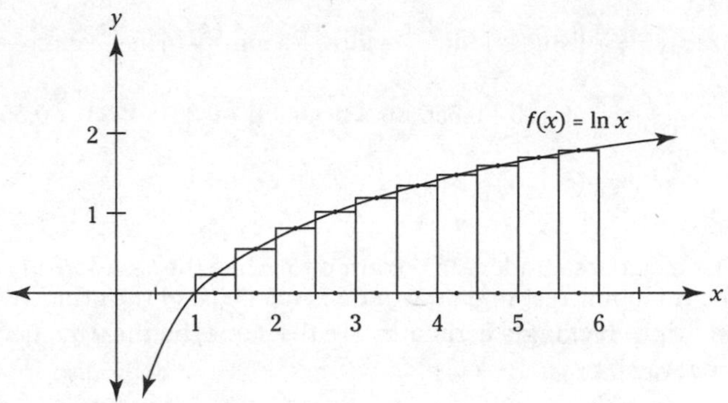

5. Compute your estimate.

$$
\begin{aligned}
Area_{10\,\text{MRs}} &= \frac{1}{2}(\ln 1.25 + \ln 1.75 + \ln 2.25 + \ln 2.75 + \ln 3.25 + \ln 3.75 + \ln 4.25 + \ln 4.75 + \ln 5.25 + \ln 5.75)\\
&\approx \frac{1}{2}(0.223 + 0.560 + 0.811 + 1.012 + 1.179 + 1.322 + 1.447 + 1.558 + 1.658 + 1.749)\\
&\approx 5.759
\end{aligned}
$$

(3) **The midpoint rectangles give the best estimate because each rectangle goes above the curve (in this sense, the estimate's too big) and also leaves an uncounted gap below the curve (in this sense, the estimate's too small).** These two errors cancel each other out to some extent. By the way, the exact area is about 5.751. The approximate area with 10 midpoint rectangles, 5.759, is only about 0.14% off.

It's harder to rank the left versus the right rectangle estimates. Kudos if you noticed that because of the shape of $f(x) = \ln x$, right rectangles give a slightly better estimate (technically, it's because $\ln x$ is concave *down* and *increasing*). It turns out that the right-rectangle approximation is off by 7.48%, and the left-rectangle estimate is off by 8.10%. If you missed this question, don't sweat it. It's basically an extra-credit type question.

(4) **The approximations are, respectively, 1.974, 1.974, and 2.013.**

Let's cut to the chase. Here are the computations for 8 left rectangles, 8 right rectangles, and 8 midpoint rectangles:

$$
\begin{aligned}
Area_{8\,\text{LR}} &= \frac{\pi}{8}\left(\sin 0 + \sin\frac{\pi}{8} + \sin\frac{2\pi}{8} + \sin\frac{3\pi}{8} + \sin\frac{4\pi}{8} + \sin\frac{5\pi}{8} + \sin\frac{6\pi}{8} + \sin\frac{7\pi}{8}\right)\\
&\approx \frac{\pi}{8}(0 + 0.383 + 0.707 + 0.924 + 1 + 0.924 + 0.707 + 0.383)\\
&= \frac{\pi}{8}(5.027) \approx 1.974
\end{aligned}
$$

$$
\begin{aligned}
Area_{8\,\text{RR}} &= \frac{\pi}{8}\left(\sin\frac{\pi}{8} + \sin\frac{2\pi}{8} + \sin\frac{3\pi}{8} + \sin\frac{4\pi}{8} + \sin\frac{5\pi}{8} + \sin\frac{6\pi}{8} + \sin\frac{7\pi}{8} + \sin\pi\right)\\
&\approx \frac{\pi}{8}(0.383 + 0.707 + 0.924 + 1 + 0.924 + 0.707 + 0.383 + 0)\\
&= \frac{\pi}{8}(5.027) \approx 1.974
\end{aligned}
$$

$$Area_{8\,MR} = \frac{\pi}{8}\left(\sin\frac{\pi}{16} + \sin\frac{3\pi}{16} + \sin\frac{5\pi}{16} + \sin\frac{7\pi}{16} + \sin\frac{9\pi}{16} + \sin\frac{11\pi}{16} + \sin\frac{13\pi}{16} + \sin\frac{15\pi}{16}\right)$$
$$\approx \frac{\pi}{8}(0.195 + 0.556 + 0.831 + 0.981 + 0.981 + 0.831 + 0.556 + 0.195)$$
$$= \frac{\pi}{8}(5.126) \approx 2.013$$

The exact area under $\sin x$ from 0 to π has the wonderfully simple answer of 2. The error of the midpoint rectangle estimate is 0.65%, and the other two have an error of 1.3%. The left and right rectangle estimates are the same, by the way, because of the symmetry of the sine wave.

5 **40**

As often happens with many types of problems in mathematics, this very simple version of a sigma sum problem is tricky. Here, there's no place to plug in the *i* values, so all the *i* does is work as a counter:

$$\sum_{i=1}^{10} 4 = 4+4+4+4+4+4+4+4+4+4 = 10 \cdot 4 = 40$$

6 **−55**

$$= (-1)^0(0+1)^2 + (-1)^1(1+1)^2 + (-1)^2(2+1)^2 + \ldots$$
$$= 1^2 - 2^2 + 3^2 - 4^2 + 5^2 - 6^2 + 7^2 - 8^2 + 9^2 - 10^2$$
$$= -55$$

7 **131,325**

$$= \sum_{i=1}^{50} 3i^2 + \sum_{i=1}^{50} 2i = 3\sum_{i=1}^{50} i^2 + 2\sum_{i=1}^{50} i$$
$$= 3\left(\frac{50(50+1)(2 \cdot 50+1)}{6}\right) + 2\left(\frac{50(50+1)}{2}\right) = 131{,}325$$

8 $\displaystyle\sum_{k=6}^{12} 5k$ **or** $\displaystyle\sum_{k=1}^{7} 5(k+5)$ **or** $\displaystyle\sum_{k=1}^{7}(5k+25)$

9 $\displaystyle\sum_{k=2}^{6} k^3$ **or** $\displaystyle\sum_{k=1}^{5}(k+1)^3$

Did you recognize this pattern of consecutive cubes?

10 $\displaystyle\sum_{i=1}^{10}(-1)^i 2^i$ **or** $\displaystyle\sum_{i=1}^{10}(-2)^i$

TIP

To make the terms in a sigma sum alternate between positive and negative, use a (−1) raised to a power (as you can see in the answer to problem 10). The power is usually i or $i+1$.

(11) **The notation and approximation are $\frac{1}{64}\sum_{i=1}^{20} i^2 + \frac{3}{16}\sum_{i=1}^{20} i \approx 84.2$.**

1. **Begin with the formula for a right sum of n rectangles using sigma notation:**

$$R_n = \sum_{i=1}^{n}\left[g(x_i)\cdot\frac{b-a}{n}\right]$$

2. **Determine the width of each rectangle, $\frac{b-a}{n}$:**

$$\frac{b-a}{n} = \frac{5-0}{20} = \frac{1}{4}$$

3. **Determine an expression for x_i:**

The width of each rectangle is $\frac{1}{4}$, so the right edges of the 20 rectangles fall on the first 20 multiples of $\frac{1}{4}$. These numbers are the x-coordinates of the 20 points x_1 through x_{20}. They can be generated by the expression $\frac{1}{4}i$, where i equals 1 through 20.

4. **Switch out the n, the x_i, and the $\frac{b-a}{n}$ in your formula, then pull the $\frac{1}{4}$ out to the left:**

$$\begin{aligned} R_{20} &= \sum_{i=1}^{20}\left[g\left(\frac{1}{4}i\right)\cdot\frac{1}{4}\right] \\ &= \frac{1}{4}\sum_{i=1}^{20} g\left(\frac{1}{4}i\right) \end{aligned}$$

5. **Determine $g\left(\frac{1}{4}i\right)$, then rewrite your formula:**

$$g(x) = x^2 + 3x$$

$$g\left(\frac{1}{4}i\right) = \left(\frac{1}{4}i\right)^2 + 3\left(\frac{1}{4}i\right)$$

$$R_{20} = \frac{1}{4}\sum_{i=1}^{20}\left[\left(\frac{1}{4}i\right)^2 + 3\left(\frac{1}{4}i\right)\right]$$

6. **Simplify, and pull everything to the outside except functions of i:**

$$\begin{aligned} &= \frac{1}{4}\sum_{i=1}^{20}\left(\frac{1}{4}i\right)^2 + \frac{1}{4}\sum_{i=1}^{20} 3\left(\frac{1}{4}i\right) \\ &= \frac{1}{4}\sum_{i=1}^{20}\frac{1}{16}i^2 + \frac{1}{4}\sum_{i=1}^{20}\frac{3}{4}i \\ &= \frac{1}{64}\sum_{i=1}^{20} i^2 + \frac{3}{16}\sum_{i=1}^{20} i \end{aligned}$$

7. **Compute the area, using the following rules for summing consecutive integers and consecutive squares of integers.**

The sum of the first n integers equals $\frac{n(n+1)}{2}$, and the sum of the squares of the first n integers equals $\frac{n(n+1)(2n+1)}{6}$.

So, finally, you've got the area:

$$\frac{1}{64}\sum_{i=1}^{20} i^2 + \frac{3}{16}\sum_{i=1}^{20} i$$
$$= \frac{1}{64}\left(\frac{20(20+1)(2\cdot 20+1)}{6}\right) + \frac{3}{16}\left(\frac{20(20+1)}{2}\right)$$
$$= \frac{1}{64}\left(\frac{20\cdot 21\cdot 41}{6}\right) + \frac{3}{16}\cdot 10\cdot 21$$
$$= \frac{17{,}220}{384} + \frac{630}{16} \approx 84.2$$

(12) **The formula is $R_n = \frac{475}{6} + \frac{100}{n} + \frac{125}{6n^2}$.**

1. **Express the sum of n rectangles instead of 20 rectangles.**

   Look back at Step 5 from Problem 11. The $\frac{1}{4}$ outside and the two $\frac{1}{4}$s inside come from the width of the rectangles that you got by dividing 5 (the span) by 20. So, the width of each rectangle could have been written as $\frac{5}{20}$. To add n rectangles instead of 20, just replace the 20 with an n — that's $\frac{5}{n}$. So, the three $\frac{1}{4}$s become $\frac{5}{n}$. At the same time, replace the 20 on top of the $\sum$ with an n: $R_n = \frac{5}{n}\sum_{i=1}^{n}\left[\left(\frac{5}{n}i\right)^2 + 3\left(\frac{5}{n}i\right)\right]$

2. **Simplify as in Step 6 of Problem 11.**

$$= \frac{5}{n}\sum_{i=1}^{n}\left(\frac{5}{n}i\right)^2 + \frac{5}{n}\sum_{i=1}^{n} 3\left(\frac{5}{n}i\right)$$
$$= \frac{5}{n}\sum_{i=1}^{n}\frac{25}{n^2}i^2 + \frac{5}{n}\sum_{i=1}^{n}\frac{15}{n}i$$
$$= \frac{125}{n^3}\sum_{i=1}^{n} i^2 + \frac{75}{n^2}\sum_{i=1}^{n} i$$

3. **Now replace the sigma sums with the expressions for the sums of integers and squares of integers, like you did in Step 7 of Problem 11.**

$$= \frac{125}{n^3}\left(\frac{n(n+1)(2n+1)}{6}\right) + \frac{75}{n^2}\left(\frac{n(n+1)}{2}\right)$$
$$= \frac{250n^2 + 375n + 125}{6n^2} + \frac{75n^2 + 75n}{2n^2}$$
$$= \frac{475n^2 + 600n + 125}{6n^2}$$
$$= \frac{475}{6} + \frac{100}{n} + \frac{125}{6n^2}$$

That's the formula for approximating the area under $g(x) = x^2 + 3x$ from 0 to 5 with n rectangles — the more you use, the better your estimate.

Check this result by plugging 20 into *n* to see whether you get the same answer as you did in Problem 11:

$$=\frac{475}{6}+\frac{100}{20}+\frac{125}{6\cdot 20^2}\approx 84.2$$

It checks.

13 **The approximations are, respectively, 81.175, 80.169, 79.267, 79.177.**

$$Area_{n\text{ rect}}=\frac{475}{6}+\frac{100}{n}+\frac{125}{6n^2}$$

$$Area_{50\text{ rect}}=\frac{475}{6}+\frac{100}{50}+\frac{125}{6\cdot 50^2}=81.175$$

$$Area_{100\text{ rect}}\approx 80.169$$

$$Area_{1000\text{ rect}}\approx 79.267$$

$$Area_{10,000\text{ rect}}\approx 79.177$$

These estimates are getting better and better; they appear to be headed toward something near 79. Now for the magic of calculus — actually (sort of) adding up an infinite number of rectangles.

14 **The area is $79.1\overline{6}$ or $79\frac{1}{6}$.**

$$\int_a^b g(x)dx=\lim_{n\to\infty}\sum_{i=1}^{n}\left[g(x_i)\cdot\frac{b-a}{n}\right]$$

$$\int_0^5\left(x^2+3x\right)dx=\lim_{n\to\infty}\left(\frac{475}{6}+\frac{100}{n}+\frac{125}{6n^2}\right)$$

$$=\frac{475}{6}+0+0$$

$$=\frac{475}{6}=79.1\overline{6}\text{ or }79\frac{1}{6}$$

15 **The approximate area is 1.974 and the error is 1.3%.**

1. **List the values for *a*, *b*, and *n*, and determine the *x*-values x_0 through x_8.**

   $a=0,\ b=\pi,\ n=8$

   $$x_0=0,\ x_1=\frac{\pi}{8},\ x_2=\frac{2\pi}{8},\ x_3=\frac{3\pi}{8},\ x_4=\frac{4\pi}{8},\ \ldots,\ x_8=\frac{8\pi}{8}=\pi$$

2. **Plug these values into the formula.**

   $$T_8=\frac{\pi-0}{2\cdot 8}\left(\sin 0+2\sin\frac{\pi}{8}+2\sin\frac{2\pi}{8}+2\sin\frac{3\pi}{8}+\ldots+2\sin\frac{7\pi}{8}+\sin\pi\right)$$

   $$\approx\frac{\pi}{16}(0+0.765+1.414+1.848+\ldots+0.765+0)\approx 1.974$$

   The exact area of 2 is given in Problem 4, and thus the relative error is $\frac{(2-1.974)}{2}$, which gives you a percent error of 1.3%.

16 **The approximate area for 16 trapezoids is 1.994 and the percent error is about 0.3%. The approximate area for 24 trapezoids is 1.997 and the percent error is about 0.15%.**

$$T_{16} = \frac{\pi - 0}{2 \cdot 16}\left(\sin 0 + 2\sin\frac{\pi}{16} + 2\sin\frac{2\pi}{16} + 2\sin\frac{3\pi}{16} + \ldots + 2\sin\frac{15\pi}{16} + \sin\pi\right)$$

$$\approx \frac{\pi}{32}(0 + 0.390 + 0.765 + \ldots + 0.765 + 0) \approx 1.994$$

$$T_{24} = \frac{\pi - 0}{2 \cdot 24}\left(\sin 0 + 2\sin\frac{\pi}{24} + 2\sin\frac{2\pi}{24} + 2\sin\frac{3\pi}{24} + \ldots + 2\sin\frac{23\pi}{24} + \sin\pi\right)$$

$$\approx \frac{\pi}{48}(0 + 0.261 + 0.518 + \ldots + 0) \approx 1.997$$

17 **The area for eight Simpson's "trapezoids" is 2.00001659 (an error of 0.00001659). The percent error for eight "trapezoids" is about 0.000830%.**

For eight Simpson's "trapezoids":

1. **List the values for *a*, *b*, and *n*, and determine the *x* values x_0 through x_{16}, the nine edges and the eight base midpoints of the eight curvy-topped "trapezoids."**

   $a = 0,\ b = \pi,\ n = 16$

   $x_0 = 0,\ x_1 = \frac{\pi}{16},\ x_2 = \frac{2\pi}{16},\ x_3 = \frac{3\pi}{16},\ x_4 = \frac{4\pi}{16}, \ldots,\ x_{16} = \frac{16\pi}{16} = \pi$

2. **Plug these values into the formula.**

$$S_{16} = \frac{\pi - 0}{3 \cdot 16}\left(\sin 0 + 4\sin\frac{\pi}{16} + 2\sin\frac{2\pi}{16} + 4\sin\frac{3\pi}{16} + 2\sin\frac{4\pi}{16} + \ldots + 4\sin\frac{15\pi}{16} + 2\sin\pi\right)$$

$$\approx \frac{\pi}{48}(0 + 0.7804 + 0.7654 + 2.2223 + 1.4142 + \ldots + 0.7804 + 0) \approx 2.00001659$$

18 $\mathbf{S_{20} \approx 5.750}$

Using the shortcut and the results from Problem 2 and the example problem, you get:

$$S_{2n} = \frac{M_n + M_n + T_n}{3}$$

$$S_{20} = \frac{M_{10} + M_{10} + T_{10}}{3}$$

$$\approx \frac{5.759 + 5.759 + 5.733}{3} \approx 5.750$$

This agrees (except for a small round-off error) with the result obtained the hard way in the Simpson's rule example problem.

If you're ready to test your skills a bit more, take the following chapter quiz that incorporates all the chapter topics.

Whaddya Know? Chapter 14 Quiz

Quiz time! Complete each problem to test your knowledge on the various topics covered in this chapter. You can then find the solutions and explanations in the next section.

1. Write the following sum using sigma notation: $1+6+11+16+21+\ldots+101$

2. Write the following sum using sigma notation: $5-6+7-8+9-10+\ldots+99-100$

3. Rewrite the sum $\sum_{i=1}^{10}(i^3+5i^2)$ where i begins at zero instead of 1. Simplify your result so that the argument of the sigma sum is a polynomial.

4. For the following, replace the ? with the correct number and supply the proper expression inside the parentheses: $\sum_{i=1}^{20} i^4 - \sum_{i=1}^{12} i^4 = \sum_{i=1}^{?}(\quad)^4$

For Problems 5 through 9, use the function $f(x)=x^4$.

5. Use the left rectangle rule formula to approximate the area under f from zero to 4 with 8 left rectangles.

6. Use the right rectangle rule formula to approximate the area under f from zero to 4 with 8 right rectangles.

7. Use the midpoint rectangle rule formula to approximate the area under f from zero to 4 with 8 midpoint rectangles.

8. Use the trapezoid rule formula to approximate the area under f from zero to 4 with 8 trapezoids. (Actually, if you know the shortcut, you can compute this approximation without using the trapezoid formula.)

9. Use the Simpson's rule formula to approximate the area under f from zero to 4 with 8 curvy-topped "trapezoids." (If you know the shortcut, you can do this computation without the formula.)

10. Use the definition of the definite integral (using a limit and sigma notation) to write the exact area under f from zero to 4. Simplify the sigma sum by pulling out as much as permitted to the outside of the sigma sum. Do not evaluate your final answer.

Answers to Chapter 14 Quiz

1. $\sum_{i=0}^{20}(5i+1)$

2. $\sum_{i=5}^{100}\left(i(-1)^{(i+1)}\right)$

3. $\sum_{i=0}^{9}\left((i+1)^3+5(i+1)^2\right)=\sum_{i=0}^{9}\left(i^3+8i^2+13i+6\right)$

4. $\sum_{i=1}^{20}i^4-\sum_{i=1}^{12}i^4=\sum_{i=13}^{20}i^4=\sum_{i=1}^{8}(i+12)^4$

5. **146.125**

$$L_8=\frac{4}{8}\left[\left(\frac{0}{2}\right)^4+\left(\frac{1}{2}\right)^4+\left(\frac{2}{2}\right)^4+\left(\frac{3}{2}\right)^4+\left(\frac{4}{2}\right)^4+\left(\frac{5}{2}\right)^4+\left(\frac{6}{2}\right)^4+\left(\frac{7}{2}\right)^4\right]$$
$$=0.5\left(0^4+0.5^4+1^4+1.5^4+2^4+2.5^4+3^4+3.5^4\right)$$
$$=0.5\cdot 292.25=146.125$$

6. **274.125**

$$R_8=\frac{4}{8}\left[\left(\frac{1}{2}\right)^4+\left(\frac{2}{2}\right)^4+\left(\frac{3}{2}\right)^4+\left(\frac{4}{2}\right)^4+\left(\frac{5}{2}\right)^4+\left(\frac{6}{2}\right)^4+\left(\frac{7}{2}\right)^4+\left(\frac{8}{2}\right)^4\right]$$
$$=0.5\left(0.5^4+1^4+1.5^4+2^4+2.5^4+3^4+3.5^4+4^4\right)$$
$$=0.5\cdot 548.25=274.125$$

7. **~202.141**

$$M_8=\frac{4}{8}\left[\left(\frac{1}{4}\right)^4+\left(\frac{3}{4}\right)^4+\left(\frac{5}{4}\right)^4+\left(\frac{7}{4}\right)^4+\left(\frac{9}{4}\right)^4+\left(\frac{11}{4}\right)^4+\left(\frac{13}{4}\right)^4+\left(\frac{15}{4}\right)^4\right]$$
$$=0.5\left(0.25^4+0.75^4+1.25^4+1.75^4+2.25^4+2.75^4+3.25^4+3.75^4\right)$$
$$=0.5\cdot 404.28125=202.140625$$

8. **210.125**

Use the shortcut! (Nothing wrong with using the trapezoid formula, however. It's good practice.)

$$T_8=\frac{L_8+R_8}{2}=\frac{146.125+274.125}{2}=210.125$$

9. **~204.802**

Use the shortcut.

$$S_{16}=\frac{M_8+M_8+T_8}{3}=\frac{2\cdot 202.140625+210.125}{3}\approx 204.802$$

This Simpson's rule answer is *extremely* close, by the way. The exact area under *f* from zero to 4 is 204.8. You learn how to compute that in Chapter 15.

10 $\lim_{n\to\infty}\left(\frac{1024}{n^5}\sum_{i=1}^{n} i^4\right)$

$$\int_a^b f(x)dx = \lim_{n\to\infty}\sum_{i=1}^{n}\left[f(x_i)\cdot\frac{b-a}{n}\right]$$

$$\int_0^4 x^4 dx = \lim_{n\to\infty}\sum_{i=1}^{n}\left[\left(\frac{4i}{n}\right)^4\cdot\frac{4}{n}\right]$$

$$= \lim_{n\to\infty}\sum_{i=1}^{n}\left[\frac{256i^4}{n^4}\cdot\frac{4}{n}\right]$$

$$= \lim_{n\to\infty}\left(\frac{1024}{n^5}\sum_{i=1}^{n} i^4\right)$$

IN THIS CHAPTER

» Using the area function

» Getting familiar with the Fundamental Theorem of Calculus

» Finding antiderivatives

» Figuring exact areas the easy way

Chapter 15 Integration: It's Backwards Differentiation

Chapter 14 shows you the hard way to calculate the area under a function using the formal definition of integration — the limit of a Riemann sum. In this chapter, I calculate areas the easy way, taking advantage of one of the most important and amazing discoveries in mathematics — that integration (finding areas) is just differentiation in reverse. That reverse process was a great discovery, and it's based on some difficult ideas, but before we get to that, let's talk about a related, straightforward reverse process, namely . . .

Antidifferentiation

The derivative of $\sin x$ is $\cos x$, so the antiderivative of $\cos x$ is $\sin x$; the derivative of x^3 is $3x^2$, so the antiderivative of $3x^2$ is x^3 — you just go backwards. There's a bit more to it, but that's the basic idea. Later in this chapter, I show you how to find areas by using antiderivatives. This is *much* easier than finding areas with the Riemann sum technique.

Now consider x^3 and its derivative $3x^2$ again. The derivative of x^3+10 is also $3x^2$, as is the derivative of x^3-5. Any function of the form x^3+C, where C is any number, has a derivative of $3x^2$. So, every such function is an antiderivative of $3x^2$.

MATH RULES

Definition of the *indefinite integral*: The indefinite integral of a function $f(x)$, written as $\int f(x)dx$, is the family of *all* antiderivatives of the function. For example, because the derivative of x^3 is $3x^2$, the indefinite integral of $3x^2$ is x^3+C, and you write

$$\int 3x^2dx = x^3+C$$

You probably recognize this integration symbol, $\int$, from the discussion of the *definite* integral in Chapter 14. The definite integral symbol, however, contains two little numbers like $\int_4^{10}$ that tell you to compute the area under a function between those two numbers, called the *limits of integration*. The naked version of the symbol, $\int$, indicates an *indefinite* integral or an *antiderivative*. This chapter is all about the intimate connection between these two symbols, these two ideas.

Figure 15-1 shows the family of antiderivatives of $3x^2$, namely x^3+C. Note that this family of curves has an infinite number of curves. They go up and down forever and are infinitely dense. The vertical gap of 2 units between each curve in Figure 15-1 is just a visual aid.

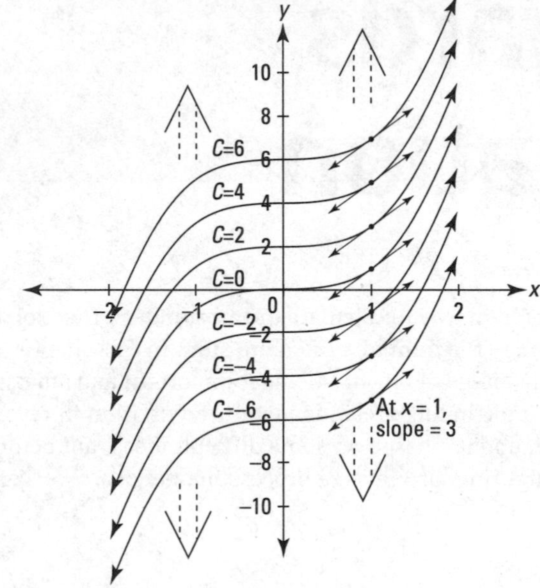

FIGURE 15-1: The family of curves x^3+C. All these functions have the same derivative, $3x^2$.

Consider a few things about Figure 15-1. The top curve on the graph is $y=x^3+6$; the one below it is $y=x^3+4$; the bottom one is $y=x^3-6$. By the power rule, these three functions, as well as all the others in this family of functions, have a derivative of $3x^2$. Now, consider the slope of each of the curves where x equals 1 (see the tangent lines drawn on the curves). The derivative of each function is $3x^2$, so when x equals 1, the slope of each curve is $3\cdot 1^2$, or 3. Thus, all these little tangent lines are parallel. Next, notice that all the functions in Figure 15-1 are identical except for being slid up or down (remember vertical shifts from Chapter 5?). Because they differ only by a vertical shift, the steepness at any x-value, like at $x=1$, is the same for all the curves.

This is the visual way to understand why each of these curves has the same derivative, and, thus, why each curve is an antiderivative of the same function.

Vocabulary, Voshmabulary: What Difference Does It Make?

In general, definitions and vocabulary are very important in mathematics, and it's a good idea to use them correctly. But with the current topic, I'm going to be a bit lazy about precise terminology, and I hereby give you permission to do so as well.

If you're a stickler, you should say that *the* indefinite integral of $3x^2$ is $x^3 + C$ and that $x^3 + C$ is the family or set of *all* antiderivatives of $3x^2$ (you don't say that $x^3 + C$ is *the* antiderivative), and you say that $x^3 + 10$, for instance, is *an* antiderivative of $3x^2$. And on a test, you should definitely write $\int 3x^2 dx = x^3 + C$. If you leave the *C* off, you'll likely lose some points.

But, when discussing these matters, no one will care or be confused if you get tired of saying "+ *C*" after every indefinite integral and just say, for example, that the indefinite integral of $3x^2$ is x^3, and you can skip the *indefinite* and just say that the *integral* of $3x^2$ is x^3. And instead of always talking about that family-of-functions business, you can just say that *the* antiderivative of $3x^2$ is $x^3 + C$ or that the antiderivative of $3x^2$ is x^3. Everyone will know what you mean. It may cost me my membership in the National Council of Teachers of Mathematics, but at least occasionally, I use this loose approach.

The Annoying Area Function

This is a tough one — gird your loins. Say you've got any old function, $f(t)$. Imagine that at some *t*-value, call it *s*, you draw a fixed vertical line. See Figure 15-2.

Then you take a moveable vertical line, starting at the same point, *s* ("*s*" is for *starting* point), and drag it to the right. As you drag the line, you sweep out a larger and larger area under the curve. This area is a function of *x*, the position of the moving line. In symbols, you write

$$A_f(x) = \int_s^x f(t)\,dt$$

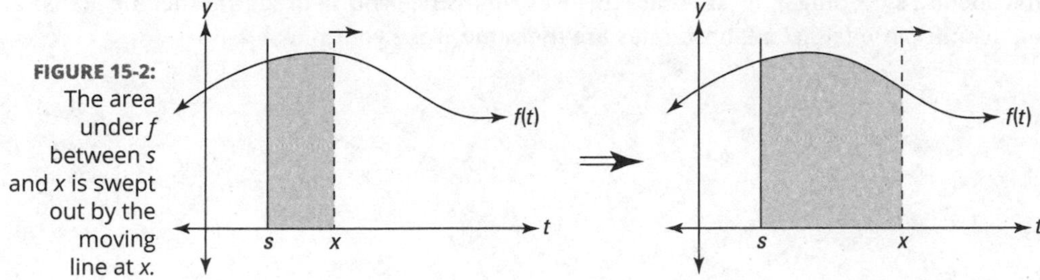

FIGURE 15-2: The area under *f* between *s* and *x* is swept out by the moving line at *x*.

Note that t is the input variable in $f(t)$ instead of x because x is already taken — it's the input variable in $A_f(x)$. The subscript f in A_f indicates that $A_f(x)$ is the area function for the particular curve f or $f(t)$. The dt is a little increment along the t-axis — actually an infinitesimally small increment.

Here's a simple example to make sure you've got a handle on how an area function works. By the way, don't feel bad if you find this extremely hard to grasp — you've got lots of company. Say you've got the simple function, $f(t)=10$, that's a horizontal line at $y=10$. If you sweep out area beginning at $s=3$, you get the following area function:

$$A_f(x)=\int_3^x 10\,dt$$

You can see that the area swept out from 3 to 4 is 10 because, in dragging the line from 3 to 4, you sweep out a rectangle with a width of 1 and a height of 10, which has an area of 1 times 10, or 10. See Figure 15-3.

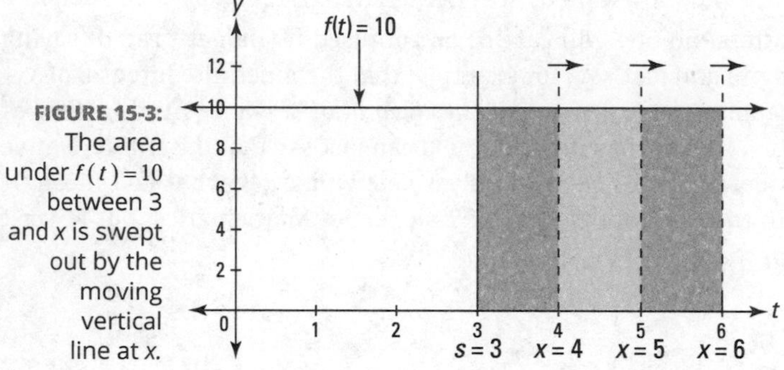

FIGURE 15-3: The area under $f(t)=10$ between 3 and x is swept out by the moving vertical line at x.

So, $A_f(4)$, the area swept out as you hit 4, equals 10. $A_f(5)$ equals 20 because when you drag the line to 5, you've swept out a rectangle with a width of 2 and a height of 10, which has an area of 2 times 10, or 20. $A_f(6)$ equals 30, and so on.

Now, imagine that you drag the line across at a rate of one unit per second. You start at $x=3$, and you hit 4 at 1 second, 5 at 2 seconds, 6 at 3 seconds, and so on. How much area are you sweeping out per second? Ten square units per second, because each second, you sweep out another 1-by-10 rectangle. Notice — this is huge — that because the width of each rectangle you sweep out is 1, the area of each rectangle, which is given by *height* times *width*, is the same as its height because anything times 1 equals itself. You see why this is huge in a minute. (By the way, the real rate you care about here is not area swept out per second, but, rather, area swept out per unit change on the x-axis. I explain it in terms of per second because it's easier to think about a sweeping-out-area rate this way. And since you're dragging the line across at *one* x-axis unit per *one* second, both rates are the same. Take your pick.)

The derivative of an area function equals the rate of area being swept out. Okay, are you sitting down? You've reached another one of the big *Ah ha!* moments in the history of mathematics. Recall that *a derivative is a rate*. So, because the rate at which the previous area function grows is 10 square units per second, you can say its derivative equals 10. Thus, you can write

$$\frac{d}{dx}A_f(x)=10$$

Again, this just tells you that with each 1-unit increase in x, A_f (the area function) goes up 10. Now here's the critical thing: Notice that this rate or derivative of 10 is the same as the height of the original function $f(t)=10$, because as you go across 1 unit, you sweep out a rectangle that's 1 by 10, which has an area of 10, the height of the function.

And the rate works out to 10 regardless of the width of the rectangle. Imagine that you drag the vertical line from $x=4$ to $x=4.001$. At a rate of one unit per second, that'll take you 1/1000th of a second, and you'll sweep out a skinny rectangle with a width of 1/1000, a height of 10, and thus an area of 10 times 1/1000, or 1/100 square units. The rate of area being swept out would therefore be $\frac{1/100 \text{ square units}}{1/1000 \text{ second}}$, which equals 10 square units per second. So, you see that with every small increment along the x-axis, the rate of area being swept out equals the function's height.

This works for any function, not just horizontal lines. Look at Figure 15-4 which shows the function $g(t)$ and its area function $A_g(x)$ that sweeps out area beginning at $s=2$.

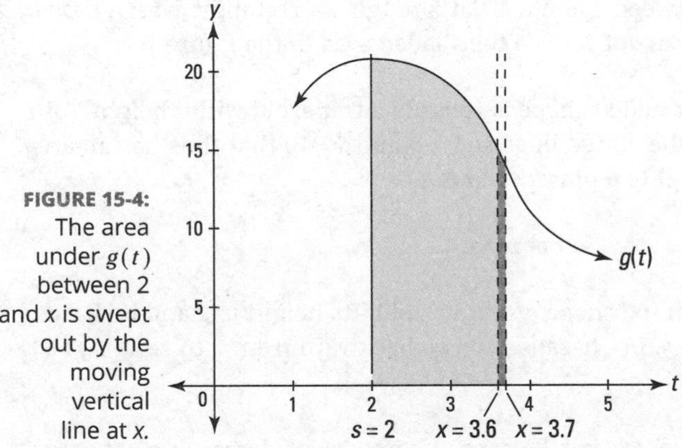

FIGURE 15-4: The area under $g(t)$ between 2 and x is swept out by the moving vertical line at x.

Between $x=3.6$ and $x=3.7$, $A_g(x)$ grows by the area of that skinny, dark-shaded "rectangle" with a width of 0.1 and a height of about 15. (As you can see, it's not really a rectangle; it's closer to a trapezoid, but it's not that either, because its tiny top is curving slightly. But, in the limit, as the width gets smaller and smaller, the skinny "rectangle" behaves precisely like a real rectangle.) So, to repeat, $A_g(x)$ grows by the area of that dark "rectangle," which has an area extremely close to 0.1 times 15, or 1.5. That area is swept out in 0.1 second, so the rate of area being swept out is $\frac{1.5 \text{ square units}}{0.1 \text{ second}}$, or 15 square units per second, the height of the function. This idea is so important that it deserves an icon.

MATH RULES

The sweeping-out area rate equals the height. The *rate* of area being swept out under a curve by an area function at a given x-value is equal to the *height* of the curve at that x-value.

EXAMPLE

Q. Consider $f(t)$, shown in the following figure. Given the area function $A_f(x)=\int_2^x f(t)dt$, approximate $A_f(4)$, $A_f(5)$, $A_f(2)$, and $A_f(0)$. Also, is A_f increasing or decreasing between $x=5$ and $x=6$? Between $x=8$ and $x=9$?

Remember: When using an area function (or a definite integral — stay tuned), **area below the horizontal axis counts as *negative* area.**

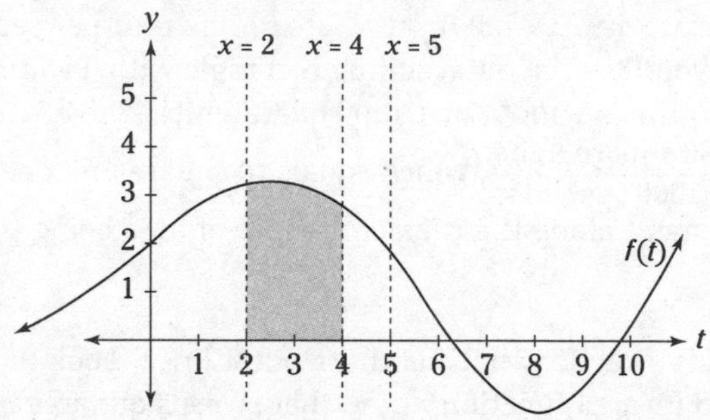

A. $A_f(4)$ is the area under $f(t)$ between 2 and 4. That's roughly a rectangle with a base of 2 and a height of 3, so **the area is about 6.** (See the shaded area in the figure.)

$A_f(5)$ adds a bit to $A_f(4)$ — the added shape is roughly a trapezoid with "height" of 1 and "bases" of 2 and 3 (along the dotted lines at $x=4$ and $x=5$) that thus has an area of about 2.5 — **so $A_f(5)$ is roughly 6 plus 2.5, or 8.5.**

$A_f(2)$ is the area between 2 and 2, which is zero.

$A_f(0)$ is another area roughly in the shape of a trapezoid. Its height is 2 and its bases are 2 and 3, so its area is about 5. But because you go backward from 2 to zero, **$A_f(0)$ equals about −5.**

Between $x=5$ and $x=6$, A_f is *increasing*. Be careful here: $f(t)$ is decreasing between 5 and 6, but as you go from 5 to 6, A_f sweeps out more and more area so it's increasing.

Between $x=8$ and $x=9$, while $f(t)$ is increasing, A_f is decreasing. Area below the t-axis counts as negative area, so in moving from 8 to 9, A_f sweeps out more and more negative area, and it thus grows more and more negative. A_f is therefore decreasing.

YOUR TURN

For Problems 1 through 4, use the area function $A_g(x)=\int_{1/2}^x g(t)dt$ and the following figure. Most answers will be approximations.

1 Where (from $x = 0$ to $x = 8$) does A_g equal zero?

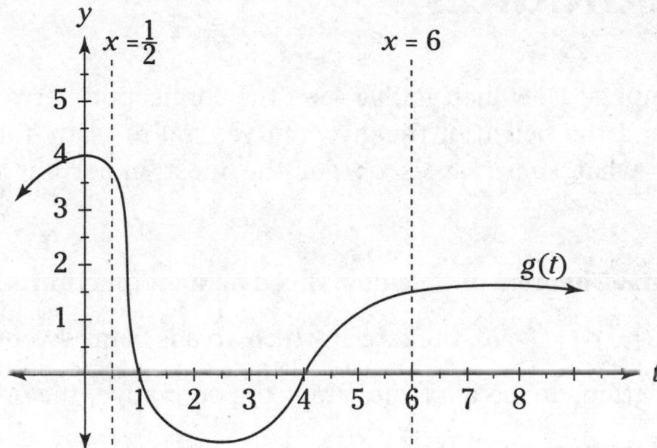

2 Where (from $x = 0$ to $x = 8$) does A_g reach

(a) its maximum value?

(b) its minimum value?

3 In what intervals between 0 and 8 is A_g

(a) increasing?

(b) decreasing?

4 Approximate $A_g(1)$, $A_g(3)$, and $A_g(5)$.

The Power and the Glory of the Fundamental Theorem of Calculus

Sound the trumpets! Now that you've seen the connection between the rate of growth of an area function and the height of the given curve, you're ready for the Fundamental Theorem of Calculus — what some say is one of the most important theorems in the history of mathematics.

MATH RULES

The Fundamental Theorem of Calculus: Given an area function A_f that sweeps out area under $f(t)$, namely, $A_f(x)=\int_s^x f(t)dt$, the rate at which area is being swept out is equal to the height of the original function. So, because the rate is the derivative, the derivative of the area function equals the original function:

$$\frac{d}{dx}A_f(x)=f(x)$$

Because $A_f(x)=\int_s^x f(t)dt$, you can also write this equation as follows:

$$\frac{d}{dx}\int_s^x f(t)dt=f(x)$$

Break out the smelling salts.

Now, because the derivative of $A_f(x)$ is $f(x)$, $A_f(x)$ is by definition an *antiderivative* of $f(x)$. Check out how this works by returning to the simple function from the previous section, $f(t)=10$, and its area function, $A_f(x)=\int_s^x 10\,dt$.

According to the Fundamental Theorem of Calculus, $\frac{d}{dx}A_f(x)=10$. Thus, A_f must be an antiderivative of 10; in other words, A_f is a function whose derivative is 10. Because any function of the form $10x+C$, where C is a number, has a derivative of 10, the antiderivative of 10 is $10x+C$. The particular number C depends on your choice of s, the point where you start sweeping out area. For a particular choice of s, the area function will be the one function (out of all the functions in the family of curves $10x+C$) that crosses the x-axis at s. To figure out C, set the antiderivative equal to zero, plug the value of s into x, and solve for C.

For this function with an antiderivative of $10x+C$, if you start sweeping out area at, say, $s=0$, then $10\cdot 0+C=0$, so $C=0$, and thus, $A_f(x)=\int_0^x 10\,dt=10x+0$, or just $10x$. (Note that C does not necessarily equal s. In fact, it usually doesn't [especially when $s\neq 0$]. When $s=0$, C often also equals zero, but not for all functions.)

Figure 15-5 shows why $A_f(x)=10x$ is the correct area function if you start sweeping out area at zero. In the top graph in the figure, the area under the curve from 0 to 3 is 30, and that's given by $A_f(3)=10\cdot 3=30$. And you can see that the area from 0 to 5 is 50, which agrees with the fact that $A_f(5)=10\cdot 5=50$.

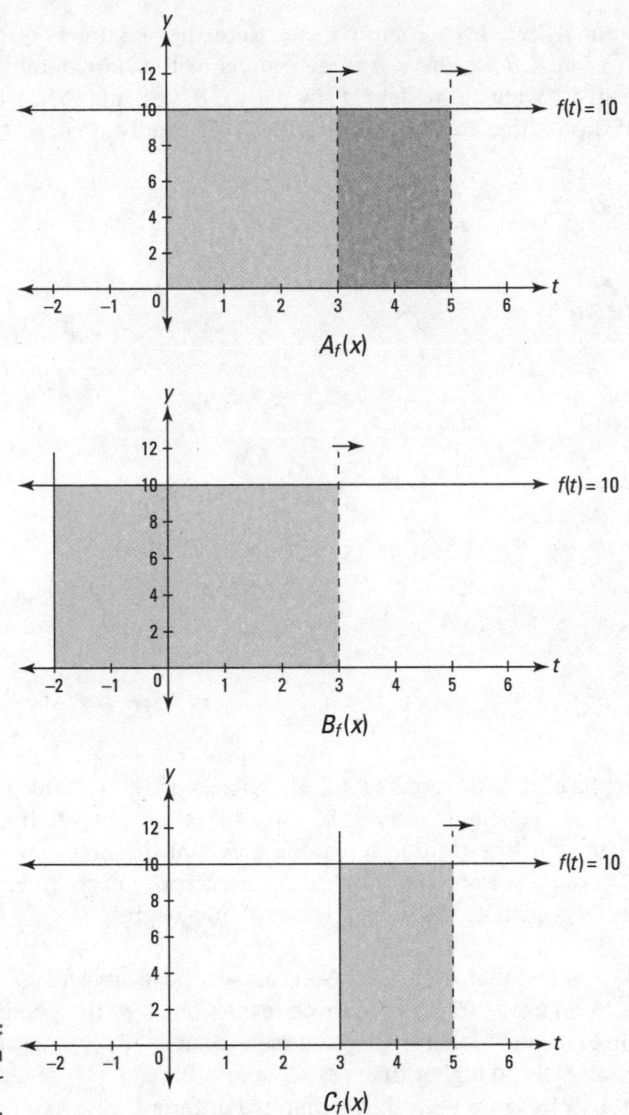

FIGURE 15-5: Three area functions for $f(t)=10$.

If instead you start sweeping out area at $s=-2$ and define a new area function, $B_f(x)=\int_{-2}^{x} 10\,dt$, then $10\cdot(-2)+C=0$, so C equals 20 and $B_f(x)$ is thus $10x+20$. This area function is 20 more than $A_f(x)$, which starts at $s=0$, because if you start at $s=-2$, you've already swept out an area of 20 by the time you get to zero. Figure 15-5 shows why $B_f(3)$ is 20 more than $A_f(3)$.

And if you start sweeping out area at $s=3$, $10\cdot 3+C=0$, so $C=-30$ and the area function is $C_f(x)=\int_{3}^{x} 10\,dt=10x-30$. This function is 30 *less* than $A_f(x)$ because with $C_f(x)$, you lose the 3-by-10 rectangle between 0 and 3 that $A_f(x)$ has (see the bottom graph in Figure 15-5).

MATH RULES

An area function is an antiderivative. The area swept out under the horizontal line $f(t)=10$, from some number s to x, is given by an antiderivative of 10, namely $10x+C$, where the value of C depends on where you start sweeping out area.

Now let's look at graphs of $A_f(x)$, $B_f(x)$, and $C_f(x)$. (Note that Figure 15-5 doesn't show the graphs of $A_f(x)$, $B_f(x)$, or $C_f(x)$. You see three graphs of the horizontal line function, $f(t)=10$; and you see the areas swept out under $f(t)$ by $A_f(x)$, $B_f(x)$, and $C_f(x)$, but you don't actually see the graphs of these three area functions.) Check out Figure 15-6.

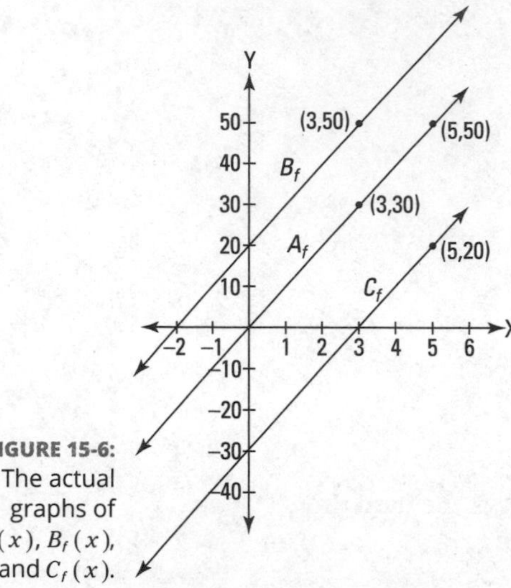

FIGURE 15-6: The actual graphs of $A_f(x)$, $B_f(x)$, and $C_f(x)$.

Figure 15-6 shows the graphs of the equations of $A_f(x)$, $B_f(x)$, and $C_f(x)$, which I worked out before: $A_f(x)=10x$, $B_f(x)=10x+20$, and $C_f(x)=10x-30$. (As you can see, all three are simple, $y=mx+b$ lines.) The y-values of these three functions give you the areas swept out under $f(t)=10$ that you see in Figure 15-5. Note that the three x-intercepts you see in Figure 15-6 are the three x-values in Figure 15-5 where the sweeping out of area begins.

We worked out that $A_f(3)=30$ and that $A_f(5)=50$. You can see those areas of 30 and 50 in the top graph of Figure 15-5. In Figure 15-6, you see these results on A_f at the points $(3,\ 30)$ and $(5,\ 50)$. You can also see in Figure 15-5 that $B_f(3)$ is 20 more than $A_f(3)$; you see that result in Figure 15-6 where $(3,\ 50)$ on B_f is 20 higher than $(3,\ 30)$ on A_f. Finally, you see in Figure 15-5 that $C_f(x)$ is 30 less than $A_f(x)$. Figure 15-6 shows that in a different way: At any x-value, the C_f line is 30 units below the A_f line.

A few observations. You already know from the Fundamental Theorem of Calculus that $\frac{d}{dx}A_f(x)=f(x)=10$ (and the same for $B_f(x)$ and $C_f(x)$). That was explained a minute ago in terms of rates: For A_f, B_f, and C_f, the rate of area being swept out under $f(t)=10$ equals 10. Figure 15-6 also shows that $\frac{d}{dx}A_f(x)=10$ (and the same for B_f and C_f), but here you see the derivative as a slope. The slopes, of course, of all three lines equal 10. Finally, note that — like you see in Figure 15-1 — the three lines in Figure 15-6 differ from each other only by a vertical translation. These three lines (and the infinity of all other vertically translated lines) are all members of the class of functions, $10x+C$, the family of antiderivatives of $f(x)=10$.

For the next example, look again at the parabola $y=x^2+1$, our friend from Chapter 14 which we analyzed in terms of the sum of the areas of rectangles (Riemann sums). Flip back to Figure 14-4, and check out the shaded region under $y=x^2+1$. Now you can finally compute the exact area of the shaded region the easy way.

The area function for sweeping out area under x^2+1 is $A_f(x)=\int_s^x (t^2+1)dt$. By the Fundamental Theorem of Calculus, $\frac{d}{dx}A_f(x)=x^2+1$, and so A_f is an antiderivative of x^2+1. Any function of the form $\frac{1}{3}x^3+x+C$ has a derivative of x^2+1 (try it), so that's the antiderivative. For Figure 14-4, you want to sweep out area beginning at zero, so $s=0$. Set the antiderivative equal to zero, plug the value of s into x, and solve for C: $\frac{1}{3}\cdot 0^3+0+C=0$, so $C=0$, and thus

$$A_f(x)=\int_0^x (t^2+1)dt=\frac{1}{3}x^3+x+0$$

The area swept out from 0 to 3 — which we do the hard way in Chapter 14 by computing the limit of a Riemann sum — is simply $A_f(3)$:

$$A_f(x)=\frac{1}{3}x^3+x$$
$$A_f(3)=\frac{1}{3}\cdot 3^3+3=9+3=12$$

Piece o' cake. That was *much* less work than doing it the hard way.

And after you know that the area function that starts at zero, $\int_0^x (t^2+1)dt$, equals $\frac{1}{3}x^3+x$, it's a snap to figure the area of other sections under the parabola that don't start at zero. Say, for example, you want the area under the parabola between 2 and 3. You can compute that area by subtracting the area between 0 and 2 from the area between 0 and 3. We just figured the area between 0 and 3 — that's 12. And the area between 0 and 2 is $A_f(2)=\frac{1}{3}\cdot 2^3+2=4\frac{2}{3}$. So, the area between 2 and 3 is $12-4\frac{2}{3}$, or $7\frac{1}{3}$. This subtraction method brings us to the next topic — the second version of the Fundamental Theorem of Calculus.

The Fundamental Theorem of Calculus: Take Two

Now you finally arrive at the super-duper shortcut integration theorem that you'll use for the rest of your natural born days — or at least till the end of your stint with calculus.

MATH RULES

The Fundamental Theorem of Calculus (second version or shortcut version): Let *F* be any antiderivative of the function *f*; then

$$\int_a^b f(x)dx=F(b)-F(a)$$

This theorem gives you the super shortcut for computing a definite integral like $\int_2^3 (x^2+1)dx$, the area under the parabola $y=x^2+1$ between 2 and 3. As I show in the previous section, you can get this area by subtracting the area between 0 and 2 from the area between 0 and 3, but to do that, you need to know that the particular area function sweeping out area beginning at zero, $\int_0^x (t^2+1)dt$, is $\frac{1}{3}x^3+x$; and to get that, you need to calculate that the value of C is zero.

The beauty of the shortcut theorem is that you don't have to even use an area function like $A_f(x)=\int_0^x (t^2+1)dt$. You just find any antiderivative, $F(x)$, of your function, and do the subtraction, $F(b)-F(a)$. The simplest antiderivative to use is the one where $C=0$. So, here's how you use the theorem to find the area under our parabola from 2 to 3. You begin with the simplest antiderivative of x^2+1; that's $F(x)=\frac{1}{3}x^3+x$. Then the theorem gives you:

$$\int_2^3 (x^2+1)dx = F(3)-F(2)$$

$F(3)-F(2)$ can be written as $\left[\frac{1}{3}x^3+x\right]_2^3$, and thus,

$$\begin{aligned}\int_2^3 (x^2+1)dx &= \left[\frac{1}{3}x^3+x\right]_2^3 \\ &= \left(\frac{1}{3}\cdot 3^3+3\right)-\left(\frac{1}{3}\cdot 2^3+2\right) \\ &= 12-4\frac{2}{3}=7\frac{1}{3}\end{aligned}$$

Granted, this is the same computation I did in the previous section using the area function with $s=0$, but that's only because for the $y=x^2+1$ function, when s is zero, C is also zero. It's sort of a coincidence, and it's not true for all functions. But regardless of the function, the shortcut works, and you don't have to worry about area functions or s or C. All you do is $F(b)-F(a)$.

Here's another problem: What's the area under $f(x)=e^x$ between $x=3$ and $x=5$? The derivative of e^x is e^x, so e^x is an antiderivative of e^x, and thus

$$\begin{aligned}\int_3^5 e^x dx &= \left[e^x\right]_3^5 \\ &= e^5-e^3 \\ &\approx 148.4-20.1\approx 128.3\end{aligned}$$

What could be simpler?

WARNING

Areas *above* the curve and *below* the x-axis count as *negative* areas. Before going on, I'd be remiss if I didn't touch on negative areas (this is virtually the same caution made at the very end of Chapter 14). Note that with the two examples, the parabola, $y=x^2+1$, and the exponential function, $y=e^x$, the areas you're computing are *under* the curves and *above* the x-axis. These areas count as ordinary, *positive* areas. But, if a function goes below the x-axis, areas above the curve and below the x-axis count as *negative* areas. This is the case whether you're using an area function, the first version of the Fundamental Theorem of Calculus, or the shortcut version. Don't worry about this for now. You see how this works in Chapter 17.

Okay, so now you've got the super shortcut for computing the area under a curve. And if one big shortcut wasn't enough to make your day, Table 15-1 lists some rules about definite integrals that can make your life much easier.

Table 15-1 Five Easy Rules for Definite Integrals

1) $\int_a^a f(x)dx = 0$ (Well, duh — there's no area "between" a and a.)

2) $\int_b^a f(x)dx = -\int_a^b f(x)dx$

3) $\int_a^b f(x)dx = \int_a^c f(x)dx + \int_c^b f(x)dx$

4) $\int_a^b kf(x)dx = k\int_a^b f(x)dx$ (k is a constant; you can pull a constant out of the integral.)

5) $\int_a^b [f(x) \pm g(x)]dx = \int_a^b f(x)dx \pm \int_a^b g(x)dx$

EXAMPLE

Q. (a) For the area function, $A_f(x) = \int_{10}^{x} (t^2 - 5t)dt$, what's $\frac{d}{dx}A_f(x)$?

(b) For the area function, $B_f(x) = \int_{-4}^{3x^2} \sin t\, dt$, what's $\frac{d}{dx}B_f(x)$?

A. (a) No work needed here. **The answer is simply $x^2 - 5x$ by the Fundamental Theorem of Calculus.**

(b) $6x \sin 3x^2$.

The *argument* of an area function is the expression at the top of the integral symbol — not the integrand. Because the argument of this area function, $3x^2$, is something other than a plain old x, this is a chain rule problem. Thus,

$$\frac{d}{dx}B_f(x) = \sin(3x^2) \cdot 6x, \text{ or } 6x\sin(3x^2)$$

Q. What's the area under $2x^2 + 5$ from 0 to 4? *Note:* this is the same question that you work on in Chapter 10 with the difficult sigma-sum-rectangle method.

A. $\frac{188}{3}$

Using the second version of the Fundamental Theorem of Calculus, $\int_0^4 (2x^2+5)dx = F(4)-F(0)$, where F is any antiderivative of $2x^2+5$. Anything of the form $\frac{2}{3}x^3+5x+C$ is an antiderivative of $2x^2+5$. You should use the simplest antiderivative where $C=0$, namely, $\frac{2}{3}x^3+5x$. Thus,

$$\begin{aligned}\int_0^4 (2x^2+5)dx &= \left[\frac{2}{3}x^3+5x\right]_0^4 \\ &= \left(\frac{2}{3}\cdot 4^3+5\cdot 4\right)-\left(\frac{2}{3}\cdot 0^3+5\cdot 0\right) \\ &= \frac{188}{3}\end{aligned}$$

You get the same answer with *much* less work than adding up all those rectangles!

YOUR TURN

5 **(a)** If $A_f(x)=\int_0^x \sin t\,dt$, what's $\frac{d}{dx}A_f(x)$?

(b) If $A_g(x)=\int_{\pi/4}^x \sin t\,dt$, what's $\frac{d}{dx}A_g(x)$?

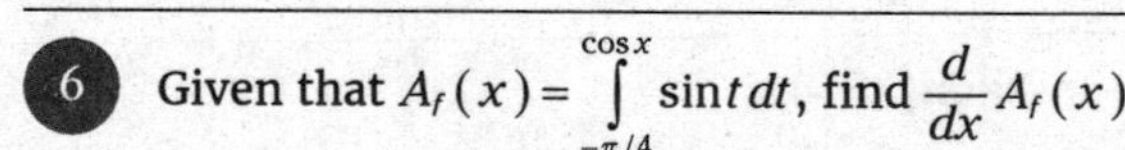

---

6 Given that $A_f(x)=\int_{-\pi/4}^{\cos x} \sin t\,dt$, find $\frac{d}{dx}A_f(x)$.

7 For $A_f(x)$ from Problem 5a, where does $\frac{d}{dx}A_f$ equal zero?

8 For $A_f(x)$ from Problem 6, evaluate $A_f'\left(\frac{\pi}{4}\right)$.

9 What's the area under $y = \sin x$ from 0 to π?

10 Evaluate $\int_0^{2\pi} \sin x\, dx$.

11 Evaluate $\int_2^3 \left(x^3 - 4x^2 + 5x - 10\right)dx$.

12 Evaluate $\int_{-1}^{2} e^x dx$.

Now that you know how to use the shortcut version of the Fundamental Theorem of Calculus (as you just did for Problems 9 to 12), you won't have to compute the area under a curve the hard way (with an area function or a Riemann sum). But that doesn't mean you're off the hook. Following are three different ways to understand why the shortcut version of the theorem works. This is difficult stuff — brace yourself.

Alternatively, you can skip these explanations if all you want to know is how to compute an area: Forget about C and just subtract $F(a)$ from $F(b)$. I include these explanations because I suspect you're dying to learn extra math just for the love of learning — right? Other books just give you the rules; I explain why they work and the underlying principles — that's why they pay me the big bucks.

Actually, in all seriousness, you should read at least some of this material. The Fundamental Theorem of Calculus is one of the most important theorems in all of mathematics, so you ought to spend some time trying hard to understand what it's all about. It's worth the effort. Of the three explanations, the first is the easiest. But if you only want to read one or two of the three, I'd read just the third, or the second and the third. Or, you could begin with the figures accompanying the three explanations, because the figures really show you what's going on. Finally, if you can't digest all of this in one sitting — no worries — you can revisit it later.

Why the theorem works: Area functions explained

One way to understand the shortcut version of the Fundamental Theorem of Calculus is by looking at area functions. As you can see in Figure 15-7, the dark-shaded area between a and b can be figured by starting with the area between s and b, then cutting away (subtracting) the area between s and a. And it doesn't matter whether you use 0 as the left edge of the areas or any other value of s. Do you see that you'd get the same result whether you use the graph on the left or the graph on the right?

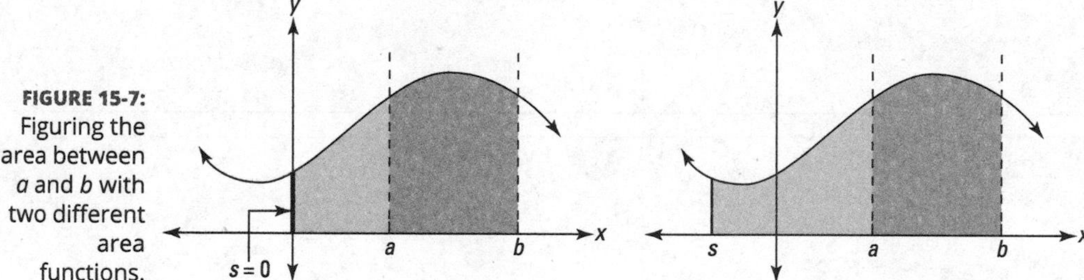

FIGURE 15-7: Figuring the area between a and b with two different area functions.

Take a look at $f(t) = 10$ (see Figure 15-8). Say you want the area between 5 and 8 under the horizontal line $f(t) = 10$, and you are forced to use calculus.

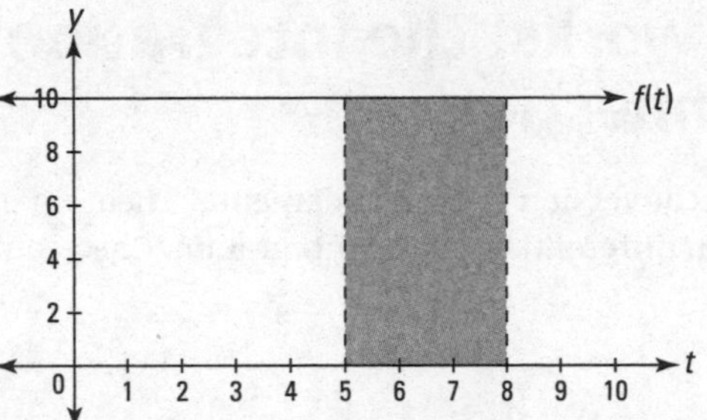

FIGURE 15-8: The shaded area equals 30 — well, duh, it's a 3-by-10 rectangle.

Look back at two of the area functions for $f(t)=10$ in Figure 15-5: $A_f(x)$ starting at zero (in which $C=0$) and $B_f(x)$ starting at –2 (where $C=20$):

$$A_f(x)=\int_0^x 10dt=10x$$

$$B_f(x)=\int_{-2}^x 10dt=10x+20$$

If you use $A_f(x)$ to compute the area between 5 and 8 in Figure 15-8, you get the following:

$$\begin{aligned}\int_5^8 10dx &= A_f(8)-A_f(5)\\ &=10\cdot 8-10\cdot 5\\ &=80-50 \qquad \text{(80 is the area of the rectangle from 0 to 8; 50 is the area of the rectangle from 0 to 5.)}\\ &=30\end{aligned}$$

If, on the other hand, you use $B_f(x)$ to compute the same area, you get the same result:

$$\begin{aligned}\int_5^8 10dx &= B_f(8)-A_f(5)\\ &=10\cdot 8+20-(10\cdot 5+20)\\ &=80+20-(50+20) \qquad \text{(This is } 100-70\text{, of course; 100 is the area of the rectangle from } -2 \text{ to 8; 70 is the area of the rectangle from } -2 \text{ to 5.)}\\ &=30\end{aligned}$$

Notice that the two 20's in the second line from the bottom cancel. Recall that all antiderivatives of $f(t)=10$ are of the form $10x+C$. Regardless of the value of C, it cancels out as in this example. Thus, you can use any antiderivative with any value of C. For convenience, everyone just uses the antiderivative with $C=0$, so that you don't mess with C at all. And the choice of s (the point where the area function begins) is irrelevant. So, when you're using the shortcut version of the Fundamental Theorem of Calculus, and computing an area with $F(b)-F(a)$, you're sort of using a mystery area function with a C value of zero and an unknown starting point, s (recall that when C is zero, s might or might not be zero.). Get it?

Why the theorem works: The integration-differentiation connection

The next explanation of the shortcut version of the Fundamental Theorem of Calculus involves the yin/yang relationship between differentiation and integration. Check out Figure 15-9.

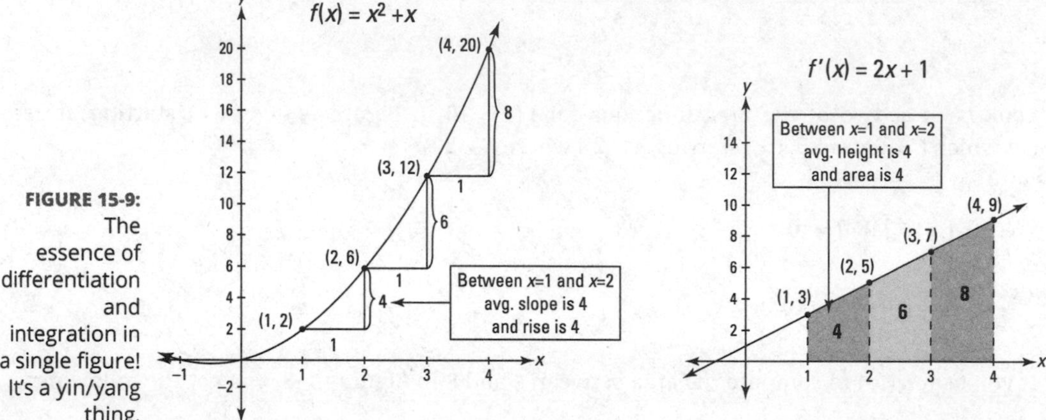

FIGURE 15-9: The essence of differentiation and integration in a single figure! It's a yin/yang thing.

The figure shows a function, $f(x) = x^2 + x$, and its derivative, $f'(x) = 2x + 1$. Look carefully at the numbers 4, 6, and 8 on both graphs. The connection between 4, 6, and 8 on the graph of f — which are the amounts of *rise* between consecutive points on the curve — and 4, 6, and 8 on the graph of f' — which are the *areas* of the trapezoids under f' — shows the intimate relationship between integration and differentiation. Figure 15-9 is a picture worth a thousand symbols and equations, encapsulating the essence of integration in a single snapshot. It shows how the shortcut version of the Fundamental Theorem of Calculus works because it shows that the *area* under $f'(x)$ between 1 and 4 equals the total *rise* on $f(x)$ between $(1, 2)$ and $(4, 20)$, in other words that

$$\int_1^4 f'(x) = f(4) - f(1)$$

Note that I've called the two functions in Figure 15-9 and in this equation f and f' to emphasize that $2x + 1$ is the *derivative* of $x^2 + x$. I could have instead referred to $x^2 + x$ as F and referred to $2x + 1$ as f, which would emphasize that $x^2 + x$ is an *antiderivative* of $2x + 1$. In that case, you would write this area equation in the standard way,

$$\int_1^4 f(x)dx = F(4) - F(1)$$

Either way, the meaning's the same. I use the derivative version to point out how finding area is differentiation in reverse. Going from left to right in Figure 15-9 is differentiation: The slopes of f correspond to heights on f'. Going from right to left is integration: Areas under f' correspond to the change in height between two points on f.

Okay, here's how it works. Imagine you're going up along f from $(1, 2)$ to $(2, 6)$. Every point along the way has a certain steepness, a slope. This slope is plotted as the y-coordinate, or height, on the graph of f'. The fact that f' goes up from $(1, 3)$ to $(2, 5)$ tells you that the slope of f goes up from 3 to 5 as you travel between $(1, 2)$ and $(2, 6)$. This all follows from basic differentiation.

Now, as you go along f from $(1, 2)$ to $(2, 6)$, the slope is constantly changing. But it turns out that because you go up a total *rise* of 4 as you *run* across 1, the average of all the slopes on f between $(1, 2)$ and $(2, 6)$ is $\frac{4}{1}$, or 4. Because each of these slopes is plotted as a y-coordinate or height on f', it follows that the average height of f' between $(1, 3)$ and $(2, 5)$ is also 4. Thus, between two given points, average slope on f equals average height on f'.

Hold on, you're almost there. *Slope* equals $\frac{rise}{run}$, so when the run is 1, the slope equals the rise. For example, from $(1, 2)$ to $(2, 6)$ on f, the curve rises up 4 and the average slope between those points is also 4. Thus, between any two points on f whose x-coordinates differ by 1, the average slope *is* the rise.

The area of a trapezoid like the ones on the right in Figure 15-9 equals its width times its average height. (This is true of any other similar shape that has a bottom like a rectangle; the top can be any crooked line or funky curve you like.) So, because the width of each trapezoid is 1, and because anything times 1 is itself, the average height of each trapezoid under f' *is* its area; for instance, the area of that first trapezoid is 4 and its average height is also 4.

Are you ready for the grand finale? Here's the whole argument in a nutshell. On f, *rise* = *average slope*; going from f to f', *average slope* = *average height*; on f', *average height* = *area*. So that gives you *rise* = *slope* = *height* = *area*, and thus, finally, *rise* = *area*. And that's what the second version of the Fundamental Theorem of Calculus says:

$$f(b) - f(a) = \int_a^b f'(x)\,dx$$

$$rise = area$$

These ideas are unavoidably difficult. You might have to read it two or three times for it to really sink in.

Notice that it makes no difference to the relationship between slope and area if you use any other function of the form $x^2 + x + C$ instead of $x^2 + x$. Any parabola like $x^2 + x + 10$ or $x^2 + x - 5$ is exactly the same shape as $x^2 + x$; it's just been slid up or down vertically. Any such parabola rises up between $x = 1$ and $x = 4$ in precisely the same way as the parabola in Figure 15-9. From 1 to 2 these parabolas go over 1, up 4. From 2 to 3 they go over 1, up 6, and so on. This is why any antiderivative can be used to find area. The total *area* under f' between 1 to 4, namely 18, corresponds to the total *rise* on any of these parabolas from 1 to 4, namely $4 + 6 + 8$, or 18.

At the risk of beating a dead horse, I've got a third explanation of the Fundamental Theorem of Calculus for you. You might prefer it to the first two because it's less abstract — it's connected to simple, commonsense ideas encountered in our day-to-day world. This explanation has a lot in common with the previous one, but the ideas are presented from a different angle.

Why the theorem works: A connection to — egad! — statistics

Don't let the title of this section put you off. I realize that many readers of this calculus book may not have studied statistics. No worries; the statistics connection I explain here involves a very simple thing covered in statistics courses, but you don't need to know any statistics at all to understand this idea. The simple idea is the relationship between a frequency distribution graph and a cumulative frequency distribution graph (you might have run across such graphs in a newspaper or magazine). Consider Figure 15-10.

The upper graph in the figure shows a frequency distribution histogram of the annual profits of Widgets-R-Us from January 1, 2001 through December 31, 2013. The rectangle marked '07, for example, shows that the company's profit for 2007 was $2,000,000 (their best year during the period 2001–2013).

The lower graph in the figure is a cumulative frequency distribution histogram for the same data used for the upper graph. The difference is simply that in the cumulative graph, the height of each column shows the total profits earned since 1/1/2001. Look at the '02 column in the lower graph and the '01 and '02 rectangles in the upper graph, for example. You can see that the '02 column shows the '02 rectangle sitting on top of the '01 rectangle, which gives that '02 column a height equal to the total of the profits from '01 and '02. Got it? As you go to the right on the cumulative graph, the height of each successive column simply grows by the amount of profits earned in the corresponding single year shown in the upper graph.

Okay. So, here's the calculus connection. (Bear with me; it takes a while to walk through all this.) Look at the top rectangle of the '08 column on the cumulative graph (let's call that graph C for short). At that point on C, you *run* across 1 year and *rise* up $1,250,000, the '08 profit you see on the frequency distribution graph (F for short). *Slope* = *rise* / *run*, so, since the run equals 1, the slope equals 1,250,000/1, or just 1,250,000, which is, of course, the same as the rise. Thus, the slope on C (at '08 or any other year) can be read as a height on F for the corresponding year. (Make sure you see how this works.) Since the heights (or function values) on F are the *slopes* of C, F is the *derivative* of C. In short, F, the derivative, tells you about the slope of C.

The next idea is that since F is the derivative of C, C, by definition, is the antiderivative of F (for example, C might equal $5x^3$ and F would equal $15x^2$). Now, what does C, the antiderivative of F, tell you about F? Imagine dragging a vertical line from left to right over F. As you sweep over the rectangles on F — year by year — the total profit you're sweeping over is shown climbing up along C.

Look at the '01 through '08 rectangles on F. You can see those same rectangles climbing up stair-step fashion along C (see the rectangles labeled A, B, C, and so on, on both graphs). The heights of the rectangles from F keep adding up on C as you climb up the stair-step shape. And I've shown how the same '01 through '08 rectangles that lie along the stair-step top of C can also be seen in a vertical stack at year '08 on C. I've drawn the cumulative graph this way so it's even more obvious how the heights of the rectangles add up. (*Note:* Most cumulative histograms are not drawn this way.)

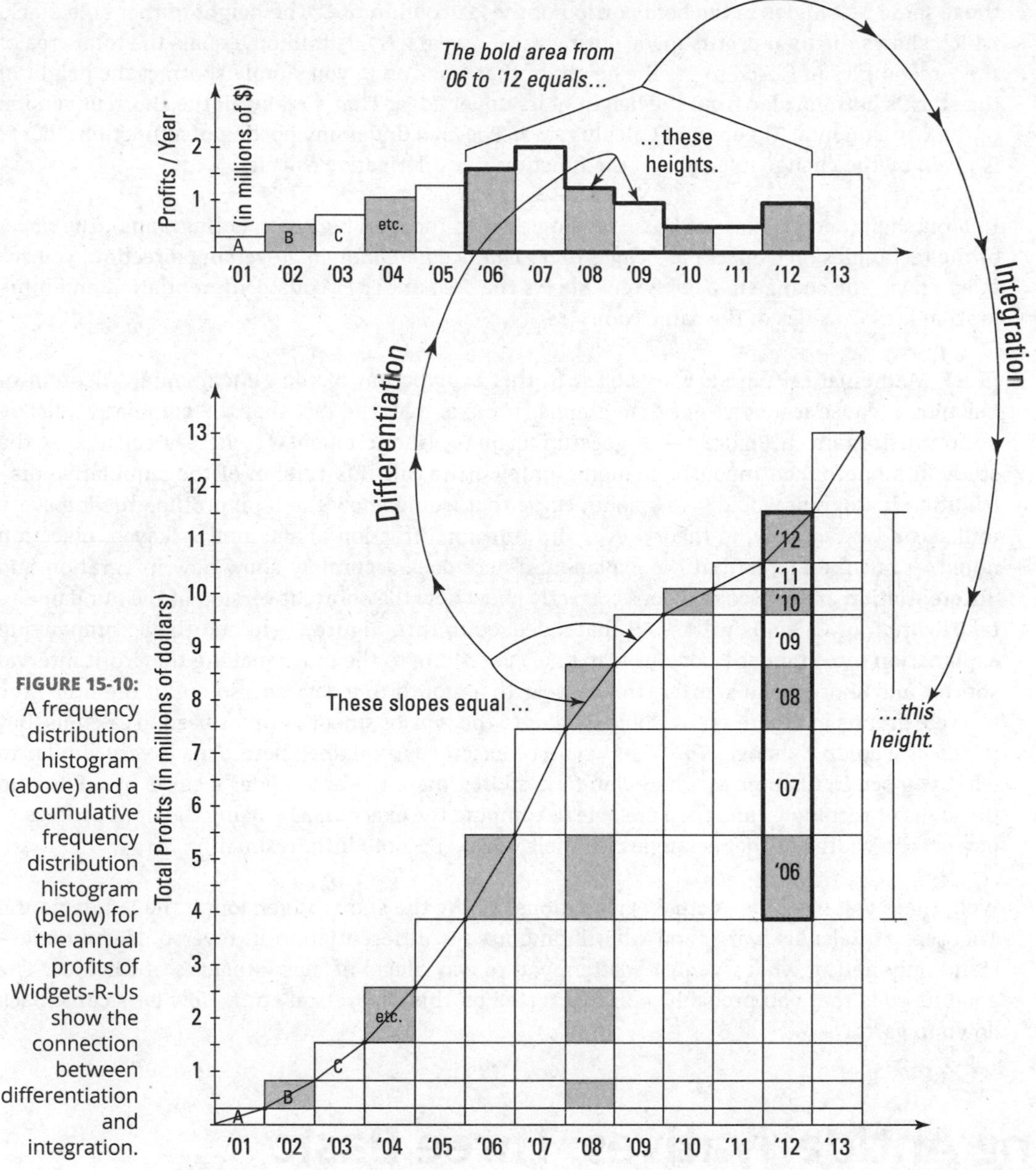

FIGURE 15-10: A frequency distribution histogram (above) and a cumulative frequency distribution histogram (below) for the annual profits of Widgets-R-Us show the connection between differentiation and integration.

Each rectangle on F has a base of 1 year, so, since $area = base \times height$, the area of each rectangle equals its height. So, as you stack up rectangles on C, you're adding up the areas of those rectangles from F. For example, the height of the '01 through '08 stack of rectangles on C ($8.5 million) equals the total area of the '01 through '08 rectangles on F. And, therefore, the heights or function values of C — which is the antiderivative of F — give you the area under the top edge of F. That's how integration works.

Okay, we're just about done. Now let's go through how these two graphs explain the shortcut version of the Fundamental Theorem of Calculus and the relationship between differentiation and integration. Look at the '06 through '12 rectangles on F (with the bold border). You can see those same rectangles in the bold portion of the '12 column of *C*. The height of that bold stack, which shows the total profits made during those 7 years, \$7.75 million, equals the total area of the 7 rectangles in *F*. And to get the height of that stack on *C*, you simply subtract the height of the stack's bottom edge from the height of its upper edge. That's really all the shortcut version of the Fundamental Theorem of Calculus says: The *area* under any portion of a function (like *F*) is given by the change in *height* on the function's antiderivative (like *C*).

In a nutshell (keep looking at those rectangles with the bold border in both graphs), the *slopes* of the rectangles on *C* appear as *heights* on *F*. That's *differentiation*. Reversing direction, you see *integration*: The change in *heights* on *C* shows the *area* under *F*. Voilà: Differentiation and integration are two sides of the same coin.

(*Note:* Mathematical purists may object to this explanation of the Fundamental Theorem of Calculus because it involves *discrete* graphs [for example, the fact that the cumulative distribution histogram in Figure 15-10 goes up at one-year increments], whereas calculus is the study of smooth, continuously changing graphs [the calculus version of the cumulative distribution histogram would be a smooth curve that would show the total profits growing every millisecond — actually, in theory, every infinitesimal fraction of a second]. Okay — objection noted — but the fact is that the explanation here does accurately show how integration and differentiation are related and does correctly show how the shortcut version of the Fundamental Theorem of Calculus works. All that's needed to turn Figure 15-10 and the accompanying explanation into standard calculus is to take everything to the limit, making the profit interval shorter and shorter and shorter: from a year to a month to a day, and so on. In the limit, the discrete graphs in Figure 15-10 would meld into the type of smooth graphs used in calculus. But the *ideas* wouldn't change. The *ideas* would be exactly as explained here. This is very similar to what you see in Chapter 14 where you first approximate the area under a curve by adding up the areas of rectangles and then are able to compute the exact area by using the limit process to narrow the widths of the rectangles till their widths become infinitesimal.)

Well, there you have it — actual explanations of why the shortcut version of the Fundamental Theorem of Calculus works and why finding area is differentiation in reverse. If you understand only half of what I've just written, you're way ahead of most students of calculus. The good news is that you probably won't be tested on this theoretical stuff. Now let's come back down to earth.

Finding Antiderivatives: Three Basic Techniques

I've been talking a lot about antiderivatives, but just how do you find them? In this section, I give you three easy techniques. Then, in Chapter 16, I give you four advanced techniques. By the way, you *will* be tested on this stuff.

Reverse rules for antiderivatives

The easiest antiderivative rules are the ones that are the reverse of derivative rules you already know. (You can brush up on derivative rules in Chapter 10 if you need to.) These are automatic, one-step antiderivatives with the exception of the reverse power rule, which is only slightly harder.

No-brainer reverse rules

You know that the derivative of $\sin x$ is $\cos x$, so reversing that tells you that an antiderivative of $\cos x$ is $\sin x$. What could be simpler? But don't forget that all functions of the form $\sin x + C$ are antiderivatives of $\cos x$. In symbols, you write

$$\frac{d}{dx}\sin x = \cos x, \text{ and therefore}$$

$$\int \cos x dx = \sin x + C$$

Table 15-2 lists the reverse rules for antiderivatives.

Table 15-2 Basic Antiderivative Formulas

1) $\int 1dx$ (or just $\int dx$) $= x + C$ (because the derivative of x is 1)

2) $\int x^n dx = \frac{x^{n+1}}{n+1} + C \quad (n \neq -1)$ 3) $\int \frac{dx}{x} = \ln|x| + C$ (rule 2 for $n = -1$)

4) $\int e^x dx = e^x + C$ 5) $\int a^x dx = \frac{1}{\ln a} a^x + C$

6) $\int \sin x dx = -\cos x + C$ 7) $\int \cos x dx = \sin x + C$

8) $\int \sec^2 x dx = \tan x + C$ 9) $\int \csc^2 x dx = -\cot x + C$

10) $\int \sec x \tan x dx = \sec x + C$ 11) $\int \csc x \cot x dx = -\csc x + C$

12) $\int \frac{dx}{\sqrt{a^2 - x^2}} = \arcsin \frac{x}{a} + C$ 13) $\int \frac{dx}{a^2 + x^2} = \frac{1}{a} \arctan \frac{x}{a} + C$

14) $\int \frac{dx}{x\sqrt{x^2 - a^2}} = \frac{1}{a} \operatorname{arcsec} \frac{|x|}{a} + C$

The slightly more difficult reverse power rule

By the power rule for differentiation, you know that

$$\frac{d}{dx}x^3 = 3x^2, \text{ and therefore}$$

$$\int 3x^2 dx = x^3 + C$$

Here's the simple method for reversing the power rule. Use $y = 5x^4$ for your function. Recall that the power rule says to do the following:

1. **Bring the power in front where it will *multiply* the rest of the derivative.**

   $5x^4 \rightarrow 4 \cdot 5x^4$

2. ***Reduce* the power by one and simplify.**

   $4 \cdot 5x^4 \rightarrow 4 \cdot 5x^3 = 20x^3$

   Thus, $y' = 20x^3$.

To reverse this process, you reverse the order of the two steps and reverse the math within each step. Here's how that works for the same problem:

1. ***Increase* the power by one.**

   The 3 becomes a 4.

   $20x^3 \rightarrow 20x^4$

2. ***Divide* by the new power and simplify.**

   $20x^4 \rightarrow \frac{20}{4}x^4 = 5x^4$

   And thus, you write $\int 20x^3 dx = 5x^4 + C$.

WARNING

The reverse power rule does not work for a power of negative one. The reverse power rule works for all powers (including negative and decimal powers) except for a power of negative one. Instead of using the reverse power rule, you should just memorize that the antiderivative of x^{-1} is $\ln|x| + C$ (rule 3 in Table 15-2).

TIP

Test your antiderivatives by differentiating them. Especially when you're new to antidifferentiation, it's a good idea to test your antiderivatives by differentiating them — you can ignore the C. If you get back to your original function, you know your antiderivative is correct.

With the antiderivative you just found and the shortcut version of the Fundamental Theorem of Calculus, you can determine the area under $20x^3$ between, say, 1 and 2:

$$\begin{aligned}
&\int 20x^3 dx = 5x^4 + C, \text{ thus} \\
&\int_1^2 20x^3 dx = \left[5x^4\right]_1^2 \\
&\qquad = 5 \cdot 2^4 - 5 \cdot 1^4 \\
&\qquad = 80 - 5 = 75
\end{aligned}$$

YOUR TURN

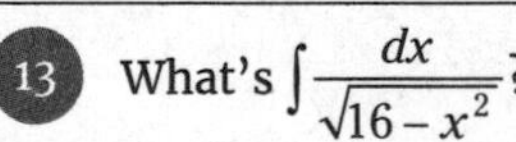

13 What's $\int \frac{dx}{\sqrt{16 - x^2}}$?

14 What's $\int 5\sqrt{x}dx \quad (x \geq 0)$?

Guessing and checking

The guess-and-check method works when the *integrand* (that's the expression after the integral symbol not counting the *dx*, and it's the thing you want to antidifferentiate) is close to a function that you know the reverse rule for. For example, say you want the antiderivative of $\cos(2x)$. Well, you know that the derivative of sine is cosine. Reversing that tells you that the antiderivative of cosine is sine. So you might think that the antiderivative of $\cos(2x)$ is $\sin(2x)$. That's your *guess*. Now *check* it by differentiating it to see if you get the original function, $\cos(2x)$:

$$\begin{aligned}&\frac{d}{dx}\sin(2x)\\&=\cos(2x)\cdot 2 \quad \text{(sine rule and chain rule)}\\&=2\cos(2x)\end{aligned}$$

This result is very close to the original function, except for that extra coefficient of 2. In other words, the answer is 2 times as much as what you want. Because you want a result that's half of this, just try an antiderivative that's half of your first guess: So, your new guess is $\frac{1}{2}\sin(2x)$. Check this second guess by differentiating it, and you get the desired result.

Here's another problem. What's the antiderivative of $(3x-2)^4$?

1. **Guess the antiderivative.**

   This looks sort of like a power rule problem, so try the reverse power rule. The antiderivative of x^4 is $\frac{1}{5}x^5$ by the reverse power rule, so your guess is $\frac{1}{5}(3x-2)^5$.

2. **Check your guess by differentiating it.**

$$\begin{aligned}&\frac{d}{dx}\left[\frac{1}{5}(3x-2)^5\right]\\&=5\cdot\frac{1}{5}(3x-2)^4\cdot 3 \quad \text{(power rule and chain rule)}\\&=3(3x-2)^4\end{aligned}$$

3. **Tweak your first guess.**

   Your result, $3(3x-2)^4$, is three times too much, so make your second guess a *third* of your first guess — that's $\frac{1}{3}\cdot\frac{1}{5}(3x-2)^5$, or $\frac{1}{15}(3x-2)^5$.

4. **Check your second guess by differentiating it.**

$$\begin{aligned}&\frac{d}{dx}\left[\frac{1}{15}(3x-2)^5\right]\\&=5\cdot\frac{1}{15}(3x-2)^4\cdot 3 \quad \text{(power rule and chain rule)}\\&=(3x-2)^4\end{aligned}$$

   This checks. You're done. The antiderivative of $(3x-2)^4$ is $\frac{1}{15}(3x-2)^5+C$.

The two previous problems show that *guess and check* works well when the function you want to antidifferentiate has an argument like $3x$ or $3x+2$ (where x is raised to the *first* power) instead of a plain old x. (Recall that in a function like $\sqrt{5x}$, the $5x$ is called the *argument*.) In this case, all you have to do is tweak your guess by the *reciprocal* of the coefficient of x: for example, the 3 in $3x+2$ (the 2 in $3x+2$ has no effect on your answer). In fact, for these easy problems, you don't really have to do any guessing and checking. You can immediately see how to tweak your guess. It becomes sort of a one-step process. If the function's argument is more complicated than $3x+2$ — like the x^2 in $\cos(x^2)$ — you have to try the next method, substitution.

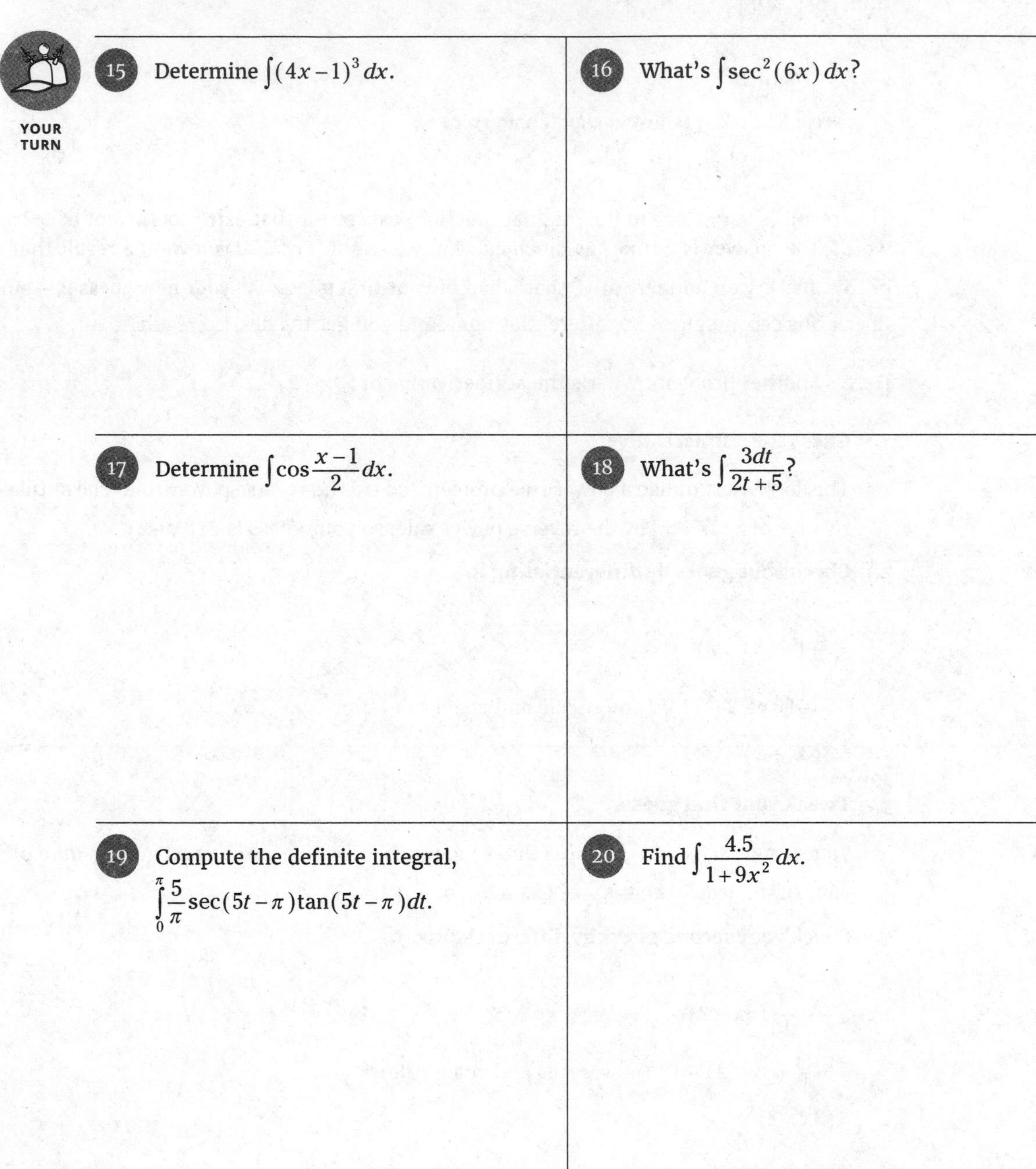

YOUR TURN

15 Determine $\int (4x-1)^3\,dx$.

16 What's $\int \sec^2(6x)\,dx$?

17 Determine $\int \cos\frac{x-1}{2}\,dx$.

18 What's $\int \frac{3dt}{2t+5}$?

19 Compute the definite integral, $\int_0^{\pi} \frac{5}{\pi}\sec(5t-\pi)\tan(5t-\pi)\,dt$.

20 Find $\int \frac{4.5}{1+9x^2}\,dx$.

The substitution method

If you look back at the examples of the guess-and-check method in the previous section, you can see why the first guess in each case didn't work. When you differentiate the guess, the chain rule produces an extra constant: 2 in the first example, 3 in the second. You then tweak the guesses with $\frac{1}{2}$ and $\frac{1}{3}$ to compensate for the extra constant.

Now say you want the antiderivative of $\cos(x^2)$ and you guess that it is $\sin(x^2)$. Watch what happens when you differentiate $\sin(x^2)$ to check it:

$$\begin{aligned}&\frac{d}{dx}\sin(x^2)\\&=\cos(x^2)\cdot 2x \quad \text{(sine rule and chain rule)}\\&=2x\cos(x)^2\end{aligned}$$

Here the chain rule produces an extra $2x$ — because the derivative of x^2 is $2x$ — but if you try to compensate for this by attaching a $\frac{1}{2x}$ to your guess, it won't work. Try it.

So, guessing and checking doesn't work for antidifferentiating $\cos(x^2)$ — actually, *no* method works for this simple-looking integrand (not all functions have antiderivatives) — but your admirable attempt at differentiation here reveals a new class of functions that you can antidifferentiate. Because the derivative of $\sin(x^2)$ is $2x\cos(x^2)$, the antiderivative of $2x\cos(x^2)$ must be $\sin(x^2)$. This function, $2x\cos(x^2)$, is the type of function you can antidifferentiate with the substitution method.

MATH RULES

Keep your eyes peeled for the derivative of the function's argument. The substitution method works when the integrand contains a function and *the derivative of the function's argument* — in other words, when it contains that extra thing produced by the chain rule — or something just like it except for a constant. And the integrand must not contain any other extra stuff.

The derivative of e^{x^3} is $e^{x^3}\cdot 3x^2$ by the e^x rule and the chain rule. So, the antiderivative of $e^{x^3}\cdot 3x^2$ is e^{x^3}. And if you were asked to find the antiderivative of $e^{x^3}\cdot 3x^2$, you would know that the substitution method would work because this expression contains $3x^2$, which is the derivative of the argument of e^{x^3}, namely x^3.

By now, you're probably wondering why this is called the substitution method. I show you why in the following step-by-step method. But first, I want to point out that you don't always have to use the step-by-step method. Assuming you understand why the antiderivative of $e^{x^3}\cdot 3x^2$ is e^{x^3}, you may encounter problems where you can just see the antiderivative without doing any work. But whether or not you can just see the answers to problems like that one, the substitution method is a good technique to learn because, for one thing, it has many uses in calculus and other areas of mathematics, and for another, your teacher may require that you know it and use it.

Okay, so here's how to find $\int 2x\cos(x^2)dx$ with substitution:

1. **Set *u* equal to the argument of the main function.**

   The argument of $\cos(x^2)$ is x^2, so you set *u* equal to x^2.

2. **Take the derivative of *u* with respect to *x*.**

   $u = x^2 \text{ so } \frac{du}{dx} = 2x$

3. **Solve for *dx*.**

   $$\frac{du}{dx} = \frac{2x}{1}$$
   $$du = 2xdx \quad \text{(cross multiplication)}$$
   $$\frac{du}{2x} = dx \quad \text{(dividing both sides by } 2x\text{)}$$

4. **Make the substitutions.**

   In $\int 2x\cos(x^2)dx$, *u* takes the place of x^2 and $\frac{du}{2x}$ takes the place of *dx*. So now you've got $\int 2x\cos u\frac{du}{2x}$. The two 2*x*'s cancel, giving you $\int \cos u\, du$.

5. **Antidifferentiate using the simple reverse rule.**

   $\int \cos u\, du = \sin u + C$

6. **Substitute x^2 back in for *u*, coming full circle.**

   u equals x^2, so x^2 goes in for the *u*:

   $\int \cos u\, du = \sin(x^2) + C$

   That's it. So $\int 2x\cos(x^2)dx = \sin(x^2) + C$.

If the original problem had been $\int 5x\cos(x^2)dx$ instead of $\int 2x\cos(x^2)dx$, you would follow the same steps except that in Step 4, after making the substitution, you would arrive at $\int 5x\cos u\frac{du}{2x}$. The *x*'s would still cancel — that's the important thing — but after canceling, you would get $\int \frac{5}{2}\cos u\, du$, which has that extra $\frac{5}{2}$ in it. No worries. You would just pull the $\frac{5}{2}$ through the $\int$ symbol, giving you $\frac{5}{2}\int \cos u\, du$. Now you would finish this problem just as you did in Steps 5 and 6, except for the extra $\frac{5}{2}$:

$$\begin{aligned}\frac{5}{2}\int \cos u\, du &= \frac{5}{2}(\sin u + C)\\ &= \frac{5}{2}\sin u + \frac{5}{2}C\\ &= \frac{5}{2}\sin(x^2) + \frac{5}{2}C\end{aligned}$$

Because *C* is any old constant, $\frac{5}{2}C$ is still any old constant, so you can get rid of the $\frac{5}{2}$ in front of the *C*. That may seem somewhat (grossly?) unmathematical, but it's right. Thus, your final answer is $\frac{5}{2}\sin(x^2) + C$. You should check this by differentiating it.

Here are a few examples of antiderivatives you can do with the substitution method so you can learn how to spot them:

- $\int 4x^2\cos\left(x^3\right)dx$

  The derivative of x^3 is $3x^2$, but you don't have to pay any attention to the 3 in $3x^2$ or the 4 in the integrand. Because the integrand contains x^2 and no other extra stuff, substitution works. Try it.

- $\int 10\sec^2 x\cdot e^{\tan x}dx$

  The integrand contains a function, $e^{\tan x}$, and the derivative of its argument $(\tan x)$ — which is $\sec^2 x$. Because the integrand doesn't contain any other extra stuff (except for the 10, which doesn't matter), substitution works. Do it.

- $\int \frac{2}{3}\cos x\sqrt{\sin x}\,dx$

  Because the integrand contains the derivative of $\sin x$, namely $\cos x$, and no other stuff except for the $\frac{2}{3}$, substitution works. Go for it.

You can do the three problems just listed with a method that combines substitution and guess-and-check (as long as your teacher doesn't insist that you show the six-step substitution solution). Try using this combo method to antidifferentiate the first example, $\int 4x^2\cos\left(x^3\right)dx$. First, you confirm that the integral fits the pattern for substitution — it does, as pointed out in the first item on the checklist. This confirmation is the only part substitution plays in the combo method. Now you finish the problem with the guess-and-check method:

EXAMPLE

Q. Find $\int 4x^2\cos\left(x^3\right)dx$.

A. $\frac{4}{3}\sin\left(x^3\right)$

1. **Make your guess.**

   The antiderivative of cosine is sine, so a good guess for the antiderivative of $4x^2\cos\left(x^3\right)$ is $\sin\left(x^3\right)$.

2. **Check your guess by differentiating it.**

   $$\begin{aligned}\frac{d}{dx}\sin\left(x^3\right)&=\cos\left(x^3\right)\cdot 3x^2 \quad \text{(sine rule and chain rule)}\\&=3x^2\cos\left(x^3\right)\end{aligned}$$

3. **Tweak your guess.**

   Your result from Step 2, $3x^2\cos\left(x^3\right)$, is $\frac{3}{4}$ of what you want, $4x^2\cos\left(x^3\right)$, so make your guess $\frac{4}{3}$ bigger (note that $\frac{4}{3}$ is the reciprocal of $\frac{3}{4}$). Your second guess is thus $\frac{4}{3}\sin\left(x^3\right)$.

4. **Check this second guess by differentiating it.**

   Oh, heck, skip this — your answer's got to work.

21 Find the antiderivative, $\int \frac{\sin x}{\sqrt{\cos x}}\,dx$, with the substitution method.

22 Find the antiderivative, $\int x^4 \sqrt[3]{2x^5+6}\,dx$, with the substitution method.

23 Use substitution to determine $\int 5x^3 e^{x^4}\,dx$.

24 Use substitution to determine $\int \frac{\sec^2 \sqrt{x}}{\sqrt{x}}\,dx$.

Finding Area with Substitution Problems

You can use the shortcut version of the Fundamental Theorem of Calculus to calculate the area under a function that you integrate with the substitution method. You can do this in two ways. In the previous section, I use substitution, setting u equal to x^2, to find the antiderivative of $2x\cos(x^2)$:

$$\int 2x\cos(x^2)dx = \sin(x^2) + C$$

If you want the area under this curve from, say, 0.5 to 1, the Fundamental Theorem of Calculus does the trick:

$$\begin{aligned}\int_{0.5}^{1} 2x\cos(x^2)dx &= \left[\sin(x^2)\right]_{0.5}^{1} \\ &= \sin(1^2) - \sin(0.5^2) \\ &= \sin 1 - \sin 0.25 \\ &\approx 0.841 - 0.247 \\ &\approx 0.594\end{aligned}$$

Another method, which amounts to the same thing, is to change the limits of integration and do the whole problem in terms of u. Refer back to the six-step solution in the section "The substitution method." What follows is very similar, except that this time you're doing definite integration rather than indefinite integration. Again, you want the area given by $\int_{0.5}^{1} 2x\cos(x^2)dx$:

1. **Set u equal to x^2.**
2. **Take the derivative of u with respect to x.**

   $$\frac{du}{dx} = 2x$$

3. **Solve for du.**

   $$du = 2x\,dx$$

   (I'm solving for du and not going on to solve for dx. It's a slight shortcut that's explained in the tip inside the solution to Problem 21. Check it out if you haven't done so already.)

4. **Determine the new limits of integration.**

   $$u = x^2, \text{ so when } x = \frac{1}{2},\ u = \frac{1}{4}$$
   $$\text{and when } x = 1,\ u = 1$$

5. **Make the substitutions, including the new limits of integration.**

   (In this problem, only one of the limits is new because when $x = 1,\ u = 1$).

   $$\int_{0.5}^{1} 2x\cos(x^2)dx = \int_{0.25}^{1} \cos u\, du$$

6. **Use the antiderivative and the Fundamental Theorem of Calculus to get the desired area *without* making the switch back to x^2.**

$$\int_{0.25}^{1} \cos u \, du = [\sin u]_{0.25}^{1}$$
$$= \sin 1 - \sin 0.25$$
$$\approx 0.594$$

It's a case of six of one, half a dozen of another with the two methods; they require about the same amount of work. So, you can take your pick — however, most teachers and textbooks emphasize the second method, so you probably should learn it.

YOUR TURN

 25 Evaluate $\int_0^2 \frac{t\,dt}{\left(t^2+5\right)^4}$. Change the indices of integration.

 26 Evaluate $\int_1^8 \frac{\left(s^{2/3}+5\right)^3}{\sqrt[3]{s}} ds$ without changing the indices of integration.

Practice Questions Answers and Explanations

1 **At about $x = 2$ or $2\frac{1}{2}$ and about $x = 6$.**

A_g equals zero twice between $x = 0$ and $x = 8$. First, somewhere between $x = 2$ and $x = 2\frac{1}{2}$, where the negative area beginning at $x = 1$ cancels out the positive area between $x = \frac{1}{2}$ and $x = 1$. The second zero of A_g is somewhere between $x = 5\frac{1}{2}$ and $x = 6$. After the first zero at about $x = 2\frac{1}{4}$, negative area is added between $2\frac{1}{4}$ and 4. The positive area from 4 to, say, $5\frac{3}{4}$ roughly cancels that out, so A_g returns to zero at about $x = 5\frac{3}{4}$.

2 (a) **A_g reaches its max at about $x = 8$.** After the zero at about $x = 5\frac{3}{4}$, A_g grows by roughly $3\frac{1}{4}$ square units by the time x gets to 8.

(b) **The minimum value of A_g is at $x = 4$, where it equals something like $-1\frac{1}{2}$.** Note that this minimum occurs at the point where all the negative area has been added (minimums often occur at points like that), and that when you move to the right past $x = 4$, the area crosses above the t-axis and the area begins to increase.

3 (a) **A_g is increasing from 0 to 1 and from 4 to 8.**

(b) **A_g is decreasing from 1 to 4.**

4 **$A_g(1)$ is slightly more than 1; $A_g(3)$ is about -1; $A_g(5)$ is also about -1.**

$A_g(1)$ is a bit bigger than the right triangle with base from $x = \frac{1}{2}$ to $x = 1$ on the t-axis and with vertex maybe at $\left(\frac{1}{2}, 4\right)$, which has an area of 1. So the area in question is slightly more than 1.

There's a zero at about $2\frac{1}{4}$. Between there and $x = 3$, the area is very roughly -1, so $A_g(3)$ is about -1.

In Problem 2b, you estimate $A_g(4)$ to be about $-1\frac{1}{2}$. Between 4 and 5, there's sort of a triangular shape with a rough area of $\frac{1}{2}$. Thus $A_g(5)$ equals about $-1\frac{1}{2} + \frac{1}{2}$ or roughly -1.

5 (a) $\frac{d}{dx}A_f(x) = \sin x$

(b) $\frac{d}{dx}A_g(x) = \sin x$

6 **The answer is $-\sin x \cdot \sin(\cos x)$.**

This is a chain rule problem. Because the derivative of $\int_{-\pi/4}^{x} \sin t\, dt$ is $\sin x$, the derivative of $\int_{-\pi/4}^{stuff} \sin t\, dt$ is $\sin(stuff) \cdot stuff'$. Thus, the derivative of $\int_{-\pi/4}^{\cos x} \sin t\, dt$ is $\sin(\cos x) \cdot (\cos x)' = -\sin x \cdot \sin(\cos x)$.

7 **$\frac{d}{dx}A_f = \sin x$, so $\frac{d}{dx}A_f$ is zero at all the zeros of $\sin x$, namely at all multiples of π, which as you probably know is written as $k\pi$ (for any integer, k).**

8 **Approximately −0.459.**

In Problem 6, you find that $A_f{}'(x) = -\sin x \cdot \sin(\cos x)$, so

$$A_f{}'\left(\frac{\pi}{4}\right) = -\sin\frac{\pi}{4} \cdot \sin\left(\cos\frac{\pi}{4}\right) = -\frac{\sqrt{2}}{2} \cdot \sin\frac{\sqrt{2}}{2} \approx -0.459$$

9 **The area is 2.** The derivative of $-\cos x$ is $\sin x$, so $-\cos x$ is an antiderivative of $\sin x$. Thus, by the Fundamental Theorem of Calculus, $\int_0^{\pi} \sin x\,dx = [-\cos x]_0^{\pi} = -(-1-1) = -(-2) = 2$.

10 $\int_0^{2\pi} \sin x\,dx = [-\cos x]_0^{2\pi} = -(1-1) = 0$**.** Do you see why the answer is zero?

11 $\int_2^3 (x^3 - 4x^2 + 5x - 10)\,dx \approx -6.58$

$$\begin{aligned}&\int_2^3 (x^3 - 4x^2 + 5x - 10)\,dx\\ &= \left[\frac{1}{4}x^4 - \frac{4}{3}x^3 + \frac{5}{2}x^2 - 10x\right]_2^3\\ &= \left(\frac{1}{4}\cdot 81 - \frac{4}{3}\cdot 27 + \frac{5}{2}\cdot 9 - 30\right) - \left(\frac{1}{4}\cdot 16 - \frac{4}{3}\cdot 8 + \frac{5}{2}\cdot 4 - 20\right)\\ &\approx -6.58\end{aligned}$$

12 $\int_{-1}^{2} e^x\,dx \approx 7.02$

$(e^x)' = e^x$, so e^x is its own antiderivative as well as its own derivative. Thus,

$$\int_{-1}^{2} e^x\,dx = \left[e^x\right]_{-1}^{2} = e^2 - e^{-1} \approx 7.02$$

13 $\int \frac{dx}{\sqrt{16 - x^2}} = \arcsin\frac{x}{4} + C$ **by rule 12 of Table 15-2.**

14 $\int 5\sqrt{x}\,dx = \frac{10}{3}x^{3/2} + C$ **or** $\frac{10\sqrt{x^3}}{3} + C$ $(x \ge 0)$**.** This is a reverse power rule problem.

15 $\int (4x-1)^3\,dx = \frac{1}{16}(4x-1)^4 + C$

1. **Guess your answer:** $\frac{1}{4}(4x-1)^4$
2. **Differentiate:** $(4x-1)^3 \cdot 4$ (by the chain rule)
   It's 4 times too much.
3. **Tweak guess:** $\frac{1}{16}(4x-1)^4$
4. **Differentiate to check:** $\frac{1}{4}(4x-1)^3 \cdot 4 = (4x-1)^3$

   Bingo.

16 $\int \sec^2 6x\,dx = \frac{1}{6}\tan(6x) + C$

Your guess at the antiderivative, $\tan(6x)$, gives you $(\tan(6x))' = \sec^2(6x)\cdot 6$. Tweak the guess to $\frac{1}{6}\tan(6x)$. Check: $\left(\frac{1}{6}\tan(6x)\right)' = \frac{1}{6}\sec^2(6x)\cdot 6 = \sec^2(6x)$

17 $\int \cos\frac{x-1}{2}\,dx = 2\sin\frac{x-1}{2} + C$

Your guess is $\sin\frac{x-1}{2}$. Differentiating that gives you $\cos\left(\frac{x-1}{2}\right)\cdot\frac{1}{2}$.

The tweaked guess is $2\sin\frac{x-1}{2}$. That's it.

(18) $\int\frac{3dt}{2t+5}=\frac{3}{2}\ln|2t+5|+C$

$\ln|2t+5|$ is your guess. Differentiating gives you $\frac{1}{2t+5}\cdot 2$.

You wanted a 3, but you got a 2, so tweak your guess by 3 over 2. (I'm a poet!)

TIP

This "poem" always works. Try it for the other problems. Often what you want is a 1. For example, for Problem 15, you'd have, "You wanted a 1 but you got $\frac{1}{2}$, so tweak your guess by 1 over $\frac{1}{2}$." That's 2, of course. It works!

Back to Problem 18. Your tweaked guess is $\frac{3}{2}\ln|2t+5|$. That's it.

(19) $\int_0^\pi \frac{5}{\pi}\sec(5t-\pi)\tan(5t-\pi)\,dt=\frac{2}{\pi}$

Don't let all those 5's and π's distract you — they're just a smoke screen.

Guess: $\sec(5t-\pi)$. Diff: $\sec(5t-\pi)\tan(5t-\pi)\cdot 5$.

Tweak: $\frac{1}{\pi}\sec(5t-\pi)$. Diff: $\frac{1}{\pi}\sec(5t-\pi)\tan(5t-\pi)\cdot 5$. Bingo. So now, $\frac{1}{\pi}[\sec(5t-\pi)]_0^\pi=$ $\frac{1}{\pi}[\sec(4\pi)-\sec(-\pi)]=\frac{2}{\pi}$.

(20) $\int\frac{4.5}{1+9x^2}dx=\frac{3}{2}\tan^{-1}(3x)+C$

I bet you've got the method down by now: Guess, diff, tweak, diff.

Guess: $\tan^{-1}(3x)$. Diff: $\frac{1}{1+(3x)^2}\cdot 3$

Tweak: $\frac{3}{2}\tan^{-1}(3x)$. Diff: $\frac{3}{2}\cdot\frac{1}{1+(3x)^2}\cdot 3$. That's it.

(21) $\int\frac{\sin x}{\sqrt{\cos x}}dx=-2\sqrt{\cos x}+C$

1. **It's not $\sqrt{\text{plain old } x}$, so substitute $u=\cos x$.**
2. **Differentiate and solve for du.**

$$\frac{du}{dx}=-\sin x$$
$$du=-\sin x\,dx$$

TIP

Shortcut for the u-substitution integration method. You can save a little time in all substitution problems by just solving for du — as I did here — and not bothering to solve for dx. You then tweak the expression inside the integral so that it contains the thing du equals and compensate for that tweaking by adding something outside the integral. In the current problem, du equals $-\sin x\,dx$. The integral contains a $\sin x\,dx$, so you multiply it by -1 to turn it into $-\sin x\,dx$ and then compensate for that -1 by multiplying the whole integral by -1. This is a wash because -1 times -1 equals 1. This may not sound like much of a shortcut, but it's a good time-saver once you get used to it. (Solving for dx is the book and classroom method, and that's why I went through that method in the text. But I like to use this tip. Either way is fine. Take your pick.)

3. **Tweak inside and outside of the integral with negative signs:** $= -\int \frac{-\sin x}{\sqrt{\cos x}} dx$

4. **Pull the switch:** $= -\int \frac{du}{\sqrt{u}}$

5. **Antidifferentiate with the reverse power rule:** $= -\int u^{-1/2} du = -2u^{1/2} + C$

6. **Get rid of *u*:** $= -2(\cos x)^{1/2} + C = -2\sqrt{\cos x} + C$

(22) $\int x^4 \sqrt[3]{2x^5+6}\, dx = \frac{3(x^5+3)\sqrt[3]{2x^5+6}}{20} + C$

1. **It's not $\sqrt{\text{plain old } x}$, so substitute $u = 2x^5 + 6$.**
2. **Differentiate and solve for *du*.**

$$\frac{du}{dx} = 10x^4$$
$$du = 10x^4 dx$$

3. **Tweak inside and outside:** $= \frac{1}{10}\int 10x^4 \sqrt[3]{2x^5+6}\, dx$

4. **Pull the switcheroo:** $= \frac{1}{10}\int \sqrt[3]{u}\, du$

5. **Apply the power rule in reverse:** $= \frac{1}{10} \cdot \frac{3}{4} u^{4/3} + C = \frac{3u\sqrt[3]{u}}{40} + C$

6. **Switch back:** $= \frac{3(2x^5+6)\sqrt[3]{2x^5+6}}{40} + C = \frac{3(x^5+3)\sqrt[3]{2x^5+6}}{20} + C$

(23) $\int 5x^3 e^{x^4} dx = \frac{5}{4} e^{x^4} + C$

1. **It's not $e^{\text{plain old } x}$, so $u = x^4$.**

2. **You know the drill:** $du = 4x^3 dx$

3. **Tweak:** $= \frac{5}{4}\int 4x^3 e^{x^4} dx$

4. **Switch:** $= \frac{5}{4}\int e^u du$

5. **Antidifferentiate:** $= \frac{5}{4} e^u + C$

6. **Switch back:** $= \frac{5}{4} e^{x^4} + C$

(24) $\int \frac{\sec^2 \sqrt{x}}{\sqrt{x}} dx = 2\tan\sqrt{x} + C$

1. **It's not $\sec^2$(plain old *x*), so $u = \sqrt{x}$.**

2. **Differentiate:** $du = \frac{1}{2} x^{-1/2} dx = \frac{1}{2\sqrt{x}} dx$

3. **Tweak:** $= 2\int \frac{\sec^2 \sqrt{x}}{2\sqrt{x}} dx$

4. **Switch:** $= 2\int \sec^2 u\, du$

5. **Antidifferentiate:** $= 2\tan u + C$

6. **Switch back:** $= 2\tan\sqrt{x} + C$

(25) $\displaystyle\int_0^2 \frac{t\,dt}{\left(t^2+5\right)^4} \approx 0.0011$

1. **Do the *U-and-Diff* (it's sweeping the nation!), and find the *u* indices of integration.**

   $u = t^2 + 5 \qquad$ when $t = 0,\ u = 5$

   $du = 2t\,dt \qquad$ when $t = 2,\ u = 9$

2. **The tweak:** $= \dfrac{1}{2}\displaystyle\int_0^2 \frac{2t\,dt}{\left(t^2+5\right)^4}$

3. **The switch:** $= \dfrac{1}{2}\displaystyle\int_5^9 \frac{du}{u^4}$

4. **Antidifferentiate and evaluate:** $= \dfrac{1}{2}\cdot\left[-\dfrac{1}{3}u^{-3}\right]_5^9 = -\dfrac{1}{6}\left(9^{-3} - 5^{-3}\right) \approx 0.0011$

(26) $\displaystyle\int_1^8 \frac{\left(s^{2/3}+5\right)^3}{\sqrt[3]{s}}\,ds = 1{,}974.375$

You know the drill: $u = s^{2/3} + 5$; $du = \dfrac{2}{3}s^{-1/3}ds = \dfrac{2}{3\sqrt[3]{s}}ds$

$$\int_1^8 \frac{\left(s^{2/3}+5\right)^3}{\sqrt[3]{s}}\,ds = \frac{3}{2}\int_1^8 \frac{2\left(s^{2/3}+5\right)^3}{3\sqrt[3]{s}}\,ds$$

$$\frac{3}{2}\int_1^8 u^3\,du$$

WARNING

You'll get a math ticket if you put an equal sign in front of the last line because it is *not* equal to the line before it. When you don't change the limits of integration, you get this mixed-up integral with an integrand in terms of *u*, but with limits of integration in terms of *x* (*s* in this problem). This may be one reason why the preferred book method includes switching the limits of integration — it's mathematically cleaner.

Now just antidifferentiate, switch back, and evaluate:

$$\frac{3}{2}\cdot\frac{1}{4}u^4$$

$$\frac{3}{2}\cdot\frac{1}{4}\left[\left(s^{2/3}+5\right)^4\right]_1^8 = \frac{3}{8}\left(9^4 - 6^4\right) = 1974.375$$

If you're ready to test your skills a bit more, take the following chapter quiz that incorporates the chapter topics.

Whaddya Know? Chapter 15 Quiz

Quiz time! Complete each problem to test your knowledge on the various topics covered in this chapter. You can then find the solutions and explanations in the next section.

For Problems 1a and 1b, use "Your Turn" Problem 1 from this chapter (with its accompanying figure), except that now, let $A_g(x) = \int_4^x g(t)\,dt$.

1 **(a)** Approximate $A_g(6)$.

(b) Approximate $A_g(1)$. (*Hint:* You'll need to use one of the rules from Table 15-1.)

2 Given that $A_f(x) = \int_e^{e^x} t^2\,dt$, find $\frac{d}{dx}A_f(x)$.

3 Evaluate $\int_{\pi/4}^{3\pi/4} \csc^2 x\,dx$.

4 What's $\int \sin x\,dx$?

5 What's $\int \frac{2}{\sqrt[5]{x^3}}\,dx$?

6 What's $\int 5(5x-8)^4\,dx$?

7 What's $\int e^{x/\pi}\,dx$?

8 Find the antiderivative, $\int \frac{e^{1/x}}{x^2}\,dx$, with the substitution method.

9 Find the antiderivative, $\int \frac{\cos\sqrt[3]{x}}{\sqrt[3]{x^2}}\,dx$, with the substitution method.

10 Evaluate $\int_1^e \frac{\cos(\ln x)}{x}\,dx$. Change the indices of integration.

Answers to Chapter 15 Quiz

1 **(a) About 2.**

$A_g(6)$ equals the area under g from 4 to 6. That shape looks sort of like a right triangle with a curvy "hypotenuse." If it were exactly a right triangle (with a base of 2 and a height that looks very close to 1.5), its area would be 1.5. Add a bit because of the curvy hypotenuse.

(b) About 2.5.

$A_g(1)$ concerns the area above g and below the x-axis between 1 and 4. That sort-of semi-circle looks like it has an area of about 2.5. Since that area is *below* the x-axis, it counts as *negative* area, and so, your answer would be about −2.5 if you were asked for the area from 1 to 4. However, $A_g(1)$ asks for the area from 4 to 1 (which is going backwards). Rule 2 from Table 15-1 tells you that the area from 4 to 1 is the opposite of the area from 1 to 4. Thus, your answer is the opposite of roughly −2.5, or, of course, roughly 2.5.

2 $\boldsymbol{e^{3x}}$

The input of any area function is the upper index of integration. So, for $A_f(x)=\int_e^{e^x} t^2 dt$, the input is e^x. Because the input of this function is not just a simple x, determining the derivative $\frac{d}{dx}A_f(x)$ is a chain rule problem. If the upper index had been just x, the derivative would equal x^2 by the fundamental theorem. So, because this is a chain rule problem, you replace the x with e^x, then multiply by the derivative of e^x (which is e^x). So, that gives you $\left(e^x\right)^2 e^x = e^{2x}\cdot e^x = e^{3x}$.

3 **2**

$$\begin{aligned}\int_{\pi/4}^{3\pi/4} \csc^2 x dx &= -[\cot x]_{\pi/4}^{3\pi/4}\\ &= -\left(\cot\frac{3\pi}{4} - \cot\frac{\pi}{4}\right)\\ &= -(-1-1) = 2\end{aligned}$$

4 $\boldsymbol{-\cos x + C}$

Be careful about negative signs when differentiating or integrating sines and cosines, and don't forget the C!

5 $\boldsymbol{5x^{2/5} + C}$ **or** $\boldsymbol{5\sqrt[5]{x^2} + C}$

Rewrite $\int \frac{2}{\sqrt[5]{x^3}}dx$ with a power rather than a radical, then use the reverse power rule:

$$\begin{aligned}\int \frac{2}{\sqrt[5]{x^3}}dx &= \int 2x^{-3/5}dx\\ &= \frac{2}{\frac{2}{5}}x^{2/5} + C = 5x^{2/5} + C\end{aligned}$$

6 $\boldsymbol{\frac{1}{5}(5x-8)^5 + C}$

Guess-and-check is a good method for this one.

(7) $\pi e^{x/\pi} + C$

This is another problem where guess-and-check works well.

(8) $-e^{1/x} + C$

$$\int \frac{e^{1/x}}{x^2} dx$$

let $u = \frac{1}{x}$

$$du = -\frac{1}{x^2} dx$$

Now rewrite the integrand in terms of u, solve, and switch back to x:

$$\int \frac{e^{1/x}}{x^2} dx = -\int -\frac{e^{1/x}}{x^2} dx = -\int e^u du = -e^u + C = -e^{1/x} + C$$

(9) $3\sin\sqrt[3]{x} + C$

$$\int \frac{\cos\sqrt[3]{x}}{\sqrt[3]{x^2}} dx$$

let $u = \sqrt[3]{x}$

$$du = \frac{1}{3} x^{-2/3} dx$$
$$= \frac{1}{3\sqrt[3]{x^2}} dx$$

Rewrite the integrand in terms of u, solve, and switch back to x:

$$\int \frac{\cos\sqrt[3]{x}}{\sqrt[3]{x^2}} dx = 3\int \frac{\cos\sqrt[3]{x}}{3\sqrt[3]{x^2}} dx = 3\int \cos u\, du = 3\sin u + C = 3\sin\sqrt[3]{x} + C$$

(10) $\sin 1 \approx 0.84147$

$$\int_1^e \frac{\cos(\ln x)}{x} dx$$

let $u = \ln x$

$$du = \frac{1}{x} dx$$

when $x = 1,\ u = 0$

when $x = e,\ u = 1$

$$\int_1^e \frac{\cos(\ln x)}{x} dx = \int_0^1 \cos u\, du = [\sin u]_0^1 = \sin 1 - \sin 0 = \sin 1$$

IN THIS CHAPTER

» Breaking down integrals into parts and finding trigonometric integrals

» Returning to your roots with *SohCahToa*

» Understanding the As, Bs, and Cs of partial fractions

» Looking at LIATE

Chapter 16 Integration Techniques for Experts

I figure it wouldn't hurt to give you a break from the kind of theoretical groundwork stuff that I lay on pretty thick in Chapter 15, so this chapter cuts to the chase and shows you just the nuts and bolts of several integration techniques. In Chapter 15, you saw three basic integration methods: the reverse rules, the guess-and-check method, and substitution. Now you graduate to four advanced techniques: integration by parts, trigonometric integrals, trigonometric substitution, and partial fractions. Ready?

Integration by Parts: Divide and Conquer

Integrating by parts is the integration version of the product rule for differentiation. Just take my word for it. The basic idea of integration by parts is to transform an integral you *can't* do into a simple product minus an integral you *can* do. Here's the formula:

MATH RULES

Integration-by-parts formula: $\int u\,dv = uv - \int v\,du$

And here's a memory aid for it: In the first two chunks, $\int u\,dv$ and uv, the u and v are in alphabetical order. If you remember that, you can remember that the integral on the right is just like the one on the left, except the u and v are reversed.

Don't try to understand the formula yet. You'll see how it works in a minute. And don't worry about understanding the first example coming up until you get to the end of it. The integration-by-parts process may seem pretty convoluted your first time through it, so you've got to be patient. After you work through a couple examples, you'll see it's really not that bad at all.

The integration-by-parts box: The integration-by-parts formula contains four things: *u*, *v*, *du*, and *dv*. To help keep everything straight, organize your problems with a box like the one in Figure 16-1.

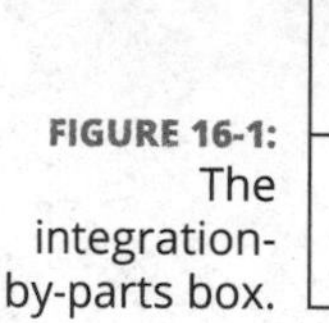

FIGURE 16-1: The integration-by-parts box.

Let's do $\int \sqrt{x} \ln x \, dx$ to see how this method works. The integration-by-parts formula converts this integral, which you can't do directly, into a simple product minus an integral you know how to do. First, you've got to split up the integrand into two chunks — one chunk becomes the *u* and the other the *dv* that you see on the left side of the formula. For this problem, the $\ln x$ becomes your *u* chunk. Then everything else is the *dv* chunk, namely $\sqrt{x}\,dx$. (In the next section, I show you how to decide what goes into the *u* chunk; then, whatever is left over is automatically the *dv* chunk.) After rewriting the integrand, you've got the following for the left side of the formula:

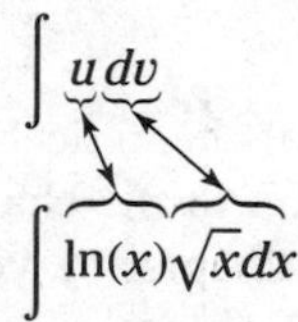

Now it's time to do the box thing. For each new problem, you should draw an empty four-square box, then put your *u* ($\ln x$ in this problem) in the upper-left square and your *dv* ($\sqrt{x}\,dx$ in this problem) in the lower-right square. See Figure 16-2.

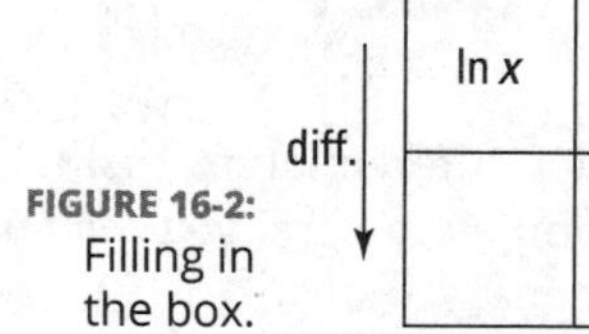

FIGURE 16-2: Filling in the box.

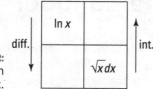

Next, you differentiate *u* to get your *du*, and you integrate *dv* to get your *v*. The arrows in Figure 16-2 remind you to differentiate on the left and to integrate on the right. Think of differentiation — the easier thing — as going down (like going downhill), and integration — the harder thing — as going up (like going uphill).

Now complete the box:

$$u = \ln x \qquad dv = \sqrt{x}\,dx$$

$$\frac{du}{dx} = \frac{1}{x} \qquad \int dv = \int \sqrt{x}\,dx$$

$$du = \frac{1}{x}dx \qquad v = \frac{2}{3}x^{3/2} \quad \text{(reverse power rule; note that you drop the } C\text{)}$$

Figure 16-3 shows the completed box.

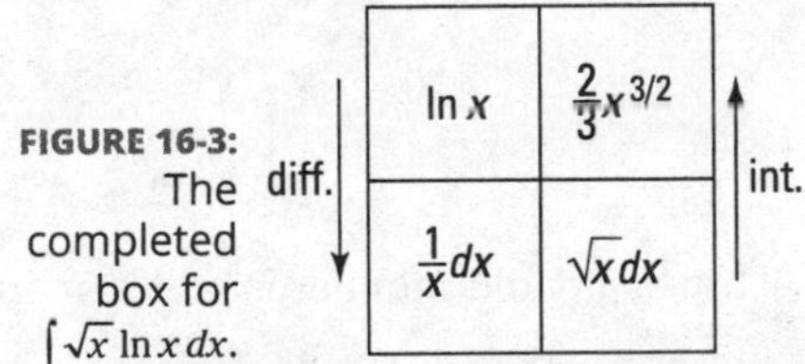

FIGURE 16-3: The completed box for $\int \sqrt{x} \ln x\, dx$.

You can also use the four-square box to help you remember the right side of the integration-by-parts formula: Start in the upper-left square and draw (or just picture) a number 7 going straight across to the right, then down diagonally to the left. See Figure 16-4.

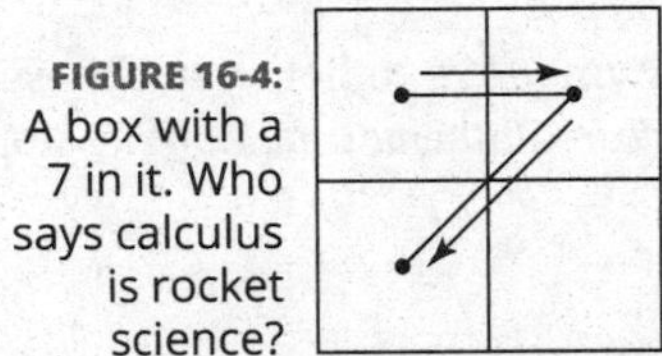

FIGURE 16-4: A box with a 7 in it. Who says calculus is rocket science?

Remembering how you "draw" the 7, look back to Figure 16-3. The right side of the integration-by-parts formula tells you to do the top part of the 7, namely $\ln x \cdot \frac{2}{3}x^{3/2}$, minus the integral of the diagonal part of the 7, $\int \frac{2}{3}x^{3/2} \cdot \frac{1}{x}dx$. By the way, all of this is *much* easier to do than to explain. Try it. You'll see how this four-square-box scheme helps you learn the formula and organize these problems.

Ready to finish? Plug everything into the formula:

$$\begin{aligned}
\int u\,dv &= uv - \int v\,du \\
\int \sqrt{x}\ln x\,dx &= \ln x \cdot \frac{2}{3}x^{3/2} - \int \frac{2}{3}x^{3/2} \cdot \frac{1}{x}dx \\
&= \frac{2}{3}x^{3/2}\ln x - \frac{2}{3}\int x^{1/2}dx \\
&= \frac{2}{3}x^{3/2}\ln x - \frac{2}{3}\left(\frac{2}{3}x^{3/2} + C\right) \quad \text{(reverse power rule)} \\
&= \frac{2}{3}x^{3/2}\ln x - \frac{4}{9}x^{3/2} - \frac{2}{3}C \\
&= \frac{2}{3}x^{3/2}\ln x - \frac{4}{9}x^{3/2} + C, \quad \text{or} \\
&= \frac{2}{3}\sqrt{x^3}\ln x - \frac{4}{9}\sqrt{x^3} + C
\end{aligned}$$

In the last step, you replace the $-\frac{2}{3}C$ with C because $-\frac{2}{3}$ times any old number is still just any old number.

Picking your *u*

Here's a great mnemonic device for how to choose your u chunk (again, once you've selected your u, everything else is automatically the dv chunk).

TIP

The LIATE mnemonic: Herbert E. Kasube came up with the acronym *LIATE* to help you choose your u (calculus nerds can check out Herb's article in the *American Mathematical Monthly* 90, 1983 issue):

L	Logarithmic	like $\log x$
I	Inverse trigonometric	like $\arctan x$
A	Algebraic	like $5x^2 + 3$
T	Trigonometric	like $\cos x$
E	Exponential	like 10^x

To pick your u chunk, go down this list in order; the first type of function on this list that appears in the integrand is the u.

Here are some helpful hints on how to remember the acronym *LIATE.* How about *Let's Integrate Another Tantalizing Example.* Or maybe you prefer *Lilliputians In Africa Tackle Elephants,* or *Lulu's Indigo And Turquoise Earrings.* The last one's not so good because it could also be *Lulu's Turquoise And Indigo Earrings* — whoops: Now you'll never remember it!

Here's a problem. Integrate $\int \arctan x\,dx$. (*Note:* Integration by parts sometimes works for integrands like this one that contain a single function.)

1. **Go down the LIATE list and pick the *u*.**

   You see that there are no logarithmic functions in $\arctan x\, dx$, but there is an inverse trigonometric function, $\arctan x$. So that's your *u*. Everything else is your *dv*, namely, plain old *dx*.

2. **Do the box thing.**

   See Figure 16-5 (and see Table 15-2 for the derivative of $\arctan x$).

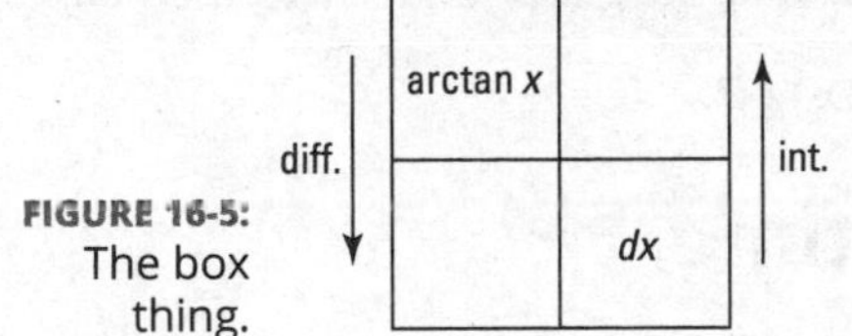

FIGURE 16-5: The box thing.

3. **Plug everything into the integration-by-parts formula (draw or imagine a 7 in the box on the right in Figure 16-5).**

   $$\int u\,dv = uv - \int v\,du$$

   $$\int \arctan x\, dx = x \arctan x - \int x \cdot \frac{1}{1+x^2} dx$$

   Now you can finish this problem by integrating $\int x \cdot \frac{1}{1+x^2} dx$ with the substitution method, setting $u = 1 + x^2$. Try it (see Chapter 15 for more on the substitution method). Note that the *u* in $u = 1 + x^2$ has nothing to do with the integration-by-parts *u*. Your final answer should be $\int \arctan x\, dx = x \arctan x - \frac{1}{2}\ln\left(1+x^2\right) + C$.

Here's another one. Integrate $\int x \sin(3x)\, dx$:

1. **Go down the LIATE list and pick the *u*.**

   Going down the LIATE list, the first type of function you find in $x\sin(3x)dx$ is a very simple algebraic one, namely *x*, so that's your *u*. Everything else is your *dv*.

2. **Do the box thing.**

   See Figure 16-6.

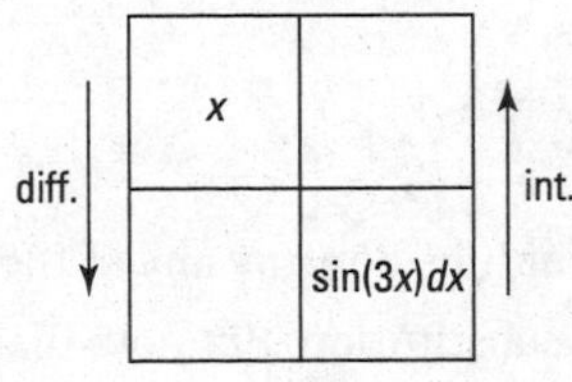

FIGURE 16-6: Yet more boxes.

3. **Plug everything into the integration-by-parts formula (draw or imagine a 7 over the box on the right in Figure 16-6).**

$$\int u\,dv = uv - \int v\,du$$

$$\int x\sin(3x)dx = -\frac{1}{3}x\cos(3x) - \int -\frac{1}{3}\cos(3x)dx$$
$$= -\frac{1}{3}x\cos(3x) + \frac{1}{3}\int \cos(3x)dx$$

You can easily integrate $\int\cos(3x)dx$ with substitution or the guess-and-check method. Go for it. Your final answer: $-\frac{1}{3}x\cos(3x) + \frac{1}{9}\sin(3x) + C$.

Integration by parts: Second time, same as the first

Sometimes you have to use the integration-by-parts method more than once because the first run through the method takes you only partway to the answer. Let's walk through a problem. Find $\int x^2 e^x dx$:

1. **Go down the LIATE list and pick the *u*.**

   You can see that $x^2 e^x dx$ contains an algebraic function, x^2, and an exponential function, e^x. (It's an exponential function because there's an *x* in the exponent.) The first on the LIATE list is x^2, so that's your *u*.

2. **Do the box thing.**

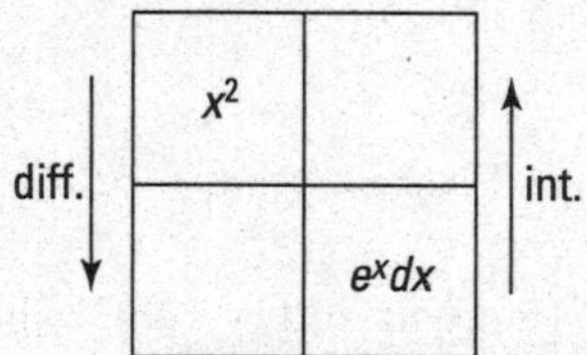

FIGURE 16-7: The boxes for $\int x^2 e^x dx$.

   See Figure 16-7.

3. **Use the integration-by-parts formula — or the "7" mnemonic.**

$$\int x^2 e^x dx = x^2 e^x - \int e^x \cdot 2x\,dx$$
$$= x^2 e^x - 2\int x e^x dx$$

   You end up with another integral, $\int xe^x dx$, that can't be done by any of the simple methods — reverse rules, guess and check, and substitution. But note that the power of *x* has been reduced from 2 to 1, so you've made some progress. When you use integration by parts again for $\int xe^x dx$, the *x* disappears entirely and you're done. Here goes:

4. **Integrate by parts again.**

I'll let you do most of this one on your own. Here's the final step:

$$\int xe^x dx = xe^x - \int e^x dx$$
$$= xe^x - e^x + C$$

5. **Take the result from Step 4 and substitute it for the $\int xe^x dx$ in the answer from Step 3 to produce the whole enchilada.**

$$\int x^2 e^x dx = x^2 e^x - 2\left(xe^x - e^x + C\right)$$
$$= x^2 e^x - 2xe^x + 2e^x - 2C$$
$$= x^2 e^x - 2xe^x + 2e^x + C$$

YOUR TURN

1 What's $\int x\cos(5x-2)dx$?

2 Evaluate $\int x \arctan x \, dx$.

3 Evaluate $\int_{-1}^{1} x 10^x dx$.

4 What's $\int x^2 e^{-x} dx$?

5 Integrate $\int e^x \sin x \, dx$. *Tip*: Sometimes you circle back to where you started from — that's a good thing!

Tricky Trig Integrals

Don't you just love trig? I'll bet you didn't realize that studying calculus was going to give you the opportunity to do so much more trig. Remember this next Thanksgiving when everyone around the dinner table is invited to mention something that they're thankful for.

In this section, you integrate powers of the six trigonometric functions, like $\int \sin^3 x\,dx$ and $\int \sec^4 x\,dx$, and products or quotients of different trig functions, like $\int \sin^2 x \cos^3 x\,dx$ and $\int \frac{\csc^2 x}{\cot x}dx$. This is pretty tedious — time to order up a double espresso.

To use the following techniques, you must have either an integrand that contains just one of the six trig functions, like $\int \csc^3 x\,dx$, or a certain pairing of trig functions, like $\int \sin^2 x \cos x\,dx$. If the integrand has two trig functions, the two must be one of these three pairs: sine with cosine, secant with tangent, or cosecant with cotangent. If you have an integrand containing something other than one of these three pairs, you can easily convert the problem into one of these pairs by using trig identities like $\sin x = \frac{1}{\csc x}$ and $\tan x = \frac{\sin x}{\cos x}$. For instance,

$$\begin{aligned}
&\int \sin^2 x \sec x \tan x\,dx\\
&= \int \sin^2 x \cdot \frac{1}{\cos x} \cdot \frac{\sin x}{\cos x}dx\\
&= \int \frac{\sin^3 x}{\cos^2 x}dx
\end{aligned}$$

After doing any needed conversions, you want to get one of the following three cases:

$$\int \sin^m x \cos^n x\,dx$$
$$\int \sec^m x \tan^n x\,dx$$
$$\int \csc^m x \cot^n x\,dx,$$

where either m or n (or both) is a positive integer.

The basic idea with most of the following trig integrals is to organize the integrand so that you can make a handy u-substitution and then integrate with the reverse power rule. You'll see what I mean in a minute.

Integrals containing sines and cosines

This section covers integrals with — can you guess? — sines and cosines.

Case 1: The power of sine is odd and positive

If the power of sine is odd and positive, lop off one sine factor and put it to the right of the rest of the expression, convert the remaining sine factors to cosines with the Pythagorean Identity, and then integrate with the substitution method where $u = \cos x$.

REMEMBER

The Pythagorean Identity: The Pythagorean Identity tells you that, for any angle x, $\sin^2 x + \cos^2 x = 1$. And thus $\sin^2 x = 1 - \cos^2 x$ and $\cos^2 x = 1 - \sin^2 x$.

Now integrate $\int \sin^3 x \cos^4 x\, dx$:

1. **Lop off one sine factor and move it to the right.**

   $$\int \sin^3 x \cos^4 x\, dx = \int \sin^2 x \cos^4 x \sin x\, dx$$

2. **Convert the remaining sines to cosines using the Pythagorean Identity and simplify.**

   $$\begin{aligned}&\int \sin^2 x \cos^4 x \sin x\, dx\\ &= \int \left(1 - \cos^2 x\right)\cos^4 x \sin x\, dx\\ &= \int \left(\cos^4 x - \cos^6 x\right)\sin x\, dx\end{aligned}$$

3. **Integrate with the substitution method, where $u = \cos x$.**

   $$\begin{aligned}&u = \cos x\\ &\frac{du}{dx} = -\sin x\\ &du = -\sin x\, dx\end{aligned}$$

   Tweak your integral so that it contains what du equals:

   $$\begin{aligned}&\int \left(\cos^4 x - \cos^6 x\right)(\sin x\, dx)\\ &= -\int \left(\cos^4 x - \cos^6 x\right)(-\sin x\, dx)\end{aligned}$$

   Now substitute and solve by the reverse power rule:

   $$\begin{aligned}&= -\int \left(u^4 - u^6\right)du\\ &= -\frac{1}{5}u^5 + \frac{1}{7}u^7 + C\\ &= -\frac{1}{5}\cos^5 x + \frac{1}{7}\cos^7 x + C \quad \text{or} \quad \frac{1}{7}\cos^7 x - \frac{1}{5}\cos^5 x + C\end{aligned}$$

Case 2: The power of cosine is odd and positive

This problem works exactly like Case 1, except that the roles of sine and cosine are reversed. Find $\int \frac{\cos^3 x}{\sqrt{\sin x}}dx$.

1. **Lop off one cosine factor and move it to the right.**

   $$\begin{aligned}\int \frac{\cos^3 x}{\sqrt{\sin x}}dx &= \int \cos^3 x\left(\sin^{-1/2} x\right)dx\\ &= \int \cos^2 x\left(\sin^{-1/2} x\right)\cos x\, dx\end{aligned}$$

2. **Convert the remaining cosines to sines with the Pythagorean Identity and simplify.**

$$\int \cos^2 x\left(\sin^{-1/2} x\right)\cos x\,dx$$
$$= \int\left(1-\sin^2 x\right)\left(\sin^{-1/2} x\right)\cos x\,dx$$
$$= \int\left(\sin^{-1/2} x - \sin^{3/2} x\right)\cos x\,dx$$

3. **Integrate with substitution, where $u = \sin x$.**

$$u = \sin x$$
$$\frac{du}{dx} = \cos x$$
$$du = \cos x\,dx$$

Now substitute:

$$= \int\left(u^{-1/2} - u^{3/2}\right)du$$

And finish integrating as in Case 1.

Case 3: The powers of both sine and cosine are even and nonnegative

Here you convert the integrand into odd powers of cosines by using the following trig identities.

REMEMBER

Two handy trig identities:

$$\sin^2 x = \frac{1-\cos(2x)}{2} \quad \text{and} \quad \cos^2 x = \frac{1+\cos(2x)}{2}$$

Then you finish the problem as in Case 2. Let's go through a problem:

$$\int \sin^4 x\cos^2 x\,dx$$
$$= \int\left(\sin^2 x\right)^2\cos^2 x\,dx$$
$$= \int\left(\frac{1-\cos(2x)}{2}\right)^2\left(\frac{1+\cos(2x)}{2}\right)dx$$
$$= \frac{1}{8}\int\left(1-\cos(2x)-\cos^2(2x)+\cos^3(2x)\right)dx \quad \text{(It's just algebra!)}$$
$$= \frac{1}{8}\int 1dx - \frac{1}{8}\int\cos(2x)dx - \frac{1}{8}\int\cos^2(2x)dx + \frac{1}{8}\int\cos^3(2x)dx$$

The first in this string of integrals is a no-brainer; the second is a simple reverse rule with a little tweak for the 2; you do the third integral by using the $\cos^2 x$ identity a second time; and the fourth integral is handled by following the steps in Case 2. Do it. Your final answer should be

$$\frac{1}{16}x - \frac{1}{64}\sin(4x) - \frac{1}{48}\sin^3(2x) + C$$

A veritable cake walk.

TIP

Don't forget your trig identities. If you get a sine-cosine problem that doesn't fit any of the three cases discussed here, try using a trig identity like $\sin^2 x + \cos^2 x = 1$ or $\cos^2 x = \frac{1+\cos(2x)}{2}$ to convert the integral into one you can handle.

For example, $\int \frac{\sin^4 x}{\cos^2 x} dx$ doesn't fit any of the three sine-cosine cases (because when you rewrite the integrand as $\sin^4 x \cos^{-2} x\, dx$, both powers are not non-negative), but you can use the Pythagorean Identity to convert it to $\int \frac{(1-\cos^2 x)^2}{\cos^2 x} dx = \int \frac{1 - 2\cos^2 x + \cos^4 x}{\cos^2 x} dx$. This splits up into $\int \sec^2 x\, dx - \int 2dx + \int \cos^2 x\, dx$, and the rest is easy. Try it. See whether you can differentiate your result and arrive back at the original problem.

YOUR TURN

 Find $\int \sqrt[3]{\sin x} \cos^3 x\, dx$.

 Evaluate $\int_0^{\pi/6} \cos^4 t \sin^2 t\, dt$.

Integrals containing secants and tangents (or cosecants and cotangents)

The method for solving integrals containing the secant-tangent pairing or the cosecant-cotangent pairing is similar to the method used for the sine-cosine problems. In this section, you'll look at two secant-tangent examples (cosecant-cotangent problems work the same way). You'll need to use the following Pythagorean Identities.

REMEMBER

The other Pythagorean Identities: For any angle x, $\tan^2 x + 1 = \sec^2 x$ and $\cot^2 x + 1 = \csc^2 x$.

Case 1: The power of tangent (or cotangent) is odd

EXAMPLE

Q. Integrate $\int \sec^3 x \tan^3 x\,dx$.

A. $= \frac{1}{5}\sec^5 x - \frac{1}{3}\sec^3 x + C$

1. Split off $\sec x \tan x$:

$$= \int \sec^2 x \tan^2 x \sec x \tan x\,dx$$

2. Use the Pythagorean Identity to convert the even number of tangents into secants:

$$= \int \sec^2 x\left(\sec^2 x - 1\right)\sec x \tan x\,dx$$
$$= \int \sec^4 x \sec x \tan x\,dx - \int \sec^2 x \sec x \tan x\,dx$$

3. Integrate with u-substitution using $u = \sec x$:

$$= \frac{1}{5}\sec^5 x - \frac{1}{3}\sec^3 x + C$$

Case 2: The power of secant (or cosecant) is even

EXAMPLE

Q. Evaluate $\int_{\pi/4}^{\pi/3} \tan^2\theta \sec^4\theta\,d\theta$.

A. $= \frac{14\sqrt{3}}{5} - \frac{8}{15}$

1. Split off a $\sec^2\theta$.

$$= \int_{\pi/4}^{\pi/3} \tan^2\theta \sec^2\theta \sec^2\theta\,d\theta$$

2. Use the Pythagorean Identity to convert the even number of secants into tangents:

$$= \int_{\pi/4}^{\pi/3} \tan^2\theta\left(\tan^2\theta + 1\right)\sec^2\theta\,d\theta = \int_{\pi/4}^{\pi/3} \tan^4\theta \sec^2\theta\,d\theta + \int_{\pi/4}^{\pi/3} \tan^2\theta \sec^2\theta\,d\theta$$

3. Do u-substitution with $u = \tan\theta$.

$$= \left[\frac{1}{5}\tan^5\theta\right]_{\pi/4}^{\pi/3} + \left[\frac{1}{3}\tan^3\theta\right]_{\pi/4}^{\pi/3}$$
$$= \frac{1}{5}\sqrt{3}^5 - \frac{1}{5}\cdot 1^5 + \frac{1}{3}\sqrt{3}^3 - \frac{1}{3}\cdot 1^3$$
$$= \frac{14\sqrt{3}}{5} - \frac{8}{15}$$

YOUR TURN

8 Find $\int \sqrt{\tan x}\sec^6 x\,dx$.

9 Determine $\int \sqrt{\csc^3 x}\cot^3 x\,dx$.

Your Worst Nightmare: Trigonometric Substitution

With the trigonometric substitution method, you can do integrals containing radicals of the following forms: $\sqrt{u^2+a^2}$, $\sqrt{a^2-u^2}$, and $\sqrt{u^2-a^2}$ (as well as powers of those roots), where a is a constant and u is an expression containing x. For instance, $\sqrt{3^2-x^2}$ is of the form $\sqrt{a^2-u^2}$.

You're going to love this technique . . . about as much as sticking a hot poker in your eye.

TIP

Desperate times call for desperate measures. Consider pulling the fire alarm on the day your teacher is presenting this topic. With any luck, your teacher will decide that they can't afford to get behind schedule and they'll just omit this topic from your final exam.

Before I show you how trigonometric substitution works, I've got some silly mnemonic tricks to help you keep the three cases of this method straight. (Remember, with mnemonic devices, silly [and vulgar] works.) First, the three cases involve three trig functions, *tangent, sine,* and *secant*. Their initial letters, *t, s,* and *s*, are the same letters as the initial letters of the name of this technique, *trigonometric substitution*. Pretty nice, eh?

Table 16-1 shows how these three trig functions pair up with the radical forms listed in the opening paragraph.

Table 16-1 A Totally Radical Table

$\tan\theta$	$\longleftrightarrow$	$\sqrt{u^2+a^2}$
$\sin\theta$	$\longleftrightarrow$	$\sqrt{a^2-u^2}$
$\sec\theta$	$\longleftrightarrow$	$\sqrt{u^2-a^2}$

To keep these pairings straight, note that the plus sign in $\sqrt{u^2+a^2}$ looks like a little *t* for *tangent*, and that the other two forms, $\sqrt{a^2-u^2}$ and $\sqrt{u^2-a^2}$, contain a *subtraction* sign — *s* is for *sine* and *secant*. To memorize what sine and secant pair up with, note that $\sqrt{a^2-u^2}$ begins with the letter *a*, and it's a *sin* to call someone an *ass*. Okay, I admit this is pretty weak. If you can come up with a better mnemonic, use it!

Ready to do some problems? I've stalled long enough.

Case 1: Tangents

Find $\int \frac{dx}{\sqrt{9x^2+4}}$. First, note that this can be rewritten as $\int \frac{dx}{\sqrt{(3x)^2+2^2}}$, so it fits the form $\sqrt{u^2+a^2}$, where $u = 3x$ and $a = 2$; you can see that this pairs up with tangent in Table 16-1.

1. **Draw a right triangle — basically a *SohCahToa* triangle — where $\tan\theta$ equals $\frac{u}{a}$, which is $\frac{3x}{2}$.**

   Because you know that $\tan\theta = \frac{\text{O}}{\text{A}}$ (from *SohCahToa* — see Chapter 6), your triangle should have $3x$ as O, the side *opposite* the angle θ, and 2 as A, the *adjacent* side. Then, your radical, $\sqrt{(3x)^2+2^2}$, or $\sqrt{9x^2+4}$, will automatically be the correct length for the hypotenuse. It's not a bad idea to confirm this with the Pythagorean Theorem, $a^2+b^2=c^2$. See Figure 16-8.

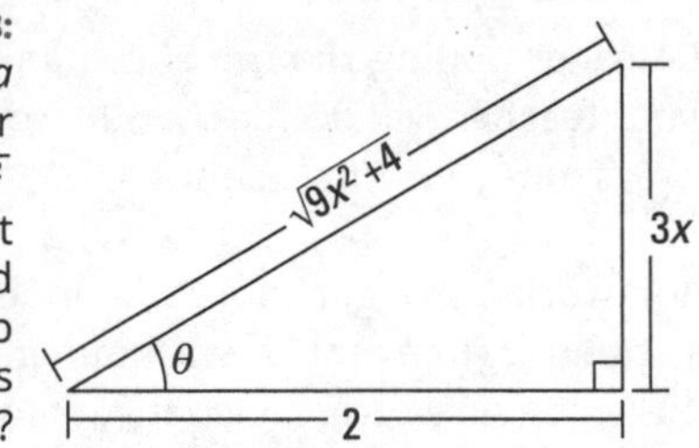

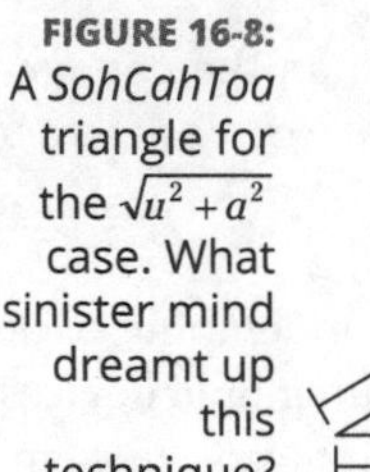
FIGURE 16-8: A *SohCahToa* triangle for the $\sqrt{u^2+a^2}$ case. What sinister mind dreamt up this technique?

2. **Solve $\tan\theta = \frac{3x}{2}$ for *x*, differentiate, and solve for *dx*.**

$$\frac{3x}{2} = \tan\theta$$
$$3x = 2\tan\theta$$
$$x = \frac{2}{3}\tan\theta$$
$$\frac{dx}{d\theta} = \frac{2}{3}\sec^2\theta$$
$$dx = \frac{2}{3}\sec^2\theta\, d\theta$$

3. **Find which trig function is represented by the radical over the *a*, and then solve for the radical.**

   Look at the triangle in Figure 16-8. The radical is the *hypotenuse* and *a* is 2, the *adjacent* side, so $\frac{\sqrt{9x^2+4}}{2}$ is $\frac{\text{H}}{\text{A}}$, which equals *secant*. So $\sec\theta = \frac{\sqrt{9x^2+4}}{2}$, and thus $\sqrt{9x^2+4} = 2\sec\theta$.

4. **Use the results from Steps 2 and 3 to make substitutions in the original problem and then integrate.**

   From Steps 2 and 3, you have $dx = \frac{2}{3}\sec^2\theta\, d\theta$ and $\sqrt{9x^2+4} = 2\sec\theta$. Now you can finally do the integration.

$$\int \frac{dx}{\sqrt{9x^2+4}} = \int \frac{\frac{2}{3}\sec^2\theta\, d\theta}{2\sec\theta}$$
$$= \frac{1}{3}\int \sec\theta\, d\theta$$
$$= \frac{1}{3}\ln|\sec\theta + \tan\theta| + C \quad \text{(an integral you should memorize or just look up)}$$

5. **Substitute the *x* expressions from Steps 1 and 3 back in for $\sec\theta$ and $\tan\theta$. You can also get the expressions from the triangle in Figure 16-8.**

$$= \frac{1}{3}\ln\left|\frac{\sqrt{9x^2+4}}{2} + \frac{3x}{2}\right| + C$$
$$= \frac{1}{3}\ln\left|\frac{\sqrt{9x^2+4}+3x}{2}\right| + C$$
$$= \frac{1}{3}\ln\left|\sqrt{9x^2+4}+3x\right| - \frac{1}{3}\ln 2 + C \quad \text{(by the log of a quotient rule, of course, and distributing the } \tfrac{1}{3}\text{)}$$
$$= \frac{1}{3}\ln\left|\sqrt{9x^2+4}+3x\right| + C \quad \text{(because } -\tfrac{1}{3}\ln 2 + C \text{ is just a constant)}$$

Now tell me, when was the last time you had so much fun? Before tackling Case 2, here are a couple tips.

TIP

Step 1 is $\frac{u}{a}$. For all three cases in trigonometric substitution, Step 1 always involves drawing a triangle in which the trig function in question equals $\frac{u}{a}$:

Case 1 is $\tan\theta = \frac{u}{a}$.

Case 2 is $\sin\theta = \frac{u}{a}$.

Case 3 is $\sec\theta = \frac{u}{a}$.

The fact that the u goes in the numerator of this $\frac{u}{a}$ fraction should be easy to remember because u is an expression in x and something like $\frac{3x}{2}$ is somewhat simpler and more natural to see than $\frac{2}{3x}$. Just remember, the x goes on top.

TIP

Step 3 is $\frac{\sqrt{}}{a}$. For all three cases, Step 3 always involves putting the radical over the a. The three cases are given here, but you don't need to memorize the trig functions in this list because you'll know which one you've got by just looking at the triangle — assuming you know *SohCahToa* and the reciprocal trig functions (flip back to Chapter 6 if you don't know them). I've left out what goes under the radicals because by the time you're doing Step 3, you've already got the right radical expression.

Case 1 is $\sec\theta = \frac{\sqrt{}}{a}$.

Case 2 is $\cos\theta = \frac{\sqrt{}}{a}$.

Case 3 is $\tan\theta = \frac{\sqrt{}}{a}$.

In a nutshell, just remember $\frac{u}{a}$ for Step 1 and $\frac{\sqrt{}}{a}$ for Step 3. How about **U̲** **A̲**re **R̲**adically **A̲**wesome?

Case 2: Sines

Integrate $\int \frac{dx}{x^2\sqrt{16-x^2}}$, rewriting it first as $\int \frac{dx}{x^2\sqrt{4^2-x^2}}$ so that it fits the form $\sqrt{a^2-u^2}$, where $a=4$ and $u=x$.

1. **Draw a right triangle where $\sin\theta = \frac{u}{a}$, which is $\frac{x}{4}$.**

   Sine equals $\frac{\text{O}}{\text{H}}$, so the *opposite* side is x and the *hypotenuse* is 4. The length of the adjacent side is then automatically equal to your radical, $\sqrt{16-x^2}$. (Confirm this with the Pythagorean Theorem.) See Figure 16-9.

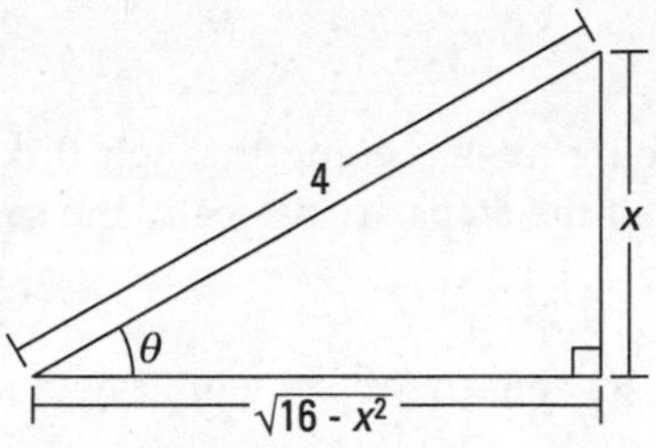

FIGURE 16-9: A *SohCahToa* triangle for the $\sqrt{a^2 - u^2}$ case.

2. **Solve $\sin\theta = \frac{x}{4}$ for x, differentiate, and solve for *dx*.**

$$\frac{x}{4} = \sin\theta$$
$$x = 4\sin\theta$$
$$\frac{dx}{d\theta} = 4\cos\theta$$
$$dx = 4\cos\theta\, d\theta$$

3. **Find which trig function equals the radical over the *a*, and then solve for the radical.**

Look at the triangle in Figure 16-9. The radical, $\sqrt{16 - x^2}$, over the *a*, 4, is $\frac{\text{A}}{\text{H}}$, which you know from *SohCahToa* equals *cosine.* That gives you

$$\cos\theta = \frac{\sqrt{16 - x^2}}{4}$$
$$\sqrt{16 - x^2} = 4\cos\theta$$

4. **Use the results from Steps 2 and 3 to make substitutions in the original problem and then integrate.**

Note that you have to make three substitutions here, not just two like in the first example. From Steps 2 and 3, you've got

$x = 4\sin\theta,\ dx = 4\cos\theta\, d\theta,$ and $\sqrt{16 - x^2} = 4\cos\theta,$ so

$$\begin{aligned}\int \frac{dx}{x^2\sqrt{16 - x^2}} &= \int \frac{4\cos\theta\, d\theta}{(4\sin\theta)^2\, 4\cos\theta} \\ &= \int \frac{d\theta}{16\sin^2\theta} \\ &= \frac{1}{16}\int \csc^2\theta\, d\theta \\ &= -\frac{1}{16}\cot\theta + C\end{aligned}$$

5. **The triangle shows that $\cot\theta = \frac{\sqrt{16 - x^2}}{x}$. Substitute back for your final answer.**

$$\begin{aligned}&= -\frac{1}{16} \cdot \frac{\sqrt{16 - x^2}}{x} + C \\ &= -\frac{\sqrt{16 - x^2}}{16x} + C\end{aligned}$$

It's a walk in the park.

Case 3: Secants

In the interest of space — and sanity — I'm going to skip going through this case in detail. But you won't have any trouble with it because all the steps are basically the same as in Cases 1 and 2.

Try this one. Integrate $\int \frac{\sqrt{x^2-9}}{x}dx$. I'll get you started. In Step 1, you draw a triangle, where $\sec\theta = \frac{u}{a}$, that's $\frac{x}{3}$. Now take it from there. Here's the answer (no peeking if you haven't done it yet): $\sqrt{x^2-9} - 3\arctan\frac{\sqrt{x^2-9}}{3} + C$, or $\sqrt{x^2-9} - 3\operatorname{arcsec}\frac{x}{3} + C$, or $\sqrt{x^2-9} - 3\arccos\frac{3}{x} + C$.

YOUR TURN

10 Integrate $\int \frac{dx}{(9x^2+4)\sqrt{9x^2+4}}$.

11 What's $\int \frac{dx}{25-x^2}$? *Hint*: This is a $\sqrt{a^2-u^2}$ problem where $\frac{u}{a} = \sin\theta$.

12 Integrate $\int \frac{dx}{\sqrt{625x^2-121}}$. *Hint*: This is a $\sqrt{u^2-a^2}$ problem where $\frac{u}{a} = \sec\theta$.

13 Last one: $\int \frac{\sqrt{4x^2-1}}{x}dx$. Same hint as in Problem 12.

The *A*'s, *B*'s, and *Cx*'s of Partial Fractions

Just when you thought it couldn't get any worse than trigonometric substitution, I give you the partial fractions technique.

You use the partial fractions method to integrate rational functions like $\frac{6x^2+3x-2}{x^3+2x^2}$. The basic idea involves "unadding" a fraction: Adding works like this: $\frac{1}{2}+\frac{1}{3}=\frac{5}{6}$. So, you can "unadd" $\frac{5}{6}$ by splitting it up into $\frac{1}{2}$ plus $\frac{1}{3}$. This is what you do with the partial fraction technique, except that you do it with complicated rational functions instead of ordinary fractions.

Before using the partial fractions technique, you have to check that your integrand is a "proper" fraction — that's one where the degree of the numerator is less than the degree of the denominator. If the integrand is "improper," like $\int \frac{2x^3+x^2-10}{x^3-3x-2}dx$, you first have to do long polynomial division to transform the improper fraction into a sum of a polynomial (which sometimes will be just a number) and a proper fraction. Here's the division for this improper fraction. Basically, it works like regular long division:

$$\begin{array}{r|l} & \quad\quad\quad\quad 2 \\ x^3-3x-2 & 2x^3+x^2+0x-10 \\ & \underline{2x^3 \quad\quad -6x\ -4} \\ & \quad\quad x^2\ +6x\ -6 \end{array}$$

With regular division, if you divide, say, 23 (the dividend) by 4 (the divisor), you get a quotient of 5 and a remainder of 3, which tells you that $\frac{23}{4}$ equals $5+\frac{3}{4}$, or $5\frac{3}{4}$. The four pieces in this polynomial division (the dividend, the divisor, the quotient, and the remainder) work the same way. The quotient is 2 and the remainder is x^2+6x-6, thus $\frac{2x^3+x^2-10}{x^3-3x-2}$ equals $2+\frac{x^2+6x-6}{x^3-3x-2}$. The original problem, $\int \frac{2x^3+x^2-10}{x^3-3x-2}dx$, therefore becomes $\int 2dx+\int \frac{x^2+6x-6}{x^3-3x-2}dx$. The first integral is just $2x+C$. You would then do the second integral with the partial fractions method. Let's walk through a basic problem and then a more advanced one.

Case 1: The denominator contains only linear factors

Integrate $\int \frac{5}{x^2+x-6}dx$. This is a Case 1 problem because the factored denominator (see Step 1) contains only *linear* factors — in other words, *first*-degree polynomials. (Also note that each factor is raised to the 1st power. If one or more factors is raised to a power greater than 1, you have a Case 3 problem.)

1. **Factor the denominator.**

$$\frac{5}{x^2+x-6}=\frac{5}{(x-2)(x+3)}$$

2. **Break up the fraction on the right into a sum of fractions, where each factor of the denominator in Step 1 becomes the denominator of a separate fraction. Then put capital-letter unknowns in the numerator of each fraction.**

$$\frac{5}{(x-2)(x+3)}=\frac{A}{(x-2)}+\frac{B}{(x+3)}$$

3. **Multiply both sides of this equation by the left side's denominator.**

   This is basic algebra, so you can't possibly want to see the steps, right?

   $5=A(x+3)+B(x-2)$

4. **Take the roots of the linear factors and plug them — one at a time — into x in the equation from Step 3, and solve for the capital-letter unknowns.**

   If $x=2$,

   $5 = A(2+3)+B(2-2)$

   $5 = 5A$

   $A=1$

   If $x=-3$,

   $5 = (-3+3)+B(-3-2)$

   $5 = -5B$

   $B=-1$

5. **Plug these results into the A and B in the equation from Step 2.**

$$\frac{5}{(x-2)(x+3)}=\frac{1}{(x-2)}+\frac{-1}{(x+3)}$$

6. **Split up the original integral into the partial fractions from Step 5 and you're home free.**

$$\begin{aligned}\int\frac{5}{x^2+x-6}dx&=\int\frac{1}{(x-2)}dx+\int\frac{-1}{(x+3)}dx\\&=\ln|x-2|-\ln|x+3|+C\\&=\ln\left|\frac{x-2}{x+3}\right|+C\qquad\text{(the log of a quotient rule)}\end{aligned}$$

Case 2: The denominator contains irreducible quadratic factors

Sometimes you can't factor a denominator all the way down to linear factors because some quadratics are irreducible — like prime numbers, they can't be factored.

TIP

Check the *discriminant*. You can easily check whether a quadratic (ax^2+bx+c) is reducible or not by checking its discriminant, b^2-4ac. If the discriminant is negative, the quadratic is irreducible. If the discriminant is a perfect square like 0, 1, 4, 9, 16, 25, and so on, the quadratic can be factored into factors you're used to seeing, like $(2x-5)(x+5)$. This is what happens in a Case 1 problem. The last possibility is that the discriminant equals a non-square positive number, as with the quadratic $x^2+10x+1$, for example, that has a discriminant of 96. In that case, the quadratic can be factored, but you get ugly factors involving square roots. You almost certainly will not get a problem like that.

Using the partial fractions technique with irreducible quadratics is a bit different. Here's a problem: Integrate $\int \frac{5x^3+9x-4}{x(x-1)(x^2+4)}dx$.

1. **Factor the denominator.**

   I did this step for you — a random act of kindness. Note that x^2+4 is irreducible because its discriminant is negative. (Like with the Case 1 example, note that the three factors in this denominator are all raised to the 1st power. If any of the powers had been greater than 1, this would be a Case 3 problem.)

2. **Break up the fraction into a sum of "partial fractions."**

   If you have an irreducible quadratic factor (like the x^2+4), the numerator for that partial fraction needs two capital-letter unknowns instead of just one. You write them in the form of $Px+Q$.

   $$\frac{5x^3+9x-4}{x(x-1)(x^2+4)}=\frac{A}{x}+\frac{B}{x-1}+\frac{Cx+D}{x^2+4}$$

3. **Multiply both sides of this equation by the left-side denominator.**

   $$5x^3+9x-4=A(x-1)(x^2+4)+B(x)(x^2+4)+(Cx+D)(x)(x-1)$$

4. **Take the roots of the linear factors and plug them — one at a time — into x in the equation from Step 3, and then solve.**

   $$\begin{array}{ll} \text{If } x=0, & \text{If } x=1, \\ -4=-4A & 10=5B \\ A=1 & B=2 \end{array}$$

   Unlike in the Case 1 example, you can't solve for all the unknowns by plugging in the roots of the linear factors, so you have more work to do.

5. **Plug into the Step 3 equation the known values of *A* and *B* and any two values for *x* not used in Step 4 (low numbers make the arithmetic easier) to get a system of two equations in *C* and *D*.**

   $A=1$ and $B=2$, so

   $$\begin{array}{ll} \text{If } x=-1, & \text{If } x=2, \\ -18=-10-10-2C+2D & 54=8+32+4C+2D \\ 2=-2C+2D & 14=4C+2D \\ 1=-C+D & 7=2C+D \end{array}$$

6. **Solve the system: $1 = -C + D$ and $7 = 2C + D$.**

   You should get $C = 2$ and $D = 3$.

7. **Split up the original integral and integrate.**

   Using the values obtained in Steps 4 and 6, $A = 1$, $B = 2$, $C = 2$, and $D = 3$, and the equation from Step 2, you can split up the original integral into three pieces:

   $$\int \frac{5x^3 + 9x - 4}{x(x-1)(x^2+4)} dx = \int \frac{1}{x} dx + \int \frac{2}{x-1} dx + \int \frac{2x+3}{x^2+4} dx$$

   And with simple algebra, you can split up the third integral on the right into two pieces, resulting in the final partial fraction decomposition:

   $$\int \frac{5x^3 + 9x - 4}{x(x-1)(x^2+4)} dx = \int \frac{1}{x} dx + \int \frac{2}{x-1} dx + \int \frac{2x}{x^2+4} dx + \int \frac{3}{x^2+4} dx$$

   The first two integrals are easy. For the third, you use substitution with $u = x^2 + 4$ and $du = 2xdx$. The fourth is done with the arctangent rule, which you should memorize: $\int \frac{dx}{a^2 + x^2} = \frac{1}{a} \arctan \frac{x}{a} + C$.

   $$\int \frac{5x^3 + 9x - 4}{x(x-1)(x^2+4)} dx = \ln|x| + 2\ln|x-1| + \ln|x^2+4| + \frac{3}{2} \arctan\left(\frac{x}{2}\right) + C$$

   $$= \ln\left|x(x-1)^2(x^2+4)\right| + \frac{3}{2} \arctan\left(\frac{x}{2}\right) + C$$

Bonus: Equating coefficients of like terms

Here's another method for finding the capital-letter unknowns that you should have in your bag of tricks. Say you get the following for your Step 3 equation (this comes from a problem with two irreducible quadratic factors):

$$2x^3 + x^2 - 5x + 4 = (Ax + B)(x^2 + 1) + (Cx + D)(x^2 + 2x + 2)$$

This equation has no linear factors, so you can't plug in the roots to get the unknowns. Instead, expand the right side of the equation:

$$2x^3 + x^2 - 5x + 4 = Ax^3 + Ax + Bx^2 + B + Cx^3 + 2Cx^2 + 2Cx + Dx^2 + 2Dx + 2D$$

And collect like terms:

$$2x^3 + x^2 - 5x + 4 = (A + C)x^3 + (B + 2C + D)x^2 + (A + 2C + 2D)x + (B + 2D)$$

Then equate the coefficients of like terms from the left and right sides of the equation:

$$\begin{aligned} 2 &= A + C \\ 1 &= B + 2C + D \\ -5 &= A + 2C + 2D \\ 4 &= B + 2D \end{aligned}$$

You then solve this system of simultaneous equations to get *A*, *B*, *C*, and *D*.

TIP

How about a shortcut? You can finish the Case 2 example by using a shortcut version of the equating-of-coefficients method. Once you have the values for *A* and *B* from Step 4, you can look back at the equation in Step 3, and equate the coefficients of the x^3 term on the left and right sides of the equation. Can you see, without actually doing the expansion, that on the right you get $(A+B+C)x^3$? So, $5x^3 = (A+B+C)x^3$, which means that $5 = A+B+C$, and because $A = 1$ and $B = 2$ (from Step 4), *C* must equal 2. Then, using these values for *A*, *B*, and *C*, and any value of *x* (other than 0 or 1), you can get *D*. How about that for a simple shortcut?

TIP

Practice makes perfect. In a nutshell, you have three ways to find your capital-letter unknowns: 1) Plugging in the roots of the linear factors of the denominator if there are any, 2) Plugging in other values of *x* and solving the resulting system of equations, and 3) Equating the coefficients of like terms. With practice, you'll get good at combining these methods to find your unknowns quickly.

Case 3: The denominator contains one or more factors raised to a power greater than 1

Say you want to integrate $\int \frac{2x+1}{x^3(x^2+1)^2}dx$ with the partial fraction technique. This is a Case 3 problem because, unlike in Cases 1 and 2, there are factors in the denominator (in this case, both factors) raised to a power greater than 1. For a problem like this, your Step 2 sum of partial fractions will look like this:

$$\frac{2x+1}{x^3(x^2+1)^2} = \frac{A}{x} + \frac{B}{x^2} + \frac{C}{x^3} + \frac{Dx+E}{(x^2+1)} + \frac{Fx+G}{(x^2+1)^2}$$

As you can see, you need a partial fraction for each different power (up to the power on the factor) of any factor that's raised to a power greater than 1.

The rest of a Case 3 solution works the same as in Cases 1 and 2. The remaining steps of the solution for this particular problem are a bit long and messy, because you've got six partial fractions. I'll spare you the gory details, but if you feel like slogging through it, here's the final solution:

$$\int \frac{2x+1}{x^3(x^2+1)^2}dx = \ln\frac{x^2+1}{x^2} - 3\arctan x - \frac{2x+1}{2(x^2+1)} - \frac{1}{2x^2} - \frac{2}{x} + C$$

YOUR TURN

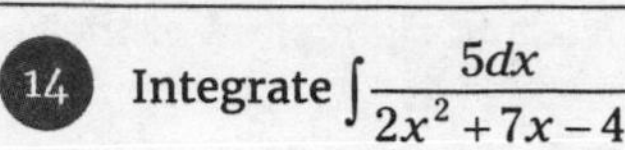

14 Integrate $\int \frac{5dx}{2x^2+7x-4}$.

15 Integrate $\int \frac{2x-3}{(3x-1)(x+4)(x+5)}dx$.

16 What's $\int \frac{x^2+x+1}{x^3-3x^2+3x-1}dx$?

17 Integrate $\int \frac{dx}{x^4+6x^2+5}$.

18 Integrate $\int \frac{4x^3+3x^2+2x+1}{x^4-1}dx$.

19 What's $\int \frac{x^2-x}{(x+1)\left(x^2+1\right)\left(x^2+2\right)}dx$?

Practice Questions Answers and Explanations

1. $\int x\cos(5x-2)\,dx = \frac{1}{5}x\sin(5x-2)+\frac{1}{25}\cos(5x-2)+C$

1. Pick x as your u, because the algebraic function x is the first on the LIATE list.

2. Fill in your box.

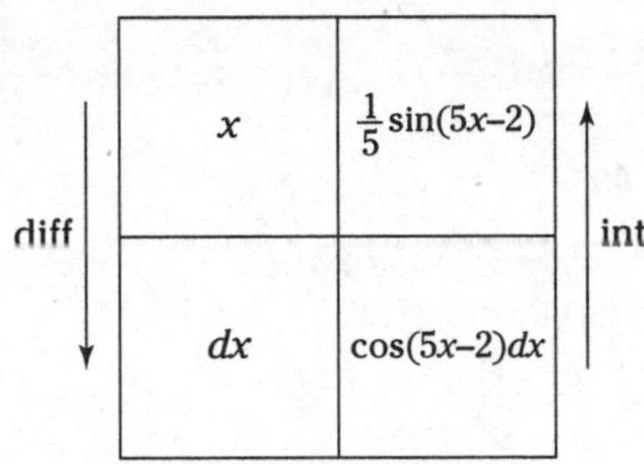

3. Use the "7" rule.

$$\int x\cos(5x-2)dx = \frac{1}{5}x\sin(5x-2)\frac{1}{5} - \sin\int(5x-2)dx$$

4. Finish by integrating.

$$= \frac{1}{5}x\sin(5x-2)+\frac{1}{25}\cos(5x-2)+C$$

2. $\int x\arctan x\,dx = \frac{1}{2}x^2\arctan x - \frac{1}{2}x + \frac{1}{2}\arctan x + C$

1. Pick $\arctan x$ as your u.

2. Do the box.

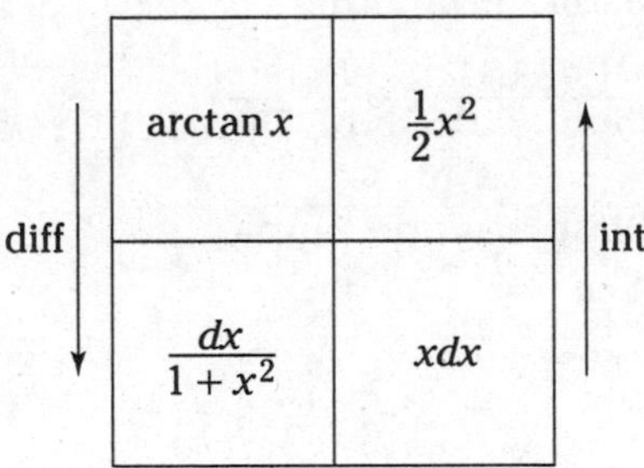

3. **Apply the "7" rule.**

$$\begin{aligned}\int x\arctan x\,dx &= \frac{1}{2}x^2\arctan x - \frac{1}{2}\int\frac{x^2dx}{1+x^2}\\ &= \frac{1}{2}x^2\arctan x - \frac{1}{2}\int\frac{x^2+1-1}{1+x^2}dx\\ &= \frac{1}{2}x^2\arctan x - \frac{1}{2}\int dx + \frac{1}{2}\int\frac{dx}{1+x^2}\\ &= \frac{1}{2}x^2\arctan x - \frac{1}{2}x + \frac{1}{2}\arctan x + C \text{ or } \frac{x^2+1}{2}\arctan x - \frac{x}{2} + C\end{aligned}$$

(3) $\displaystyle\int_{-1}^{1} x10^x\,dx = \frac{101\ln 10 - 99}{10(\ln 10)^2}$

1. **Pick the algebraic x as your u.**
2. **Box it.**

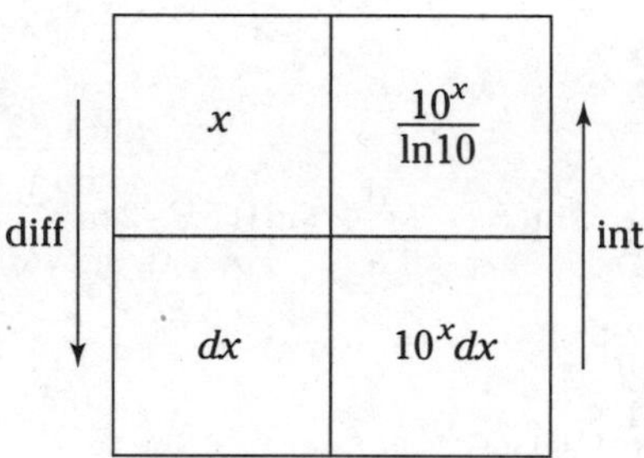

3. **Do the "7."**

$$\begin{aligned}\int_{-1}^{1} x10^x\,dx &= \left[\frac{x10^x}{\ln 10}\right]_{-1}^{1} - \frac{1}{\ln 10}\int_{-1}^{1}10^x\,dx\\ &= \frac{10}{\ln 10} + \frac{1}{10\ln 10} - \frac{1}{\ln 10}\left[\frac{10^x}{\ln 10}\right]_{-1}^{1}\\ &= \frac{10}{\ln 10} + \frac{1}{10\ln 10} - \frac{1}{\ln 10}\left(\frac{10}{\ln 10} - \frac{1}{10\ln 10}\right)\\ &= \frac{10}{\ln 10} + \frac{1}{10\ln 10} - \frac{10}{(\ln 10)^2} + \frac{1}{10(\ln 10)^2}\\ &= \frac{101\ln 10 - 99}{10(\ln 10)^2}\end{aligned}$$

4 $\int x^2 e^{-x} dx = -e^{-x}(x^2 + 2x + 2) + C$

1. **Pick x^2 as your u.**
2. **Box it.**

diff ↓			int ↑
	x^2	$-e^{-x}$	
	$2xdx$	$e^{-x}dx$	

3. **"7" it.**

$$\int x^2 e^{-x} dx = -x^2 e^{-x} + 2\int x e^{-x} dx$$

In the second integral, the power of x is reduced by 1, so you're making progress.

4. **Repeat the process for the second integral: Pick it and box it.**

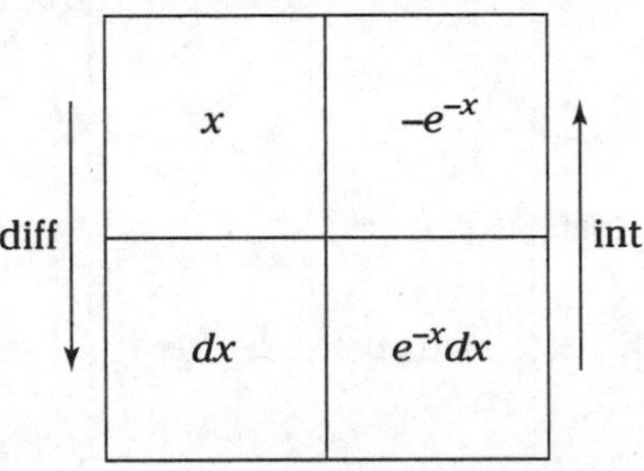

5. **Apply the "7" rule for the second integral.**

$$\int x e^{-x} dx = -xe^{-x} + \int e^{-x} dx = -xe^{-x} - e^{-x} + C$$

6. **Take this result and plug it into the second integral from Step 3.**

$$\begin{aligned}\int x^2 e^{-x} dx &= -x^2 e^{-x} + 2(-xe^{-x} - e^{-x} + C)\\ &= -x^2 e^{-x} - 2xe^{-x} - 2e^{-x} + C\\ &= -e^{-x}(x^2 + 2x + 2) + C\end{aligned}$$

(5) $\int e^x \sin x\,dx = \frac{e^x \sin x}{2} - \frac{e^x \cos x}{2} + C$

1. **Pick $\sin x$ as your u — it's a T from LIATE.**
2. **Box it.**

diff ↓			int ↑
	$\sin x$	e^x	
	$\cos x\,dx$	$e^x dx$	

3. **"7" it.**

$$\int e^x \sin x\,dx = e^x \sin x - \int e^x \cos x\,dx$$

Doesn't look like progress, but it is. Repeat this process for $\int e^x \cos x\,dx$.

4. **Pick $\cos x$ as your u and box it.**

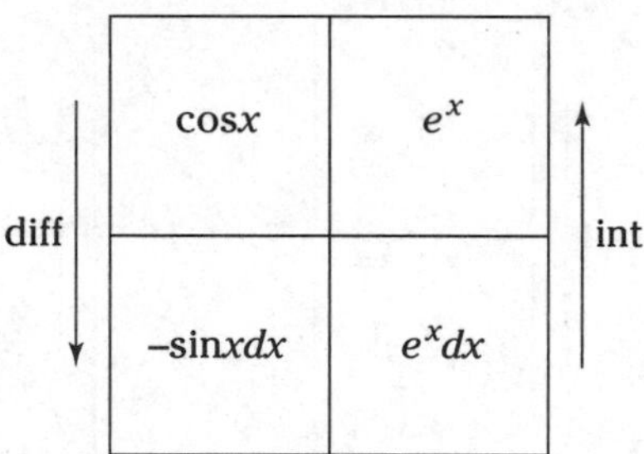

5. **"7" it.**

$$\int e^x \cos x\,dx = e^x \cos x + \int e^x \sin x\,dx$$

6. **Plug this result into the second integral from Step 3.**

$$\int e^x \sin x\,dx = e^x \sin x - e^x \cos x - \int e^x \sin x\,dx$$

7. **You want to solve for $\int e^x \sin x\,dx$, so bring them both to the left side and solve.**

$$2\int e^x \sin x\,dx = e^x \sin x - e^x \cos x + C$$

$$\int e^x \sin x\,dx = \frac{e^x \sin x}{2} - \frac{e^x \cos x}{2} + C$$

6 $\int \sqrt[3]{\sin x} \cos^3 x\,dx = \frac{3}{4}\sin^{4/3} x - \frac{3}{10}\sin^{10/3} x + C$

1. **Split off one $\cos x$.**

$$\int \sqrt[3]{\sin x} \cos^2 x \cos x\,dx$$

2. **Convert the even number of cosines into sines with the Pythagorean Identity.**

$$= \int \sqrt[3]{\sin x}\left(1-\sin^2 x\right)\cos x\,dx = \int \sin^{1/3} x \cos x\,dx - \int \sin^{7/3} x \cos x\,dx$$

3. **Integrate with u-substitution using $u = \sin x$.**

$$= \frac{3}{4}\sin^{4/3} x - \frac{3}{10}\sin^{10/3} x + C$$

7 $\int_0^{\pi/6} \cos^4 t \sin^2 t\,dt = \frac{\pi}{96}$

1. **Convert to odd powers of cosine with trig identities $\cos^2 x = \frac{1+\cos(2x)}{2}$ and $\sin^2 x = \frac{1-\cos(2x)}{2}$.**

$$= \int_0^{\pi/6} \left(\frac{1+\cos(2t)}{2}\right)^2 \frac{1-\cos(2t)}{2}\,dt$$

2. **Simplify and FOIL.**

$$= \frac{1}{8}\int_0^{\pi/6} \left(1-\cos^2(2t)\right)\left(1+\cos(2t)\right)dt = \frac{1}{8}\int_0^{\pi/6} 1dt + \frac{1}{8}\int_0^{\pi/6} \cos(2t)dt - \frac{1}{8}\int_0^{\pi/6} \cos^2(2t)dt - \frac{1}{8}\int_0^{\pi/6} \cos^3(2t)dt$$

3. **Integrate.**

The first and second are simple; for the third, you use the same trig identity again; the fourth is handled like you handled Problem 6. Here's what you should get:

$$= \frac{1}{8}\int_0^{\pi/6} 1dt + \frac{1}{8}\int_0^{\pi/6} \cos(2t)dt - \frac{1}{16}\int_0^{\pi/6} 1dt - \frac{1}{16}\int_0^{\pi/6} \cos(4t)dt - \frac{1}{8}\int_0^{\pi/6} \cos(2t)dt + \frac{1}{8}\int_0^{\pi/6} \sin^2(2t)\cos(2t)dt$$

$$= \frac{1}{16}\int_0^{\pi/6} dt - \frac{1}{16}\int_0^{\pi/6} \cos(4t)dt + \frac{1}{8}\int_0^{\pi/6} \sin^2(2t)\cos(2t)dt$$

$$= \left[\frac{1}{16}t\right]_0^{\pi/6} - \left[\frac{1}{64}\sin(4t)\right]_0^{\pi/6} + \left[\frac{1}{48}\sin^3(2t)\right]_0^{\pi/6}$$

$$= \frac{\pi}{96} - \frac{\sqrt{3}}{128} + \frac{\sqrt{3}}{128}$$

$$= \frac{\pi}{96}$$

8 $\int \sqrt{\tan\theta}\sec^6\theta\,d\theta = \frac{2}{11}\tan^{11/2}\theta + \frac{4}{7}\tan^{7/2}\theta + \frac{2}{3}\tan^{3/2}\theta + C$

1. **Split off a $\sec^2\theta$.**

$$= \int \tan^{1/2}\theta \sec^4\theta \sec^2\theta\,d\theta$$

2. **Convert to tangents.**

$$= \int \tan^{1/2}\theta\left(\tan^2\theta+1\right)^2 \sec^2\theta\, d\theta$$
$$= \int \tan^{1/2}\theta\left(\tan^4\theta+2\tan^2\theta+1\right)\sec^2\theta\, d\theta$$
$$= \int \tan^{9/2}\theta \sec^2\theta\, d\theta + 2\int \tan^{5/2}\theta\sec^2\theta\, d\theta + \int \tan^{1/2}\theta\sec^2\theta\, d\theta$$

3. **Do u-substitution with $u = \tan\theta$.**

$$= \frac{2}{11}\tan^{11/2}\theta + \frac{4}{7}\tan^{7/2}\theta + \frac{2}{3}\tan^{3/2}\theta + C$$

(9) $\int \sqrt{\csc^3 x}\cot^3 x\, dx = -\frac{2}{5}\csc^{7/2} x + \frac{2}{3}\csc^{3/2} x + C$

1. **Split off $\csc x \cot x$.**

$$= \int \csc^{1/2} x \cot^2 x \csc x \cot x\, dx$$

2. **Convert the even number of cotangents to cosecants with the Pythagorean Identity.**

$$= \int \csc^{1/2} x\left(\csc^2 x - 1\right)\csc x \cot x\, dx$$

3. **Finish with a u-substitution.**

$$= \int \csc^{5/2} x \csc x \cot x\, dx - \int \csc^{1/2} x \csc x \cot x\, dx$$
$$= \int u^{5/2}(-du) - \int u^{1/2}(-du)$$
$$= -\frac{2}{7}u^{7/2} + \frac{2}{3}u^{3/2} + C$$
$$= -\frac{2}{7}\csc^{7/2} x + \frac{2}{3}\csc^{3/2} x + C$$

(10) $\int \frac{dx}{\left(9x^2+4\right)\sqrt{9x^2+4}} = \frac{x}{4\sqrt{9x^2+4}} + C$

1. **Rewrite as $\int \frac{dx}{\sqrt{(3x)^2+2^2}^{\,3}}$.**

2. **Draw your triangle, remembering that $\tan\theta = \frac{u}{a}$.**

   See the following figure.

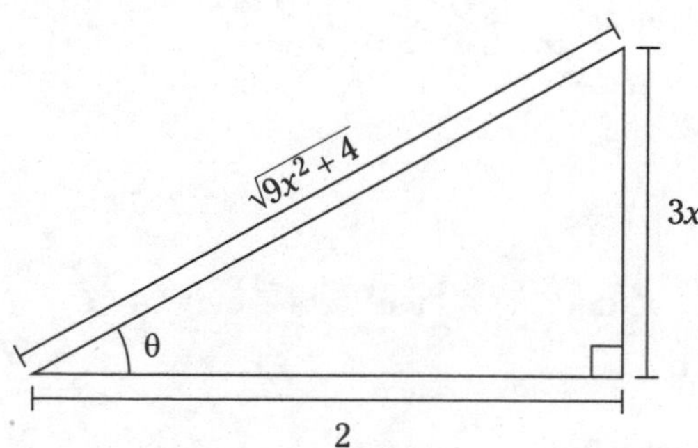

3. **Solve $\tan\theta = \frac{3x}{2}$ for x, differentiate, and solve for dx.**

$3x = 2\tan\theta \qquad x = \frac{2}{3}\tan\theta \qquad dx = \frac{2}{3}\sec^2\theta\, d\theta$

4. **Do the $\frac{\sqrt{}}{a}$ thing.**

$\frac{\sqrt{9x^2+4}}{2} = \sec\theta \qquad \sqrt{9x^2+4} = 2\sec\theta$

5. **Substitute.**

$$\int \frac{dx}{\sqrt{9x^2+4}^3}$$

$$= \int \frac{\frac{2}{3}\sec^2\theta d\theta}{(2\sec\theta)^3} = \frac{1}{12}\int \frac{d\theta}{\sec\theta} = \frac{1}{12}\int \cos\theta d\theta$$

6. **Integrate to get $\frac{1}{12}\sin\theta + C$.**
7. **Switch back to x (use the triangle).**

$$= \frac{1}{12}\left(\frac{3x}{\sqrt{9x^2+4}}\right) + C = \frac{x}{4\sqrt{9x^2+4}} + C$$

(11) $\int \frac{dx}{25-x^2} = \frac{1}{5}\ln\left|\frac{x+5}{\sqrt{25-x^2}}\right| + C$

1. **Rewrite as $\int \frac{dx}{5^2 - x^2}$.**
2. **Draw your triangle.**

For this problem, $\sin\theta = \frac{u}{a}$. Check out the figure.

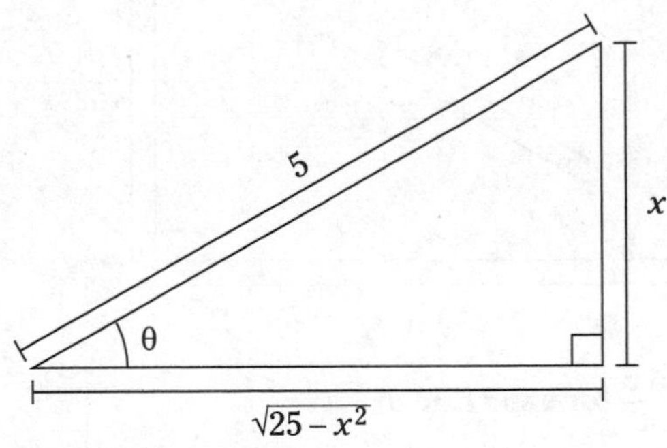

3. **Solve $\sin\theta = \frac{x}{5}$ for x, and then get dx.**

$x = 5\sin\theta \qquad dx = 5\cos\theta\, d\theta$

4. **Do the $\frac{\sqrt{\ }}{a}$ thing.**

$$\frac{\sqrt{25-x^2}}{5}=\cos\theta \qquad \sqrt{25-x^2}=5\cos\theta$$

5. **Substitute.**

$$\int\frac{dx}{25-x^2}$$
$$=\int\frac{5\cos\theta\,d\theta}{(5\cos\theta)^2}$$
$$=\frac{1}{5}\int\sec\theta d\theta$$

6. **Integrate (you may want to just look up this antiderivative in a table):**

   You should get $\frac{1}{5}\ln|\sec\theta+\tan\theta|+C$.

7. **Switch back to x (use your triangle).**

$$=\frac{1}{5}\ln\left|\frac{5}{\sqrt{25-x^2}}+\frac{x}{\sqrt{25-x^2}}\right|+C=\frac{1}{5}\ln\left|\frac{x+5}{\sqrt{25-x^2}}\right|+C$$

12 $\int\frac{dx}{\sqrt{625x^2-121}}=\frac{1}{25}\ln x\left|25x+\sqrt{625x^2-121}\right|+C$

1. **Rewrite as $\int\frac{dx}{\sqrt{(25x)^2-11^2}}$.**
2. **Do the triangle thing.**

   For this problem, $\sec\theta=\frac{u}{a}$.

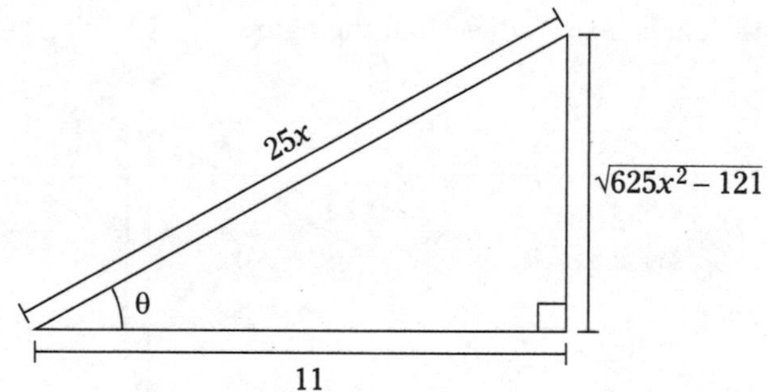

3. **Solve $\sec\theta=\frac{25x}{11}$ for x and find dx.**

$$x=\frac{11}{25}\sec\theta \qquad dx=\frac{11}{25}\sec\theta\tan\theta\,d\theta$$

4. **Do the $\frac{\sqrt{\ }}{a}$ thing.**

$$\frac{\sqrt{625x^2-121}}{11}=\tan\theta \qquad \sqrt{625x^2-121}=11\tan\theta$$

5. **Substitute.**

$$\int \frac{dx}{\sqrt{625x^2-121}} = \int \frac{\frac{11}{25}\sec\theta\tan\theta\, d\theta}{11\tan\theta} = \frac{1}{25}\int \sec\theta d\theta$$

6. **Integrate.**

$$= \frac{1}{25}\ln|\sec\theta + \tan\theta| + C$$

7. **Switch back to x (see Steps 3 and 4).**

$$= \frac{1}{25}\ln\left|\frac{25x}{11} + \frac{\sqrt{625x^2-121}}{11}\right| + C$$

$$= \frac{1}{25}\ln\left|25x + \sqrt{625x^2-121}\right| - \frac{1}{25}\ln 11 + C$$

$$= \frac{1}{25}\ln\left|25x + \sqrt{625x^2-121}\right| + C$$

(13) $\boldsymbol{\int \frac{\sqrt{4x^2-1}}{x}dx = \sqrt{4x^2-1} - \arctan\sqrt{4x^2-1} + C}$

1. **Rewrite as $\int \frac{\sqrt{(2x)^2-1^2}}{x}dx$.**
2. **Draw your triangle.**

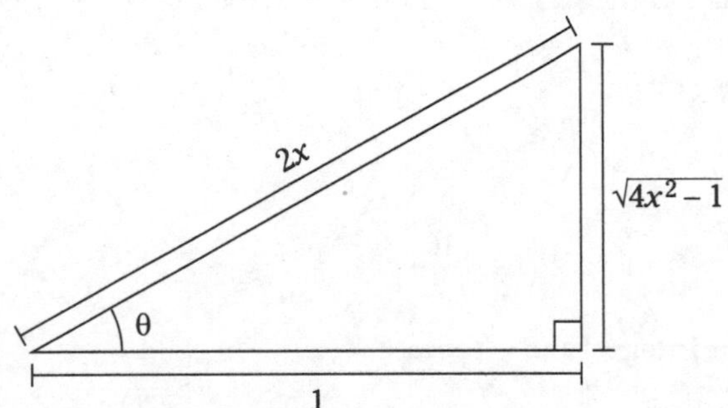

3. **Solve $\sec\theta = \frac{2x}{1}$ for x; get dx.**

$$x = \frac{1}{2}\sec\theta \qquad dx = \frac{1}{2}\sec\theta\tan\theta\, d\theta$$

4. **Do the $\frac{\sqrt{}}{a}$ thing.**

$$\sqrt{4x^2-1} = \tan\theta$$

5. **Substitute.**

$$\int \frac{\sqrt{4x^2-1}}{x}dx$$

$$= \int \frac{\tan\theta}{\frac{1}{2}\sec\theta} \cdot \frac{1}{2}\sec\theta\tan\theta\, d\theta = \int \tan^2\theta\, d\theta$$

6. **Integrate.**

$$= \int (\sec^2\theta - 1)\,d\theta = \tan\theta - \theta + C$$

7. **Switch back to *x* (see Step 4).**

$$= \sqrt{4x^2 - 1} - \arctan\sqrt{4x^2 - 1} + C$$

or

$$= \sqrt{4x^2 - 1} - \text{arcsec}\,2x + C$$

(14) $\int \frac{5dx}{2x^2 + 7x - 4} = \frac{5}{9}\ln\left|\frac{2x-1}{x+4}\right| + C$

1. **Factor the denominator.**

$$= \int \frac{5dx}{(2x-1)(x+4)}$$

2. **Break up the fraction into a sum of partial fractions.**

$$\frac{5}{(2x-1)(x+4)} = \frac{A}{2x-1} + \frac{B}{x+4}$$

3. **Multiply both sides by the least common denominator.**

$$5 = A(x+4) + B(2x-1)$$

4. **Plug the roots of the factors into *x* one at a time.**

$x = -4$ gives you $\quad x = \frac{1}{2}$ gives you

$$5 = -9B \qquad 5 = \frac{9}{2}A$$

$$B = -\frac{5}{9} \qquad A = \frac{10}{9}$$

5. **Split up your integral and integrate.**

$$\int \frac{5dx}{2x^2 + 7x - 4} = \frac{10}{9}\int \frac{dx}{2x-1} + \frac{-5}{9}\int \frac{dx}{x+4} = \frac{10}{9}\ln|2x-1| + \frac{-5}{9}\ln|x+4| + C = \frac{5}{9}\ln\left|\frac{2x-1}{x+4}\right| + C$$

(15) $\int \frac{2x-3}{(3x-1)(x+4)(x+5)}dx = \frac{-7}{208}\ln|3x-1| + \frac{11}{13}\ln|x+4| - \frac{13}{16}\ln|x+5| + C$

1. **The denominator is already factored, so go ahead and write your sum of partial fractions.**

$$\frac{2x-3}{(3x-1)(x+4)(x+5)} = \frac{A}{3x-1} + \frac{B}{x+4} + \frac{C}{x+5}$$

2. **Multiply both sides by the LCD.**

$$2x - 3 = A(x+4)(x+5) + B(3x-1)(x+5) + C(3x-1)(x+4)$$

3. **Plug the roots of the factors into x one at a time.**

$$x=\frac{1}{3}\text{ gives you}: \quad -\frac{7}{3}=\frac{208}{9}A; \quad A=-\frac{21}{208}$$

$$x=-4\text{ gives you}: \quad -11=-13B; \quad B=\frac{11}{13}$$

$$x=-5\text{ gives you}: \quad -13=16C; \quad C=-\frac{13}{16}$$

4. **Split up and integrate.**

$$\int\frac{2x-3}{(3x-1)(x+4)(x+5)}dx=\frac{-21}{208}\int\frac{dx}{3x-1}+\frac{11}{13}\int\frac{dx}{x+4}+\frac{-13}{16}\int\frac{dx}{x+5}$$
$$=\frac{-7}{208}\ln|3x-1|+\frac{11}{13}\ln|x+4|-\frac{13}{16}\ln|x+5|+C$$

16 $\displaystyle\int\frac{x^2+x+1}{x^3-3x^2+3x-1}dx=\ln|x-1|-\frac{3(2x-1)}{2(x-1)^2}+C$

1. **Factor the denominator.**

$$=\int\frac{x^2+x+1}{(x-1)^3}dx$$

2. **Write the partial fractions.**

$$\frac{x^2+x+1}{(x-1)^3}=\frac{A}{x-1}+\frac{B}{(x-1)^2}+\frac{C}{(x-1)^3}$$

3. **Multiply by the LCD.**

$$x^2+x+1=A(x-1)^2+B(x-1)+C$$

4. **Plug in the single root, which is 1, giving you $C=3$.**
5. **Equate coefficients of like terms.**

   Without multiplying out the entire right side in Step 3, you can see that the x^2 term on the right will be Ax^2. Because the coefficient of x^2 on the left is 1, A must equal 1.

6. **Plug in 0 for x in the Step 3 equation, giving you $1=A-B+C$.**

   Because you know A is 1 and C is 3, B must be 3.

   Note: You can solve for A, B, and C in many ways, but the way I did it is probably the quickest.

7. **Split up and integrate.**

$$\int\frac{x^2+x+1}{x^3-3x^2+3x-1}dx=\int\frac{dx}{x-1}+3\int\frac{dx}{(x-1)^2}+3\int\frac{dx}{(x-1)^3}=\ln|x-1|-\frac{3}{x-1}-\frac{3}{2(x-1)^2}+C$$

17 $\int \frac{dx}{x^4+6x^2+5} = \frac{1}{4}\arctan x - \frac{\sqrt{5}}{20}\arctan\frac{x\sqrt{5}}{5} + C$

1. **Factor.**

$$\int \frac{dx}{(x^2+5)(x^2+1)}$$

2. **Write the partial fractions.**

$$\frac{1}{(x^2+5)(x^2+1)} = \frac{Ax+B}{x^2+5} + \frac{Cx+D}{x^2+1}$$

3. **Multiply by the LCD.**

$$1 = (Ax+B)(x^2+1) + (Cx+D)(x^2+5)$$

4. **Plug in the easiest numbers to work with, 0 and 1, to effortlessly get two equations.**

$x = 0$: $\quad 1 = B + 5D$

$x = 1$: $\quad 1 = 2A + 2B + 6C + 6D$

5. **After FOILing out the equation in Step 3, equate coefficients of like terms to come up with two more equations.**

The x^2 term gives you $0 = B + D$.

This equation plus the first one in Step 4 give you $B = -\frac{1}{4}$, $D = \frac{1}{4}$.

The x^3 term gives you $0 = A + C$.

Now, this equation plus the second one in Step 4 plus the known values of B and D give you $A = 0$ and $C = 0$.

6. **Split up and integrate.**

$$\begin{aligned}\int \frac{dx}{x^4+6x^2+5} &= \int \frac{-\frac{1}{4}dx}{x^2+5} + \int \frac{\frac{1}{4}dx}{x^2+1}\\ &= -\frac{1}{4}\int \frac{dx}{x^2+5} + \frac{1}{4}\int \frac{dx}{x^2+1}\\ &= -\frac{1}{4\sqrt{5}}\arctan\frac{x}{\sqrt{5}} + \frac{1}{4}\arctan x + C\end{aligned}$$

18 $\int \frac{4x^3+3x^2+2x+1}{x^4-1}dx = \frac{1}{2}\ln\left[(x^2+1)|x-1|^5|x+1|\right] + \arctan x + C$

1. **Factor.**

$$\int \frac{4x^3+3x^2+2x+1}{(x-1)(x+1)(x^2+1)}dx$$

2. **Write the partial fractions.**

$$\frac{4x^3+3x^2+2x+1}{(x-1)(x+1)(x^2+1)} = \frac{A}{x-1} + \frac{B}{x+1} + \frac{Cx+D}{x^2+1}$$

3. **Multiply by the LCD.**

$$4x^3+3x^2+2x+1=A(x+1)(x^2+1)+B(x-1)(x^2+1)+(Cx+D)(x-1)(x+1)$$

4. **Plug in roots.**

$x=1: \quad 10=4A; \quad A=2.5$

$x=-1: \quad -2=-4B; \quad B=0.5$

5. **Equating the coefficients of the x^3 term gives you C.**

$4=A+B+C$

$A=2.5,\ B=0.5,\ \text{so } C=1$

6. **Plugging in zero and the known values of A, B, and C gets you D.**

$1=2.5-0.5-D$

$D=1$

7. **Integrate.**

$$\int\frac{4x^3+3x^2+2x+1}{x^4-1}dx=2.5\int\frac{dx}{x-1}+0.5\int\frac{dx}{x+1}+\int\frac{x+1}{x^2+1}dx$$
$$=2.5\ln|x-1|+0.5\ln|x+1|+0.5\ln\left|x^2+1\right|+\arctan x+C$$
$$=\frac{1}{2}\ln\left[\left(x^2+1\right)|x-1|^5|x+1|\right]+\arctan x+C$$

(19) $\displaystyle\int\frac{x^2-x}{(x+1)\left(x^2+1\right)\left(x^2+2\right)}dx=\frac{1}{6}\ln\frac{(x+1)^2}{x^2+2}-\arctan x+\frac{2\sqrt{2}}{3}\arctan\frac{x\sqrt{2}}{2}+C$

1. **Break the already-factored function into partial fractions.**

$$\frac{x^2-x}{(x+1)\left(x^2+1\right)\left(x^2+2\right)}=\frac{A}{x+1}+\frac{Bx+C}{x^2+1}+\frac{Dx+E}{x^2+2}$$

2. **Multiply by the LCD.**

$$x^2-x=A\left(x^2+1\right)\left(x^2+2\right)+(Bx+C)(x+1)\left(x^2+2\right)+(Dx+E)(x+1)\left(x^2+1\right)$$

3. **Plug in the single root (−1).**

$2=6A \qquad A=\frac{1}{3}$

4. **Plug 0, 1, and −2 into x and $\frac{1}{3}$ into A.**

$x=0: \quad 0=\frac{2}{3}+2C+E$

$x=1: \quad 0=2+6B+6C+4D+4E$

$x=-2: \quad 6=10+12B-6C+10D-5E$

5. **Equate coefficients of the x^4 terms (with $A=\frac{1}{3}$).**

$0=\frac{1}{3}+B+D$

6. **Solve the system of four equations from Steps 4 and 5. You get the following:**

$$B = 0 \qquad C = -1 \qquad D = -\frac{1}{3} \qquad E = \frac{4}{3}$$

If you find an easier way to solve for *A* through *E*, go to my website and send me an email.

7. **Integrate.**

$$\int \frac{x^2 - x}{(x+1)(x^2+1)(x^2+2)}dx = \frac{1}{3}\int \frac{dx}{x+1} - \int \frac{dx}{x^2+1} - \frac{1}{3}\int \frac{x-4}{x^2+2}dx$$
$$= \frac{1}{3}\ln|x+1| - \arctan x - \frac{1}{6}\ln\left(x^2+2\right) + \frac{2\sqrt{2}}{3}\arctan\frac{x\sqrt{2}}{2} + C$$
$$= \frac{1}{6}\ln\frac{(x+1)^2}{x^2+2} - \arctan x + \frac{2\sqrt{2}}{3}\arctan\frac{x\sqrt{2}}{2} + C$$

If you're ready to test your skills a bit more, take the following chapter quiz that incorporates all the chapter topics.

Whaddya Know? Chapter 16 Quiz

Quiz time! Complete each problem to test your knowledge on the various topics covered in this chapter. You can then find the solutions and explanations in the next section.

For this 12-question integration quiz, I'm going to cut you some slack and ask you to provide only a couple preliminary steps for each problem. You do not need to complete the integration. (I'll supply the final answers for those who feel like finishing the problems.)

For integration-by-parts Problems 1 to 3, identify/determine the u, du, v, and dv.

1 Integrate $\int x\sec^2 x\,dx$.

2 Integrate $\int \cos x\ln(\sin x)dx$.

3 Integrate $\int x^3\ln\left(x^3\right)dx$.

For the trig integrals in Problems 4 to 6, simply identify the part of the integrand that you should split off and move to the right.

4 Integrate $\int \cos^3 x\sin^4 x\,dx$.

5 Integrate $\int \csc^4 x\cot^4 x\,dx$.

6 Integrate $\int \sec^5 x\tan^3 x\,dx$.

For trigonometric substitution Problems 7 to 9, you should do two preliminary things. First, determine the u and the a, then set $\sin\theta$, or $\tan\theta$, or $\sec\theta$ equal to $\frac{u}{a}$. This is part of Step 1 in the solutions to the example problems in the text. (Your answer to this first part of the problem should look something like $\sin\theta=\frac{3x}{2}$.) Second, find which trig function is represented by the radical over the a. This is part of Step 3 in the example solutions. (Your answer to this second part of the problem should look something like $\sec\theta=\frac{\sqrt{9x^2+4}}{2}$.)

7 Integrate $\int \frac{x^2}{\sqrt{9-4x^2}}dx$.

8 Integrate $\int \frac{\sqrt{x^2-4}}{x}dx$.

9 Integrate $\int \frac{x^3}{\sqrt{1+25x^2}}dx$.

For partial-fraction Problems 10 to 12, do two things. First, write the partial fraction decomposition (Step 2 in the solutions to the example problems in the text). Second, solve for the capital-letter unknowns (this is Step 4 in the solutions in the text).

10 Integrate $\int \frac{2x^2}{(x^2+1)^2}dx$.

11 Integrate $\int \frac{4x}{(x-2)(x^2+4)}dx$.

12 Integrate $\int \frac{60}{(x-1)(x-2)(x+3)(x+4)}dx$.

Answers to Chapter 16 Quiz

1. $u = x,\ du = dx,\ v = \tan x,\ dv = \sec^2 x\, dx$

   (The final answer for go-getters is $x \tan x + \ln|\cos x| + C$.)

2. $u = \ln(\sin x),\ du = \dfrac{\cos x}{\sin x} dx,\ v = \sin x,\ dv = \cos x\, dx$

   (The final answer is $\sin x(\ln(\sin x) - 1) + C$.)

3. $u = \ln\left(x^3\right),\ du = \dfrac{3}{x} dx,\ v = \dfrac{x^4}{4},\ dv = x^3 dx$

   (The final answer is $\dfrac{x^4}{16}\left(4\ln\left(x^3\right) - 3\right) + C$.)

4. **Split off $\cos x$.**

   (The final answer is $-\dfrac{1}{7}\sin^7 x + \dfrac{1}{5}\sin^5 x + C$.)

5. **Split off $\csc^2 x$.**

   (The final answer is $-\dfrac{1}{7}\cot^7 x - \dfrac{1}{5}\cot^5 x + C$.)

6. **Split off $\sec x \tan x$.**

   (The final answer is $\dfrac{1}{7}\sec^7 x - \dfrac{1}{5}\sec^5 x + C$.)

7. $\sin\theta = \dfrac{2x}{3};\ \cos\theta = \dfrac{\sqrt{9 - 4x^2}}{3}$

   (The final answer is $\dfrac{9}{16}\sin^{-1}\left(\dfrac{2x}{3}\right) - \dfrac{1}{8}x\sqrt{9 - 4x^2} + C$.)

8. $\sec\theta = \dfrac{x}{2};\ \tan\theta = \dfrac{\sqrt{x^2 - 4}}{2}$

   (The final answer is $\sqrt{x^2 - 4} - 2\tan^{-1}\dfrac{\sqrt{x^2 - 4}}{2} + C$.)

9. $\tan\theta = 5x;\ \sec\theta = \sqrt{1 + 25x^2}$

   (The final answer is $\dfrac{1}{75}x^2\sqrt{1 + 25x^2} - \dfrac{2}{1875}\sqrt{1 + 25x^2} + C$.)

10. $\dfrac{2x^2}{\left(x^2+1\right)^2} = \dfrac{Ax + B}{\left(x^2+1\right)} + \dfrac{Cx + D}{\left(x^2+1\right)^2};\ A = 0,\ B = 2,\ C = 0,\ D = -2$

    (The final answer is $\tan^{-1} x - \dfrac{x}{x^2 + 1} + C$.)

11. $\dfrac{4x}{(x-2)\left(x^2+4\right)} = \dfrac{A}{(x-2)} + \dfrac{Bx + C}{\left(x^2+4\right)};\ A = 1,\ B = -1,\ C = 2$

    (The final answer is $\ln|x - 2| - \dfrac{1}{2}\ln\left(x^2 + 4\right) + \tan^{-1}\dfrac{x}{2} + C$.)

12. $\dfrac{60}{(x-1)(x-2)(x+3)(x+4)} = \dfrac{A}{x-1} + \dfrac{B}{x-2} + \dfrac{C}{x+3} + \dfrac{D}{x+4};\ A = -3,\ B = 2,\ C = 3,\ D = -2$

    (The final answer is $-3\ln|x-1| + 2\ln|x-2| + 3\ln|x+3| - 2\ln|x+4| + C$.)

IN THIS CHAPTER

» One mean theorem: "The Golden Rule!? Don't make me laugh."

» Adding up the area between curves

» Figuring out volumes of odd shapes with the deli meat method

» Mastering the disk and washer methods

» Finding arc length and surface area

» Other stuff you'll never need

Chapter 17
Who Needs Freud? Using the Integral to Solve Your Problems

As I say in Chapter 14, integration is basically just adding up small pieces of something to get the total for the whole thing — *really* small pieces, actually, *infinitely* small pieces. Thus, the integral

$$\int_{5\,sec.}^{20\,sec.} \textit{little piece of distance}$$

tells you to add up all the little pieces of distance traveled during the 15-second interval from 5 to 20 seconds to get the total distance traveled during that interval.

In all problems, the little piece after the integration symbol is always an expression in x (or some other variable). For the above integral, for example, the little piece of distance might be given by, say, x^2dx, Then the definite integral

$$\int_5^{20} x^2 dx$$

would give you the total distance traveled during the time interval. Because you're now an expert at computing integrals like this one, that's no longer the issue; your main challenge in this chapter is simply to come up with the algebraic expression for the little pieces you're adding up. But before I begin the adding-up problems, I want to cover another integration topic: The Mean Value Theorem and a function's average value.

The Mean Value Theorem for Integrals and Average Value

The best way to understand the Mean Value Theorem for integrals is with a diagram — look at Figure 17-1.

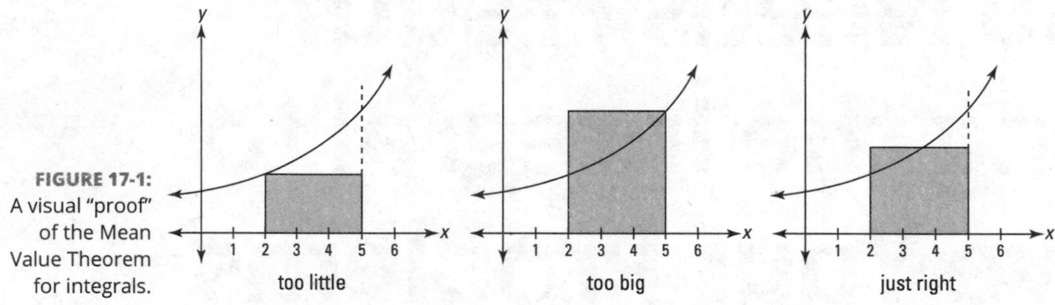

FIGURE 17-1: A visual "proof" of the Mean Value Theorem for integrals.

The graph on the left in Figure 17-1 shows a rectangle whose area is clearly *less than* the area under the curve between 2 and 5. This rectangle has a height equal to the lowest point on the curve in the interval from 2 to 5. The middle graph shows a rectangle whose height equals the highest point on the curve. Its area is clearly *greater than* the area under the curve. By now you're thinking, "Isn't there a rectangle taller than the short one and shorter than the tall one whose area is *the same as* the area under the curve?" Of course. And this rectangle obviously crosses the curve somewhere in the interval. This so-called "mean value rectangle," shown on the right, basically sums up the Mean Value Theorem for integrals. It's really just common sense. But here's the mumbo jumbo.

MATH RULES

The Mean Value Theorem for integrals: If $f(x)$ is a continuous function on the closed interval $[a, b]$, then there exists a number c in the closed interval such that

$$\int_a^b f(x)dx = f(c) \cdot (b-a)$$

The theorem basically just guarantees the existence of the mean value rectangle. (Note that there can only be one mean value rectangle, but its top will sometimes cross the function more than once. Thus, there can be more than one c value that satisfies the theorem.)

The area of the mean value rectangle — which is the same as the area under the curve — equals *length* times *width,* or *base* times *height,* right? So, if you divide its area, $\int_a^b f(x)dx$, by its base, $(b-a)$, you get its height, $f(c)$. This height is the *average value* of the function over the interval in question.

MATH RULES

Average value: The *average value* of a function $f(x)$ over a closed interval $[a, b]$ is

$$\frac{1}{b-a}\int_a^b f(x)dx$$

which is the height of the mean value rectangle.

EXAMPLE

Q. What's the average speed of a car between $t = 9$ seconds and $t = 16$ seconds whose speed in *feet per second* is given by the function $f(t) = 30\sqrt{t}$?

A. Approximately 105.7 feet per second.

The definition of average value gives you the answer in one step: The average speed is $\frac{1}{16-9}\int_9^{16} 30\sqrt{t}\,dt$. Evaluate that integral and you're done. (That's all there is to it, so the following two-step process is somewhat superfluous. However, it shows the logic underlying the average value idea.)

1. **Determine the area under the curve between 9 and 16.**

$$\int_9^{16} 30\sqrt{t}\,dt$$

$$= 30\left[\frac{2}{3}t^{3/2}\right]_9^{16}$$

$$= 30\left(\frac{128}{3} - \frac{54}{3}\right) = 740$$

This area, by the way, is the total distance traveled during the period from 9 to 16 seconds, namely 740 feet. Do you see why? Consider the mean value rectangle for this problem. Its height is a speed (because the function values, or heights, are speeds) and its base is an amount of time, so its area is *speed* times *time,* which equals *distance.* Alternatively, recall that the derivative of position is velocity (see Chapter 12). So, the antiderivative of velocity — what I just did in this step — is position, and the change of position from 9 to 16 seconds gives the total distance traveled.

2. **Divide this area, total distance, by the time interval from 9 to 16, namely 7.**

$$\textit{Average speed} = \frac{\textit{total distance}}{\textit{total time}} = \frac{740 \text{ feet}}{7 \text{ seconds}} \approx 105.7 \text{ feet per second}$$

The definition of average value tells you to multiply the total area by $\frac{1}{b-a}$, which in this problem is $\frac{1}{16-9}$, or $\frac{1}{7}$. But because dividing by 7 is the same as multiplying by $\frac{1}{7}$, you can divide like I do in this step. It makes more sense to think about these problems in terms of division: Area equals *base* times *height*, so the height of the mean value rectangle equals its area *divided* by its base.

THE MVT FOR INTEGRALS AND FOR DERIVATIVES: TWO PEAS IN A POD

Remember the Mean Value Theorem for derivatives from Chapter 11? The graph on the left in the figure shows how it works for the function $f(x) = x^3$. The basic idea is that there's a point on the curve between 0 and 2 where the slope is the same as the slope of the secant line from $(0, 0)$ to $(2, 8)$ — that's a slope of 4. When you do the math, you get $x = \frac{2\sqrt{3}}{3}$ for this point. Well, it turns out that the point guaranteed by the Mean Value Theorem for integrals — the point where the mean value rectangle crosses the derivative of this curve (shown on the right in the figure) — has the very same *x*-value. Pretty nice, eh?

If you really want to understand the intimate relationship between differentiation and integration, think long and hard about the many connections between the two graphs in the accompanying figure. This figure is a real gem, if I do say so myself. (For more on the differentiation/integration connection, check out my other favorites, Figures 15-9 and 15-10.)

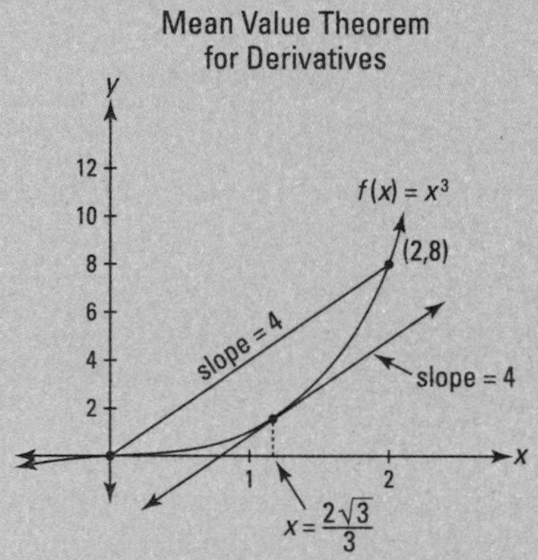

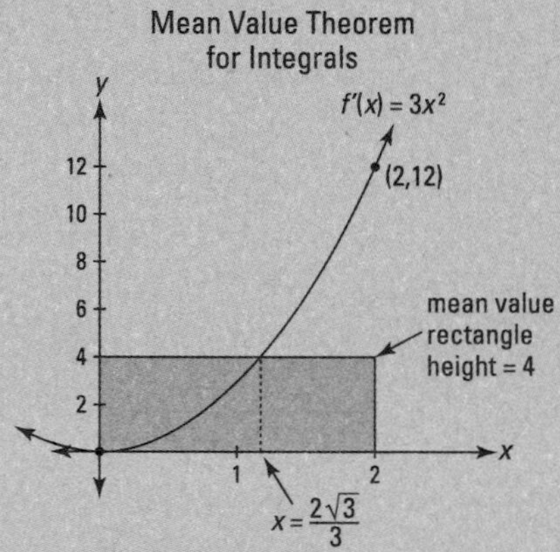

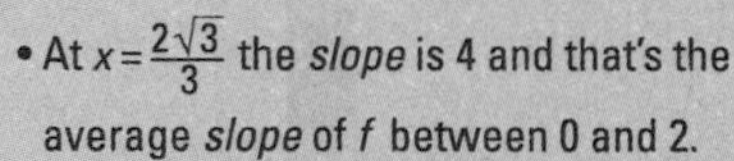

• At x = 2√3/3 the slope is 4 and that's the
average slope of f between 0 and 2.

- The least *slope* of f in the interval is 0.
- The greatest *slope* of f in the interval is 12.
- The total *rise* along f from 0 to 2 is 8.

- At $x = \frac{2\sqrt{3}}{3}$ the *height* is 4 and that's the average *height* of f' between 0 and 2.
- The least *height* of f' in the interval is 0.
- The greatest *height* of f' in the interval is 12.
- The total *area* under f' from 0 to 2 is 8.

YOUR TURN

1 What's the average value of $f(x) = \frac{x}{(x^2+1)^3}$ from 1 to 3?

2 A car's speed in *feet per second* is given by $f(t) = t^{1.7} - 6t + 80$. What's its average speed from $t = 5$ seconds to $t = 15$ seconds? What's that in *miles per hour*?

The Area between Two Curves — Double the Fun

This is the first of several topics in this chapter where your task is to come up with an expression for a little bit of something, then add up the bits by integrating. For this first problem type, the little bit is a narrow rectangle that sits on one curve and goes up to another. Let's walk through an example: Find the area between $y = 2 - x^2$ and $y = \frac{1}{2}x$ from $x = 0$ to $x = 1$. See Figure 17-2.

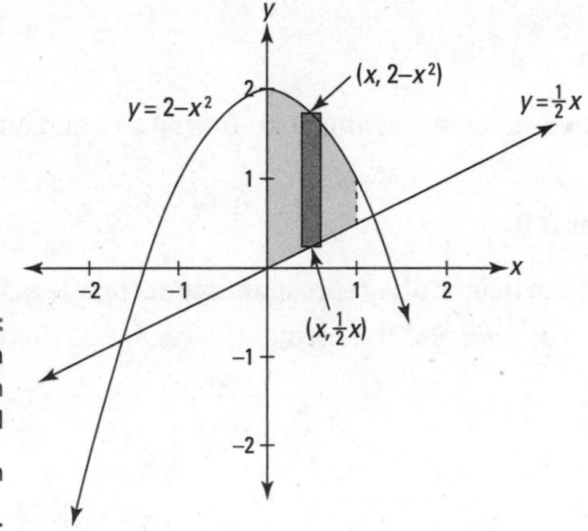

FIGURE 17-2: The area between $y = 2 - x^2$ and $y = \frac{1}{2}x$ from $x = 0$ to $x = 1$.

To get the height of the representative rectangle in Figure 17-2, subtract the y-coordinate of its bottom from the y-coordinate of its top — that's $\left(2-x^2\right)-\frac{1}{2}x$. Its base is the infinitesimal dx. So, because *area* equals *height* times *base*,

$$\textit{Area of representative rectangle} = \left(\left(2-x^2\right)-\frac{1}{2}x\right)dx$$

Now you just add up the areas of all the rectangles from 0 to 1 by integrating:

$$\int_0^1\left(\left(2-x^2\right)-\frac{1}{2}x\right)dx$$

$$=\left[2x-\frac{1}{3}x^3-\frac{1}{4}x^2\right]_0^1 \quad \text{(power rule for all 3 pieces)}$$

$$=\left(2-\frac{1}{3}-\frac{1}{4}\right)-(0-0-0)=\frac{17}{12} \text{ square units}$$

Now to make things a little more twisted, in the next problem the curves cross (see Figure 17-3). When this happens, you have to split the total shaded area into two separate regions before integrating. Try this one: Find the area between $\sqrt[3]{x}$ and x^3 from $x=0$ to $x=2$.

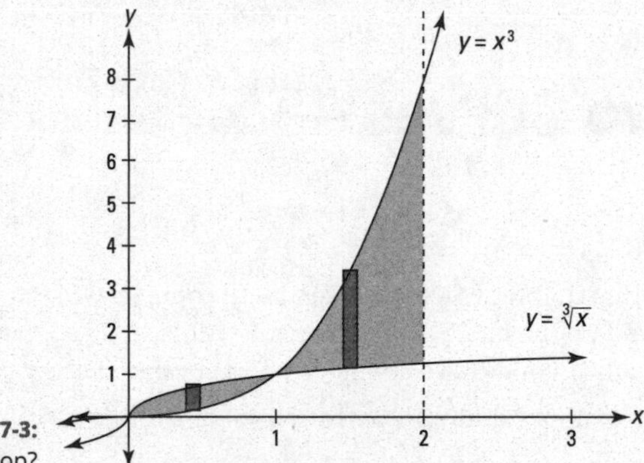

FIGURE 17-3: Who's on top?

1. **Determine where the curves cross.**

   They cross at $(1, 1)$, so you've got two separate regions: one from 0 to 1 and another from 1 to 2.

2. **Figure the area of the region on the left.**

   For this region, $\sqrt[3]{x}$ is above x^3. So, the height of a representative rectangle is $\sqrt[3]{x}-x^3$, its area is *height* times *base*, or $\left(\sqrt[3]{x}-x^3\right)dx$, and the area of the region is, therefore,

$$\int_0^1 \left(\sqrt[3]{x} - x^3\right)dx$$
$$= \left[\frac{3}{4}x^{4/3} - \frac{1}{4}x^4\right]_0^1$$
$$= \left(\frac{3}{4} - \frac{1}{4}\right) - (0-0) = \frac{1}{2}$$

3. **Figure the area of the region on the right.**

   In the right-side region, x^3 is above $\sqrt[3]{x}$, so the height of a rectangle is $x^3 - \sqrt[3]{x}$ and thus you've got

$$\int_1^2 \left(x^3 - \sqrt[3]{x}\right)dx$$
$$= \left[\frac{1}{4}x^4 - \frac{3}{4}x^{4/3}\right]_1^2$$
$$= \left(4 - \frac{3}{2}\sqrt[3]{2}\right) - \left(\frac{1}{4} - \frac{3}{4}\right)$$
$$= 4.5 - 1.5\sqrt[3]{2} \approx 2.61$$

4. **Add up the areas of the two regions to get the total area.**

$$0.5 + \sim 2.61 \approx 3.11 \text{ square units}$$

TIP

Height equals top minus bottom. Note that the height of a representative rectangle is always its *top* minus its *bottom*, regardless of whether these numbers are positive or negative. For instance, a rectangle that goes from 20 up to 30 has a height of $30 - 20$, or 10; a rectangle that goes from –3 up to 8 has a height of $8 - (-3)$, or 11; and a rectangle that goes from –15 up to –10 has a height of $-10 - (-15)$, or 5.

If you think about this top-minus-bottom method for figuring the height of a rectangle, you can now see — assuming you didn't already see it — why the definite integral of a function counts area below the x-axis as negative. (I mention this in Chapters 14 and 15.) For example, consider Figure 17-4.

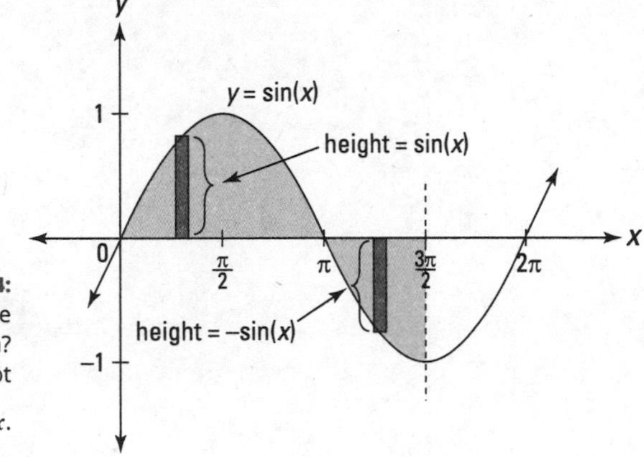

FIGURE 17-4: What's the shaded area? *Hint:* It's not $\int_0^{3\pi/2} \sin x \, dx$.

If you want the total area of the shaded region shown in Figure 17-4, you have to divide the shaded region into two separate pieces like you did in the last problem. One piece goes from 0 to π, and the other from π to $\frac{3\pi}{2}$.

For the first piece, from 0 to π, a representative rectangle has a height equal to the function itself, $y = \sin x$, because its top is on the function and its bottom is at zero — and of course, anything minus zero is itself. So, the area of this first piece is given by the ordinary definite integral $\int_0^{\pi} \sin x \, dx$.

But for the second piece from π to $\frac{3\pi}{2}$, the top of a representative rectangle is at zero — recall that the x-axis is the line $y = 0$ — and its bottom is on $y = \sin x$, so its height is $0 - \sin x$, or just $-\sin x$. So, to get the area of this second piece, you figure the definite integral of the *negative* of the function, $\int_{\pi}^{3\pi/2} -\sin x \, dx$, which is the same as $-\int_{\pi}^{3\pi/2} \sin x \, dx$.

Because this *negative* integral gives you the ordinary, *positive* area of the piece below the x-axis, the *positive* definite integral $\int_{\pi}^{3\pi/2} \sin x \, dx$ gives a *negative* area.

That's why if you figure the definite integral $\int_0^{3\pi/2} \sin x \, dx$ over the entire span, the piece below the x-axis counts as a negative area, and the answer gives you the *net* of the area above the x-axis minus the area below the axis — rather than the total shaded area.

YOUR TURN

 What's the area enclosed by $f(x) = x^2$ and $g(x) = \sqrt{x}$?

 What's the total area enclosed by $f(t) = t^3$ and $g(t) = t^5$?

5 The lines $y = x$, $y = 2x - 5$, and $y = -2x + 3$ form a triangle in the first and fourth quadrants. What's the area of this triangle?

6 What's the area of the triangular shape in the first quadrant enclosed by $\sin x$, $\cos x$, and the line $y = \frac{1}{2}$? (I'm referring to the triangular shape that begins at about $x = 0.5$ and ends at about $x = 1$.)

Volumes of Weird Solids: No, You're Never Going to Need This

In geometry, you learned how to figure the volumes of simple solids like boxes, cylinders, and spheres. Integration enables you to calculate the volumes of an endless variety of much more complicated shapes. This section shows you the meat-slicer method, the disk method, and the washer method.

The meat-slicer method

This metaphor is actually quite accurate. Picture a hunk of meat being cut into very thin slices on one of those deli meat slicers. That's the basic idea here. You slice up a three-dimensional shape, then add up the volumes of the slices to determine the total volume.

Here's a problem: What's the volume of the solid whose length runs along the x-axis from 0 to π and whose cross sections perpendicular to the x-axis are equilateral triangles such that the midpoints of their bases lie on the x-axis and their top vertices are on the curve $y = \sin x$? Is that a mouthful or what? This problem is almost harder to describe and to picture than it is to do. Take a look at this thing in Figure 17-5.

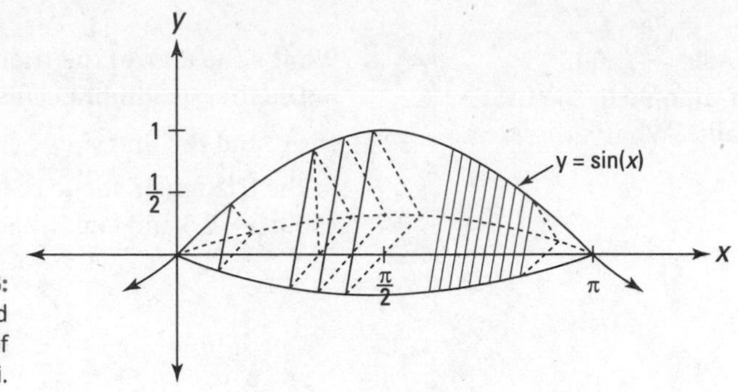

FIGURE 17-5: One weird hunk of pastrami.

So, what's the volume?

1. **Determine the area of any old cross section.**

   Each cross section is an equilateral triangle with a height of $\sin x$. (The height of the second triangle from the left is shown in Figure 17-5 with the vertical dotted segment.) If you do the geometry, you'll see that the base of each triangle is $\frac{2\sqrt{3}}{3}$ times its height, or $\frac{2\sqrt{3}}{3}\sin x$. (*Hint:* Half of an equilateral triangle is a $30° - 60° - 90°$ triangle.) So, the triangle's area, given by $A = \frac{1}{2}bh$ is $\frac{1}{2}\left(\frac{2\sqrt{3}}{3}\sin x\right)\sin x$, or $\frac{\sqrt{3}}{3}\sin^2 x$.

2. **Find the volume of a representative slice.**

   The volume of a slice is just its cross-sectional area times its infinitesimal thickness, *dx*. So, you've got the volume of a slice:

   $$\textit{Volume of representative slice} = \frac{\sqrt{3}}{3}\sin^2 x\,dx$$

3. **Add up the volumes of the slices from 0 to π by integrating.**

   If the following seems a bit difficult, well, heck, you better get used to it. This is calculus after all. (Actually, it's not really that bad if you go through it patiently, step by step.)

   $$\int_0^{\pi}\frac{\sqrt{3}}{3}\sin^2 x\,dx$$

   $$= \frac{\sqrt{3}}{3}\int_0^{\pi}\sin^2 x\,dx$$

   $$= \frac{\sqrt{3}}{3}\int_0^{\pi}\frac{1-\cos(2x)}{2}dx$$ (trig integrals with sines and cosines, Case 3, from Chapter 16)

   $$= \frac{\sqrt{3}}{6}\left(\int_0^{\pi}1\,dx - \int_0^{\pi}\cos(2x)dx\right)$$

$$= \frac{\sqrt{3}}{6}\left([x]_0^\pi - \left[\frac{\sin(2x)}{2} \right]_0^\pi \right)$$

$$= \frac{\sqrt{3}}{6}\left(\pi - 0 - \left(\frac{\sin(2\pi)}{2} - \frac{\sin(0)}{2} \right) \right)$$

$$= \frac{\sqrt{3}}{6}(\pi - 0 - (0 - 0))$$

$$= \frac{\pi\sqrt{3}}{6} \approx 0.91 \text{ cubic units}$$

It's a ~~piece o' cake~~ slice o' meat.

EXAMPLE

Q. What's the volume of the shape shown in the following figure? Its base is formed by the functions $f(x) = \sqrt{x}$ and $g(x) = -\sqrt{x}$. Its cross sections are isosceles triangles whose heights grow linearly from zero at the origin to 1 when $x = 1$.

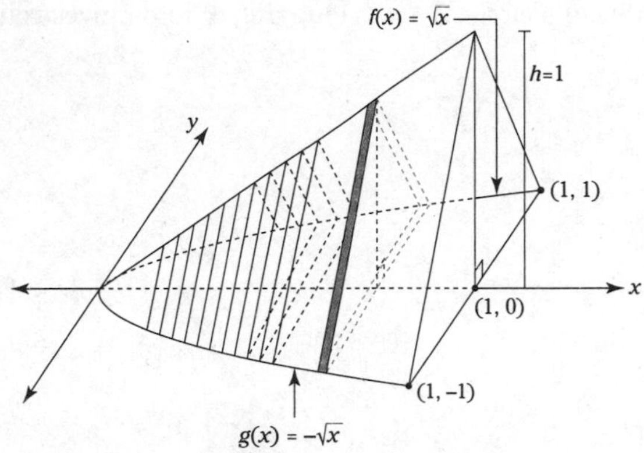

A. **The volume is $\frac{2}{5}$ cubic units.**

1. **Always try to sketch the figure first (of course, I've done it for you here).**

2. **Indicate on your sketch a representative thin slice of the volume in question.**

   This slice should always be perpendicular to the axis or direction along which you are integrating. In other words, if your integrand contains, say, a *dx*, your slice should be perpendicular to the *x*-axis. Also, the slice should not be at either end of the three-dimensional figure or at any other special place. Rather, it should be at some arbitrary, nondescript location within the shape.

3. **Express the volume of this slice.**

   It's easy to show — trust me — that the height of each triangle is the same as its *x*-coordinate. Its base goes from $-\sqrt{x}$ up to $\sqrt{x}$ and is thus $2\sqrt{x}$. And its thickness is *dx*.

   Therefore, $Volume_{\text{slice}} = \frac{1}{2}(2\sqrt{x})x \cdot dx = x\sqrt{x}dx$.

4. **Add up the slices from 0 to 1 by integrating.**

$$\int_0^1 x\sqrt{x}\,dx = \int_0^1 x^{3/2}\,dx = \left[\frac{2}{5}x^{5/2}\right]_0^1 = \frac{2}{5} \text{ cubic units}$$

The disk method

The disk method is sort of a special case of the meat slicer method that you use when the cross-section slices are all circles. So, the two methods are related, but your approach with the disk method is quite different. Here's how it works. Find the volume of the solid — between $x = 2$ and $x = 3$ — generated by rotating the curve $y = e^x$ about the x-axis. See Figure 17-6.

1. **Determine the area of any old cross section.**

   Each cross section is a circle with radius e^x. So, its area is given by the formula for the area of a circle, $A = \pi r^2$. Plugging e^x into r gives you

   $$A = \pi\left(e^x\right)^2 = \pi e^{2x}$$

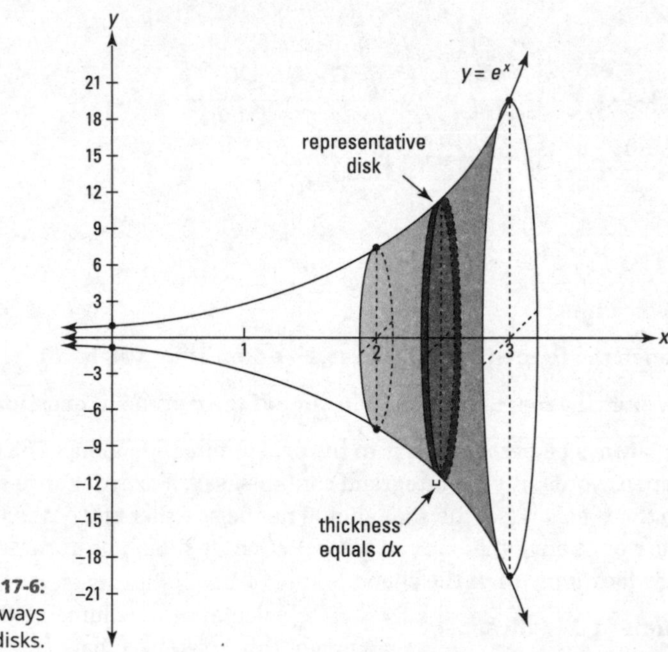

FIGURE 17-6: A sideways stack of disks.

2. **Tack on dx to get the volume of an infinitely thin representative disk.**

   $$\text{Volume of disk} = \overbrace{\pi e^{2x}}^{\text{area}} \cdot \overbrace{dx}^{\text{thickness}}$$

3. **Add up the volumes of the disks from 2 to 3 by integrating.**

$$Total\ volume = \int_2^3 \pi e^{2x}dx$$
$$= \pi \int_2^3 e^{2x}dx$$
$$= \frac{\pi}{2}\int_2^3 e^{2x}2dx$$ (The two new 2's are to tweak the integral for the *u*-substitution; see next line of equation.)
$$= \frac{\pi}{2}\int_4^6 e^u du$$ (by substitution with $u = 2x$ and $du = 2dx$; when $x = 2$, $u = 4$; when $x = 3$, $u = 6$)
$$= \frac{\pi}{2}\left[e^u\right]_4^6$$
$$= \frac{\pi}{2}\left(e^6 - e^4\right) \approx 548 \text{ cubic units}$$

TIP

A representative disk is located at no particular place. Note that Step 1 refers to "any old" cross section. I call it that because when you consider a representative disk like the one shown in Figure 17-6, you should focus on a disk that's in no particular place. The one shown in Figure 17-6 is located at an *unknown* position on the x-axis, and its radius goes from the x-axis up to the curve $y = e^x$. Thus, its radius is the *unknown* length of e^x. If, instead, you use some special disk like the left-most disk at $x = 2$, you're more likely to make the mistake of thinking that a representative disk has some *known* radius like e^2. (This tip also applies to the meat-slicer method in the previous section and the washer method in the next section.)

The washer method

I could have put the washer method and the disk method in one section and called the section "The disk/washer method," because the two methods are based on the very same idea. The only difference with the washer method is that each slice has a hole in its middle that you have to subtract. There's nothing to it.

Here you go. Take the area bounded by $y = x^2$ and $y = \sqrt{x}$, and generate a solid by revolving that area about the x-axis. See Figure 17-7.

Just think: All the forces of the evolving universe and all the twists and turns of your life have brought you to *this* moment when you are finally able to calculate the volume of this weird solid — something for your diary. So, what's the volume of this bowl-like shape?

1. **Determine where the two curves intersect.**

   It should take very little trial and error to see that $y = x^2$ and $y = \sqrt{x}$ intersect at $x = 0$ and $x = 1$ — how nice is that? So, the solid in question spans the interval on the x-axis from 0 to 1.

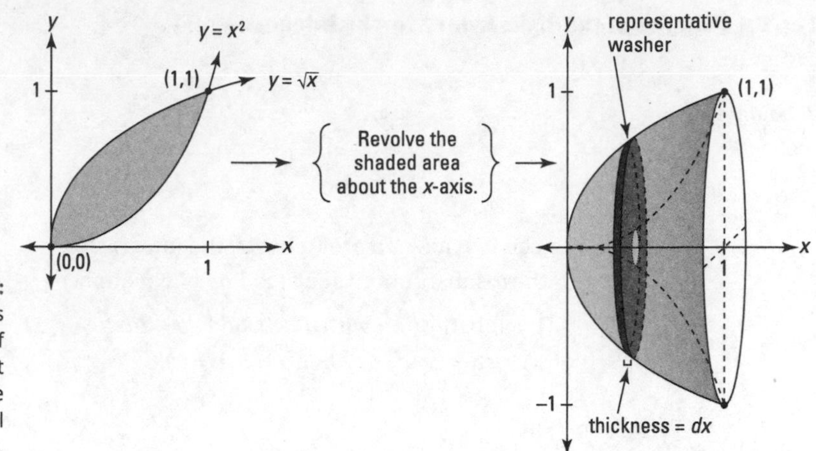

FIGURE 17-7: A sideways stack of washers — just add up the volumes of all the washers.

2. **Figure the cross-sectional area of a thin representative washer.**

   Each slice has the shape of a washer — see Figure 17-8 — so its cross-sectional area equals the area of the entire circle minus the area of the hole.

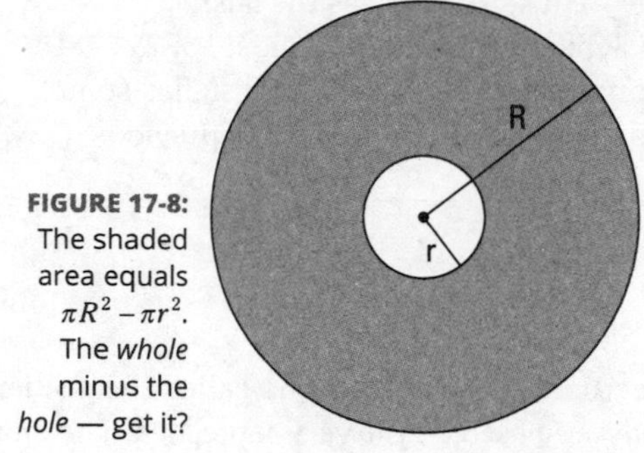

FIGURE 17-8: The shaded area equals $\pi R^2 - \pi r^2$. The *whole* minus the *hole* — get it?

   The area of the circle minus the hole is $\pi R^2 - \pi r^2$, where R is the outer radius (the big radius) and r is the hole's radius (the little radius). For this problem, the outer radius is $\sqrt{x}$ and the hole's radius is x^2, giving you

$$\begin{aligned} Area &= \pi\left(\sqrt{x}\right)^2 - \pi\left(x^2\right)^2 \\ &= \pi x - \pi x^4 \end{aligned}$$

3. **Multiply this area by the thickness, dx, to get the volume of a representative washer.**

$$Volume = \left(\pi x - \pi x^4\right)dx$$

4. Add up the volumes of the even-thinner-than-paper-thin washers from 0 to 1 by integrating.

$$\begin{aligned}
Volume &= \int_0^1 \left(\pi x - \pi x^4\right) dx \\
&= \pi \int_0^1 \left(x - x^4\right) dx \\
&= \pi \left[\frac{1}{2}x^2 - \frac{1}{5}x^5\right]_0^1 \\
&= \pi \left[\left(\frac{1}{2} - \frac{1}{5}\right) - (0 - 0)\right] \\
&= \frac{3}{10}\pi \approx 0.94 \text{ cubic units}
\end{aligned}$$

TIP

Area equals big circle minus little circle. Focus on the simple fact that the area of a washer is the area of the entire disk, πR^2, minus the area of the hole, πr^2: Thus, $Area = \pi R^2 - \pi r^2$. When you integrate, you get $\int_a^b \left(\pi R^2 - \pi r^2\right) dx$. If you factor out the pi, and bring it to the outside of the integral, you get $\pi \int_a^b \left(R^2 - r^2\right) dx$ which is the formula given in most books. But if you just learn that formula by rote, you might forget it. You're more likely to remember the formula and how to do these problems if you understand the simple big-circle-minus-little-circle idea.

YOUR TURN

7 Use the meat-slicer method to determine the volume of the following solid. The solid's base is on the x-y coordinate plane; it's the area surrounded by $y = x^2$, $y = -x^2$, and the line x = 1. Cross sections of the solid are squares perpendicular to the x-axis that rise up from the x-y plane. The base of each square runs from $y = -x^2$ to $y = x^2$.

8 Use the meat-slicer method to derive the formula for the volume of a pyramid with a square base (see the following figure). *Hint:* Integrate from 0 to h along the positive side of the upside-down y-axis. (I set the problem up this way because it simplifies it. You can draw the y-axis the regular way if you like, but then you get an upside-down pyramid.) Your formula should be in terms of s and h.

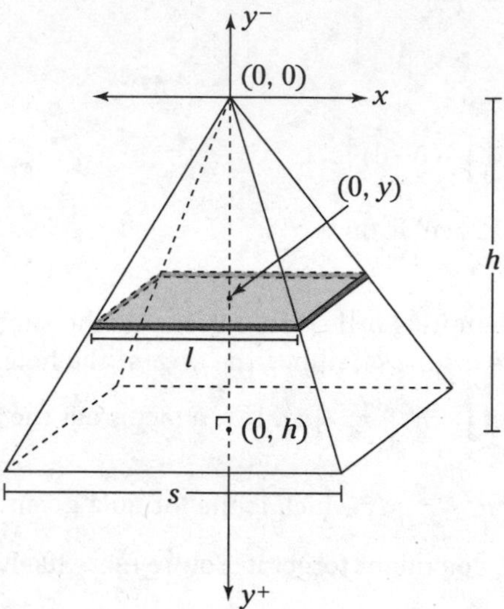

9 Use the washer method to find the volume of the solid that results when the area enclosed by $f(x) = x$ and $g(x) = \sqrt{x}$ is revolved around the x-axis.

10 Same as Problem 9, but with $f(x)=x^2$ and $g(x)=4x$.

11 Use the disk method to derive the formula for the volume of a cone. *Hint:* What's your function? See the following figure. Your formula should be in terms of *r* and *h*.

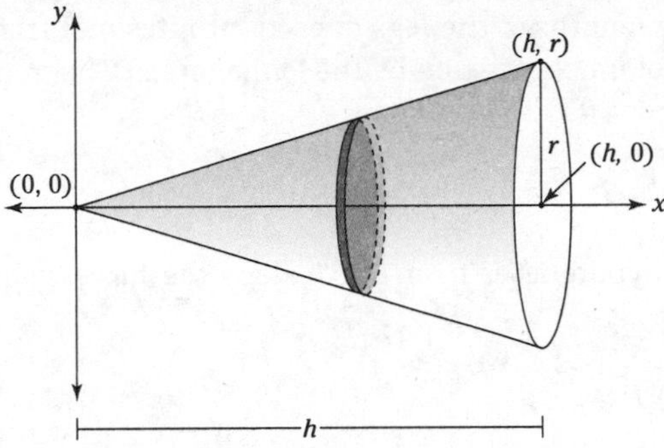

Analyzing Arc Length

So far in this chapter, you've added up the areas of thin rectangles to get total area and the volumes of thin slices to get total volume. Now, you're going to add up minute lengths along a curve to get the whole length.

I could just give you the formula for *arc length* (the length along a curve), but I'd rather show you why it works and how to derive it. Lucky you. The idea is to divide a length of curve into tiny sections, figure the length of each section, and then add up all the lengths. Figure 17-9 shows how each section of a curve can be approximated by the hypotenuse of a tiny right triangle.

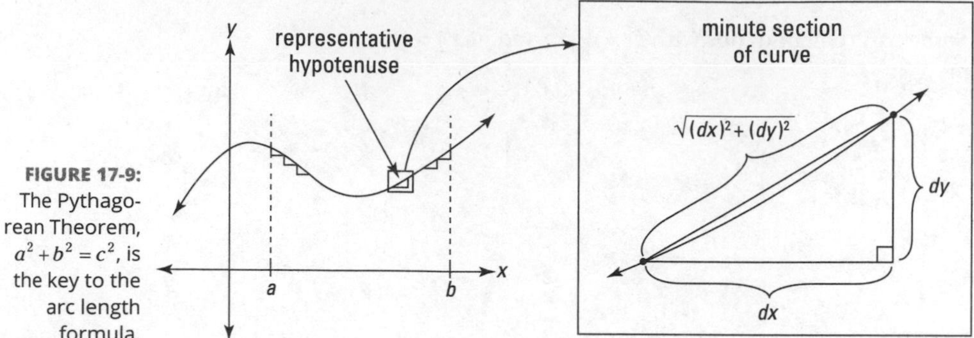

FIGURE 17-9: The Pythagorean Theorem, $a^2+b^2=c^2$, is the key to the arc length formula.

You can imagine that as you zoom in further and further, dividing the curve into more and more sections, the minute sections of the curve get straighter and straighter, and thus the perfectly straight hypotenuses become better and better approximations of the curve. That's why — when this process of adding up smaller and smaller sections is taken to the limit — you get the precise length of the curve.

So, all you have to do is add up all the hypotenuses along the curve between your start and finish points. The lengths of the legs of each infinitesimal triangle are *dx* and *dy*, and thus the length of the hypotenuse — given by the Pythagorean Theorem — is

$$\sqrt{(dx)^2+(dy)^2}$$

To add up all the hypotenuses from *a* to *b* along the curve, you just integrate:

$$\int_a^b \sqrt{(dx)^2+(dy)^2}$$

A little tweaking and you have the formula for arc length. First, factor out a $(dx)^2$ under the square root and simplify:

$$\int_a^b \sqrt{(dx)^2\left[1+\frac{(dy)^2}{(dx)^2}\right]} = \int_a^b \sqrt{(dx)^2\left[1+\left(\frac{dy}{dx}\right)^2\right]}$$

Now you can take the square root of $(dx)^2$ — that's *dx*, of course — and bring it outside the radical, and, voilà, you've got the formula.

MATH RULES

Arc length formula: The arc length along a curve, $y=f(x)$, from *a* to *b*, is given by the following integral:

$$Arc\ length = \int_a^b \sqrt{1+\left(\frac{dy}{dx}\right)^2}\,dx$$

The expression inside this integral is simply the length of a representative hypotenuse.

Try this one: What's the length along $y=(x-1)^{3/2}$ from $x=1$ to $x=5$?

1. **Take the derivative of your function.**

$$y=(x-1)^{3/2}$$
$$\frac{dy}{dx}=\frac{3}{2}(x-1)^{1/2}$$

2. **Plug this into the formula and integrate.**

$$\int_a^b \sqrt{1+\left(\frac{dy}{dx}\right)^2}\,dx$$
$$=\int_1^5 \sqrt{1+\left(\frac{3}{2}(x-1)^{1/2}\right)^2}\,dx$$
$$=\int_1^5 \sqrt{1+\frac{9}{4}(x-1)}dx$$
$$=\int_1^5 \left(\frac{9}{4}x-\frac{5}{4}\right)^{1/2}dx$$
$$=\left[\frac{4}{9}\cdot\frac{2}{3}\left(\frac{9}{4}x-\frac{5}{4}\right)^{3/2}\right]_1^5$$

(See how I got that? It's the guess-and-check integration technique with the reverse power rule. The $\frac{4}{9}$ is the tweak amount you need because of the coefficient $\frac{9}{4}$.)

$$=\left[\frac{1}{27}(9x-5)^{3/2}\right]_1^5 \quad \text{(Algebra questions are strictly prohibited!)}$$
$$=\frac{1}{27}\sqrt{40}^3-\frac{1}{27}\sqrt{4}^3$$
$$=\frac{8}{27}\left(\sqrt{10}^3-1\right)\approx 9.07 \text{ units}$$

Now if you ever find yourself on a road with the shape of $y=(x-1)^{3/2}$ and your odometer is broken, you can figure the exact length of your drive. Your friends will be very impressed — or very concerned.

YOUR TURN

12 Find the distance from $(2, 1)$ to $(5, 10)$ with the arc length formula.

13 Confirm your answer to Problem 12 with the distance formula.

Surfaces of Revolution — Pass the Bottle 'Round

A surface of revolution is a three-dimensional surface with circular cross sections, like a vase or a bell or a wine bottle. For these problems, you divide the surface into narrow circular bands, figure the surface area of a representative band, and then just add up the areas of all the bands to get the total surface area. Figure 17-10 shows such a shape with a representative band.

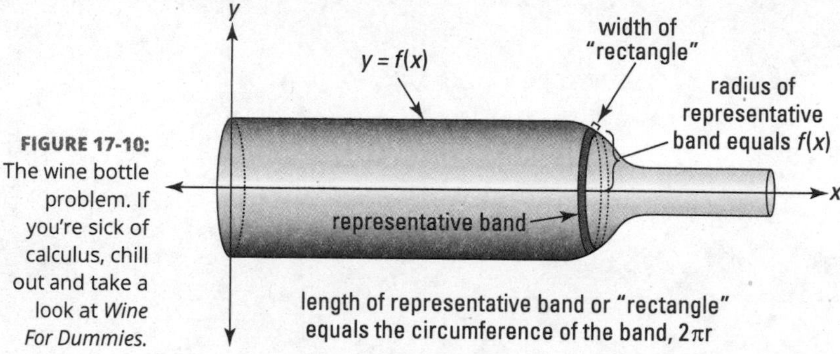

FIGURE 17-10: The wine bottle problem. If you're sick of calculus, chill out and take a look at *Wine For Dummies.*

What's the surface area of a representative band? Well, if you cut the band and unroll it, you get sort of a long, narrow rectangle whose area, of course, is length times width. The rectangle wraps around the whole circular surface, so its length is the circumference of the circular cross section, or $2\pi r$, where r is the height of the function (for garden-variety problems anyway). The width of the rectangle or band is the same as the length of the infinitesimal hypotenuse you used in the section on arc length, namely $\sqrt{1+\left(\frac{dy}{dx}\right)^2}\,dx$. Thus, the surface area of a representative band, from length times width, is $2\pi r\sqrt{1+\left(\frac{dy}{dx}\right)^2}\,dx$, which brings you to the formula.

MATH RULES

Surface of revolution formula: A surface generated by revolving a function, $y = f(x)$, about an axis has a surface area — between a and b — given by the following integral.

$$\textit{Surface area} = \int_a^b 2\pi r\sqrt{1+\left(\frac{dy}{dx}\right)^2}\,dx = 2\pi\int_a^b r\sqrt{1+\left(\frac{dy}{dx}\right)^2}\,dx$$

If the axis of revolution is the x-axis, r will equal $f(x)$ — as shown in Figure 17-10. If the axis of revolution is some other line, like $y = 5$, it's a bit more complicated — something to look forward to.

Now try one: What's the surface area — between $x = 1$ and $x = 2$ — of the surface generated by revolving $y = x^3$ about the x-axis? See Figure 17-11.

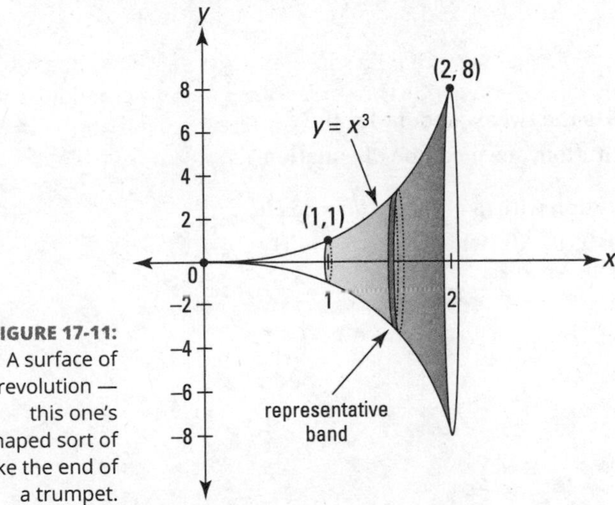

FIGURE 17-11: A surface of revolution — this one's shaped sort of like the end of a trumpet.

1. **Take the derivative of your function.**

$$y = x^3$$
$$\frac{dy}{dx} = 3x^2$$

Now you can finish the problem by just plugging everything into the formula, but I'll do it step by step to reinforce the idea that whenever you integrate, you write down a representative little bit of something — that's the integrand — then you add up all the little bits by integrating.

2. **Figure the surface area of a representative narrow band.**

The radius of the band is x^3, so its circumference is $2\pi x^3$ — that's the band's "length." Its width, a tiny hypotenuse, is $\sqrt{1+\left(\frac{dy}{dx}\right)^2}\,dx = \sqrt{1+\left(3x^2\right)^2}\,dx$. And, thus, its area — *length* times *width* — is $2\pi x^3\sqrt{1+\left(3x^2\right)^2}\,dx$.

3. **Add up the areas of all the bands from 1 to 2 by integrating.**

$$\int_1^2 2\pi x^3\sqrt{1+\left(3x^2\right)^2}\,dx$$

$$=2\pi\int_1^2 x^3\sqrt{1+9x^4}\,dx$$

$$=\frac{2\pi}{36}\int_1^2 36x^3\sqrt{1+9x^4}\,dx$$ (The 36 is the tweak amount for the *u*-substitution; see next line of equation.)

$$=\frac{\pi}{18}\int_{10}^{145} u^{1/2}du$$ (substitution with $u=1+9x^4$, $du=36x^3dx$; when $x=1$, $u=10$; when $x=2$, $u=145$)

$$=\frac{\pi}{18}\left[\frac{2}{3}u^{3/2}\right]_{10}^{145}$$

$$=\frac{\pi}{18}\left(\frac{2}{3}\cdot 145^{3/2}-\frac{2}{3}\cdot 10^{3/2}\right)$$

$$\approx 199.5 \text{ square units}$$

That's a wrap.

YOUR TURN

14 What's the surface area generated by revolving $f(x)=\frac{3}{4}x$ from $x=0$ to $x=4$ about the x-axis?

15 Confirm your answer to Problem 14 with the formula for the lateral area of a cone, $LA = \pi r\ell$, where ℓ is the slant height of the cone.

16 What's the surface area generated by revolving $f(x) = \sqrt{x}$ from $x = 0$ to $x = 9$ about the x-axis?

Practice Questions Answers and Explanations

① **The average value is 0.03.**

$$Average\ value = \frac{total\ area}{base} = \frac{\int_1^3 \frac{x}{(x^2+1)^3}dx}{3-1}$$

Do this with a u-substitution.

$$u = x^2 + 1 \quad \text{when } x = 1,\ u = 2$$
$$du = 2xdx \quad \text{when } x = 3,\ u = 10$$

$$= \frac{\frac{1}{2}\int_1^3 \frac{2x}{(x^2+1)^3}dx}{2}$$
$$= \frac{1}{4}\int_2^{10} \frac{du}{u^3}$$
$$= -\frac{1}{8}\left[u^{-2}\right]_2^{10}$$
$$= -\frac{1}{8}\left(10^{-2} - 2^{-2}\right) = 0.03$$

② **Its average speed is about 72.62 feet per second or 49.51 miles per hour.**

$$Average\ speed = \frac{total\ distance}{total\ time}$$
$$= \frac{\int_5^{15}\left(t^{1.7} - 6t + 80\right)dt}{15-5}$$
$$= \frac{\left[\frac{1}{2.7}t^{2.7} - 3t^2 + 80t\right]_5^{15}}{10}$$
$$\approx \frac{554.73 - 675 + 1200 - (28.57 - 75 + 400)}{10}$$
$$\approx 72.62 \text{ feet per second} \approx 49.51 \text{ miles per hour}$$

③ **The area is $\frac{1}{3}$.**

1. **Graph the functions.**

   You're on your own with this graph.

2. **Find the points of intersection.**

   They're nice and simple: $(0,\ 0)$ and $(1,\ 1)$.

3. **Find the area.**

REMEMBER

The rectangular slices have a height given by *top minus bottom*.

$$Area = \int_0^1\left(\sqrt{x} - x^2\right)dx = \left[\frac{2}{3}x^{3/2} - \frac{1}{3}x^3\right]_0^1 = \frac{2}{3} - \frac{1}{3} = \frac{1}{3}$$

④ **The area is $\frac{1}{6}$.**

1. **Graph the functions.**

   You should see three points of intersection.

2. **Find the points.**

   The points are $(-1,\ -1)$, $(0,\ 0)$, and $(1,\ 1)$.

3. **Find the area on the left.**

   t^5 is above t^3, so $Area = \int_{-1}^{0}\left(t^5 - t^3\right)dt = \left[\frac{1}{6}t^6 - \frac{1}{4}t^4\right]_{-1}^{0} = 0 - \left(\frac{1}{6} - \frac{1}{4}\right) = \frac{1}{12}$

4. **Find the area on the right.**

   t^3 is on top for this chunk; find the area, then add it to the left-side area.

   $Area = \int_{0}^{1}\left(t^3 - t^5\right)dt = \left[\frac{1}{4}t^4 - \frac{1}{6}t^6\right]_{0}^{1} = \frac{1}{4} - \frac{1}{6} = \frac{1}{12}$

   Therefore, the total area is $\frac{1}{12} + \frac{1}{12}$, or $\frac{1}{6}$.

   Note that had you observed that both t^3 and t^5 are odd functions, you could have reasoned that the two areas are the same, and then calculated just one of them and doubled the result.

⑤ **The area is 6.**

1. **Graph the three lines.**

2. **Find the three points of intersection.**

   a. $y = x$ intersects $y = 2x - 5$ at $x = 2x - 5$; $x = 5$ and, thus, $y = 5$.

   b. $y = x$ intersects $y = -2x + 3$ at $x = -2x + 3$; $x = 1$ and, thus, $y = 1$.

   c. $y = 2x - 5$ intersects $y = -2x + 3$ at $2x - 5 = -2x + 3$; $x = 2$ and, thus, $y = -1$.

3. **Integrate to find the area from $x = 1$ to $x = 2$.**

   $y = x$ is on the top and $y = -2x + 3$ is on the bottom, so

$$\begin{aligned}
Area &= \int_{1}^{2}(x - (-2x + 3))dx \\
&= 3\int_{1}^{3}(x - 1)dx \\
&= 3\left[\frac{1}{2}x^2 - x\right]_{1}^{2} \\
&= 3\left[(2-2) - \left(\frac{1}{2} - 1\right)\right] = \frac{3}{2}
\end{aligned}$$

4. Integrate to find the area from $x = 2$ to $x = 5$.

$y = x$ is on the top again, but, for this chunk, $y = 2x - 5$ is on the bottom, thus

$$\begin{aligned}Area &= \int_2^5 (x-(2x-5))dx \\ &= \int_2^5 (-x+5)dx \\ &= \left[-\frac{1}{2}x^2+5x\right]_2^5 \\ &= -\frac{25}{2}+25-(-2+10) = \frac{9}{2}\end{aligned}$$

The grand total from Steps 3 and 4 equals 6.

Note that you didn't need the y-coordinates of the three points of intersection of the three lines, but it's nice to know them, because then you can see exactly where the triangle is.

Granted, using calculus for this problem is loads of fun, but it's totally unnecessary. If you cut the triangle into two triangles — corresponding to Steps 3 and 4 — you can get the total area with simple coordinate geometry.

6 **The area is $\sqrt{3} - \sqrt{2} - \frac{\pi}{12}$.**

1. Do the graph and find the intersections.

a. From the example problem, you know that $\sin x$ and $\cos x$ intersect at $x = \frac{\pi}{4}$.

b. $y = \frac{1}{2}$ intersects $\sin x$ at $\sin x = \frac{1}{2}$, so $x = \frac{\pi}{6}$.

c. $y = \frac{1}{2}$ intersects $\cos x$ at $\cos x = \frac{1}{2}$, so $x = \frac{\pi}{3}$.

2. Integrate to find the area from $\frac{\pi}{6}$ to $\frac{\pi}{4}$ and from $\frac{\pi}{4}$ to $\frac{\pi}{3}$.

$$\begin{aligned}Area &= \int_{\pi/6}^{\pi/4}\left(\sin x - \frac{1}{2}\right)dx + \int_{\pi/4}^{\pi/3}\left(\cos x - \frac{1}{2}\right)dx \\ &= \left[-\cos x - \frac{1}{2}x\right]_{\pi/6}^{\pi/4} + \left[\sin x - \frac{1}{2}x\right]_{\pi/4}^{\pi/3} \\ &= -\frac{\sqrt{2}}{2} - \frac{\pi}{8} - \left(-\frac{\sqrt{3}}{2} - \frac{\pi}{12}\right) + \frac{\sqrt{3}}{2} - \frac{\pi}{6} - \left(\frac{\sqrt{2}}{2} - \frac{\pi}{8}\right) \\ &= \sqrt{3} - \sqrt{2} - \frac{\pi}{12} \quad \text{Cool answer, eh?}\end{aligned}$$

7 **The area is $\frac{4}{5}$.**

You have to add up the square cross sections of this funny shape from 0 to 1. Each square has a base of $x^2 - (-x^2)$, or $2x^2$. The area of a square is the square of its base, of course, so each square cross section has an area of $4x^4$. Finish by integrating:

$$\begin{aligned}&\int_0^1 (4x^4)dx \\ &= \left[\frac{4}{5}x^5\right]_0^1 = \frac{4}{5}\end{aligned}$$

8 **The volume formula is $\frac{1}{3}s^2h$.**

Using similar triangles, you can establish the following proportion: $\frac{y}{h} = \frac{l}{s}$.

You want to express the side of your representative slice as a function of y (and the constants, s and h), so that's $l = \frac{ys}{h}$.

The volume of your representative square slice equals its cross-sectional area times its thickness, dy, so now you have

$$Volume_{\text{slice}} = \left(\frac{ys}{h}\right)^2 dy$$

WARNING

Don't forget that when integrating, constants behave just like ordinary numbers.

$$Volume_{\text{pyramid}} = \int_0^h \left(\frac{ys}{h}\right)^2 dy = \frac{s^2}{h^2}\int_0^h y^2 dy = \frac{s^2}{h^2} \cdot \left[\frac{1}{3}y^3\right]_0^h = \frac{s^2}{h^2} \cdot \frac{1}{3}h^3 = \frac{1}{3}s^2h$$

That's the old familiar pyramid volume formula: $\frac{1}{3} \cdot base \cdot height$ — the hard way.

9 **The volume is $\frac{\pi}{6}$.**

1. **Sketch the solid, including a representative slice.**

   See the following figure.

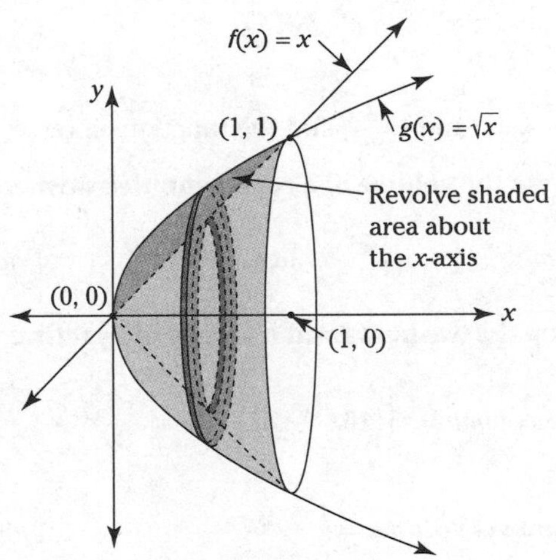

2. **Express the volume of your representative slice.**

$$Volume_{\text{washer}} = \pi\left(R^2 - r^2\right)dx = \pi\left(\sqrt{x}^2 - x^2\right)dx = \pi\left(x - x^2\right)dx$$

3. **Add up the infinite number of infinitely thin washers from 0 to 1 by integrating.**

$$Total\ volume = \int_0^1 \pi\left(x - x^2\right)dx = \pi\left[\frac{1}{2}x^2 - \frac{1}{3}x^3\right]_0^1 = \pi\left(\frac{1}{2} - \frac{1}{3}\right) = \frac{\pi}{6}$$

(10) The volume is $\frac{2048\pi}{15}$ cubic units.

1. **Sketch the solid and a representative slice.**

   See the following figure.

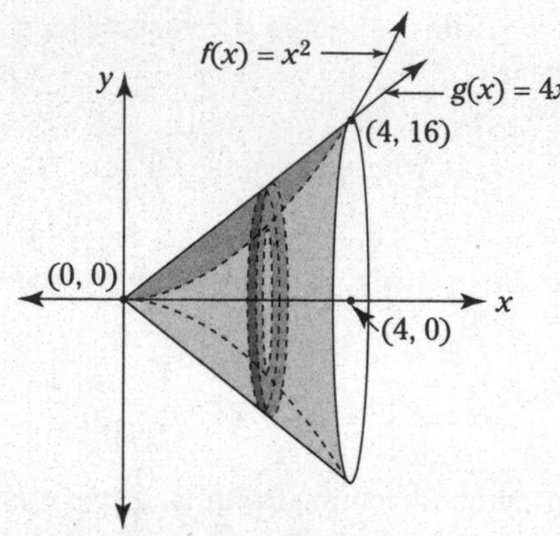

2. **Determine where the functions intersect.**

   The functions intersect at where $f(x) = g(x)$, so

$$x^2 = 4x$$
$$x^2 - 4x = 0$$
$$x(x-4) = 0$$

   Thus, $x = 0$ and $x = 4$, and the functions intersect at $(0, 0)$ and $(4, 16)$.

3. **Express the volume of a representative washer.**

$$Volume_{\text{washer}} = \pi\left(R^2 - r^2\right)dx = \pi\left((4x)^2 - \left(x^2\right)^2\right)dx = \pi\left(16x^2 - x^4\right)dx$$

4. **Add up the washers from 0 to 4 by integrating.**

$$Total\ volume = \pi\int_0^4 \left(16x^2 - x^4\right)dx = \pi\left[\frac{16}{3}x^3 - \frac{1}{5}x^5\right]_0^4 = \pi\left(\frac{1024}{3} - \frac{1024}{5}\right) = \frac{2048\pi}{15}$$

(11) The formula is $Volume = \frac{1}{3}\pi r^2 h$.

1. **Find the function that revolves about the x-axis to generate the cone.**

   The function is the line that goes through $(0, 0)$ and (h, r). Its slope is thus $\frac{r}{h}$, and its equation is therefore $f(x) = \frac{r}{h}x$.

2. **Express the volume of a representative disk.**

   The radius of your representative disk is $f(x)$ and its thickness is dx. Thus, its volume is given by

$$Volume_{\text{disk}} = \pi(f(x))^2\,dx = \pi\left(\frac{r}{h}x\right)^2 dx$$

3. **Add up the disks from $x = 0$ to $x = h$ by integrating.**

Don't forget that *r* and *h* are constants that behave like ordinary numbers.

$$Volume_{cone} = \int_0^h \pi\left(\frac{r}{h}x\right)^2 dx = \frac{\pi r^2}{h^2}\int_0^h x^2 dx = \frac{\pi r^2}{h^2}\left[\frac{1}{3}x^3\right]_0^h = \frac{\pi r^2}{h^2}\cdot\frac{1}{3}h^3 = \frac{1}{3}\pi r^2 h$$

(12) **The distance is $3\sqrt{10}$.**

1. **Find a function for the "arc."**

It's really a line, of course — that connects the two points. I'm sure you remember the point-slope formula from your algebra days:

$$y - y_1 = m(x - x_1)$$
$$y - 1 = 3(x - 2)$$
$$y = 3x - 5$$

2. **Find y'.**

I hope you don't have to look very far: $y' = 3$.

3. **Plug into the formula.**

$$Arc\ length = \int_2^5 \sqrt{1+3^2}dx = \left[x\sqrt{10}\right]_2^5 = 3\sqrt{10}$$

(13) You should get the following:

$$Distance = \sqrt{(x_2 - x_1)^2 + (y_2 - y_1)^2} = \sqrt{(5-2)^2 + (10-1)^2} = 3\sqrt{10}$$

(14) **The surface area is 15π.**

1. **Sketch the function and the surface.**
2. **Plug the function and its derivative into the formula.**

$$Surface\ area = 2\pi\int_0^4 \frac{3}{4}x\sqrt{1+\left(\frac{3}{4}\right)^2}dx = \frac{3\pi}{2}\int_0^4 x\sqrt{\frac{25}{16}}dx = \frac{15\pi}{8}\left[\frac{1}{2}x^2\right]_0^4 = 15\pi$$

(15)

1. **Determine the radius and slant height of the cone.**

From your sketch and the function, you can easily determine that the function goes through $(4, 3)$, and that, therefore, the radius is 3 and the slant height is 5 (it's the hypotenuse of a 3-4-5 triangle).

2. **Plug into the formula.**

$$Lateral\ area = \pi r\ell = 15\pi$$

It checks.

(16) **The surface area is $\frac{\pi}{6}(37\sqrt{37}-1)$.**

1. **Plug the function and its derivative into the formula.**

$$f(x)=\sqrt{x}$$
$$f'(x)=\frac{1}{2\sqrt{x}}$$

$$\textit{Surface area}=2\pi\int_0^9\sqrt{x}\sqrt{1+\left(\frac{1}{2\sqrt{x}}\right)^2}\,dx=2\pi\int_0^9\sqrt{x}\sqrt{1+\frac{1}{4x}}\,dx=2\pi\int_0^9\sqrt{x+\frac{1}{4}}\,dx$$

2. **Integrate.**

$$=2\pi\left[\frac{2}{3}\left(x+\frac{1}{4}\right)^{3/2}\right]_0^9=\frac{4\pi}{3}\left(\left(\frac{37}{4}\right)^{3/2}-\left(\frac{1}{4}\right)^{3/2}\right)=\frac{\pi}{6}(37\sqrt{37}-1)$$

If you're ready to test your skills a bit more, take the following chapter quiz that incorporates all the chapter topics.

Whaddya Know? Chapter 17 Quiz

Quiz time! Complete each problem to test your knowledge on the various topics covered in this chapter. You can then find the solutions and explanations in the next section.

For Problems 1 and 2, use the function $f(x)=4x-2x^2$.

1. Find the average value of the function over the interval $[0, 2]$.

2. Find all points on f in the interval $[0, 2]$ that have a height equal to f's average value over that interval.

3. Find the area enclosed by $f(x)=\tan x$ and $g(x)=\sin(2x)$. (I'm only interested in the total area of the two enclosed areas that meet at the origin.)

4. Find the area enclosed by $f(x)=\ln x$ and $g(x)=\dfrac{e^{((x-1)/(e-1))}-1}{e-1}$ (Is that a cool function or what?) *Hint 1:* You can find the two points where f and g intersect by trial and error. Try plugging in the two values of x that produce very simple values for g. *Hint 2:* $e-1$ is just a number. Congrats to anyone who can get through the messiness of this problem to arrive at the beautifully simple answer.

5. Use the meat-slicer method to determine the volume of the following solid. The solid's base is on the x-y coordinate plane; it's the area in the 1st and 4th quadrants surrounded by $f(x)=x^2$, $g(x)=-x$, and $x=2$. Cross sections of the solid are rectangles perpendicular to the x-axis that rise up from the x-y plane. The base of each rectangle runs from $g(x)=-x$ up to $f(x)=x^2$; the height of each rectangle equals the square of its base.

6. Consider the area in the 1st quadrant enclosed by $f(x)=x^3$ and $g(x)=\sqrt[3]{x}$. What's the volume of the solid generated by revolving that area about the x-axis?

7. Consider the area in the 1st quadrant enclosed by $f(x)=x^2$ and $g(x)=\sqrt{x}$. What's the volume of the solid generated by revolving that area about the line $y=-1$?

8. Use the arc length formula to determine the distance from $(-1, 0)$ to $(1, 0)$ in the x-y coordinate plane. The answer is 2, duh! So, why in the world would you calculate this with the arc length formula? Just because.

9. Use the arc length formula to derive the formula for the circumference of a circle: $C=2\pi r$. *Hint:* Use a circle centered at the origin, and determine the arc length of the semicircle in the 1st and 2nd quadrants.

10. Use the surface of revolution formula to derive the formula for the lateral surface area of a right circular cylinder with a radius of r and a height of h: $LSA=2\pi rh$.

Answers to Chapter 17 Quiz

1. **The average value over the interval $[0,\ 2]$ is $\frac{4}{3}$.**

   Piece o' cake. Just use the average value formula:

$$\begin{aligned}
\textit{Average value} &= \frac{1}{b-a}\int_a^b f(x)\,dx \\
&= \frac{1}{2-0}\int_0^2 \left(4x-2x^2\right)dx \\
&= \frac{1}{2}\left[2x^2-\frac{2}{3}x^3\right]_0^2 \\
&= \frac{1}{2}\left(8-\frac{16}{3}-(0-0)\right)=\frac{4}{3}
\end{aligned}$$

2. **In the interval $[0,\ 2]$, f reaches its average value at two points: $\left(1-\frac{\sqrt{3}}{3},\ \frac{4}{3}\right)$ and $\left(1+\frac{\sqrt{3}}{3},\ \frac{4}{3}\right)$.**

   Set *f* equal to the average value, and solve.

$$\begin{aligned}
4x-2x^2 &= \frac{4}{3} \\
-2x^2+4x-\frac{4}{3} &= 0 \\
6x^2-12x+4 &= 0 \\
3x^2-6x+2 &= 0
\end{aligned}$$

   Finish with the quadratic formula:

$$\begin{aligned}
x &= \frac{6\pm\sqrt{36-4\cdot 3\cdot 2}}{6} \\
&= \frac{6\pm\sqrt{12}}{6}=1\pm\frac{\sqrt{3}}{3}
\end{aligned}$$

   You must check that these two *x*-values are within the given interval of $[0,\ 2]$. Check. Done.

3. **The total area is $1-\ln 2$.**

   Set the functions equal to each other and solve to find out where they intersect:

$$\begin{aligned}
\tan x &= \sin(2x) \\
\frac{\sin x}{\cos x} &= 2\sin x\cos x \\
2\sin x\cos^2 x-\sin x &= 0 \\
\sin x\left(2\cos^2 x-1\right) &= 0
\end{aligned}$$

Finish with the zero product property:

$$2\cos^2 x - 1 = 0$$

$$\sin x = 0 \quad \text{or} \quad \cos^2 x = \frac{1}{2}$$

$$\cos x = \pm\frac{\sqrt{2}}{2}$$

The first equation gives you $x = 0$; the second gives you $x = -\frac{\pi}{4}$ and $x = \frac{\pi}{4}$. (Well, each of those two equations actually has an infinite number of solutions, but the three solutions just mentioned are the only ones you need for the two enclosed areas that meet at the origin.) So, you've got an enclosed area in the 3rd quadrant from $x = -\frac{\pi}{4}$ to $x = 0$ where $\tan x$ is on top, and another enclosed area in the 1st quadrant from $x = 0$ to $x = \frac{\pi}{4}$ where $\sin(2x)$ is on top. You're all set to finish:

$$\begin{aligned}
\textit{Total area} &= \int_{-\pi/4}^{0} (\tan x - \sin(2x))dx + \int_{0}^{\pi/4} (\sin(2x) - \tan x)dx \\
&= \left[\ln|\sec x| + \frac{1}{2}\cos(2x)\right]_{-\pi/4}^{0} + \left[-\frac{1}{2}\cos(2x) - \ln|\sec x| +\right]_{0}^{\pi/4} \\
&= \left(0 + \frac{1}{2} - \left(\ln\sqrt{2} + 0\right)\right) + \left(0 - \ln\sqrt{2} - \left(-\frac{1}{2} - 0\right)\right) \\
&= 1 - 2\ln\sqrt{2} = 1 - \ln 2
\end{aligned}$$

4 **The area is 3 − *e*.**

$f(1)$ and $g(1)$ both equal zero; and $f(e)$ and $g(e)$ both equal 1. So, f and g intersect at $(1,\ 0)$ and $(e,\ 1)$. Between $x = 1$ and $x = e$, $\ln x$ is above $\frac{e^{((x-1)/(e-1))} - 1}{e-1}$, so the area enclosed by f and g is given by

$$\begin{aligned}
&\int_1^e \left(\ln x - \frac{e^{((x-1)/(e-1))} - 1}{e-1}\right)dx \\
&= \int_1^e \ln x\,dx - \int_1^e \frac{e^{((x-1)/(e-1))} - 1}{e-1}dx \\
&= \int_1^e \ln x\,dx - \frac{1}{e-1}\int_1^e \left(e^{((x-1)/(e-1))} - 1\right)dx \\
&= [x\ln x - x]_1^e - \frac{1}{e-1}\left[(e-1)e^{((x-1)/(e-1))} - x\right]_1^e \\
&= (e - e - (0-1)) - \frac{1}{e-1}((e-1)e - e - ((e-1) - 1)) \\
&= 1 - \frac{1}{e-1}((e-1)e - 2e + 2) \\
&= 1 - (e-2) = 3 - e
\end{aligned}$$

5 **The volume is $73\frac{17}{35}$.**

You just have to add up the rectangular cross sections of this odd shape from 0 to 2. Each rectangle has a base of $x^2-(-x)$, or x^2+x. The height of each rectangle is the square of its base, so each rectangle's height is given by $\left(x^2+x\right)^2$, or $x^4+2x^3+x^2$. Area equals base times height, of course, so each rectangular cross section has an area of $\left(x^2+x\right)\left(x^4+2x^3+x^2\right)$, or $x^6+3x^5+3x^4+x^3$. Finish by integrating:

$$\int_0^2\left(x^6+3x^5+3x^4+x^3\right)dx$$
$$=\left[\frac{1}{7}x^7+\frac{1}{2}x^6+\frac{3}{5}x^5+\frac{1}{4}x^4\right]_0^2$$
$$=\frac{128}{7}+32+\frac{96}{5}+4=73\frac{17}{35}$$

6 **The volume is $\frac{16\pi}{35}$.**

$f(x)=x^3$ and $g(x)=\sqrt[3]{x}$ intersect at $(0, 0)$ and $(1, 1)$. The washer method gives you the volume:

$$Volume=\pi\int_a^b\left(R^2-r^2\right)dx$$
$$=\pi\int_0^1\left(\sqrt[3]{x}^2-\left(x^3\right)^2\right)dx$$
$$=\pi\int_0^1\left(x^{2/3}-x^6\right)dx$$
$$=\pi\left[\frac{3}{5}x^{5/3}-\frac{1}{7}x^7\right]_0^1$$
$$=\pi\left(\frac{3}{5}-\frac{1}{7}\right)=\frac{16\pi}{35}$$

7 **The volume is $\frac{29\pi}{30}$.**

Another washer method problem, but because you're revolving the area about the line $y=-1$ instead of the x-axis, both R and r will have a length 1 greater than they would have had.

$$Volume=\pi\int_a^b\left(R^2-r^2\right)dx$$
$$=\pi\int_0^1\left(\left(\sqrt{x}+1\right)^2-\left(x^2+1\right)^2\right)dx$$
$$=\pi\int_0^1\left(x+2\sqrt{x}+1-\left(x^4+2x^2+1\right)\right)dx$$
$$=\pi\int_0^1\left(-x^4-2x^2+x+2x^{1/2}\right)dx$$
$$=\pi\left[-\frac{1}{5}x^5-\frac{2}{3}x^3+\frac{1}{2}x^2+\frac{4}{3}x^{3/2}\right]_0^1$$
$$=\pi\left(-\frac{1}{5}-\frac{2}{3}+\frac{1}{2}+\frac{4}{3}\right)=\frac{29\pi}{30}$$

8 **The length is 2.**

You need the function which gives you the x-axis. That's $f(x)=0$. Its derivative is $f'(x)=0$. Now, just plug everything into the arc length formula:

$$\textit{Arc length} = \int_{-1}^{1}\sqrt{1+\left(\frac{dy}{dx}\right)^2}\,dx = \int_{-1}^{1}\sqrt{1+0}\,dx = \int_{-1}^{1}dx = [x]_{-1}^{1} = 1-(-1) = 2$$

Big surprise.

9 $C = 2\pi r$

The equation of a circle with radius r centered at the origin is $x^2+y^2=r^2$. Solve for y to obtain a function of x: $y=\sqrt{r^2-x^2}$. That's the equation of the top half of the circle. You need its derivative for the arc length formula:

$$y=\left(r^2-x^2\right)^{1/2}$$

$$y' = \frac{1}{2}\left(r^2-x^2\right)^{-1/2}(-2x)$$

$$= \frac{-x}{\sqrt{r^2-x^2}}$$

Plug this into the arc length formula, and integrate from $-r$ to r to get the length of the top half of the circle (you'll need to do some not-so-simple algebra):

$$\textit{Arc length} = \int_{-r}^{r}\sqrt{1+\left(\frac{-x}{\sqrt{r^2-x^2}}\right)^2}\,dx$$

$$= \int_{-r}^{r}\sqrt{1+\frac{x^2}{r^2-x^2}}\,dx$$

$$= \int_{-r}^{r}\sqrt{\frac{r^2-x^2}{r^2-x^2}+\frac{x^2}{r^2-x^2}}\,dx$$

$$= \int_{-r}^{r}\sqrt{\frac{r^2}{r^2-x^2}}\,dx$$

$$= r\int_{-r}^{r}\sqrt{\frac{1}{r^2-x^2}}\,dx$$

$$= r\int_{-r}^{r}\frac{1}{\sqrt{r^2-x^2}}\,dx$$

$$= r\left[\arcsin\left(\frac{x}{r}\right)\right]_{-r}^{r}$$

$$= r\left(\frac{\pi}{2}-\left(-\frac{\pi}{2}\right)\right) = \pi r$$

That's the arc length of your semicircle, so that gives you $2\pi r$ for the circle's circumference.

10 **See the following explanation.**

To create a right circular cylinder with radius r, just take the line $y = r$ and revolve it about the x-axis. You need the derivative of this horizontal line for the surface of revolution formula. The slope and derivative of any horizontal line is, of course, zero. The cylinder has a height of h, so you need to integrate from 0 to h. Now you've got everything you need for the surface of revolution formula:

$$\begin{aligned}
\textit{Surface area} &= 2\pi \int_a^b r\sqrt{1+\left(\frac{dy}{dx}\right)^2}\,dx \\
&= 2\pi r \int_0^h \sqrt{1+0^2}\,dx \\
&= 2\pi r \int_0^h dx \\
&= 2\pi r [x]_0^h = 2\pi rh
\end{aligned}$$

IN THIS CHAPTER

» The hospital rule — in case studying calculus makes you ill

» Meeting integrals without manners

» The paradox of Gabriel's horn

Chapter 18
Taming the Infinite with Improper Integrals

The main topic of this chapter is really amazing when you stop and think about it: calculating the area (or volume) of shapes that are *infinitely* long. The word *infinity* comes up in mathematics so often that perhaps we become jaded about the concept and forget how truly incredible it is. It's about 93 million miles from the Earth to the Sun. That distance is so great that it's nearly impossible to wrap our minds around it, but it's nothing compared to the distance to Alpha Centauri A (the nearest star), which is 4.24 light-years away — about 268,000 times as far as the distance to the Sun. Our Milky Way Galaxy is about 100,000 light-years across, and it's about 4.5 million light-years to our nearest spiral galaxy neighbor, the Andromeda Galaxy. Go out about 10,000 times that far and you reach the "edge" of the observable universe at about 46 or 47 billion light-years away. That's definitely quite a ways out there, but it's *nothing* compared to infinity.

The shapes you deal with in this chapter are not just bigger than the entire universe; they're so big that they make the universe seem like a speck of dust by comparison. And, yet, using the powerful tools of calculus (including L'Hôpital's rule), you're able to compute the area of these gargantuan shapes. And some of them turn out to have nice, manageable areas like, say, 10 square inches! It's time to get started. Next stop: the twilight zone.

L'Hôpital's Rule: Calculus for the Sick

L'Hôpital's rule is a great shortcut for doing limit problems. Remember limits — from way back in Chapters 7 and 8 — like $\lim_{x \to 3} \frac{x^2 - 9}{x - 3}$? By the way, if you're wondering why I'm showing you this limit shortcut now, it's because (a) you may need it to solve some improper integral problems (the topic of the next section in this chapter), and (b) you'll need it for some of the infinite series problems in Chapter 19.

As with most limit problems — not counting no-brainer problems — you can't do $\lim_{x \to 3} \frac{x^2 - 9}{x - 3}$ with direct substitution: Plugging 3 into x gives you $\frac{0}{0}$, which is undefined. In Chapter 8, you learn to do this problem by factoring the numerator into $(x-3)(x+3)$ and then canceling the $(x-3)$. That leaves you with $\lim_{x \to 3}(x+3)$, which equals 6.

Now watch how easy it is to do this limit with L'Hôpital's rule. Simply take the derivative of the numerator and the derivative of the denominator. Don't use the quotient rule; just take the derivatives of the numerator and denominator separately. The derivative of $x^2 - 9$ is $2x$ and the derivative of $x - 3$ is 1. L'Hôpital's rule lets you replace the numerator and denominator by their derivatives like this:

$$\lim_{x \to 3} \frac{x^2 - 9}{x - 3} = \lim_{x \to 3} \frac{2x}{1}$$

The new limit is a no-brainer: $\lim_{x \to 3} \frac{2x}{1} = \frac{2 \cdot 3}{1} = 6$

That's all there is to it. L'Hôpital's rule transforms a limit you can't do with direct substitution into one you can do with substitution. That's what makes it such a great shortcut.

Here's the mumbo jumbo.

MATH RULES

L'Hôpital's rule: Let f and g be differentiable functions. If the limit of $\frac{f(x)}{g(x)}$ as x approaches c produces $\frac{0}{0}$ or $\frac{\pm\infty}{\pm\infty}$ when you substitute the value of c into x, then

$$\lim_{x \to c} \frac{f(x)}{g(x)} = \lim_{x \to c} \frac{f'(x)}{g'(x)}$$

Note that c can be a number or $\pm\infty$. And note that in the $\pm\infty$ over $\pm\infty$ case, both infinities can be of the same sign or one can be positive and the other negative.

Here's a problem involving $\frac{\infty}{\infty}$: What's $\lim_{x \to \infty} \frac{\ln x}{x}$? Direct substitution gives you $\frac{\infty}{\infty}$, so you can use L'Hôpital's rule. The derivative of $\ln x$ is $\frac{1}{x}$, and the derivative of x is 1, so

$$\lim_{x \to \infty} \frac{\ln x}{x} = \lim_{x \to \infty} \frac{\frac{1}{x}}{1} = \frac{\frac{1}{\infty}}{1} = \frac{0}{1} = 0$$

Here's another one: Evaluate $\lim_{x\to 0}\frac{e^{3x}-1}{x}$. Substitution gives you $\frac{0}{0}$ so L'Hôpital's rule applies. The derivative of $e^{3x}-1$ is $3e^{3x}$ and the derivative of x is 1, thus

$$\lim_{x\to 0}\frac{e^{3x}-1}{x}=\lim_{x\to 0}\frac{3e^{3x}}{1}=\frac{3\cdot 1}{1}=3$$

WARNING

You must have zero over zero or infinity over infinity. The mumbo jumbo says that to use L'Hôpital's rule, substitution must produce either $\frac{0}{0}$ or $\frac{\pm\infty}{\pm\infty}$. You must get one of these *acceptable* "indeterminate" forms in order to apply the shortcut. Don't forget to check this.

Getting unacceptable forms into shape

If substitution produces one of the *unacceptable* forms, $\pm\infty\cdot 0$ or $\infty-\infty$, you first have to tweak the problem to get an acceptable form before using L'Hôpital's rule.

For instance, find $\lim_{x\to\infty}\left(e^{-x}\sqrt{x}\right)$. Substituting infinity into x gives you $0\cdot\infty$ so you've got to tweak it:

$$\lim_{x\to\infty}\left(e^{-x}\sqrt{x}\right)=\lim_{x\to\infty}\left(\frac{\sqrt{x}}{e^{x}}\right)$$

Now you've got the $\frac{\infty}{\infty}$ case, so you're all set to use L'Hôpital's rule. The derivative of $\sqrt{x}$ is $\frac{1}{2\sqrt{x}}$, and the derivative of e^x is e^x, so

$$\lim_{x\to\infty}\left(\frac{\sqrt{x}}{e^{x}}\right)=\lim_{x\to\infty}\left(\frac{\frac{1}{2\sqrt{x}}}{e^{x}}\right)=\frac{\frac{1}{2\sqrt{\infty}}}{e^{\infty}}=\frac{\frac{1}{\infty}}{\infty}=\frac{0}{\infty}=0$$

Here's another problem: What's $\lim_{x\to 0^+}\left(\frac{1}{1-\cos x}-\frac{1}{x}\right)$? (Recall from Chapter 7 that $\lim_{x\to 0^+}$ means that x approaches 0 from the right only; this is a *one-sided* limit.) First, substitute zero into x (actually, since x is approaching zero from the right, you must imagine plugging a tiny positive number into x, or you can sort of think of it as plugging a "positive" zero into x). Substitution gives you $\left(\frac{1}{1-0.999999...}-\frac{1}{0^+}\right)$, which results in $\infty-\infty$, one of the unacceptable forms. So, tweak the limit expression with some algebra:

$$\begin{aligned}&\lim_{x\to 0^+}\left(\frac{1}{1-\cos x}-\frac{1}{x}\right)\\ &=\lim_{x\to 0^+}\left(\frac{x}{x(1-\cos x)}-\frac{1-\cos x}{x(1-\cos x)}\right)\\ &=\lim_{x\to 0^+}\left(\frac{x-1+\cos x}{x(1-\cos x)}\right)\end{aligned}$$

Now substitution gives you $\frac{0}{0}$, so you can finish with L'Hôpital's rule:

$$\begin{aligned}
&\lim_{x\to 0^+}\left(\frac{x-1+\cos x}{x(1-\cos x)}\right)\\
&=\lim_{x\to 0^+}\left(\frac{1-\sin x}{1(1-\cos x)+x(\sin x)}\right)\\
&=\frac{1-0^+}{1\cdot 0^+ + 0^+\cdot 0^+}\\
&=\frac{1}{0^+}\\
&=+\infty
\end{aligned}$$

That's it.

Looking at three more unacceptable forms

When substitution of the arrow-number into the limit expression produces one of the unacceptable forms $1^{\pm\infty}$, 0^0, or ∞^0, you use the following logarithm trick to turn it into an acceptable form.

EXAMPLE

Q. Find $\lim_{x\to 0^+}(\sin x)^x$.

A. **The limit equals 1.**

Substitution gives you $(\sin 0)^0$, which equals 0^0, so you do the following:

1. **Set the limit equal to *y*.**

$$y=\lim_{x\to 0^+}(\sin x)^x$$

2. **Take the log of both sides.**

$$\begin{aligned}
\ln(y)&=\ln\left(\lim_{x\to 0^+}(\sin x)^x\right)\\
&=\lim_{x\to 0^+}\left(\ln(\sin x)^x\right) && \text{(Take my word for it.)}\\
&=\lim_{x\to 0^+}(x\ln(\sin x)) && \text{(Better review the log rules in Chapter 4 if you don't get this.)}
\end{aligned}$$

3. **This limit is a $0\cdot(-\infty)$ case, so tweak it.**

$$=\lim_{x\to 0^+}\left(\frac{\ln(\sin x)}{\frac{1}{x}}\right)$$

4. Now you've got a $\frac{-\infty}{\infty}$ case, so you can use L'Hôpital's rule.

The derivative of $\ln(\sin x)$ is $\frac{1}{\sin x} \cdot \cos x$, or $\cot x$, and the derivative of $\frac{1}{x}$ is $-\frac{1}{x^2}$, so

$$\lim_{x \to 0^+} \left(\frac{\ln(\sin x)}{\frac{1}{x}} \right)$$

$$= \lim_{x \to 0^+} \left(\frac{\cot x}{-\frac{1}{x^2}} \right)$$

$$= \lim_{x \to 0^+} \left(\frac{-x^2}{\tan x} \right)$$

5. This is a $\frac{0}{0}$ case, so use L'Hôpital's rule again.

$$= \lim_{x \to 0^+} \left(\frac{-2x}{\sec^2 x} \right) = \frac{0}{1} = 0$$

Hold your horses! This is *not* the answer.

6. Solve for *y*.

Do you see that the answer of 0 in Step 5 is the answer to the equation from way back in Step 2: $\ln y = \ln\left(\lim_{x \to 0^+} (\sin x)^x \right)$? So, the 0 in Step 5 tells you that $\ln y = 0$. Now solve for *y*:

$$\ln y = 0$$
$$y = 1$$

Because you set your limit equal to *y* in Step 1, this, finally, is your answer:

$$\mathbf{\lim_{x \to 0^+} (\sin x)^x = 1}$$

WARNING

Ordinary math doesn't work with infinity (or zero to the zero power). Don't make the mistake of thinking that you can use ordinary arithmetic or the laws of exponents when dealing with any of the acceptable or unacceptable indeterminate forms. It might look like $\infty - \infty$ should equal zero, for example, but it doesn't. By the same token, $0 \cdot \infty \neq 0$, $\frac{0}{0} \neq 1$, $\frac{\infty}{\infty} \neq 1$, $0^0 \neq 1$, $\infty^0 \neq 1$, and $1^\infty \neq 1$.

YOUR TURN

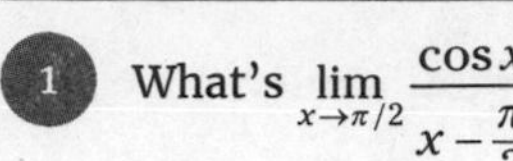

1 What's $\lim_{x \to \pi/2} \frac{\cos x}{x - \frac{\pi}{2}}$?

2 $\lim_{x \to 0} \frac{1 - \cos x}{x^2} = ?$

3 Evaluate $\lim_{x \to \pi/4} ((\tan x - 1)\sec(6x))$.

4 What's $\lim_{x \to 0^-} \left(\frac{1}{\sin x} - \frac{1}{x} \right)$?

5 Evaluate $\lim_{x \to 0^+} (\csc x - \log x)$.

6 What's $\lim_{x \to 0} (1 + x)^{1/x}$?

Improper Integrals: Just Look at the Way That Integral Is Holding Its Fork!

Definite integrals are *improper* when they go infinitely far up, down, right, or left. They go up or down infinitely far in problems like $\int_2^4 \frac{1}{x-3}dx$ that have one or more vertical asymptotes. They go infinitely far to the right or left in problems like $\int_5^\infty \frac{1}{x^2}dx$ or $\int_{-\infty}^\infty \frac{1}{x^4+1}dx$, where one or both of the limits of integration are infinite. (There are a couple other weird types of improper integrals, but they're rare — don't worry about them.) It would seem to make sense to just use the term *infinite* instead of *improper* to describe these integrals, except for the remarkable fact that many of these "infinite" integrals give you a *finite* area. More about this in a minute.

You solve both types of improper integrals by turning them into limit problems. Take a look at how it works.

Improper integrals with vertical asymptotes

There are two cases to consider here: problems where there's a vertical asymptote at one of the edges of the area in question and problems where there's a vertical asymptote somewhere in the middle of the area.

A vertical asymptote at one of the limits of integration

What's the area under $y=\frac{1}{x^2}$ from 0 to 1? This function is undefined at $x=0$, and it has a vertical asymptote there. So, you've got to turn the definite integral into a limit where c approaches the x-value of the asymptote:

$$\int_0^1 \frac{1}{x^2}dx = \lim_{c\to 0^+}\int_c^1 \frac{1}{x^2}dx$$ (The area in question is to the right of zero, so c approaches zero from the right.)

$$= \lim_{c\to 0^+}\left[-\frac{1}{x}\right]_c^1$$ (reverse power rule)

$$= \lim_{c\to 0^+}\left((-1)-\left(-\frac{1}{c}\right)\right)$$

$$= -1-(-\infty) = -1+\infty = \infty$$

This area is infinite, which probably doesn't surprise you because the curve goes up to infinity. But hold on to your hat — the next function also goes up to infinity at $x=0$, but its area is finite!

Find the area under $y = \frac{1}{\sqrt[3]{x}}$ from 0 to 1. This function is also undefined at $x = 0$, so the process is the same as in the previous problem:

$$\begin{aligned}\int_0^1 \frac{1}{\sqrt[3]{x}}dx &= \lim_{c\to 0^+} \int_c^1 \frac{1}{\sqrt[3]{x}}dx \\ &= \lim_{c\to 0^+} \left[\frac{3}{2}x^{2/3}\right]_c^1 \quad \text{(reverse power rule)} \\ &= \lim_{c\to 0^+} \left(\frac{3}{2} - \frac{3}{2}c^{2/3}\right) \\ &= \frac{3}{2} - 0 = \frac{3}{2}\end{aligned}$$

MATH RULES

Convergence and divergence: You say that an improper integral *converges* if the limit exists — that is, if the limit equals a finite number like in the second problem. Otherwise, an improper integral is said to *diverge* — like in the first problem. When an improper integral diverges, the area in question (or part of it) usually (but not always) equals infinity or negative infinity.

A vertical asymptote between the limits of integration

If the undefined point of the integrand is somewhere in between the limits of integration, you split the integral in two — at the undefined point — then turn each integral into a limit and go from there. Evaluate $\int_{-1}^{8} \frac{1}{\sqrt[3]{x}}dx$. This integrand is undefined at $x = 0$.

1. **Split the integral in two at the undefined point.**

$$\int_{-1}^{8} \frac{1}{\sqrt[3]{x}}dx = \int_{-1}^{0} \frac{1}{\sqrt[3]{x}}dx + \int_{0}^{8} \frac{1}{\sqrt[3]{x}}dx$$

2. **Turn each integral into a limit and evaluate.**

For the $\int_{-1}^{0}$ integral, the area is to the left of zero, so *c* approaches zero from the left. For the $\int_{0}^{8}$ integral, the area is to the right of zero, so *c* approaches zero from the right.

$$\begin{aligned}&= \lim_{c\to 0^-} \int_{-1}^{c} \frac{1}{\sqrt[3]{x}}dx + \lim_{c\to 0^+} \int_{c}^{8} \frac{1}{\sqrt[3]{x}}dx \\ &= \lim_{c\to 0^-} \left[\frac{3}{2}x^{2/3}\right]_{-1}^{c} + \lim_{c\to 0^+} \left[\frac{3}{2}x^{2/3}\right]_{c}^{8} \\ &= \lim_{c\to 0^-} \left(\frac{3}{2}c^{2/3} - \frac{3}{2}\right) + \lim_{c\to 0^+} \left(6 - \frac{3}{2}c^{2/3}\right) \\ &= -\frac{3}{2} + 6 = 4.5\end{aligned}$$

WARNING

Keep your eyes peeled for x-values where an integrand is undefined. If you fail to notice that an integrand is undefined at an x-value between the limits of integration, and you integrate the ordinary way, you may get the wrong answer. The above problem, $\int_{-1}^{8} \frac{1}{\sqrt[3]{x}}dx$ (undefined at $x = 0$), happens to work out correctly if you do it the ordinary way. However, if you do $\int_{-1}^{1} \frac{1}{x^2}dx$ (also undefined at $x = 0$) the ordinary way, not only do you get the wrong answer, but you get the totally absurd answer of *negative* 2, despite the fact that the area in question is above the x-axis and is therefore a positive area. The moral: *Don't risk it.*

If a part diverges, the whole diverges. If either part of the split-up integral diverges, the original integral diverges. You can't get, say, $-\infty$ for one part and ∞ for the other part and add them up to get zero.

Improper integrals with one or two infinite limits of integration

You do these improper integrals by turning them into limits where c approaches infinity or negative infinity.

Integrals with one infinite limit of integration

Let's look at two somewhat-similar problems: $\int_1^{\infty} \frac{1}{x^2}dx$ and $\int_1^{\infty} \frac{1}{x}dx$.

$$\begin{aligned}
\int_1^{\infty} \frac{1}{x^2}dx &= \lim_{c\to\infty} \int_1^{c} \frac{1}{x^2}dx \\
&= \lim_{c\to\infty} \left[-\frac{1}{x} \right]_1^c \\
&= \lim_{c\to\infty} \left(-\frac{1}{c} - \left(-\frac{1}{1} \right) \right) \\
&= 0 - (-1) = 1
\end{aligned}$$

So, this improper integral *converges*.

In the next improper integral, the denominator is smaller, x instead of x^2, and thus the fraction is *bigger*, so you'd expect $\int_1^{\infty} \frac{1}{x}dx$ to be bigger than $\int_1^{\infty} \frac{1}{x^2}dx$, which it is. But it's not just bigger, it's *way* bigger:

$$\begin{aligned}
\int_1^{\infty} \frac{1}{x}dx &= \lim_{c\to\infty} \int_1^{c} \frac{1}{x}dx \\
&= \lim_{c\to\infty} [\ln x]_1^c \\
&= \lim_{c\to\infty} (\ln c - \ln 1) \\
&= \infty - 0 = \infty
\end{aligned}$$

This improper integral *diverges*.

Figure 18-1 shows these two functions. The area under $\frac{1}{x^2}$ from 1 to ∞ is exactly the same as the area of the 1-by-1 square to its left: 1 square unit. The area under $\frac{1}{x}$ from 1 to ∞ is *much, much* bigger — actually, it's infinitely bigger than a square large enough to enclose the Milky Way Galaxy. Their shapes are quite similar, but their areas couldn't be more different.

FIGURE 18-1: The area under $\frac{1}{x^2}$ from 1 to ∞ and the area under $\frac{1}{x}$ from 1 to ∞.

By the way, these two functions make another appearance in Chapter 19, which covers infinite series. Deciding whether an infinite series converges or diverges — a distinction quite similar to the difference between these two functions — is one of the main topics in Chapter 19.

Integrals with two infinite limits of integration

When both of the limits of integration are infinite, you split the integral in two and turn each part into a limit. Splitting up the integral at $x = 0$ is convenient because zero's an easy number to deal with, but you can split it up anywhere you like. Zero may also seem like a good choice because it looks like it's in the middle between $-\infty$ and ∞. But that's an illusion because there is no middle between $-\infty$ and ∞, or you could say that any point on the x-axis is the middle.

Same warning as the one in the section, "A vertical asymptote between the limits of integration." **If either "half" of the split-up integral diverges, the whole, original integral diverges.**

Q. Evaluate $\int_{-\infty}^{\infty} \frac{1}{x^2+1} dx$.

A. **The area equals π.**

1. **Split the integral in two.**

$$\int_{-\infty}^{\infty} \frac{1}{x^2+1} dx = \int_{-\infty}^{0} \frac{1}{x^2+1} dx + \int_{0}^{\infty} \frac{1}{x^2+1} dx$$

2. **Turn each part into a limit.**

$$= \lim_{c \to -\infty} \int_{c}^{0} \frac{1}{x^2+1} dx + \lim_{c \to \infty} \int_{0}^{c} \frac{1}{x^2+1} dx$$

3. **Evaluate each part and add up the results.**

$$= \lim_{c \to -\infty} \left[\tan^{-1} x \right]_c^0 + \lim_{c \to \infty} \left[\tan^{-1} x \right]_0^c$$

$$= \lim_{c \to -\infty} \left(\tan^{-1} 0 - \tan^{-1} c \right) + \lim_{c \to \infty} \left(\tan^{-1} c - \tan^{-1} 0 \right)$$

$$= \left(0 - \left(-\frac{\pi}{2} \right) \right) + \left(\frac{\pi}{2} - 0 \right)$$

$$= \pi$$

Why don't you do this problem again, splitting up the integral somewhere other than at $x = 0$, to confirm that you get the same result.

YOUR TURN

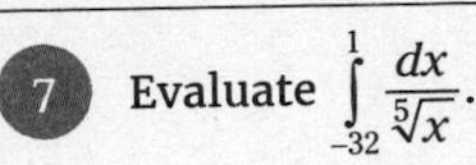

7 Evaluate $\int_{-32}^{1} \frac{dx}{\sqrt[5]{x}}$.

8 Compute $\int_{0}^{6} x \ln x \, dx$.

9 $\int_{1}^{\infty} \frac{dx}{x\sqrt{x^2 - 1}} = ?$ *Hint:* Split up at $x = 2$.

10 What's $\int_{1}^{\infty} \frac{dx}{\sqrt{x^3}}$?

11 $\int_{1}^{\infty} \frac{1}{x}\sqrt{\arctan x}\, dx = ?$ *Hint:* Use the example problem from a couple pages back, $\int_{1}^{\infty} \frac{1}{x} dx$.

12 $\int_{-\infty}^{\infty} \frac{1}{x} dx = ?$ *Hint:* Break into four parts.

Blowing Gabriel's horn

WARNING

This horn problem may blow your mind.

Gabriel's horn is the solid generated by revolving about the x-axis the unbounded region between $y = \frac{1}{x}$ and the x-axis (for $x \geq 1$). See Figure 18-2. Playing this instrument poses several not-insignificant challenges: 1) It has no end for you to put in your mouth; 2) Even if it did, it would take you till the end of time to reach the end; 3) Even if you could reach the end and put it in your mouth, you couldn't force any air through it because the hole is infinitely small; 4) Even if you could blow the horn, it'd be kind of pointless because it would take an infinite amount of time for the sound to come out the other end. There are additional difficulties — infinite weight, doesn't fit in universe, and so on — but I suspect you get the picture.

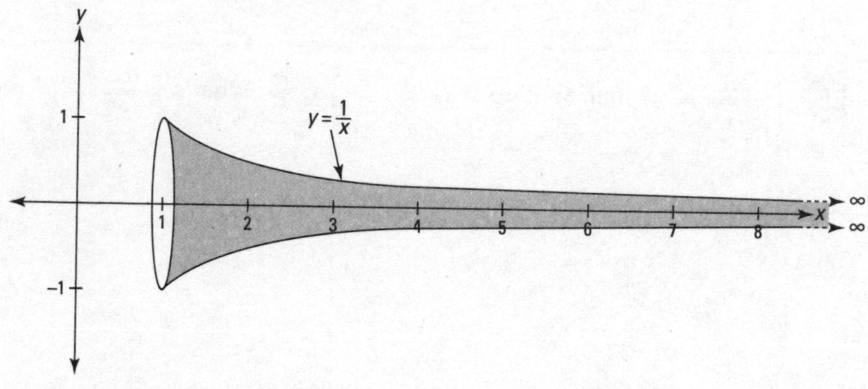

FIGURE 18-2: Gabriel's horn.

Believe it or not, Gabriel's horn has a finite volume, but an *infinite* surface area! You use the disk method to figure its volume (see Chapter 17). Recall that the volume of each representative disk is $\pi r^2 dx$. For this problem, the radius is $\frac{1}{x}$, so the little bit of volume is $\pi \left(\frac{1}{x} \right)^2 dx$. You find the total volume by adding up the little bits from 1 to infinity:

$$Volume = \int_1^{\infty} \pi \left(\frac{1}{x} \right)^2 dx = \pi \int_1^{\infty} \frac{1}{x^2} dx$$

In the section on improper integrals, we calculated that $\int_1^{\infty} \frac{1}{x^2} dx = 1$, so the volume is $\pi \cdot 1$, or just π.

To determine the surface area, you first need the function's derivative (the method for calculating surface area is covered in Chapter 17):

$$y = \frac{1}{x}$$
$$\frac{dy}{dx} = -\frac{1}{x^2}$$

Now plug everything into the surface area formula:

$$\begin{aligned}\textit{Surface area} &= 2\pi\int_1^{\infty}\frac{1}{x}\sqrt{1+\left(-\frac{1}{x^2}\right)^2}\,dx\\ &= 2\pi\int_1^{\infty}\frac{1}{x}\sqrt{1+\frac{1}{x^4}}\,dx\end{aligned}$$

In the previous section, we determined that $\int_1^{\infty}\frac{1}{x}dx=\infty$, and because $\frac{1}{x}\sqrt{1+\frac{1}{x^4}}$ is always greater than $\frac{1}{x}$ in the interval $[1, \infty)$, $\int_1^{\infty}\frac{1}{x}\sqrt{1+\frac{1}{x^4}}\,dx$ must also equal ∞. Finally, 2π times ∞ is still ∞, of course, so the surface area is infinite.

Bonus question for those with a philosophical bent: Assuming Gabriel is omnipotent, could he overcome the above-mentioned difficulties and blow this horn? *Hint:* All the calculus in the world won't help you with this one.

Practice Questions Answers and Explanations

1 $\lim\limits_{x\to\pi/2}\frac{\cos x}{x-\frac{\pi}{2}}=-1$

1. **Plug in: $\frac{0}{0}$ — onward!**
2. **Replace the numerator and denominator with their derivatives: $=\lim\limits_{x\to\pi/2}\frac{-\sin x}{1}$.**
3. **Plug in again: $=\frac{-\sin\frac{\pi}{2}}{1}=-1$**

2 $\lim\limits_{x\to 0}\frac{1-\cos x}{x^2}=\frac{1}{2}$

1. **Plug in: $\frac{0}{0}$; no worries.**
2. **Replace with derivatives: $=\lim\limits_{x\to 0}\frac{\sin x}{2x}$**
3. **Plug in: $\frac{0}{0}$ again, so repeat.**
4. **Replace with derivatives again: $=\lim\limits_{x\to 0}\frac{\cos x}{2}$**
5. **Finish: $=\frac{\cos 0}{2}=\frac{1}{2}$**

3 $\lim\limits_{x\to\pi/4}((\tan x-1)\sec(6x))=\frac{1}{3}$

1. **Plugging in gives you $0\cdot\infty$, so on to Step 2.**
2. **Rewrite: $=\lim\limits_{x\to\pi/4}\frac{\tan x-1}{\cos(6x)}=\frac{0}{0}$. Copasetic.**
3. **Replace with derivatives: $=\lim\limits_{x\to\pi/4}\frac{\sec^2 x}{-6\sin(6x)}$**
4. **Plug in to finish: $=\frac{\sec^2\frac{\pi}{4}}{-6\sin\frac{3\pi}{2}}=\frac{2}{6}=\frac{1}{3}$**

4 $\lim\limits_{x\to 0^-}\left(\frac{1}{\sin x}-\frac{1}{x}\right)=0$

1. **Plugging in gives you $-\infty-(-\infty)=-\infty+\infty$, so you have to tweak it.**
2. **Rewrite by adding the fractions: $=\lim\limits_{x\to 0^-}\frac{x-\sin x}{x\sin x}$**

   That's a good bingo: $\frac{0}{0}$
3. **Replace with derivatives: $=\lim\limits_{x\to 0^-}\frac{1-\cos x}{\sin x+x\cos x}$**

   That's $\frac{0}{0}$ again, so use L'Hôpital's rule a second time.
4. **Replace with derivatives: $=\lim\limits_{x\to 0^-}\frac{\sin x}{\cos x+\cos x-x\sin x}$**
5. **Plug in to finish: $=\frac{0}{2-0}=0$**

5 $\lim\limits_{x\to 0^+}(\csc x - \log x) = \infty$

This limit equals $\infty - (-\infty)$, which equals $\infty + \infty = \infty$. You're done! L'Hôpital's rule isn't needed. You gotta be on your toes.

6 $\lim\limits_{x\to 0}(1+x)^{1/x} = e$

1. **This is a 1^∞ case.**
2. **Set your limit equal to y and take the natural log of both sides.**

$$y = \lim_{x\to 0}(1+x)^{1/x}$$
$$\ln y = \ln\left(\lim_{x\to 0}(1+x)^{1/x}\right)$$

3. **I give you permission to pull the limit to the outside.**

$$\ln y = \lim_{x\to 0}\left(\ln(1+x)^{1/x}\right)$$

4. **Use the log of a power rule.**

$$\ln y = \lim_{x\to 0}\left(\frac{1}{x}\ln(1+x)\right)$$

5. **Plugging in gives you a $\infty \cdot 0$ case, so rewrite.**

$$\ln y = \lim_{x\to 0}\frac{\ln(1+x)}{x}$$

6. **Now you've got a $\frac{0}{0}$ case — I'm down with it.**
7. **Replace with derivatives.**

$$\ln y = \lim_{x\to 0}\frac{\frac{1}{1+x}}{1} = 1$$

8. **Your original limit equals y, so you have to solve for y.**

$$\ln y = 1$$
$$y = e$$

7 $\int_{-32}^{1}\frac{dx}{\sqrt[5]{x}} = -18.75$

1. **The integrand is undefined at $x = 0$, so break in two.**

$$\int_{-32}^{1}\frac{dx}{\sqrt[5]{x}} = \int_{-32}^{0}\frac{dx}{\sqrt[5]{x}} + \int_{0}^{1}\frac{dx}{\sqrt[5]{x}}$$

2. **Turn into one-sided limits.**

$$= \lim_{a\to 0^-}\int_{-32}^{a}\frac{dx}{\sqrt[5]{x}} + \lim_{b\to 0^+}\int_{b}^{1}\frac{dx}{\sqrt[5]{x}}$$

3. **Integrate.**

$$= \lim_{a\to 0^-}\left[\frac{5}{4}x^{4/5}\right]_{-32}^{a} + \lim_{b\to 0^+}\left[\frac{5}{4}x^{4/5}\right]_{b}^{1} = 0-\frac{5}{4}\cdot 16+\frac{5}{4}-0=-18.75$$

(8) $\int_0^6 x\ln x\,dx = 18\ln 6 - 9$

1. **The integral is improper because it's undefined at $x = 0$, so turn it into a limit.**

$$= \lim_{c\to 0^+}\int_c^6 x\ln x\,dx$$

2. **Integrate by parts.**

Hint: $\ln x$ is L from LIATE. You should obtain:

$$= \lim_{c\to 0^+}\left[\frac{1}{2}\cdot x^2\ln x - \frac{1}{4}x^2\right]_c^6$$

$$= \lim_{c\to 0^+}\left(\frac{1}{2}\cdot 36\ln 6 - 9 - \frac{1}{2}c^2\ln c + \frac{1}{4}c^2\right)$$

$$= 18\ln 6 - 9 - \frac{1}{2}\lim_{c\to 0^+}\left(c^2\ln c\right)$$

3. **Time to practice L'Hôpital's rule.**

This is a $0\cdot(-\infty)$ limit, so turn it into a $\frac{-\infty}{\infty}$ one:

$$= 18\ln 6 - 9 - \frac{1}{2}\lim_{c\to 0^+}\frac{\ln c}{\frac{1}{c^2}}$$

4. **Replace the numerator and denominator with derivatives and finish.**

$$= 18\ln 6 - 9 - \frac{1}{2}\lim_{c\to 0^+}\frac{\frac{1}{c}}{-\frac{2}{c^3}} = 18\ln 6 - 9 - \frac{1}{2}\lim_{c\to 0^+}\left(-\frac{c^2}{2}\right) = 18\ln 6 - 9$$

(9) $\int_1^\infty \frac{dx}{x\sqrt{x^2-1}} = \frac{\pi}{2}$

This is a doubly improper integral because it goes up to infinity *and* right to infinity. You have to split it up and tackle each infinite integral separately.

1. **It doesn't matter where you split it up; how about splitting it at 2, a nice, easy-to-deal-with number.**

$$= \int_1^2 \frac{dx}{x\sqrt{x^2-1}} + \int_2^\infty \frac{dx}{x\sqrt{x^2-1}}$$

2. **Turn each integral into a limit.**

$$= \lim_{a\to 1^+}\int_a^2 \frac{dx}{x\sqrt{x^2-1}} + \lim_{b\to\infty}\int_2^b \frac{dx}{x\sqrt{x^2-1}}$$

3. **Integrate.**

$$= \lim_{a\to 1^+}[\operatorname{arcsec} x]_a^2 + \lim_{b\to\infty}[\operatorname{arcsec} x]_2^b$$
$$= \lim_{a\to 1^+}(\operatorname{arcsec} 2 - \operatorname{arcsec} a) + \lim_{b\to\infty}(\operatorname{arcsec} b - \operatorname{arcsec} 2)$$
$$= \operatorname{arcsec} 2 - 0 + \frac{\pi}{2} - \operatorname{arcsec} 2 = \frac{\pi}{2}$$

10 $\displaystyle\int_1^\infty \frac{dx}{\sqrt{x^3}} = 2$

1. **Turn into a limit:** $= \lim_{c\to\infty}\int_1^c \frac{1}{\sqrt{x^3}}dx = \lim_{c\to\infty}\int_1^c x^{-3/2}dx$

2. **Integrate and finish:** $= \lim_{c\to\infty}\left[-2x^{-1/2}\right]_1^c = \lim_{c\to\infty}\left(-2c^{-1/2}+2\right) = 2$

11 $\displaystyle\int_1^\infty \frac{1}{x}\sqrt{\arctan x}\,dx = \infty$

No work is required for this one, "just" logic. You know from Problem 10 that $\int_1^\infty \frac{1}{x}dx = \infty$. Now, compare $\int_1^\infty \frac{1}{x}\sqrt{\arctan x}\,dx$ to $\int_1^\infty \frac{1}{x}dx$. But first note that because $\int_1^\infty \frac{1}{x}dx$ equals infinity, so will $\int_{10}^\infty \frac{1}{x}dx$, $\int_{100}^\infty \frac{1}{x}dx$, or $\int_{1{,}000{,}000}^\infty \frac{1}{x}dx$, because the area under $\frac{1}{x}$ from 1 to any other number must be finite.

From $\sqrt{3}$ to ∞, $\arctan x > 1$; therefore, $\sqrt{\arctan x} > 1$, and thus $\frac{1}{x}\sqrt{\arctan x} > \frac{1}{x}$. Because $\int_{\sqrt{3}}^\infty \frac{1}{x}dx = \infty$ and because between $\sqrt{3}$ and ∞, $\frac{1}{x}\sqrt{\arctan x}$ is always greater than $\frac{1}{x}$, $\int_{\sqrt{3}}^\infty \frac{1}{x}\sqrt{\arctan x}\,dx$ must also equal ∞. Finally, because $\int_{\sqrt{3}}^\infty \frac{1}{x}\sqrt{\arctan x}\,dx$ equals ∞, $\int_1^\infty \frac{1}{x}\sqrt{\arctan x}\,dx$ must as well.

Aren't you glad no work was required for this problem?

12 $\displaystyle\int_{-\infty}^\infty \frac{1}{x}dx$ **is undefined.**

Quadruply improper!

1. **Split into four parts.**

$$\int_{-\infty}^\infty \frac{1}{x}dx = \int_{-\infty}^{-1}\frac{1}{x}dx + \int_{-1}^0 \frac{1}{x}dx + \int_0^1 \frac{1}{x}dx + \int_1^\infty \frac{1}{x}dx$$

2. **Turn into limits.**

$$= \lim_{a \to -\infty} \int_a^{-1} \frac{1}{x} dx + \lim_{b \to 0^-} \int_{-1}^{b} \frac{1}{x} dx + \lim_{c \to 0^+} \int_c^{1} \frac{1}{x} dx + \lim_{d \to \infty} \int_1^{d} \frac{1}{x} dx$$

3. **Integrate.**

$$= \lim_{a \to -\infty} [\ln|x|]_a^{-1} + \lim_{b \to 0^-} [\ln|x|]_{-1}^{b} + \lim_{c \to 0^+} [\ln|x|]_c^{1} + \lim_{d \to \infty} [\ln|x|]_1^{d}$$

$$= \lim_{a \to -\infty} (\ln 1 - \ln|a|) + \lim_{b \to 0^-} (\ln|b| - \ln 1) + \lim_{c \to 0^+} (\ln 1 - \ln|c|) + \lim_{d \to \infty} (\ln|d| - \ln 1)$$

4. **Finish:** $= -\infty + (-\infty) + \infty + \infty$

Therefore, the limit doesn't exist, and the definite integral is thus undefined.

WARNING

Looks can be deceiving. If you look at the graph of $y = \frac{1}{x}$, its perfect symmetry may make you think that $\int_{-\infty}^{\infty} \frac{1}{x} dx$ would equal zero. But — strange as it seems — it doesn't work that way. And, to repeat the warning from earlier in this chapter, you can't simplify Step 4 to $-\infty + \infty$ and sum that up to zero.

If you're ready to test your skills a bit more, take the following chapter quiz that incorporates all the chapter topics.

Whaddya Know? Chapter 18 Quiz

Quiz time! Complete each problem to test your knowledge on the various topics covered in this chapter. You can then find the solutions and explanations in the next section.

1 Find $\lim_{x \to \infty} \frac{e^x}{10^x}$.

2 Evaluate $\lim_{x \to \pi/2^-} \left(x \tan x - \frac{\pi}{2} \tan x \right)$.

3 What's $\lim_{x \to e} \frac{\ln x - 1}{x^2 - e^2}$?

4 Evaluate $\lim_{x \to 1} \frac{x - 1 + \sin\left(x^3 - 1\right)}{x^2 - 1}$.

5 Evaluate the following improper integrals.

(a) $\int_{-\infty}^{0} 3x^2 e^{x^3} dx$

(b) $\int_{-\infty}^{0} 4x^3 e^{x^4} dx$

(c) $\int_{-\infty}^{0} 5x^4 e^{x^5} dx$

6 Evaluate $\int_{0}^{e} \ln x \, dx$.

7 Find $\int_{1}^{\infty} \ln x \, dx$.

8 $\int_{0}^{1} \frac{dx}{x \ln x} = ?$

9 Evaluate $\int_{\pi/4}^{3\pi/4} \sec^2 x \, dx$.

10 Evaluate $\int_{1}^{10} \frac{dx}{\sqrt[3]{x - 2}}$.

Answers to Chapter 18 Quiz

1. $\lim_{x\to\infty}\frac{e^x}{10^x}=0$

Plugging infinity into x gives you $\frac{\infty}{\infty}$, so you're good to go for L'Hôpital's rule. Replacing the numerator and denominator with their derivatives gives you $\lim_{x\to\infty}\frac{e^x}{10^x\ln 10}$, which, again, results in $\frac{\infty}{\infty}$. No problem; from time to time, you have to use L'Hôpital's rule more than once. With this problem, however, repeated use of L'Hôpital's rule gets you nowhere. Try it.

Tricked you! L'Hôpital's rule is not needed here. Algebra I is all you need: $\lim_{x\to\infty}\frac{e^x}{10^x}=\lim_{x\to\infty}\left(\frac{e}{10}\right)^x$. Any number, like $\frac{e}{10}$, that's between −1 and 1 gets smaller and smaller as you raise it to higher and higher powers, so the limit as x approaches infinity is zero.

When doing calculus (or any other type of math), don't forget to try low-level ideas from pre-algebra, algebra I and II, geometry, trig, and so on.

2. $\lim_{x\to\pi/2^-}\left(x\tan x-\frac{\pi}{2}\tan x\right)=-1$

If you plug $\frac{\pi}{2}$ in for x, you'll have something minus itself, so you might think the answer should be zero. But it doesn't work that way. Be careful: $\tan x$ as x approaches $\frac{\pi}{2}$ from the left equals infinity. And, don't forget: $\infty-\infty\neq 0$!

You've got to do a little work so you'll be able to apply L'Hôpital's rule. First, factor the limit expression: $\lim_{x\to\pi/2^-}\left(\tan x\left(x-\frac{\pi}{2}\right)\right)$. Now you've got an $\infty\cdot 0$ case, which is usually quite easy to turn into one of the acceptable forms: $\frac{0}{0}$ or $\frac{\pm\infty}{\pm\infty}$, to wit: $\lim_{x\to\pi/2^-}\frac{x-\frac{\pi}{2}}{\cot x}$ is a $\frac{0}{0}$ case. Now you're all set to finish with L'Hôpital's rule:

$$\lim_{x\to\pi/2^-}\frac{x-\frac{\pi}{2}}{\cot x}=\lim_{x\to\pi/2^-}\frac{1}{-\csc^2 x}=-1$$

3. $\lim_{x\to e}\frac{\ln x-1}{x^2-e^2}=\frac{1}{2e^2}$

Plugging e into x gives you $\frac{0}{0}$, so you're all set to use L'Hôpital's rule:

$$\lim_{x\to e}\frac{\ln x-1}{x^2-e^2}=\lim_{x\to e}\frac{\frac{1}{x}}{2x}=\lim_{x\to e}\frac{1}{2x^2}=\frac{1}{2e^2}$$

4) $\lim_{x\to1}\frac{x-1+\sin\left(x^3-1\right)}{x^2-1}=2$

This is another $\frac{0}{0}$ case, so replace the numerator and denominator with their derivatives, and then finish:

$$\lim_{x\to1}\frac{x-1+\sin\left(x^3-1\right)}{x^2-1}=\lim_{x\to1}\frac{1+\cos\left(x^3-1\right)\cdot 3x^2}{2x}=\frac{4}{2}=2$$

5) (a) $\int_{-\infty}^{0}3x^2e^{x^3}dx=1$

$$\int_{-\infty}^{0}3x^2e^{x^3}dx=\lim_{c\to-\infty}\int_{c}^{0}3x^2e^{x^3}dx=\lim_{c\to-\infty}\left[e^{x^3}\right]_c^0=\lim_{c\to-\infty}\left(1-e^{c^3}\right)=1-0=1$$

(b) $\int_{-\infty}^{0}4x^3e^{x^4}dx$ **diverges.**

$$\int_{-\infty}^{0}4x^3e^{x^4}dx=\lim_{c\to-\infty}\int_{c}^{0}4x^3e^{x^4}dx=\lim_{c\to-\infty}\left[e^{x^4}\right]_c^0=\lim_{c\to-\infty}\left(1-e^{c^4}\right)=1-\infty=-\infty$$

(c) $\int_{-\infty}^{0}5x^4e^{x^5}dx=1$

$$\int_{-\infty}^{0}5x^4e^{x^5}dx=\lim_{c\to-\infty}\int_{c}^{0}5x^4e^{x^5}dx=\lim_{c\to-\infty}\left[e^{x^5}\right]_c^0=\lim_{c\to-\infty}\left(1-e^{c^5}\right)=1-0=1$$

6) $\int_{0}^{e}\ln x\,dx=0$

$$\int_{0}^{e}\ln x\,dx=\lim_{c\to0^+}\int_{c}^{e}\ln x\,dx=\lim_{c\to0^+}\left[x\ln x-x\right]_c^e=\lim_{c\to0^+}\left(e\cdot1-e-(c\ln c-c)\right)$$

In the final limit expression, everything equals zero or cancels, leaving only $-\lim_{c\to0^+}(c\ln c)$. Plugging in gives you $0\cdot(-\infty)$ for the limit expression, so you've got to rewrite the limit as $-\lim_{c\to0^+}\frac{\ln c}{\frac{1}{c}}$, which is a $\frac{-\infty}{\infty}$ case that you can finish with L'Hôpital's rule:

$$-\lim_{c\to0^+}\frac{\ln c}{\frac{1}{c}}=-\lim_{c\to0^+}\frac{\frac{1}{c}}{\frac{-1}{c^2}}=-\lim_{c\to0^+}(-c)=0$$

⑦ $\int_1^{\infty} \ln x dx$ **diverges.**

$$\int_1^{\infty} \ln x dx = \lim_{c\to\infty}\int_1^{c} \ln x dx = \lim_{c\to\infty}[x\ln x - x]_1^c = \lim_{c\to\infty}(c\ln c - c - (0-1))$$

Plugging infinity into c in $c\ln c - c$ gives you $\infty - \infty$, so this looks like it might be a L'Hôpital's rule problem. But all you need is a little algebra:

$$\lim_{c\to\infty}(c\ln c - c - (0-1)) = \lim_{c\to\infty}(c(\ln c - 1) + 1) = \infty\cdot\infty + 1 = \infty$$

By the way, if you look at the graph of $y = \ln x$, it should be obvious that the area under the curve from 1 to infinity is infinite. The natural logarithm function rises from $(1, 0)$, going to the right toward infinity and going infinitely high. The area, of course, is infinite. (*Caution:* While this argument is correct, your professor may not buy it.)

⑧ $\int_0^1 \frac{dx}{x\ln x}$ **diverges.**

$$\begin{aligned}
\int_0^1 \frac{dx}{x\ln x} &= \int_0^{1/2}\frac{dx}{x\ln x} + \int_{1/2}^{1}\frac{dx}{x\ln x}\\
&= \lim_{c\to 0^+}\int_c^{1/2}\frac{dx}{x\ln x} + \lim_{c\to 1^-}\int_{1/2}^{c}\frac{dx}{x\ln x}\\
&= \lim_{c\to 0^+}[\ln|\ln x|]_c^{1/2} + \lim_{c\to 1^-}[\ln|\ln x|]_{1/2}^c\\
&= \lim_{c\to 0^+}\left(\ln\left|\ln\frac{1}{2}\right| - \ln|\ln c|\right) + \lim_{c\to 1^-}\left(\ln|\ln c| - \ln\left|\ln\frac{1}{2}\right|\right)\\
&= \left(\ln\left|\ln\frac{1}{2}\right| - \infty\right) + \left(-\infty - \ln\left|\ln\frac{1}{2}\right|\right)\\
&= -\infty - \infty\\
&= -\infty
\end{aligned}$$

⑨ $\int_{\pi/4}^{3\pi/4} \sec^2 x\, dx$ **diverges.**

The function $y = \sec^2 x$ has a vertical asymptote at $x = \frac{\pi}{2}$ so you have to split the integral in two there. To wit —

$$\begin{aligned}
\int_{\pi/4}^{3\pi/4} \sec^2 x dx &= \lim_{c\to\pi/2^-}\int_{\pi/4}^{c}\sec^2 x dx + \lim_{c\to\pi/2^+}\int_{c}^{3\pi/4}\sec^2 x dx\\
&= \lim_{c\to\pi/2^-}[\tan x]_{\pi/4}^{c} + \lim_{c\to\pi/2^+}[\tan x]_c^{3\pi/4}\\
&= \lim_{c\to\pi/2^-}(\tan c - 1) + \lim_{c\to\pi/2^+}(-1-\tan c)\\
&= \infty - 1 + (-1) - (-\infty)\\
&= \infty
\end{aligned}$$

Note that if you fail to notice that there's a vertical asymptote between the limits of integration, and you do the definite integral the ordinary way, you'll get the absurd answer of *negative* 2 despite the fact that the function is everywhere above the x-axis, which means that the area in question would have to be *positive*.

10 $\displaystyle\int_1^{10} \frac{dx}{\sqrt[3]{x-2}}$

This problem is quite similar to Problem 9; there's a vertical asymptote between the limits of integration. So, split the integral in two at the asymptote:

$$\begin{aligned}
\int_1^{10} \frac{dx}{\sqrt[3]{x-2}} &= \lim_{c\to 2^-} \int_1^{c} \frac{dx}{\sqrt[3]{x-2}} + \lim_{c\to 2^+} \int_c^{10} \frac{dx}{\sqrt[3]{x-2}} \\
&= \lim_{c\to 2^-} \left[\frac{3}{2}(x-2)^{2/3}\right]_1^c + \lim_{c\to 2^+} \left[\frac{3}{2}(x-2)^{2/3}\right]_c^{10} \\
&= \lim_{c\to 2^-} \left(\frac{3}{2}(c-2)^{2/3} - \frac{3}{2}\right) + \lim_{c\to 2^+} \left(6 - \frac{3}{2}(c-2)^{2/3}\right) \\
&= -\frac{3}{2} + 6 = 4.5
\end{aligned}$$

Unlike with Problem 9, if you fail to notice the asymptote here and you do the definite integral the regular way, you'll luck out and still get the correct answer. (But — unless your professor is the world's most lenient grader — you will certainly lose some points.)

IN THIS CHAPTER

» Segueing from sequences into series

» An infinite series — the rain delays just wouldn't end

» Getting musical with the harmonic series

» Taking a close look at telescoping series

» Rooting for the root test

» Testing for convergence

» Analyzing alternating series

Chapter 19

Infinite Series: Welcome to the Outer Limits

As with just about every topic in calculus, the subject of this chapter involves the idea of infinity — specifically, series that continue to infinity. An infinite series is the sum of an endless list of numbers like $\frac{1}{2}+\frac{1}{3}+\frac{1}{4}+\frac{1}{5}+\ldots$. Because the list is unending, it's not surprising that such a sum can be infinite. What's remarkable is that some infinite series add up to a *finite* number. This chapter covers ten tests for deciding whether the sum of a series is finite or infinite.

What you do in this chapter is quite fantastic when you think about it. Consider the series $0.1+0.01+0.001+0.0001+\ldots$. If you go out far enough, you'll find a number that has so many zeros to the right of the decimal point that even if each zero were as small as a proton, there wouldn't be enough room in the entire universe just to write it down! As vast as our universe is, anything in it — say, the number of elementary particles — is a proverbial drop in the bucket next to the things you look at in this chapter. Actually, not even a drop in the bucket, because next to infinity, any finite thing amounts to *nothing.* Maybe you've heard about the late astronomer, Carl Sagan, getting emotional about the "billions and billions" of stars in our galaxy. "Billions and billions" — *pffffftt.*

Sequences and Series: What They're All About

Here's a *sequence*: $\frac{1}{2}, \frac{1}{4}, \frac{1}{8}, \frac{1}{16}, \ldots$. Change the commas to addition signs and you've got a *series*: $\frac{1}{2}+\frac{1}{4}+\frac{1}{8}+\frac{1}{16}+\ldots$. Pretty simple, eh? Investigating series is what this chapter is all about, but I need to briefly discuss sequences to lay the groundwork for series.

Stringing sequences

A *sequence* is simply a list of numbers. An *infinite sequence* is an *unending* list of numbers. That's the only kind we're interested in here, and whenever the term *sequence* (or *series*) is used alone in this chapter, it means an infinite sequence (or infinite series).

Here's the general form for an infinite sequence:

$$a_1,\ a_2,\ a_3,\ a_4, \ldots, a_n, \ldots$$

where n runs from 1 (usually) to infinity (sometimes n starts at zero or another number). The fourth *term* of this sequence, for example, is a_4 (read "a sub 4"); the nth term is a_n (read "a sub n"). The thing we care about is what happens to a sequence infinitely far out to the right, or as mathematicians say, "in the limit." A shorthand notation for this sequence is $\{a_n\}$.

A few paragraphs back, I mentioned the following sequence. It's defined by the formula $a_n = \frac{1}{2^n}$:

$$\frac{1}{2}, \frac{1}{4}, \frac{1}{8}, \frac{1}{16}, \ldots, \frac{1}{2^n}, \ldots$$

What happens to this sequence as n approaches infinity should be pretty easy to see. Each term gets smaller and smaller, right? And if you go out far enough, you can find a term as close to zero as you want, right? So,

$$\lim_{n\to\infty} a_n = \lim_{n\to\infty} \frac{1}{2^n} = \frac{1}{2^\infty} = \frac{1}{\infty} = 0$$

Recall from Chapters 7 and 8 how to interpret this limit: As n approaches infinity (but never gets there), a_n gets closer and closer to zero (but never gets there).

Convergence and divergence of sequences

Because the limit of the previous sequence is a *finite* number, you say that the sequence *converges*.

MATH RULES

Convergence and divergence: For any sequence $\{a_n\}$, if $\lim_{n\to\infty} a_n = L$, where L is a real number (which does not include infinity or negative infinity), then the sequence *converges* to L. Otherwise, the sequence is said to *diverge*. (*Note:* These definitions are very similar to the definitions of these two terms that you saw in Chapter 18.)

Sequences that converge sort of settle down to some particular number — plus or minus some miniscule amount — after you go out to the right far enough. Sequences that diverge never settle down. Instead, diverging sequences might . . .

» Increase forever, in which case $\lim_{n\to\infty} a_n = \infty$. Such a sequence is said to "blow up." A sequence can also have a limit of negative infinity.

» Oscillate (go up and down) like the sequence $1, -1, 1, -1, 1, -1, 1, -1\ldots$

» Exhibit no pattern at all — this is rare.

Sequences and functions go hand in hand

The sequence $\left\{\frac{1}{2^n}\right\} = \frac{1}{2}, \frac{1}{4}, \frac{1}{8}, \frac{1}{16}, \ldots, \frac{1}{2^n}, \ldots$ can be thought of as an infinite set of discrete points (*discrete* is a fancy math word for *separate*) along the continuous function $f(x) = \frac{1}{2^x}$. Figure 19-1 shows the curve $f(x) = \frac{1}{2^x}$ and the points on the curve that make up the sequence.

The sequence is made up of the outputs (the y-values) of the function where the inputs (the x-values) are the positive integers (1, 2, 3, 4, . . .).

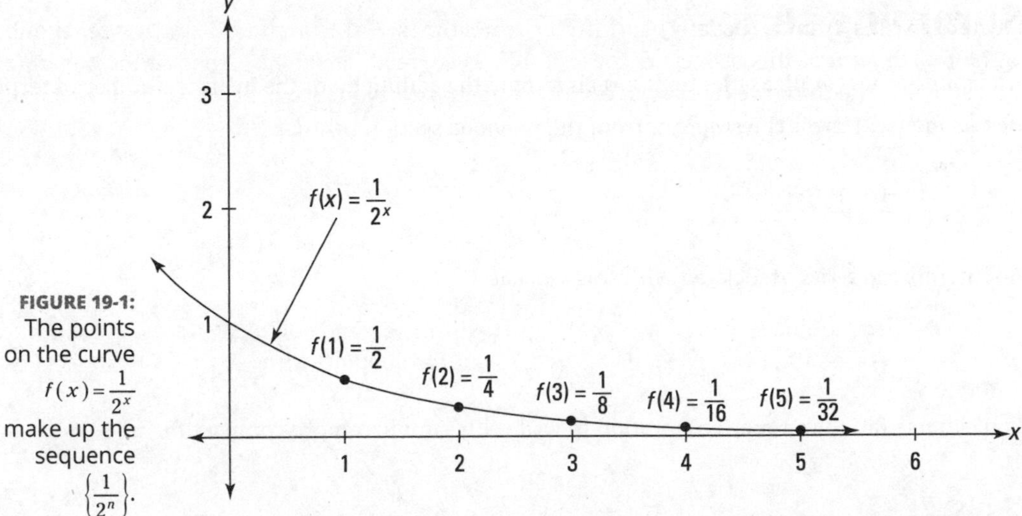

FIGURE 19-1: The points on the curve $f(x) = \frac{1}{2^x}$ make up the sequence $\left\{\frac{1}{2^n}\right\}$.

A sequence and the related function go hand in hand. If the limit of the function as x approaches infinity is some finite number, L, then the limit of the sequence is also L, and thus, the sequence converges to L. Also, the graph of such a convergent function/sequence pair has a horizontal asymptote at $y = L$; the graph in Figure 19-1 has an asymptote with the equation $y = 0$.

Determining limits with L'Hôpital's rule

Remember L'Hôpital's rule from Chapter 18? You're going to use it now to find limits of sequences. Does the sequence $a_n = \frac{n^2}{2^n}$ converge or diverge? By plugging in 1, then 2, then 3, and so on into $\frac{n^2}{2^n}$, you generate the first few terms of the sequence:

$$\frac{1}{2}, 1, \frac{9}{8}, 1, \frac{25}{32}, \frac{36}{64}, \frac{49}{128}, \frac{64}{256}, \ldots$$

What do you think? After going up for a couple terms, the sequence goes down and it appears that it'll keep going down — looks like it will converge to zero. L'Hôpital's rule proves it. You use the rule to determine the limit of the function $f(x) = \frac{x^2}{2^x}$, which goes hand in hand with the sequence $\frac{n^2}{2^n}$.

REMEMBER

Take two separate derivatives. To use L'Hôpital's rule, take the derivative of the numerator and the derivative of the denominator separately; you do not use the quotient rule.

For this problem, you have to use L'Hôpital's rule twice:

$$\lim_{x\to\infty} \frac{x^2}{2^x} = \lim_{x\to\infty} \frac{2x}{2^x \ln 2} = \lim_{x\to\infty} \frac{2}{2^x \ln 2 \ln 2} = \frac{2}{\infty} = 0$$

Because the limit of the function is zero, so is the limit of the sequence, and thus the sequence $\frac{n^2}{2^n}$ converges to zero.

Summing series

An *infinite series* (or just *series* for short) is simply the adding up of the infinite number of terms of a sequence. Here's the *sequence* from the previous section again, $a_n = \frac{1}{2^n}$:

$$\frac{1}{2}, \frac{1}{4}, \frac{1}{8}, \frac{1}{16}, \ldots$$

And here's the *series* associated with this *sequence:*

$$\frac{1}{2} + \frac{1}{4} + \frac{1}{8} + \frac{1}{16} + \ldots$$

You can use fancy summation notation to write this sum in a more compact form:

$$\sum_{n=1}^{\infty} \frac{1}{2^n}$$

The summation symbol tells you to plug 1 in for n, then 2, then 3, and so on, and then to add up all the terms (more on summation notation in Chapter 14). Nitpickers may point out that you can't actually add up an infinite number of terms. Okay, so here's the fine print for the nitpickers. An infinite sum is technically a limit. In other words,

$$\sum_{n=1}^{\infty} \frac{1}{2^n} = \lim_{b\to\infty} \sum_{n=1}^{b} \frac{1}{2^n}$$

To find an infinite sum, you take a limit — just like you do for improper (infinite) integrals (see Chapter 18). From here on, though, I just write infinite sums like $\sum_{n=1}^{\infty} \frac{1}{2^n}$ and dispense with the limit mumbo jumbo.

Partial sums

Continuing with the same series, take a look at how the sum grows by listing the "sum" of one term (kind of like the sound of one hand clapping), the sum of two terms, three terms, four, and so on:

$$S_1 = \frac{1}{2}$$
$$S_2 = \frac{1}{2} + \frac{1}{4} = \frac{3}{4}$$
$$S_3 = \frac{1}{2} + \frac{1}{4} + \frac{1}{8} = \frac{7}{8}$$
$$S_4 = \frac{1}{2} + \frac{1}{4} + \frac{1}{8} + \frac{1}{16} = \frac{15}{16}$$
$$\vdots$$
$$S_n = \frac{1}{2} + \frac{1}{4} + \frac{1}{8} + \frac{1}{16} + \frac{1}{32} + \frac{1}{64} + \ldots + \frac{1}{2^n} = \frac{2^n - 1}{2^n}$$

Each of these sums is called a *partial sum* of the series.

MATH RULES

Definition of *partial sum*: The *n*th partial sum, S_n, of an infinite series is the sum of the first *n* terms of the series.

The convergence or divergence of a series — the main event

If you now list the preceding partial sums, you have the following *sequence* of partial sums:

$$\frac{1}{2}, \frac{3}{4}, \frac{7}{8}, \frac{15}{16}, \ldots$$

The main point of this chapter is figuring out whether such a sequence of partial sums *converges* — homes in on a finite number — or *diverges*. If the sequence of partial sums converges, you say that the series converges; otherwise, the sequence of partial sums diverges and you say that the series diverges. The rest of this chapter is devoted to the many techniques used in making this determination.

By the way, if you're getting a bit confused by the terms *sequence* and *series* and the connection between them, you're not alone. Keeping these terms straight can be tricky. For starters, note that there are two sequences associated with every series. With the series $\frac{1}{2} + \frac{1}{4} + \frac{1}{8} + \frac{1}{16} + \ldots$, for example, you have the underlying sequence, $\frac{1}{2}, \frac{1}{4}, \frac{1}{8}, \frac{1}{16}, \ldots$, and also the sequence of partial sums, $\frac{1}{2}, \frac{3}{4}, \frac{7}{8}, \frac{15}{16}, \ldots$. It's not a bad idea to try to keep these things straight, but all you really need to worry about is whether the *series* adds up to some finite number or not. If it does, it *converges*; if not, it *diverges*. The reason for getting into the somewhat confusing notion of a *sequence* of partial sums is that the definitions of convergence and divergence are based on the behavior of sequences, not series. Now — don't get me wrong — in mathematics, terminology is important, but it's the ideas that really matter, and, again, the important *idea* you need to focus on is whether or not a series sums up to a finite number.

What about the previous series? Does it converge or diverge? It shouldn't take too much imagination to see the following pattern:

$$S_1 = \frac{1}{2} = 1 - \frac{1}{2}$$
$$S_2 = \frac{3}{4} = 1 - \frac{1}{4}$$
$$S_3 = \frac{7}{8} = 1 - \frac{1}{8}$$
$$S_4 = \frac{15}{16} = 1 - \frac{1}{16}$$
$$\vdots$$
$$S_n \quad = 1 - \frac{1}{2^n}$$

Finding the limit of this sequence of partial sums is a no-brainer:

$$\lim_{n\to\infty} S_n = \lim_{n\to\infty}\left(1 - \frac{1}{2^n}\right) = 1 - \frac{1}{\infty} = 1 - 0 = 1$$

So, this series converges to 1. In symbols,

$$\sum_{n=1}^{\infty} \frac{1}{2^n} = \frac{1}{2} + \frac{1}{4} + \frac{1}{8} + \frac{1}{16} + \ldots = 1$$

By the way, this may remind you of that paradox about walking toward a wall, where your first step is halfway to the wall, your second step is half of the remaining distance, your third step is half the remaining distance, and so on. Will you ever get to the wall? Answer: It depends. More about that later.

Convergence or Divergence? That Is the Question

This section contains nine ways of determining whether a series converges or diverges. (Then, in the next section on alternating series, you look at a final tenth test for convergence/divergence.) Note that all of the series you investigate in this section are made up of positive terms.

A no-brainer divergence test: The nth term test

If the individual terms of a series (in other words, the terms of the series' underlying sequence) do not converge to zero, then the series must diverge. This is the *n*th term test for divergence.

MATH RULES

The *n*th term test: If $\lim_{n\to\infty} a_n \neq 0$, then $\sum a_n$ diverges. (I presume you figured out that with this naked summation symbol, n runs from 1 to infinity.)

(*Note:* The *n*th term test not only works for ordinary positive series like the ones in this section, but it also works for series with positive and negative terms. More about this at the end of this chapter in the section, "Alternating Series.")

If you think about it, the *n*th term test is just common sense. When a series converges, the sum is homing in on a certain number. The only way this can happen is when the numbers being added are getting infinitesimally small — like in the series I've been talking about: $\frac{1}{2}+\frac{1}{4}+\frac{1}{8}+\frac{1}{16}+\ldots$. Imagine, instead, that the terms of a series are converging, say, to 1, like in the series $\frac{1}{2}+\frac{2}{3}+\frac{3}{4}+\frac{4}{5}+\frac{5}{6}+\ldots$, generated by the formula $a_n = \frac{n}{n+1}$. In that case, when you add up the terms, you are adding numbers extremely close to 1 over and over forever — and this must add up to infinity. So, in order for a series to converge, the terms of the series must converge to zero. But make sure you understand what this *n*th term test does *not* say.

WARNING

When the terms of a series converge to zero, that does *not* guarantee that the series converges. In hifalutin logicianese — the fact that the terms of a series converge to zero is a *necessary* but *not sufficient* condition for concluding that the series converges to a finite sum.

Because this test is often very easy to apply, it should be one of the first things you check when trying to determine whether a series converges or diverges.

If you're asked to determine whether $\sum_{n=1}^{\infty}\left(1+\frac{1}{n}\right)^n$ converges or diverges, note that every term of this series is a number greater than 1 being raised to a positive power. This always results in a number greater than 1, and thus, the terms of this series do not converge to zero, and the series must therefore diverge.

EXAMPLE

Q. Does $\sum_{n=1}^{\infty}\sqrt[n]{1-\frac{1}{n}}$ converge or diverge?

A. **The series diverges.** You can answer this question with common sense if your calc teacher allows such a thing. As n gets larger and larger, $1-\frac{1}{n}$ increases and gets closer and closer to one. And when you take any root of a number like 0.9, the root is *larger* than the original number — and the higher the root index, the larger the answer is. So $\sqrt[n]{1-\frac{1}{n}}$ has to get larger as n increases, and thus $\lim_{n\to\infty}\sqrt[n]{1-\frac{1}{n}}$ cannot possibly equal zero. The series, therefore, diverges by the *n*th term test.

If your teacher doesn't like that approach, you can do the following: Plugging ∞ into the limit produces $\left(1-\frac{1}{\infty}\right)^{1/\infty}$, which is 1^0, and that equals 1 — you're done. (Note that 1^0 is *not* one of the forms that give you a L'Hôpital's rule problem — see Chapter 18.) Because $\lim_{n\to\infty}\sqrt[n]{1-\frac{1}{n}}$ equals 1, $\sum_{n=1}^{\infty}\sqrt[n]{1-\frac{1}{n}}$ diverges by the *n*th term test. (If your teacher is a real stickler for rigor, they might not like this approach either because, technically, you're not supposed to plug ∞ into n even though it works just fine. Oh, well.)

YOUR TURN

 Does $\sum_{n=1}^{\infty} \frac{2n^2 - 9n - 8}{5n^2 + 20n + 12}$ converge or diverge?

 By the nth term test, does $\sum_{n=1}^{\infty} \frac{1}{n}$ converge or diverge?

Three basic series and their convergence/divergence tests

In this subsection, you look at geometric series, p-series, and telescoping series. Geometric series and p-series are relatively simple but important series that, in addition to being interesting in their own right, can be used as benchmarks when determining the convergence or divergence of more complicated series. Telescoping series don't come up much, but many calculus texts describe them, so who am I to buck tradition?

Geometric series

A geometric series is a series of the form

$$a + ar + ar^2 + ar^3 + ar^4 + \ldots = \sum_{n=0}^{\infty} ar^n$$

The first term, a, is called the *leading term*. Each term after the first equals the preceding term multiplied by r, which is called the *ratio*.

For example, if a is 5 and r is 3, you get

$$5 + 5 \cdot 3 + 5 \cdot 3^2 + 5 \cdot 3^3 + \ldots \quad = 5 + 15 + 45 + 135 + \ldots$$

You just multiply each term by 3 to get the next term. By the way, the 3 in this example is called the *ratio* because the ratio of any term *divided* by its preceding term equals 3, but I think it makes a lot more sense to think of the 3 as your *multiplier*.

If a is 100 and r is 0.1, you get

$$100+100\cdot 0.1+100\cdot 0.1^2+100\cdot 0.1^3+100\cdot 0.1^4+\ldots$$
$$=100+10+1+0.1+0.01+\ldots$$

If that rings a bell, you've got a good memory. It's the series for the "Achilles versus the tortoise" paradox (go way back to Chapter 2).

And if a is $\frac{1}{2}$ and r is also $\frac{1}{2}$, you get the series I've been talking so much about:

$$\frac{1}{2}+\frac{1}{4}+\frac{1}{8}+\frac{1}{16}+\ldots$$

The convergence/divergence rule for geometric series is a snap.

MATH RULES

Geometric series rule: If $0<|r|<1$, the geometric series $\sum_{n=0}^{\infty} ar^n$ converges to $\frac{a}{1-r}$. If $|r|\geq 1$, the series diverges. (Note that this rule works when $-1<r<0$, in which case you get an *alternating series;* more about that at the end of this chapter.)

In the first example (a half a page back), $a=5$ and $r=3$, so the series diverges. In the second example, a is 100 and r is 0.1, so the series converges to $\frac{100}{1-0.1}=\frac{100}{0.9}=111\frac{1}{9}$. That's the answer to the "Achilles versus the tortoise" problem: Achilles passes the tortoise after running $111\frac{1}{9}$ meters. And in the third example, $a=\frac{1}{2}$ and $r=\frac{1}{2}$, so the series converges to $\frac{\frac{1}{2}}{1-\frac{1}{2}}=1$. This is how far you walk if you start 1 yard from the wall, then step halfway to the wall, then half of the remaining distance, and so on and so on. You take an infinite number of steps, but travel a mere yard. And how long will it take you to get to the wall? Well, if you keep up a constant speed and don't pause between steps (which, of course, is impossible), you'll get there in the same amount of time it would take you to walk any old yard. If you do pause between each step, even for a billionth of a second, you'll *never* get to the wall.

p-series

A p-series is of the form

$$\sum_{n=1}^{\infty}\frac{1}{n^p}=\frac{1}{1^p}+\frac{1}{2^p}+\frac{1}{3^p}+\frac{1}{4^p}+\ldots$$

(where p is a *positive* power). The p-series for $p=1$ is called the *harmonic* series. Here it is:

$$\frac{1}{1}+\frac{1}{2}+\frac{1}{3}+\frac{1}{4}+\frac{1}{5}+\frac{1}{6}+\ldots$$

Although this grows *very* slowly — after 10,000 terms, the sum is only about 9.79! — the harmonic series in fact diverges to infinity.

By the way, this is called a *harmonic* series because the numbers in the series have something to do with the way a musical string like a guitar string vibrates — don't ask. For history buffs, in the sixth century B.C., Pythagoras investigated the harmonic series and its connection to the notes of the lyre.

Here's the convergence/divergence rule for p-series.

MATH RULES

The p-series rule: The p-series $\sum_{n=1}^{\infty} \frac{1}{n^p}$ converges if $p > 1$ and diverges if $p \leq 1$.

As you can see from this rule, the harmonic series forms the convergence/divergence borderline for p-series. Any p-series with terms *larger* than the terms of the harmonic series *diverges*, and any p-series with terms *smaller* than the terms of the harmonic series *converges*.

The p-series for $p = 2$ is another common series:

$$1 + \frac{1}{2^2} + \frac{1}{3^2} + \frac{1}{4^2} + \frac{1}{5^2} + \frac{1}{6^2} + \ldots$$
$$= 1 + \frac{1}{4} + \frac{1}{9} + \frac{1}{16} + \frac{1}{25} + \frac{1}{36} + \ldots$$

The p-series rule tells you that this series converges. Note, however, that the p-series rule can't tell you what number this series converges to. (Contrast that to the geometric series rule, which can answer both questions.) By other means — beyond the scope of this book — it can be shown that this sum converges to $\frac{\pi^2}{6}$.

TIP

Multiplying by a constant doesn't matter. When analyzing the series in this section and the rest of the chapter, remember that multiplying a series by any non-zero constant, big or small and positive or negative, never affects whether it converges or diverges. For example, if $\sum_{n=1}^{\infty} u_n$ converges, then so will $1000 \cdot \sum_{n=1}^{\infty} u_n$.

Telescoping series

You don't see many telescoping series (and there aren't any in the practice problems), but the telescoping series rule is a good one to keep in your bag of tricks — you never know when it might come in handy. Consider the following series:

$$\sum_{n=1}^{\infty} \frac{1}{n(n+1)} = \frac{1}{2} + \frac{1}{6} + \frac{1}{12} + \frac{1}{20} + \frac{1}{30} + \ldots$$

To see that this is a telescoping series, you have to use the partial fractions technique from Chapter 16 — sorry to have to bring that up again — to rewrite $\frac{1}{n(n+1)}$ as $\frac{1}{n} - \frac{1}{n+1}$. Now you've got

$$\sum_{n=1}^{\infty} \left(\frac{1}{n} - \frac{1}{n+1} \right) = \left(1 - \frac{1}{2}\right) + \left(\frac{1}{2} - \frac{1}{3}\right) + \left(\frac{1}{3} - \frac{1}{4}\right) + \left(\frac{1}{4} - \frac{1}{5}\right) + \ldots + \left(\frac{1}{n} - \frac{1}{n+1}\right)$$

Do you see how all these terms will now collapse, or *telescope*? The $\frac{1}{2}$s cancel, the $\frac{1}{3}$s cancel, the $\frac{1}{4}$s cancel, and so on. All that's left is the first term, 1 (actually, it's only half a term), and the "last" half-term, $-\frac{1}{n+1}$. So, the sum of the first n terms is simply $1 - \frac{1}{n+1}$. In the limit, as n approaches infinity, $\frac{1}{n+1}$ converges to zero, and thus the sum converges to $1 - 0$, or 1.

Each term in a telescoping series can be written as the difference of two half-terms — call them h-terms. The telescoping series can then be written as

$$(h_1 - h_2) + (h_2 - h_3) + (h_3 - h_4) + (h_4 - h_5) + \ldots + (h_n - h_{n+1}) + \ldots$$

I bet you're dying for another rule, so here's the next one.

Telescoping series rule: A telescoping series of the form immediately above converges if h_{n+1} converges to a finite number. In that case, the series converges to $h_1 - \lim_{n\to\infty} h_{n+1}$. If h_{n+1} diverges, the series diverges.

Note that this rule, like the rule for geometric series, lets you determine what number a convergent telescoping series converges to. These are the only two rules I cover where you can do this. The other eight rules for determining convergence or divergence don't allow you to determine what a convergent series converges to. But hey, you know what they say, "two out of ten ain't bad."

3. Does $0.008 - 0.006 + 0.0045 - 0.003375 + 0.00253125 - \ldots$ converge or diverge? If it converges, what's the sum?

---

4. Does $\sum_{n=1}^{\infty} \frac{1}{10n}$ converge or diverge?

---

5. Does $1 + \frac{\sqrt[4]{2}}{2} + \frac{\sqrt[4]{3}}{3} + \frac{\sqrt[4]{4}}{4} + \ldots + \frac{\sqrt[4]{n}}{n}$ converge or diverge?

---

6. Does $\frac{1}{2} + \frac{1}{4} + \frac{1}{8} + \frac{1}{12} + \frac{1}{16} + \frac{1}{20} + \ldots$ converge or diverge?

Three comparison tests for convergence/divergence

Say you're trying to figure out whether a series converges or diverges, but it doesn't fit any of the tests you know. No worries. You find a benchmark series that you know converges or diverges and then compare your new series to the known benchmark. For the next three tests, if the benchmark converges, your series converges; and if the benchmark diverges, your series diverges.

The direct comparison test

This is a simple, common-sense rule. If you've got a series with terms that are *less than or equal to* the terms of a *convergent* benchmark series, then your series must also converge. And if your series has terms that are *greater than or equal to* the terms of a *divergent* benchmark series, then your series must also diverge. Here's the mumbo jumbo.

MATH RULES

Direct comparison test: Let $0 \le a_n \le b_n$ for all n.

If $\sum_{n=1}^{\infty} b_n$ converges, then $\sum_{n=1}^{\infty} a_n$ converges.

If $\sum_{n=1}^{\infty} a_n$ diverges, then $\sum_{n=1}^{\infty} b_n$ diverges.

Let's walk though a couple problems. Determine whether $\sum_{n=1}^{\infty} \frac{1}{5+3^n}$ converges or diverges. Piece o' cake. This series resembles $\sum_{n=1}^{\infty} \frac{1}{3^n}$, which is a geometric series with r equal to $\frac{1}{3}$. (Note that you can rewrite this in the standard geometric series form as $\sum_{n=0}^{\infty} \frac{1}{3}\left(\frac{1}{3}\right)^n$.) Because $0 < |r| < 1$, this series converges. And because $\frac{1}{5+3^n}$ is *less* than $\frac{1}{3^n}$ for all values of n, $\sum_{n=1}^{\infty} \frac{1}{5+3^n}$ must also converge.

Here's another one: Does $\sum_{n=1}^{\infty} \frac{\ln n}{n}$ converge or diverge? This series resembles $\sum_{n=1}^{\infty} \frac{1}{n}$, the harmonic p-series that is known to diverge. Because $\frac{\ln n}{n}$ is *greater* than $\frac{1}{n}$ for all values of $n \ge 3$, $\sum_{n=1}^{\infty} \frac{\ln n}{n}$ must also diverge. By the way, if you're wondering why I'm allowed to consider only the terms where $n \ge 3$, here's why:

MATH RULES

Feel free to ignore initial terms. For any of the convergence/divergence tests, you can disregard *any* number of terms at the beginning of a series. And if you're comparing two series, you can ignore any number of terms from the beginning of either or both of the series — and you can ignore a different number of terms in each of the two series.

This utter disregard of innocent beginning terms is allowed because the first, say, 10 or 1000 or 1,000,000 terms of a series always sum to a finite number and thus never have any effect on whether the series converges or diverges. Note, however, that disregarding a number of terms *would* affect the total that a convergent series converges to.

(Are you wondering why this disregard of beginning terms doesn't violate the direct comparison test's requirement that $0 \le a_n \le b_n$ for *all n*? Everything's copacetic because you can lop off any number of terms at the beginning of each series and let the counter, n, start at 1 anywhere in each series. Thus the "first" terms a_1 and b_1 can actually be located anywhere along each series. See what I mean?)

WARNING

Fore! (That was a joke.) The direct comparison test tells you *nothing* if the series you're investigating is *greater* than a known *convergent* series or *less* than a known *divergent* series.

For example, say you want to determine whether $\sum_{n=1}^{\infty} \frac{1}{10+\sqrt{n}}$ converges. This series resembles $\sum_{n=1}^{\infty} \frac{1}{\sqrt{n}}$, which is a p-series with p equal to $\frac{1}{2}$. The p-series test says that this series diverges, but that doesn't help you because the terms of your series are *less* than the terms of this known divergent benchmark.

Instead, you should compare your series to the divergent harmonic series, $\sum_{n=1}^{\infty} \frac{1}{n}$. Your series, $\frac{1}{10+\sqrt{n}}$, is greater than $\frac{1}{n}$ for all $n \ge 14$ (it takes a little work to show this; give it a try). Because your series is *greater* than the *divergent* harmonic series, your series must also diverge.

The limit comparison test

The idea behind this test is related to the tip just before the subsection "Telescoping series." If you take a known convergent series and multiply each of its terms by any non-zero number, big or small, then that new series also converges. The same thing goes for a divergent series multiplied by any non-zero number. That new series also diverges. This is an oversimplified explanation — with the limit comparison test, it's only in the limit that one series is sort of a multiple of the other — but it conveys the basic principle.

You can discover whether such a connection exists between two series by looking at the ratio of the *n*th terms of the two series as n approaches infinity. Here's the test.

MATH RULES

Limit comparison test: For two series, $\sum a_n$ and $\sum b_n$, if $a_n > 0$, $b_n > 0$, and $\lim_{n\to\infty}\left(\frac{a_n}{b_n}\right) = L$, where L is *finite* and *positive*, then either both series converge or both diverge.

TIP

Use this test when your series goes the wrong way. This is a good test to use when you can't use the direct comparison test for your series because it goes the wrong way — in other words, your series is *larger* than a known *convergent* series or *smaller* than a known *divergent* series.

EXAMPLE

Q. Does $\sum_{n=2}^{\infty} \frac{1}{n^2 - \ln n}$ converge or diverge?

A. **The series converges.**

This series resembles the convergent p-series, $\frac{1}{n^2}$, so that's your benchmark. But you can't use the direct comparison test because the terms of your series are *greater* than $\frac{1}{n^2}$. Instead, you use the limit comparison test.

Take the limit of the ratio of the nth terms of the two series. It doesn't matter which series you put in the numerator and which in the denominator, but putting the known, benchmark series in the denominator makes it a little easier to do these problems and grasp the results.

$$\begin{aligned}
&\lim_{n\to\infty}\frac{\frac{1}{n^2-\ln n}}{\frac{1}{n^2}}\\
&=\lim_{n\to\infty}\frac{n^2}{n^2-\ln n}\\
&=\lim_{n\to\infty}\frac{2n}{2n-\frac{1}{n}} \quad \text{(L'Hôpital's rule)}\\
&=\lim_{n\to\infty}\frac{2}{2+\frac{1}{n^2}} \quad \text{(L'Hôpital's rule again)}\\
&=\frac{2}{2+\frac{1}{\infty}}=\frac{2}{2+0}=1
\end{aligned}$$

Because the limit is finite and positive and because the benchmark series converges, your series must also converge. Thus, $\sum_{n=2}^{\infty}\frac{1}{n^2-\ln n}$ converges.

TIP

Use this test for rational functions. The limit comparison test is a good one for series where the general term is a *rational function* — in other words, where the general term is a quotient of two polynomials.

EXAMPLE

Q. Determine the convergence or divergence of $\sum_{n=1}^{\infty}\frac{5n^2-n+1}{n^3+4n+3}$.

A. **The series diverges.**

1. **Determine the benchmark series.**

   Take the highest power of n in the numerator and the denominator — ignoring any coefficients and all other terms — then simplify. Like this:

   $$\frac{5\textcircled{n^2}-n+1}{\textcircled{n^3}+4n+3}\to\frac{n^2}{n^3}=\frac{1}{n}$$

   That's the benchmark series, $\frac{1}{n}$, the *divergent* harmonic series.

2. **Take the limit of the ratio of the *n*th terms of the two series.**

$$\lim_{n\to\infty}\frac{\dfrac{5n^2-n+1}{n^3+4n+3}}{\dfrac{1}{n}}$$

$$=\lim_{n\to\infty}\frac{5n^3-n^2+n}{n^3+4n+3}$$

$$=\lim_{n\to\infty}\frac{5-\dfrac{1}{n}+\dfrac{1}{n^2}}{1+\dfrac{4}{n^2}+\dfrac{3}{n^3}} \quad \text{(dividing numerator and denominator by } n^3\text{)}$$

$$=\frac{5-\dfrac{1}{\infty}+\dfrac{1}{\infty}}{1+\dfrac{4}{\infty}+\dfrac{3}{\infty}}=\frac{5-0+0}{1+0+0}=5$$

3. **Because the limit from Step 2 is finite and positive and because the benchmark series diverges, your series must also diverge.**

   Thus, $\sum_{n=1}^{\infty}\frac{5n^2-n+1}{n^3+4n+3}$ diverges.

TIP

Okay, so I'm a rebel. The limit comparison test is always stated as it appears at the beginning of this section, but I want to point out — recklessly ignoring the noble tradition of calculus textbook authors — that in a sense it's incomplete. The limit, *L*, doesn't have to be finite and positive for the test to work. First, if the benchmark series is convergent, and you put it in the denominator of the limit, and the limit is *zero*, then your series must also converge. Note that if the limit is infinity, you can't conclude anything. And second, if the benchmark series is divergent, and you put it in the denominator, and the limit is *infinity*, then your series must also diverge. If the limit is zero, you can't conclude anything.

The integral comparison test

The third benchmark test involves comparing the series you're investigating to its companion improper integral (see Chapter 18 for more on improper integrals). If the integral converges, your series converges; and if the integral diverges, so does your series. By the way, to the best of my knowledge, no one else calls this the integral comparison test — but they should because that's the way it works.

EXAMPLE

Q. Determine the convergence or divergence of $\sum_{n=2}^{\infty}\frac{1}{n\ln n}$.

A. **The series diverges.**

The direct comparison test doesn't work because the terms of this series are *less than* the terms of the *divergent* harmonic series, $\frac{1}{n}$. Trying the limit comparison test is the next natural choice, but it doesn't work either — try it. But if you notice that the series is an expression you know how to integrate, you're home free (you did notice that, right?). Just compute the companion improper integral with the same limits of integration as the index numbers of the summation — like this:

$$\begin{aligned}
&\int_2^\infty \frac{1}{x\ln x}dx \\
&= \lim_{b\to\infty}\int_2^b \frac{1}{x\ln x}dx \\
&= \lim_{b\to\infty}\int_{\ln 2}^{\ln b} \frac{1}{u}du \qquad \text{(substitution with } u=\ln x \text{ and } du=\frac{1}{x}dx; \\
&\qquad\qquad\qquad\qquad\quad \text{when } x=2,\ u=\ln 2, \text{ and when } x=b,\ u=\ln b) \\
&= \lim_{b\to\infty}\left[\ln u\right]_{\ln 2}^{\ln b} \\
&= \lim_{b\to\infty}(\ln(\ln b)-\ln(\ln 2)) \\
&= (\ln(\ln\infty)-\ln(\ln 2)) \\
&= \infty-\ln(\ln 2)=\infty
\end{aligned}$$

Because the integral diverges, your series diverges.

After you've determined the convergence or divergence of a series with the integral comparison test, you can then use that series as a benchmark for investigating other series with the direct comparison or limit comparison tests.

For instance, the integral test just told you that $\sum_{n=2}^{\infty}\frac{1}{n\ln n}$ diverges. Now you can use this series to investigate $\sum_{n=3}^{\infty}\frac{1}{n\ln n-\sqrt{n}}$ with the direct comparison test. Do you see how? Or you can investigate, say, $\sum_{n=1}^{\infty}\frac{1}{n\ln n+\sqrt{n}}$ with the limit comparison test. Try it.

TIP

Don't forget the integral test. The integral comparison test is fairly easy to use, so don't neglect to ask yourself whether you can integrate the series expression or something close to it. If you can, it's a BINGO.

By the way, in Chapter 18, you saw the following two improper integrals: $\int_1^\infty \frac{1}{x}dx$, which diverges, and $\int_1^\infty \frac{1}{x^2}dx$, which converges. Look back at Figure 18-1. Now that you know the integral comparison test, you can appreciate the connection between those integrals and their companion p-series: the divergent harmonic series, $\sum_{n=1}^{\infty}\frac{1}{n}$, and the convergent p-series, $\sum_{n=1}^{\infty}\frac{1}{n^2}$.

Here's the mumbo jumbo for the integral comparison test. Note the fine print.

MATH RULES

Integral comparison test: If $f(x)$ is positive, continuous, and decreasing for all $x\geq 1$ and if $a_n=f(n)$, then $\sum_{n=1}^{\infty}a_n$ and $\int_1^\infty f(x)dx$ either both converge or both diverge.

YOUR TURN

7 For Problems 7 through 14, determine whether the series converges or diverges. $\sum_{n=1}^{\infty} \frac{10(0.9)^n}{\sqrt{n}}$

8 $\sum_{n=1}^{\infty} \frac{1.1^n}{10n}$

9 $\frac{1}{1001} + \frac{1}{2001} + \frac{1}{3001} + \frac{1}{4001} + \ldots$

10 $\sum_{n=1}^{\infty} \frac{1}{n + \sqrt{n} + \ln n}$

11 $\sum_{n=1}^{\infty} \frac{1}{n^3 - \ln^3 n}$

12 $\sum_{n=2}^{\infty} \frac{1}{n \ln n + \sin n}$

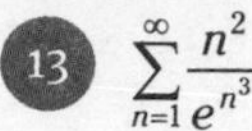

13 $\sum_{n=1}^{\infty} \frac{n^2}{e^{n^3}}$

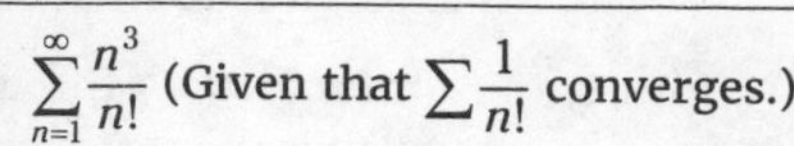

14 $\sum_{n=1}^{\infty} \frac{n^3}{n!}$ (Given that $\sum \frac{1}{n!}$ converges.)

The two "R" tests: Ratios and roots

Unlike the three benchmark tests from the previous section, the ratio and root tests don't compare a new series to a known benchmark. They work by looking only at the nature of the series you're trying to figure out. They form a cohesive pair because the results of both tests tell you the same thing. If the result is less than 1, the series converges; if it's more than 1, the series diverges; and if it's exactly 1, you learn nothing and must try a different test. (As presented here, the ratio and root tests are used for series of *positive* terms. In other books, you may see a different version of each test that uses the absolute value of the terms. These absolute value versions can be used for series made up of both positive and negative terms. Don't sweat this; the different versions amount to the same thing.)

MATH RULES

The ratio test: Given a series $\sum u_n$, consider the limit of the ratio of a term to the previous term, $\lim_{n\to\infty} \frac{u_{n+1}}{u_n}$. If this limit is less than 1, the series converges. If it's greater than 1 (this includes ∞), the series diverges. And if it equals 1, the ratio test tells you nothing.

TIP

When to use the ratio test. The ratio test works especially well with series involving *factorials* like $n!$ or where n is in the power, like 3^n.

REMEMBER

Definition of the *factorial* symbol. The factorial symbol (!), tells you to multiply like this: $6! = 6 \cdot 5 \cdot 4 \cdot 3 \cdot 2 \cdot 1$. And notice how things cancel when you have factorials in the numerator and denominator of a fraction: $\frac{6!}{5!} = \frac{6 \cdot \cancel{5} \cdot \cancel{4} \cdot \cancel{3} \cdot \cancel{2} \cdot \cancel{1}}{\cancel{5} \cdot \cancel{4} \cdot \cancel{3} \cdot \cancel{2} \cdot \cancel{1}} = 6$ and $\frac{5!}{6!} = \frac{\cancel{5} \cdot \cancel{4} \cdot \cancel{3} \cdot \cancel{2} \cdot \cancel{1}}{6 \cdot \cancel{5} \cdot \cancel{4} \cdot \cancel{3} \cdot \cancel{2} \cdot \cancel{1}} = \frac{1}{6}$. In both cases, everything cancels but the 6. In the same way, $\frac{(n+1)!}{n!} = n+1$ and $\frac{n!}{(n+1)!} = \frac{1}{n+1}$; everything cancels but the $(n+1)$. Lastly, it seems weird, but $0! = 1$ — just take my word for it.

Q. Does $\sum_{n=0}^{\infty} \frac{3^n}{n!}$ converge or diverge?

EXAMPLE **A. The series converges.**

Here's what you do. You look at the limit of the ratio of the $(n+1)$st term to the nth term:

$$\begin{aligned}
&\lim_{n\to\infty} \frac{\frac{3^{n+1}}{(n+1)!}}{\frac{3^n}{n!}} \\
&= \lim_{n\to\infty} \frac{3^{n+1}\cdot n!}{(n+1)!\cdot 3^n} \\
&= \lim_{n\to\infty} \frac{3}{n+1} \\
&= \frac{3}{\infty+1} = 0
\end{aligned}$$

Because this limit is less than 1, $\sum_{n=0}^{\infty} \frac{3^n}{n!}$ converges.

Q. Here's another series: $\sum_{n=1}^{\infty} \frac{n^n}{n!}$. What's your guess — does it converge or diverge?

EXAMPLE **A. The series diverges.**

Look at the limit of the $(n+1)$st term over the nth term:

$$\begin{aligned}
&\lim_{n\to\infty} \frac{\frac{(n+1)^{n+1}}{(n+1)!}}{\frac{n^n}{n!}} \\
&= \lim_{n\to\infty} \frac{(n+1)^{n+1}\cdot n!}{(n+1)!\cdot n^n} \\
&= \lim_{n\to\infty} \frac{(n+1)^{n+1}}{(n+1)\cdot n^n} \\
&= \lim_{n\to\infty} \frac{(n+1)^n}{n^n} \\
&= \lim_{n\to\infty} \left(\frac{n+1}{n}\right)^n \\
&= \lim_{n\to\infty} \left(1+\frac{1}{n}\right)^n \\
&= e \quad \text{(} \lim_{n\to\infty}\left(1+\frac{1}{n}\right)^n = e \text{ is one of the limits you should memorize, as discussed in Chapter 8.)} \\
&\approx 2.718
\end{aligned}$$

Because the limit is greater than 1, $\sum_{n=1}^{\infty} \frac{n^n}{n!}$ diverges.

MATH RULES

The root test: Note its similarity to the ratio test. Given a series $\sum u_n$, consider the limit of the nth root of the nth term, $\lim_{n\to\infty} \sqrt[n]{u_n}$. If this limit is less than 1, the series converges. If it's greater than one (including ∞), the series diverges. And if it equals 1, the root test says nothing.

TIP

When to use the root test. The root test (like the ratio test) is a good one to try if the series involves nth powers. (If you've got an nth power problem, and you're not sure which of the two tests to try first, start with the ratio test — it's often the easier one to use.)

EXAMPLE

Q. Does $\sum_{n=1}^{\infty} \frac{e^{2n}}{n^n}$ converge or diverge?

A. The series converges.

Here's what you do:

$$\begin{aligned}
&\lim_{n\to\infty} \sqrt[n]{\frac{e^{2n}}{n^n}} \\
&= \lim_{n\to\infty} \frac{e^{2n/n}}{n^{n/n}} \\
&= \lim_{n\to\infty} \frac{e^2}{n} \\
&= \frac{e^2}{\infty} = 0
\end{aligned}$$

Because the limit is less than 1, the series converges. (By the way, you can also do this series with the ratio test, but it's harder — an exception to the rule of thumb in the previous tip.)

TIP

Making a good guess about convergence/divergence: Sometimes it's useful to make an educated guess about the convergence or divergence of a series before you launch into one or more of the convergence/divergence tests. Here's a tip that helps you make a good guess with some series. (Much of this tip is virtually identical to the tip in Chapter 8 about evaluating limits at infinity.) The following expressions are listed from "smallest" to "largest": n^{10}, 5^n, $n!$, and n^n. A series with a "smaller" expression over a "larger" one converges, for example, $\sum_{n=1}^{\infty} \frac{n^{50}}{n!}$ or $\sum_{n=1}^{\infty} \frac{n!}{n^n}$; and a series with a "larger" expression over a "smaller" one diverges, for instance, $\sum_{n=1}^{\infty} \frac{n^n}{100^n}$ or $\sum_{n=1}^{\infty} \frac{25^n}{n^{100}}$. Note a few things:

- Coefficients don't change the order — for example, $1000n^{10}$, $12 \cdot 5^n$, $300n!$, and $0.065n^n$.
- For the first expression (n^{10}), the power can be any number; for the second expression (5^n), the number must be greater than one.
- Replacing n with a multiple of n (like 5^{10n} or $(6n)^{6n}$) doesn't change the order — with one important exception: With the factorial expression, replacing the n with kn (if $k \geq 1$ like with $(4n)!$ or $(1.8n)!$) makes the expression the largest of the four; if $0 < k < 1$, the order doesn't change.

YOUR TURN

15 For Problems 15 through 20, determine whether the series converges or diverges. $\sum_{n=1}^{\infty} \frac{1}{\ln^n (n+2)}$

16 $\sum_{n=1}^{\infty} \frac{n^{\sqrt{n}}}{\sqrt{n}^n}$

17 $\sum_{n=1}^{\infty} \frac{n!}{n^n}$

18 $\sum_{n=1}^{\infty} n\left(\frac{3}{4}\right)^n$

19 $\sum_{n=1}^{\infty} \frac{n^{\sqrt{n}}}{n!}$

20 $\sum_{n=1}^{\infty} \frac{n!}{4^n}$

Alternating Series

In the previous sections, you've been looking at series of *positive* terms. Now you look at *alternating series* — series where the terms alternate between positive and negative — like this:

$$1-\frac{1}{2}+\frac{1}{4}-\frac{1}{8}+\frac{1}{16}-\frac{1}{32}+\frac{1}{64}-\ldots$$

Finding absolute versus conditional convergence

Many divergent series of positive terms converge if you change the signs of their terms so they alternate between positive and negative. For example, you know that the harmonic series diverges:

$$1+\frac{1}{2}+\frac{1}{3}+\frac{1}{4}+\frac{1}{5}+\frac{1}{6}+\ldots$$

But, if you change every other sign to negative, you obtain the *alternating harmonic series,* which *converges*:

$$1-\frac{1}{2}+\frac{1}{3}-\frac{1}{4}+\frac{1}{5}-\frac{1}{6}+\ldots$$

By the way, although I'm not going to show you how to compute it, this series converges to $\ln 2$, which equals about 0.6931.

Definition of *conditional convergence:* An alternating series is said to be conditionally convergent if it's convergent as it is but would become divergent if all its terms were made positive.

Definition of *absolute convergence:* An alternating series is said to be absolutely convergent if it would be convergent even if all its terms were made positive. And any such absolutely convergent alternating series is also automatically convergent as it is.

Q. Determine the convergence or divergence of the following alternating series:

$$\sum_{n=0}^{\infty}\left((-1)^n\frac{1}{2^n}\right)=1-\frac{1}{2}+\frac{1}{4}-\frac{1}{8}+\frac{1}{16}-\ldots$$

A. **The series converges.** If all these terms were positive, you'd have the familiar geometric series,

$$\sum_{n=0}^{\infty}\frac{1}{2^n}=1+\frac{1}{2}+\frac{1}{4}+\frac{1}{8}+\frac{1}{16}+\ldots$$

which, by the geometric series rule, converges to 2. Because the positive series converges, the alternating series must also converge (though to a different result — see the following) and you say that the alternating series is *absolutely convergent.*

The fact that absolute convergence implies ordinary convergence is just common sense if you think about it. The previous geometric series of positive terms converges to 2. If you made all the terms negative, it would sum to −2, right? So, if some of the terms are positive and some negative, the series must converge to something between −2 and 2.

Did you notice that the alternating series is a geometric series *as it is* with $r = -\frac{1}{2}$? (Recall that the geometric series rule works for alternating series as well as for positive series.) The geometric series rule gives its sum: $\frac{a}{1-r} = \frac{1}{1-\left(-\frac{1}{2}\right)} = \frac{2}{3}$.

The alternating series test

MATH RULES

Alternating series test: An alternating series converges if two conditions are met:

1. Its nth term converges to zero.
2. Its terms are non-increasing — in other words, each term is less than or equal to its predecessor (ignoring the minus signs).

   (Note that you are free to ignore any number of initial terms when checking whether condition 2 is satisfied.)

Using this simple test, you can easily show many alternating series to be convergent. The terms just have to converge to zero and get smaller and smaller (they rarely stay the same). The alternating harmonic series converges by this test:

$$\sum_{n=1}^{\infty}\left((-1)^{n+1}\frac{1}{n}\right) = 1 - \frac{1}{2} + \frac{1}{3} - \frac{1}{4} + \frac{1}{5} - \frac{1}{6} + \ldots$$

So does the alternating geometric series I discussed at the end of the previous section. And so do the following two series:

$$\sum_{n=1}^{\infty}\left((-1)^{n+1}\frac{1}{\sqrt{n}}\right) = 1 - \frac{1}{\sqrt{2}} + \frac{1}{\sqrt{3}} - \frac{1}{\sqrt{4}} + \frac{1}{\sqrt{5}} - \frac{1}{\sqrt{6}} + \ldots$$

$$\sum_{n=1}^{\infty}\left((-1)^{n+1}\frac{1}{n^2}\right) = 1 - \frac{1}{2^2} + \frac{1}{3^2} - \frac{1}{4^2} + \frac{1}{5^2} - \frac{1}{6^2} + \ldots$$

WARNING

The alternating series test can't tell you whether a series is absolutely or conditionally convergent. The alternating series test can only tell you whether an alternating series itself converges. The test says nothing about the corresponding positive-term series. In other words, the test cannot tell you whether a series is absolutely convergent or conditionally convergent. To answer that, you must investigate the positive series with a different test.

EXAMPLE

Q. Determine the convergence or divergence of the following alternating series. If convergent, determine whether the convergence is conditional or absolute.

$$\sum_{n=3}^{\infty}\left((-1)^{n+1}\frac{\ln n}{n}\right)$$

A. **The series is conditionally convergent.**

1. Check that the *n*th term converges to zero.

$$\lim_{n\to\infty}\frac{\ln n}{n}$$
$$=\lim_{n\to\infty}\frac{\frac{1}{n}}{1}\quad\text{(by L'Hôpital's rule)}$$
$$=0$$

TIP

Consider the *n*th term. Always check the *n*th term first because if it doesn't converge to zero, you're done — the alternating series *and* the positive series will both diverge. Note that the *n*th term test of *divergence* (see the section on the *n*th term test) applies to alternating series as well as positive series.

2. Check that the terms decrease or stay the same (ignoring the minus signs).

To show that $\frac{\ln n}{n}$ decreases, take the derivative of the function $f(x)=\frac{\ln x}{x}$. Remember differentiation? I know it's been a while.

$$f'(x)=\frac{\frac{1}{x}\cdot x-\ln x}{x^2}\quad\text{(quotient rule)}$$
$$=\frac{1-\ln x}{x^2}$$

This is negative for all $x\geq 3$ (because the natural log of anything 3 or greater is more than 1, and x^2 for $x\geq 3$ is always positive), so the derivative and thus the slope of the function are negative, and therefore the function is decreasing. Finally, because the function is decreasing, the terms of the series are also decreasing (when $n\geq 3$). That does it: $\sum_{n=3}^{\infty}\left((-1)^{n+1}\frac{\ln n}{n}\right)$ converges by the alternating series test.

3. Determine the type of convergence.

You can see that for $n\geq 3$ the positive series, $\frac{\ln n}{n}$, is greater than the divergent harmonic series, $\frac{1}{n}$, so the positive series diverges by the direct comparison test. Thus, the alternating series is *conditionally* convergent.

WARNING

I can't think of a good title for this warning. If the alternating series fails to satisfy the second requirement of the alternating series test, it does *not* follow that your series diverges, only that this test fails to show convergence.

You're getting so good at this, so how about another problem?

Q. Test the convergence of $\sum_{n=4}^{\infty}\left((-1)^n \frac{\ln n}{n^3}\right)$.

EXAMPLE

TIP

You might want to consider the positive series first. If you think you can show that the *positive* series converges or diverges, you may want to try that before using the alternating series test, because . . .

You may have to do this later anyway to determine the type of convergence, and

If you can show that the positive series *converges,* you're done in one step, and you've shown that the alternating series is *absolutely* convergent.

A. **The alternating series converges absolutely.**

Because the *positive* series $\frac{\ln n}{n^3}$ resembles the convergent *p*-series, $\frac{1}{n^3}$, you guess that it converges. So, try to show the convergence of the positive series $\sum_{n=4}^{\infty}\frac{\ln n}{n^3}$. The limit comparison test seems appropriate here, and $\sum_{n=4}^{\infty}\frac{1}{n^3}$ is the natural choice for the benchmark series, but with that benchmark, the test fails — try it. When this happens, you can sometimes get home by trying a larger convergent series. So, try the limit comparison test with the convergent *p*-series, $\sum_{n=4}^{\infty}\frac{1}{n^2}$:

$$\lim_{n\to\infty}\frac{\frac{\ln n}{n^3}}{\frac{1}{n^2}}$$

$$= \lim_{n\to\infty}\frac{\ln n}{n}$$

$= 0$ (I did this in the previous problem with L'Hôpital's rule.)

Because this limit is zero, the positive series $\sum_{n=4}^{\infty}\frac{\ln n}{n^3}$ converges (see the section, "The limit comparison test"); and because the positive series converges, so does the given alternating series. Thus, $\sum_{n=4}^{\infty}\left((-1)^n \frac{\ln n}{n^3}\right)$ converges *absolutely*.

Q. One last problem and I'll let you go. Test the convergence of $\sum_{n=1}^{\infty}\left((-1)^{n+1}\frac{n}{n+1}\right) = \frac{1}{2}-\frac{2}{3}+\frac{3}{4}-\frac{4}{5}+\frac{5}{6}-\ldots$. This is an easy one.

EXAMPLE

A. **The series diverges.** The *n*th term of this series (ignoring the minus signs) converges to 1 (it's a L'Hôpital's rule no-brainer), so you're done. Because the *n*th term does not converge to zero, the series diverges by the *n*th term test.

YOUR TURN

For Problems 21 and 22, determine whether the series converges or diverges. If the series converges, determine whether the convergence is absolute or conditional.

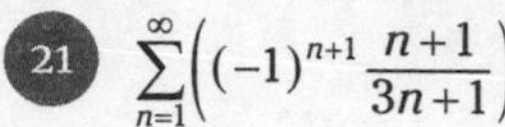

21 $\sum_{n=1}^{\infty}\left((-1)^{n+1}\frac{n+1}{3n+1}\right)$

22 $\sum_{n=3}^{\infty}\left((-1)^{n}\frac{n+1}{n^2-2}\right)$

Keeping All the Tests Straight

You now probably feel like you know — have a vague recollection of? — a gazillion convergence/divergence tests and are wondering how to keep track of all of them. Actually, I've given you only ten tests in all — that's a nice, easy-to-remember round number. Here's how you can keep the tests straight.

First are the three series with names: the geometric series, *p*-series, and telescoping series. A geometric series converges if $0<|r|<1$. A *p*-series converges if $p>1$. A telescoping series converges if the second "half-term" converges to a finite number.

Next are the three comparison tests: the direct comparison, limit comparison, and integral comparison tests. All three compare a new series to a known benchmark series. If the benchmark series converges, so does the series you're investigating; if the benchmark diverges, so does your new series.

And then you have the two "*R*" tests: the ratio test and the root test. Both analyze just the series in question instead of comparing it to a benchmark series. Both involve taking a limit, and the results of both are interpreted the same way. If the limit is less than 1, the series converges; if the limit is greater than 1, the series diverges; and if the limit equals 1, the test is inconclusive.

Finally, you have two tests that form bookends for the other eight — the *n*th term test of divergence and the alternating series test. These two form a coherent pair. You can remember them as the *n*th term test of divergence and the *n*th term test of convergence. The alternating series test involves more than just testing the *n*th term, but this is a good memory aid.

Well, there you have it: Calculus, schmalculus.

Practice Questions Answers and Explanations

1. $\sum_{n=1}^{\infty}\frac{2n^2-9n-8}{5n^2+20n+12}$ **diverges.** You know (vaguely remember?) from Chapter 8 on limits that $\lim_{x\to\infty}\frac{2x^2-9x-8}{5x^2+20x+12}=\frac{2}{5}$ by the horizontal asymptote rule. Because this limit doesn't converge to zero, neither does the underlying sequence of the series. And, therefore, the *n*th term test tells you that the series must diverge.

2. $\sum_{n=1}^{\infty}\frac{1}{n}$ **converges . . . . NOT.** It should be obvious that $\lim_{n\to\infty}\frac{1}{n}=0$. If you conclude that the series, $\sum_{n=1}^{\infty}\frac{1}{n}$, must therefore converge by the *n*th term test, I've got some good news and some bad news for you. The bad news is that you're wrong — you have to use the *p*-series test to find out whether this converges or not (check out the solution to Problem 4). The good news is that you made this mistake here instead of on a test.

TIP

Don't forget that the *n*th term test is no help in determining the convergence or divergence of a series when the underlying sequence converges to zero.

3. $0.008-0.006+0.0045-0.003375+0.00253125-\ldots$ **converges to** $\frac{4}{875}$**.**

   1. **Determine the ratio of the second term to the first term.**

   $$\frac{-0.006}{0.008}=-\frac{3}{4}$$

   2. **Check to see whether all the other ratios of the other pairs of consecutive terms equal** $-\frac{3}{4}$**.**

   $$\frac{0.0045}{-0.006}=-\frac{3}{4}?\text{ Check.}\qquad\frac{-0.003375}{0.0045}=-\frac{3}{4}?\text{ Check.}\qquad\frac{0.00253125}{-0.003375}=-\frac{3}{4}?\text{ Check.}$$

   Voilà! A geometric series with $r=-\frac{3}{4}$.

   3. **Apply the geometric series rule.**

   Because $-1<|r|<1$, the series converges to

   $$\frac{a}{1-r}=\frac{0.008}{1-\left(-\frac{3}{4}\right)}=\frac{4}{875}$$

4. $\sum_{n=1}^{\infty}\frac{1}{10n}$ **diverges.**

   There are a couple ways to solve this one. Probably the easiest way is to, first, simply pull the 10 (actually it's $\frac{1}{10}$) to the outside of the sigma symbol (you learn in Chapter 14 that this is allowed). That gives you: $\frac{1}{10}\sum_{n=1}^{\infty}\frac{1}{n}$. And that's $\frac{1}{10}$ times the divergent harmonic series (it's divergent because it's a *p*-series with $p=1$). That does it. The tip just above the "Telescoping series" subsection says that multiplication by a constant doesn't affect the convergence or divergence of a series, so, since $\sum_{n=1}^{\infty}\frac{1}{n}$ is known to diverge, $\sum_{n=1}^{\infty}\frac{1}{10n}=\frac{1}{10}\sum_{n=1}^{\infty}\frac{1}{n}$ must diverge as well.

5 $1+\frac{\sqrt[4]{2}}{2}+\frac{\sqrt[4]{3}}{3}+\frac{\sqrt[4]{4}}{4}+\ldots+\frac{\sqrt[4]{n}}{n}$ **diverges.**

This may not look like a p-series, but you can't always judge a book by its cover.

1. **Rewrite the terms with exponents instead of roots.**

   $$1+\frac{2^{1/4}}{2}+\frac{3^{1/4}}{3}+\frac{4^{1/4}}{4}+\ldots+\frac{n^{1/4}}{n}$$

2. **Use ordinary laws of exponents to simplify.**

   $$1+\frac{1}{2^{3/4}}+\frac{1}{3^{3/4}}+\frac{1}{3^{3/4}}+\ldots+\frac{1}{n^{3/4}}$$

3. **Apply the p-series rule.**

   You've got a p-series with $p=\frac{3}{4}$, so this series diverges.

6 $\frac{1}{2}+\frac{1}{4}+\frac{1}{8}+\frac{1}{12}+\frac{1}{16}+\frac{1}{20}+\ldots$ **diverges.**

This looks like it might be a geometric series, so:

1. **Find the first ratio.**

   $$\frac{\frac{1}{4}}{\frac{1}{2}}=\frac{1}{2}$$

2. **Test the other pairs.**

   $$\frac{\frac{1}{8}}{\frac{1}{4}}=\frac{1}{2}\text{? Check.}\qquad \frac{\frac{1}{12}}{\frac{1}{8}}=\frac{1}{2}\text{? No.}$$

   Thus, this is *not* a geometric series, and therefore the geometric series rule does not apply.

3. **Try something else.**

   The key to this problem is to notice the simple pattern in the denominators (ignoring the first term): namely, 4, 8, 12, 16, ... They're the multiples of 4, of course. Thus, you can write the sum (again, ignoring the first term) as $\sum_{n=1}^{\infty}\frac{1}{4n}$. And that equals $\frac{1}{4}\sum_{n=1}^{\infty}\frac{1}{n}$. This is a constant times the divergent harmonic series, and (just like you saw in the solution to Problem 4) this multiplication doesn't affect divergence. Therefore, $\frac{1}{4}\sum_{n=1}^{\infty}\frac{1}{n}$ diverges.

   Finally, if all the terms beginning with $\frac{1}{4}$ sum to infinity, adding back the $\frac{1}{2}$ still gives you infinity.

7 $\sum_{n=1}^{\infty}\frac{10(0.9)^n}{\sqrt{n}}$ **converges.**

1. **Look in the summation expression for a series you recognize that can be used for your benchmark series.**

   You should recognize $\sum 0.9^n$ as a convergent geometric series, because r, namely 0.9, is between 0 and 1.

2. **Use the direct comparison test to compare $\sum_{n=1}^{\infty}\frac{10(0.9)^n}{\sqrt{n}}$ to $\sum_{n=1}^{\infty}0.9^n$.**

First, you can pull the 10 out and ignore it because multiplying a series by a constant has no effect on its convergence or divergence. That gives you $\sum_{n=1}^{\infty}\frac{0.9^n}{\sqrt{n}}$.

Now, because each term of $\sum_{n=1}^{\infty}\frac{0.9^n}{\sqrt{n}}$ is less than or equal to the corresponding term of the *convergent* series $\sum_{n=1}^{\infty}0.9^n$, $\sum_{n=1}^{\infty}\frac{0.9^n}{\sqrt{n}}$ has to converge as well. Finally, because $\sum_{n=1}^{\infty}\frac{0.9^n}{\sqrt{n}}$ converges, so does $\sum_{n=1}^{\infty}\frac{10(0.9)^n}{\sqrt{n}}$.

(8) **$\sum_{n=1}^{\infty}\frac{1.1^n}{10n}$ diverges.**

1. **Find an appropriate benchmark series.**

Like in Problem 7, there is a geometric series in the numerator, $\sum_{n=1}^{\infty}1.1^n$. By the geometric series rule, it diverges. But unlike Problem 7, this doesn't help you, because the given series is *less* than that *divergent* geometric series. Use the series in the denominator instead:

$$\sum_{n=1}^{\infty}\frac{1.1^n}{10n}=\frac{1}{10}\sum_{n=1}^{\infty}\frac{1.1^n}{n}$$

The denominator of $\sum_{n=1}^{\infty}\frac{1.1^n}{n}$ is the divergent p-series $\sum_{n=1}^{\infty}\frac{1}{n}$.

2. **Apply the direct comparison test.**

Because each term of $\sum_{n=1}^{\infty}\frac{1.1^n}{n}$ is *greater* than the corresponding term of the *divergent* series $\sum_{n=1}^{\infty}\frac{1}{n}$, $\sum_{n=1}^{\infty}\frac{1.1^n}{n}$ diverges as well — and therefore so does $\sum_{n=1}^{\infty}\frac{1.1^n}{10n}$.

(9) **$\frac{1}{1001}+\frac{1}{2001}+\frac{1}{3001}+\frac{1}{4001}+\ldots$ diverges.**

1. **Ask yourself what this series resembles.**

It's sort of like the divergent harmonic series, $\frac{1}{1}+\frac{1}{2}+\frac{1}{3}+\frac{1}{4}+\ldots$, right?

2. **Multiply the given series by 1001 so that you can compare it to the harmonic series.**

$$1001\left(\frac{1}{1001}+\frac{1}{2001}+\frac{1}{3001}+\frac{1}{4001}+\ldots\right)=\frac{1001}{1001}+\frac{1001}{2001}+\frac{1001}{3001}+\frac{1001}{4001}+\ldots$$

3. **Use the direct comparison test.**

It's easy to show that the terms of the series in Step 2 are greater than or equal to the terms of the divergent p-series, so it, and thus your given series, diverges as well.

(10) $\sum_{n=1}^{\infty} \frac{1}{n+\sqrt{n}+\ln n}$ **diverges.**

Try the limit comparison test: Use the divergent harmonic series $\sum_{n=1}^{\infty} \frac{1}{n}$ as your benchmark.

$$\lim_{n\to\infty} \frac{\frac{1}{n+\sqrt{n}+\ln n}}{\frac{1}{n}}$$

$$= \lim_{n\to\infty} \frac{n}{n+\sqrt{n}+\ln n}$$

$$= \lim_{n\to\infty} \frac{1}{1+\frac{1}{2\sqrt{n}}+\frac{1}{n}} \quad \text{(by L'Hôpital's rule)}$$

$$= 1$$

Because the limit is finite and positive, the limit comparison test tells you that $\sum_{n=1}^{\infty} \frac{1}{n+\sqrt{n}+\ln n}$ diverges along with the benchmark series. By the way, you could do this problem with the direct comparison test as well. Do you see how? *Hint:* You can use the harmonic series as your benchmark, but you have to tweak it first.

(11) $\sum_{n=1}^{\infty} \frac{1}{n^3-\ln^3 n}$ **converges.**

1. **Do a quick check to see whether the direct comparison test will give you an immediate answer.**

   It doesn't because $\sum_{n=1}^{\infty} \frac{1}{n^3-\ln^3 n}$ is *greater* than the known *convergent* p-series $\sum_{n=1}^{\infty} \frac{1}{n^3}$.

2. **Try the limit comparison test with $\sum_{n=1}^{\infty} \frac{1}{n^3}$ as your benchmark.**

$$\lim_{n\to\infty} \frac{\frac{1}{n^3-\ln^3 n}}{\frac{1}{n^3}}$$

$$= \lim_{n\to\infty} \frac{n^3}{n^3-\ln^3 n}$$

$$= \lim_{n\to\infty} \frac{1}{1-\frac{\ln^3 n}{n^3}}$$

$$= \lim_{n\to\infty} \frac{1}{1-\left(\frac{\ln n}{n}\right)^3}$$

$$= \frac{1}{1 - \lim_{n\to\infty}\left(\frac{\ln n}{n}\right)^3} \quad \text{(Just take my word for it.)}$$

$$= \frac{1}{1 - \left(\lim_{n\to\infty}\frac{\ln n}{n}\right)^3} \quad \text{(Just take my word for it.)}$$

$$= \frac{1}{1 - \left(\lim_{n\to\infty}\frac{\frac{1}{n}}{1}\right)^3} \quad \text{(L'Hôpital's rule)}$$

$$= 1$$

Because this is finite and positive, the limit comparison test tells you that since the benchmark series converges, $\sum_{n=1}^{\infty}\frac{1}{n^3 - \ln^3 n}$ must converge as well.

12 $\sum_{n=2}^{\infty}\frac{1}{n\ln n + \sin n}$ **diverges.**

1. **You know you can integrate $\int \frac{1}{x\ln x}dx$ with a simple u-substitution, so do it, and then you'll be able to use the integral comparison test.**

$$\int_2^{\infty}\frac{dx}{x\ln x}$$

$$= \lim_{c\to\infty}\int_2^{c}\frac{dx}{x\ln x} \qquad u = \ln x \quad \text{when } x = 2,\ u = \ln 2$$

$$du = \frac{dx}{x} \quad \text{when } x = c,\ u = \ln c$$

$$= \lim_{c\to\infty}\int_{\ln 2}^{\ln c}\frac{du}{u}$$

$$= \lim_{c\to\infty}\left[\ln u\right]_{\ln 2}^{\ln c}$$

$$= \lim_{c\to\infty}(\ln(\ln c) - \ln(\ln 2))$$

$$= \infty$$

By the integral comparison test, $\sum_{n=2}^{\infty}\frac{1}{n\ln n}$ diverges along with its companion improper integral, $\int_2^{\infty}\frac{dx}{x\ln x}$.

2. **Try the direct comparison test.**

It won't work yet because $\frac{1}{n\ln n + \sin n}$ is sometimes *less* than the *divergent* series $\frac{1}{n\ln n}$.

3. **Try multiplication by a constant (always easy to do and always a good thing to try).**

$\sum_{n=2}^{\infty}\frac{1}{n\ln n}$ diverges; thus, so does $\frac{1}{2}\sum_{n=2}^{\infty}\frac{1}{n\ln n} = \sum_{n=2}^{\infty}\frac{1}{2n\ln n}$.

4. **Now try the direct comparison test again.**

It's easy to show that $\frac{1}{n\ln n + \sin n}$ is always *greater* than $\frac{1}{2n\ln n}$ (for $n \geq 2$), and thus the direct comparison test tells you that $\sum_{n=2}^{\infty}\frac{1}{n\ln n + \sin n}$ must diverge along with $\sum_{n=2}^{\infty}\frac{1}{2n\ln n}$.

(13) $\sum_{n=1}^{\infty} \frac{n^2}{e^{n^3}}$ **converges.**

This is tailor-made for the integral test:

$$\int_1^{\infty} \frac{x^2}{e^{x^3}} dx = \lim_{c\to\infty} \int_1^{c} \frac{x^2}{e^{x^3}} dx = \lim_{c\to\infty} \frac{1}{3} \int_1^{c^3} \frac{du}{e^u} = \frac{1}{3} \lim_{c\to\infty} \left[-e^{-u} \right]_1^{c^3} = -\frac{1}{3} \lim_{c\to\infty} \left(\frac{1}{e^{c^3}} - \frac{1}{e} \right) = \frac{1}{3e}$$

Because the integral converges, so does the series.

(14) $\sum_{n=1}^{\infty} \frac{n^3}{n!}$ **converges.**

1. **Try the limit comparison test with the convergent series, $\sum_{n=1}^{\infty} \frac{1}{n!}$, as the benchmark.**

   $$\lim_{n\to\infty} \frac{\frac{n^3}{n!}}{\frac{1}{n!}} = \lim_{n\to\infty} \frac{n! n^3}{n!} = \infty.$$ No good. This result tells you nothing.

2. **Try the following nifty trick.**

   Ignore the first three terms of $\sum_{n=1}^{\infty} \frac{n^3}{n!}$, which doesn't affect the convergence or divergence of the series. (You ignore three terms because the power on n is 3; that's what makes this trick work.) The series is now $\frac{4^3}{4!} + \frac{5^3}{5!} + \frac{6^3}{6!} + \ldots$, which can be written as $\sum_{n=1}^{\infty} \frac{(n+3)^3}{(n+3)!}$.

3. **Try the limit comparison test again.**

   $$\lim_{n\to\infty} \frac{\frac{(n+3)^3}{(n+3)!}}{\frac{1}{n!}}$$
   $$= \lim_{n\to\infty} \frac{n!(n+3)^3}{(n+3)!}$$
   $$= \lim_{n\to\infty} \frac{(n+3)^3}{(n+3)(n+2)(n+1)}$$
   $$= \lim_{n\to\infty} \frac{n^3 + \text{lesser powers of } n}{n^3 + \text{lesser powers of } n}$$
   $$= 1 \quad \text{(by the horizontal asymptote rule)}$$

   Thus, $\sum_{n=1}^{\infty} \frac{(n+3)^3}{(n+3)!}$ converges by the limit comparison test. And because $\sum_{n=1}^{\infty} \frac{n^3}{n!}$ is the same series except for its first three terms, it converges as well.

(15) $\sum_{n=1}^{\infty} \frac{1}{\ln^n (n+2)}$ **converges.**

Try the root test:

$$\lim_{n\to\infty} \left(\frac{1}{\ln^n (n+2)} \right)^{1/n}$$
$$= \lim_{n\to\infty} \frac{1}{\ln(n+2)} = 0$$

This is less than 1, so the series converges.

16 $\sum_{n=1}^{\infty} \frac{n^{\sqrt{n}}}{\sqrt{n}^n}$ **converges.**

Try the root test again:

$$\lim_{n\to\infty}\left(\frac{n^{\sqrt{n}}}{\sqrt{n}^n}\right)^{1/n} = \lim_{n\to\infty}\frac{n^{\sqrt{n}/n}}{n^{1/2}} = \lim_{n\to\infty}\frac{1}{n^{1/2-\sqrt{n}/n}} = \lim_{n\to\infty}\frac{1}{n^{1/2-1/\sqrt{n}}} = 0$$

Thus the series converges.

17 $\sum_{n=1}^{\infty} \frac{n!}{n^n}$ **converges.**

There's a factorial, so try the ratio test:

$$\lim_{n\to\infty}\frac{\dfrac{(n+1)!}{(n+1)^{n+1}}}{\dfrac{n!}{n^n}}$$

$$= \lim_{n\to\infty}\frac{(n+1)!n^n}{n!(n+1)^{n+1}}$$

$$= \lim_{n\to\infty}\frac{(n+1)n^n}{(n+1)^{n+1}}$$

$$= \lim_{n\to\infty}\frac{n^n}{(n+1)^n}$$

$$= \lim_{n\to\infty}\left(\frac{n}{n+1}\right)^n$$

$$= \lim_{n\to\infty}\left(1-\frac{1}{n+1}\right)^n$$

This is extremely close to one of the limits you should memorize (from the beginning of Chapter 8). It needs just a little work: Set $u = n+1$ then substitute:

$$\lim_{n\to\infty}\left(1-\frac{1}{n+1}\right)^n = \lim_{u\to\infty}\left(1-\frac{1}{u}\right)^{u-1} = \lim_{u\to\infty}\left(1-\frac{1}{u}\right)^u \cdot \lim_{u\to\infty}\left(1-\frac{1}{u}\right)^{-1} = \frac{1}{e}\cdot 1 = \frac{1}{e}$$

Because this is less than one, the series converges.

18 $\sum_{n=1}^{\infty} n\left(\frac{3}{4}\right)^n$ **converges.**

Rewrite this so it's one big nth power: $\sum_{n=1}^{\infty}\left(n^{1/n}\cdot\frac{3}{4}\right)^n$. Now look at the limit of the nth root.

$$\lim_{n\to\infty}\left(\left(\frac{3}{4}n^{1/n}\right)^n\right)^{1/n}$$

$$= \lim_{n\to\infty}\frac{3}{4}n^{1/n}$$

$$= \frac{3}{4}\lim_{n\to\infty} n^{1/n} \quad \text{(an unacceptable L'Hôpital's rule case: } \infty^0\text{)}$$

Now, set the limit equal to y, and take the log of both sides:

$$y = \frac{3}{4}\lim_{n\to\infty} n^{1/n}$$

$$\ln y = \ln\left(\frac{3}{4}\lim_{n\to\infty} n^{1/n}\right)$$

$$= \ln\frac{3}{4} + \ln\left(\lim_{n\to\infty} n^{1/n}\right)$$

$$= \ln\frac{3}{4} + \lim_{n\to\infty}\left(\ln n^{1/n}\right)$$

$$= \ln\frac{3}{4} + \lim_{n\to\infty}\frac{\ln n}{n}$$

$$= \ln\frac{3}{4} + \lim_{n\to\infty}\frac{\frac{1}{n}}{1} \quad \text{(L'Hôpital's rule)}$$

$$\ln y = \ln\frac{3}{4}$$

$$y = \frac{3}{4}$$

Thus, the limit of the nth root is $\frac{3}{4}$, and therefore the series converges.

(19) $\sum_{n=1}^{\infty}\frac{n^{\sqrt{n}}}{n!}$ **converges.**

Try the ratio test: $\lim_{n\to\infty}\frac{\frac{(n+1)^{\sqrt{n+1}}}{(n+1)!}}{\frac{n^{\sqrt{n}}}{n!}} = \lim_{n\to\infty}\frac{n!(n+1)^{\sqrt{n+1}}}{(n+1)!n^{\sqrt{n}}} = \lim_{n\to\infty}\frac{(n+1)^{\sqrt{n+1}}}{(n+1)n^{\sqrt{n}}} = \lim_{n\to\infty}\frac{(n+1)^{\sqrt{n+1}-1}}{n^{\sqrt{n}}} = 0$

(Okay, I admit it, I used my calculator to get that last limit.)

By the ratio test, the series converges.

(20) $\sum_{n=1}^{\infty}\frac{n!}{4^n}$ **diverges.**

Try the ratio test: $\lim_{n\to\infty}\frac{\frac{(n+1)!}{4^{n+1}}}{\frac{n!}{4^n}} = \lim_{n\to\infty}\frac{(n+1)!4^n}{n!4^{n+1}} = \lim_{n\to\infty}\frac{n+1}{4} = \infty$

Thus the series diverges.

(21) $\sum_{n=1}^{\infty}\left((-1)^{n+1}\frac{n+1}{3n+1}\right)$ **diverges.**

This one is a no-brainer, because $\lim_{n\to\infty}\frac{n+1}{3n+1} = \frac{1}{3}$, the first condition of the alternating series test, is not satisfied, which means that both the alternating series and the series of positive terms are divergent.

22 $\sum_{n=3}^{\infty}\left((-1)^n \frac{n+1}{n^2-2}\right)$ **converges conditionally.**

Check the two conditions of the alternating series test:

1. **Does the limit equal zero?**

$$\begin{aligned}&\lim_{n\to\infty}\frac{n+1}{n^2-2}\\&=\lim_{n\to\infty}\frac{1}{2n} \quad \text{(L'Hôpital's rule)}\\&=0 \quad \text{Check.}\end{aligned}$$

2. **Are the terms non-increasing?**

$$\begin{aligned}&\frac{n+1}{n^2-2}\overset{?}{\geq}\frac{(n+1)+1}{(n+1)^2-2}\\&\frac{n+1}{n^2-2}\overset{?}{\geq}\frac{n+2}{n^2+2n-1}\\&(n+1)(n^2+2n-1)\overset{?}{\geq}(n+2)(n^2-2)\\&n^3+2n^2-n+n^2+2n-1\overset{?}{\geq}n^3-2n+2n^2-4\\&n^3+3n^2+n-1\overset{?}{\geq}n^3+2n^2-2n-4\\&n^2+3n+3\geq 0 \quad \text{Check.}\end{aligned}$$

(Extra credit question: Do you see why the above inequality math is a bit loose? It is valid, however.) Thus, the series is at least conditionally convergent. And it is easy to show that it is only conditionally convergent and not absolutely convergent by the direct comparison test. Each term of the given series, $\sum_{n=3}^{\infty}\frac{n+1}{n^2-2}$, is greater than the corresponding term of the series $\sum_{n=3}^{\infty}\frac{n}{n^2}$, because each term of $\sum_{n=3}^{\infty}\frac{n+1}{n^2-2}$ has a larger numerator and a smaller denominator. Since $\sum_{n=3}^{\infty}\frac{n}{n^2}$ is the same as the divergent harmonic series, $\sum_{n=3}^{\infty}\frac{1}{n}$, it follows that $\sum_{n=3}^{\infty}\frac{n+1}{n^2-2}$ is divergent as well.

If you're ready to test your skills a bit more, take the following chapter quiz that incorporates all the chapter topics.

Whaddya Know? Chapter 19 Quiz

Quiz time! Complete each problem to test your knowledge on the various topics covered in this chapter. You can then find the solutions and explanations in the next section.

For the problems in this quiz, determine 1) whether the given series converges or diverges, 2) (only for convergent geometric or telescoping series) what number the series converges to, and 3) (when appropriate) whether the convergence is conditional or absolute.

1. $\sum_{n=1}^{\infty} \frac{n-\sqrt{n}}{n^2}$

2. $\sum_{n=1}^{\infty} \frac{2n}{n^2+8n+9}$

3. $\sum_{n=1}^{\infty} \frac{n^3 n!}{n^n}$

4. $\sum_{n=1}^{\infty} \frac{1}{n \sin^2 n}$

5. $\sum_{n=1}^{\infty} \frac{1+\ln n}{n^2+\ln n}$

6. $\sum_{n=1}^{\infty} \frac{1+e^n}{n^2+e^n}$

7. $\sum_{n=2}^{\infty} \frac{1}{\sqrt{n} \cdot \ln n}$

8. $\sum_{n=2}^{\infty} \frac{(-1)^n}{\log n}$

9. $\sum_{n=1}^{\infty} \frac{\log n}{n}$

10. $\sum_{n=2}^{\infty} \frac{\log n}{\ln n}$

11. $\sum_{n=1}^{\infty} \frac{\log^3(n^3)}{n^3}$

12. $\sum_{n=1}^{\infty} \frac{\pi}{e^{n-1}}$

Answers to Chapter 19 Quiz

1. $\sum_{n=1}^{\infty} \frac{n-\sqrt{n}}{n^2}$ **diverges to infinity.**

   Use some simple algebra to rewrite the fraction (always consider taking a fraction apart like this):

$$\begin{aligned}\sum_{n=1}^{\infty} \frac{n-\sqrt{n}}{n^2} &= \sum_{n=1}^{\infty}\left(\frac{n}{n^2}-\frac{\sqrt{n}}{n^2}\right)\\ &= \sum_{n=1}^{\infty}\frac{n}{n^2}-\sum_{n=1}^{\infty}\frac{\sqrt{n}}{n^2}\\ &= \sum_{n=1}^{\infty}\frac{1}{n}-\sum_{n=1}^{\infty}\frac{1}{n^{3/2}}\end{aligned}$$

   The first series is the harmonic series, which diverges to infinity; the second is a p-series with $p=\frac{3}{2}$, so it converges. Thus, the original series diverges.

2. $\sum_{n=1}^{\infty} \frac{2n}{n^2+8n+9}$ **diverges to infinity.**

   Always ask yourself whether you can easily integrate the series expression, because if you can, the integral comparison test is a snap. The derivative of the denominator is $2n+8$. If that were in the numerator, you'd be all set for a simple u-substitution. Like with Problem 1, some simple algebra does the trick — just add and subtract the same thing from the numerator (a handy trick to remember):

$$\sum_{n=1}^{\infty}\frac{2n}{n^2+8n+9}=\sum_{n=1}^{\infty}\frac{2n+8-8}{n^2+8n+9}=\sum_{n=1}^{\infty}\left(\frac{2n+8}{n^2+8n+9}-\frac{8}{n^2+8n+9}\right)=\sum_{n=1}^{\infty}\frac{2n+8}{n^2+8n+9}-\sum_{n=1}^{\infty}\frac{8}{n^2+8n+9}$$

   For the first series, compare it to the improper integral $\int_1^{\infty}\frac{2x+8}{x^2+8x+9}dx$, which you can solve with a u-substitution ($u=x^2+8x+9$):

$$\begin{aligned}&\int_1^{\infty}\frac{2x+8}{x^2+8x+9}dx\\ &=\int_{18}^{\infty}\frac{du}{u}\\ &=\lim_{c\to\infty}\int_{18}^{c}\frac{du}{u} \quad \text{(If you remember this improper integral from Ch. 18, you can stop here; you know it diverges.)}\\ &=\lim_{c\to\infty}[\ln u]_{18}^{c}\\ &=\lim_{c\to\infty}(\ln c-\ln 18)=\infty\end{aligned}$$

   Because $\int_1^{\infty}\frac{2x+8}{x^2+8x+9}dx$ diverges to infinity, so does $\sum_{n=1}^{\infty}\frac{2n+8}{n^2+8n+9}$.

   For the second series, pull the 8 to the outside: $\sum_{n=1}^{\infty}\frac{8}{n^2+8n+9}=8\sum_{n=1}^{\infty}\frac{1}{n^2+8n+9}$. Now you've got a series where each term is less than the convergent p-series (with $p=2$), so your series converges as well, and 8 times that series still converges.

Infinity plus a finite number equals infinity, so the original series diverges.

I wanted to show you the above algebra trick, so I did the problem with the integral comparison test. But you can do the problem more quickly with the limit comparison test. Give it a try.

(3) $\sum_{n=1}^{\infty}\frac{n^3 n!}{n^n}$ **converges.**

Use the ratio test:

$$\lim_{n\to\infty}\frac{\frac{(n+1)^3(n+1)!}{(n+1)^{n+1}}}{\frac{n^3 n!}{n^n}}$$

$$=\lim_{n\to\infty}\frac{(n+1)^3(n+1)!n^n}{(n+1)^{n+1}n^3 n!}$$

$$=\lim_{n\to\infty}\frac{(n+1)^3(n+1)n^{n-3}}{(n+1)^{n+1}}$$

$$=\lim_{n\to\infty}\frac{(n+1)^4 n^{n-3}}{(n+1)^{n+1}}$$

$$=\lim_{n\to\infty}\frac{n^{n-3}}{(n+1)^{n-3}}$$

$$=\lim_{n\to\infty}\left(\frac{n}{n+1}\right)^{n-3}$$

$$=\lim_{n\to\infty}\left(1-\frac{1}{n+1}\right)^{n-3}$$

$$=\lim_{n\to\infty}\left(1-\frac{1}{n+1}\right)^{n+1}\cdot\lim_{n\to\infty}\left(1-\frac{1}{n+1}\right)^{-4}$$

The first limit is from the list of limits to memorize (Chapter 8).

$$=\frac{1}{e}\cdot 1=\frac{1}{e}$$

Because the limit of the ratio is less than 1, the series converges.

(4) $\sum_{n=1}^{\infty}\frac{1}{n\sin^2 n}$ **diverges to infinity.**

Use the direct comparison test:

$\sin^2 n$ is always greater than or equal to zero and less than or equal to 1. Thus, each term of the given series is greater than or equal to each term of the divergent harmonic series, and, therefore, it also diverges.

Congrats if you noticed the potential problem that the denominator might equal zero, which would throw a wrench into the works (always be on the lookout for this). No worries: $\sin^2 x$ is only zero at multiples of pi, and no positive integer value of n can ever be equal to a multiple of pi. (Multiples of pi, of course, are always irrational numbers which, when written in decimal form, never end.)

5. $\sum_{n=1}^{\infty} \frac{1+\ln n}{n^2+\ln n}$ **converges.**

This resembles a series, $\frac{1}{n^2}$, that you should know converges, so you might guess that it converges as well. But the direct comparison test fails because the given series is *greater* than the *convergent* $\frac{1}{n^2}$ series. The ratio test also fails, because the limit equals 1. Next, you might try the limit comparison test with $\frac{1}{n^2}$ as your benchmark series. The limit equals infinity. Strike three — but you're not out! Try the limit comparison test again, but with a larger convergent series as your benchmark series, $\frac{1}{n^{3/2}}$. (Using a larger convergent series in this situation can make your life easier.)

$$\lim_{n\to\infty} \frac{\frac{1+\ln n}{n^2+\ln n}}{\frac{1}{n^{3/2}}}$$

$$= \lim_{n\to\infty} \frac{(1+\ln n)n^{3/2}}{n^2+\ln n}$$

$$= \lim_{n\to\infty} \frac{n^{3/2}+n^{3/2}\ln n}{n^2+\ln n}$$

$$= \lim_{n\to\infty} \frac{\frac{3}{2}n^{1/2}+\frac{3}{2}n^{1/2}\ln n+n^{1/2}}{2n+\frac{1}{n}} \quad \text{(by L'Hôpital's rule)}$$

$$= \lim_{n\to\infty} \frac{\frac{5}{2}n^{1/2}+\frac{3}{2}n^{1/2}\ln n}{2n+\frac{1}{n}}$$

$$= \lim_{n\to\infty} \frac{\frac{5}{4}n^{-1/2}+\frac{3}{4}n^{-1/2}\ln n+\frac{3}{2}n^{-1/2}}{2-\frac{1}{n^2}} \quad \text{(by L'Hôpital's rule again)}$$

I'll spare you the rest of the gory details, but you can show that the limit of the middle term in the numerator equals zero (also by using L'Hôpital's rule), so the limit of the whole enchilada ends up equaling $\frac{0+0+0}{2}=0$.

Finally, the fact that this limit is zero means that the given series is infinitely *smaller* than the *convergent* benchmark series, so the given series must converge as well. (I had to use the "Okay, so I'm a rebel" tip at the end of the subsection, "The limit comparison test." This common-sense tip is perfectly sound, but your calc prof might not buy it.)

6. $\sum_{n=1}^{\infty} \frac{1+e^n}{n^2+e^n}$ **diverges to infinity.**

Don't forget to use the *n*th term test!

It's easy to show that $\lim_{n\to\infty} \frac{1+e^n}{n^2+e^n}=1$ (use L'Hôpital's rule three times). Because the *n*th term does not converge to zero, the series diverges.

7. $\displaystyle\sum_{n=2}^{\infty}\frac{1}{\sqrt{n}\cdot\ln n}$ **diverges to infinity.**

Use the limit comparison test (sort of) with the harmonic series as your benchmark series:

$$\lim_{n\to\infty}\frac{\dfrac{1}{\sqrt{n}\cdot\ln n}}{\dfrac{1}{n}}$$
$$=\lim_{n\to\infty}\frac{n}{\sqrt{n}\cdot\ln n}$$
$$=\lim_{n\to\infty}\frac{\sqrt{n}}{\ln n}$$
$$=\lim_{n\to\infty}\frac{\dfrac{1}{2\sqrt{n}}}{\dfrac{1}{n}}\quad\text{(by L'Hôpital's rule)}$$
$$=\lim_{n\to\infty}\frac{\sqrt{n}}{2}=\infty$$

The given series is, thus, infinitely *greater* than the *divergent* harmonic series, so it must diverge as well. The reason I wrote "sort of" is that to get this result, I had to use the "Okay, so I'm a rebel" tip again.

8. $\displaystyle\sum_{n=2}^{\infty}\frac{(-1)^n}{\log n}$ **converges conditionally.**

This series meets the requirements of the alternating series test, so it converges. Next, you have to consider the ordinary, positive series: $\displaystyle\sum_{n=2}^{\infty}\frac{1}{\log n}$. That series diverges by the direct comparison test — using the harmonic series as your benchmark series. (As you have probably noticed, the harmonic series is one of the most useful — perhaps the most useful — benchmark series.) Because the series of positive terms diverges, the given alternating series converges conditionally.

9. $\displaystyle\sum_{n=1}^{\infty}\frac{\log n}{n}$ **diverges to infinity.**

This is a snap with the direct comparison test.

Each term of the given series is greater than or equal to the corresponding term of the divergent harmonic series (for all $n\geq 10$), so this series diverges as well. Don't forget: You are always free to disregard any number of initial terms.

10. $\displaystyle\sum_{n=2}^{\infty}\frac{\log n}{\ln n}$ **diverges to infinity.**

Don't forget to use pre-calc!

$\dfrac{\log n}{\ln n}=\log e\approx 0.434$ for all values of n by the change of base rule. Thus, the series sums to infinity times that value, which, of course, is infinity. The series also diverges by the nth term test.

11. $\displaystyle\sum_{n=1}^{\infty}\frac{\log^3\left(n^3\right)}{n^3}$ **converges.**

Use the very same approach as with Problem 5:

You can use the limit comparison test with the known benchmark series, $\frac{1}{n^2}$. (The test will fail if you use the smaller benchmark series, $\frac{1}{n^3}$.)

$$\lim_{n\to\infty}\frac{\frac{\log^3(n^3)}{n^3}}{\frac{1}{n^2}}$$

$$=\lim_{n\to\infty}\frac{\log^3(n^3)}{n}$$

$$=\lim_{n\to\infty}\frac{3\log^2(n^3)\frac{1}{n^3}3n^2}{1} \quad \text{(by L'Hôpital's rule; the derivative is a nested chain rule problem)}$$

$$=\lim_{n\to\infty}\frac{9\log^2(n^3)}{n}$$

$$=\lim_{n\to\infty}\frac{18\log(n^3)\frac{1}{n^3}3n^2}{1} \quad \text{(L'Hôpital's rule again)}$$

$$=\lim_{n\to\infty}\frac{54\log(n^3)}{n}$$

$$=\lim_{n\to\infty}\frac{54\frac{1}{n^3}3n^2}{1} \quad \text{(L'Hôpital's rule yet again)}$$

$$=\lim_{n\to\infty}\frac{162}{n}=0$$

Just like with Problem 5, this zero result tells you that the given series converges. And the same caveat as you see at the end of the Problem 5 solution applies here.

12 $\sum_{n=1}^{\infty}\frac{\pi}{e^{n-1}}$ **converges to** $\frac{e\pi}{e-1}\approx 4.97$**.**

Congrats if you saw that this one is a geometric series in disguise.

You can rewrite this in standard geometric series format, $\sum_{n=0}^{\infty}ar^n$:

$$\sum_{n=1}^{\infty}\frac{\pi}{e^{n-1}}=\sum_{n=0}^{\infty}\pi\left(\frac{1}{e}\right)^n$$

$0<\frac{1}{e}<1$, so, by the geometric series rule, this series converges to $\frac{\pi}{1-\frac{1}{e}}=\frac{e\pi}{e-1}\approx 4.97$.

Index

Symbols and Numerics

± infinity
  limits of rational functions at, 158–159
  solving limits at, with calculators, 160–161
30°-60°-90° triangles, 45, 97
45°-45°-90° triangles, 45, 96–97

A

absolute convergence, 602–603
absolute extrema
  finding on closed intervals, 273–277
  finding over function's entire domain, 278–281
absolute maximum, 261
acceleration
  about, 326–328
  negative, 333
  positive, 333
  second squared, 334
  slowing down, 333
  speed, velocity and, 328–329
  speeding up, 333
acceleration function, 327
adding
  fractions, 29–30
  series, 584–586
algebra
  about, 27
  absolute value, 34
  example questions, 33, 41
  factoring, 38–39
  fractions, 28–34
  logarithms, 37
  power rules, 34–35
  practice questions, 33–34, 42–43
  practice questions answers and explanations, 52–56
  quiz question answers, 61–65
  quiz questions, 59–60
  roots, 35–37
  solving
    limit problems with, 147–152
    limits at infinity with, 161–163
    quadratic equations, 39–41
Algebra II For Dummies (Sterling), 39
algebraic expression, 31
alternating series
  about, 602
  alternating series test, 603–606
  finding absolute *versus* conditional convergence, 602–603
alternating series test, 603–606
American Mathematical Monthly 90, 1983 issue, 482
analyzing arc length, 537–539
angles
  measuring with radians, 98–99
  in unit circles, 98
answers
  practice questions
    algebra, 52–56
    differentiation, 205–208, 292–309, 346–361, 385–392
    differentiation rules, 240–252
    functions, 87–90
    geometry, 56–58
    improper integrals, 570–574
    infinite series, 607–615
    integrals, 544–550
    integration, 427–434, 471–475, 503–516
    limits, 135–136, 165–174
    transformations, 90
    trigonometry, 109–112

answers *(continued)*
quiz questions
algebra, 61–65
differentiation, 211–213, 311–319, 363–365, 394–395
differentiation rules, 254–258
functions, 92–94
geometry, 61–65
improper integrals, 576–579
infinite series, 617–621
integrals, 552–556
integration, 436–437, 477–478, 519
limits, 139–140, 176–180
transformations, 92–94
trigonometry, 114–116
antiderivatives
finding, 460–468
reverse rules for, 460–462
antidifferentiation, 439–441
approximations
of area
about, 403–411
with Simpson's rule, 421, 423–425
with trapezoid rule, 421–422
linear, 374–378
arc length, analyzing, 537–539
area
approximating
about, 403–411
with Simpson's rule, 421, 423–425
with trapezoid rule, 421–422
finding
under a curve, 401–403
with definite integrals, 417–420
with substitution problems, 469–470
formulas for, 44
negative, 426, 450
between two curves, 525–529
area functions, 441–445, 454–455
arrow-number, 120
asymptotes
horizontal, 123–124
vertical, 107, 123, 563–565
average rate, 201
average speed, 126–127, 332
average value, Mean Value Theorem for, 522–525
average velocity, 330

B

boxes, maximum volume of, 322–323
business problems, 378–384

C

calculating
angles with radians, 98–99
instantaneous speed with limits, 126–128
related rates, 343–344
calculators
about, 143–146
solving limits at ± infinity with, 160–161
calculus. *See also specific topics*
about, 7–9, 21
differentiation, 13–15
infinite series, 17–19
integration, 15–17
limit concept, 21–22
precision, 24
real-world examples of, 9–11
zooming, 22–24
canceling, in fractions, 31–33
chain rule, 225–231
Cheat Sheet (website), 3
closed intervals, finding absolute extrema on, 273–277
coefficients, equating of like terms, 500–501
comparison tests, for convergence/divergence, 592–596, 606
completing the square, for solving quadratic equations, 41
composite functions, 225–227

concave down interval, 287
concavity points
about, 260–261
finding, 281–284
conditional convergence, 602–603
constant multiple rule, 217–218
constant rule, 216
constants, multiplying by, 590
continuity. *See also* limits
definition of, 131
linking limits and, 129–131
continuous function, 129
convergence
about, 564
absolute, 602–603
comparison tests for, 592–596, 606
conditional, 602–603
guessing about, 600
nth term test, 586–588
of sequences, 582–583
of series, 585–586
tests for, 588–591
convergent series, 17–19
coordinate geometry formulas, 44
corrals, maximum area of, 323–325
cosecants
about, 220
integrals containing, 489–491
cosines
graphing, 106–107
integrals containing, 486–489
cost, marginal, 378–380
cotangents, integrals containing, 489–491
critical numbers, 263–264
cubes, difference/sum of, 38
curves
area between two, 525–529
derivative of, 191–194
differentiation and, 259–319
finding area under, 401–403
curving incline problem, 8–9

D

decreasing intervals, 286
definite integrals, finding area with, 417–420
degrees, radians compared with, 99
demand function, 379
denominators
containing irreducible quadratic factors, 498–500
containing linear factors, 497–498
containing one or more factors raised to a power greater than 1, 501–502
derivative-hole connection, 131
derivatives
about, 188
of area function, 443
of a curve, 191–194
defined, 13, 184
first derivative test, 264–268
of the function's argument, 465
graphs of, 285–288
higher-order, 237–239
of inverse functions, 235–236
of a line, 188
Mean Value Theorem for, 524
meaning of, 198
non-existence of, 202–203
of position, 327
as a rate, 15, 188–191
second derivative test, 268–272
as a slope, 14
of velocity, 327
determining limits, with L'Hôpital's rule, 583–584
difference of cubes, 38
difference of squares, 38
difference quotient, 28, 195–201
difference rule, 218
differentiation
about, 13, 183–184, 259–260, 262, 321, 367
absolute maximum, 261
acceleration problems, 326–336

differentiation *(continued)*
answers to quiz questions, 211–213, 311–319, 363–365, 394–395
antidifferentiation, 439–441
average rate, 201
business problems, 378–384
concavity, 260–261
critical numbers, 263–264
curve shapes and, 259–319
derivative
of a curve, 191–194
of a line, 188
as a rate, 15, 188–191
as a slope, 14
difference quotient, 195–201
displacement and velocity, 330–331
downhill, 261–262
economics problems, 378–384
example questions, 193, 202–203, 266–267, 270–271, 276–277, 279–280, 290–291, 376–377
finding
absolute extrema on closed intervals, 273–277
absolute extrema over function's entire domain, 278–281
concavity points, 281–284
inflection points, 281–284
local extrema, 262–272
first derivative test, 264–268
graphs of derivatives, 285–288
inflection points, 260–261
instantaneous rate, 201
linear approximations, 374–377
local minimum, 261
marginals, 378–381
maximum area of corrals, 323–325
maximum height, 329–330
maximum profit, 382–383
maximum volume of boxes, 322–323
Mean Value Theorem, 289–291
minimum height, 329–330
negative slopes, 260
non-existence of derivatives, 202–204
normal line problem, 368, 370–371
optimization problems, 321–326
position problems, 326–336
positive slopes, 260
practice questions, 193–194, 203–204, 268, 272, 277, 280–281, 283–284, 288, 291, 325–326, 334–336, 345, 372–373, 377–378, 383–384
practice questions answers and explanations, 205–208, 292–309, 346–361, 385–392
quiz questions, 209–210, 310, 362, 393
rate-slope connection, 190–191
related rates, 336–345
second derivative test, 268–272
second squared, 334
slope and, 184–188
slope of a line, 186–187
speed, 190
speed and distance traveled, 331–332
tangent line problem, 368–369
teetering on corners, 261
velocity problems, 326–336
differentiation rules
about, 215–216
answers to quiz questions, 254–258
chain rule, 225–229
constant multiple rule, 217–218
constant rule, 216
difference rule, 218
differentiating
exponential functions, 220–221
implicitly, 231–233
inverse functions, 234–237
logarithmic functions, 221
trig functions, 220
example questions, 218, 223–224, 232–233, 236
higher-order derivatives, 237–239
power rule, 216–217
practice questions, 219, 224–225, 230–231, 233–234, 237, 238–239
practice questions answers and explanations, 240–252

product rule, 222
quiz questions, 253
quotient rule, 222–223
sum rule, 218
direct comparison test, 592–593
discriminant, 40, 499
disk method, 532–533
displacement, velocity and, 330–331
distance
formula for, 44, 126
speed and distance traveled, 331–332
distributing powers, 35
divergence
about, 564
comparison tests for, 592–596, 606
guessing about, 600
nth term test, 586–588
of sequences, 582–583
of series, 585–586
tests for, 588–591
divergent series, 17
dividing fractions, 29
domain, finding absolute extrema over function's entire, 278–281

E

economics problems, 378–384
endpoints, testing, 323
entire domain, finding absolute extrema over function's, 278–281
equations, 31
evaluating
limits, 141–180
limits at infinity, 156–164
Example icon, 3
example questions
algebra, 33, 41
differentiation, 193, 202–203, 266–267, 270–271, 276–277, 279–280, 290–291, 376–377
differentiation rules, 218, 223–224, 232–233, 236
functions, 73, 80, 82
geometry, 45–46
improper integrals, 560–561, 566
infinite series, 587, 592, 593–596, 599, 600, 602–603, 604, 605
integrals, 523–524, 531–532
integration, 409–410, 412–413, 424–425, 444, 451–452, 467, 490
limits, 124, 133, 146, 148–150, 155, 161–163
transformations, 85
trigonometry, 102, 108
exponential functions, differentiating, 220–221
expressions, 31
extra cost, 378

F

factorial symbol, 598
factoring
about, 38–39
for solving quadratic equations, 39–40
trinomial, 39
finding
absolute extrema
on closed intervals, 273–277
over function's entire domain, 278–281
absolute *versus* conditional convergence, 602–603
antiderivatives, 460–468
area
under a curve, 401–403
with definite integrals, 417–420
with substitution problems, 469–470
concavity points, 281–284
inflection points, 281–284
local extrema, 262–272
finite number, 122
first derivative test, 264–268
first-degree polynomials, 497–498
formulas
for arc length, 538
for average speed, 126–127

formulas *(continued)*
coordinate geometry, 44
for distance, 44, 126
for geometry, 44
for instantaneous speed, 128
midpoint, 44
quadratic, 40–41
for surface area, 44
for surface of revolution, 540
for three-dimensional shapes, 44
for two-dimensional shapes, 44
for volumes, 44
45°-45°-90° triangles, 45, 96–97
fractions
about, 28
adding, 29–30
canceling in, 31–33
dividing, 29
multiplying, 28–29
partial, 497–502
rules for, 28
subtracting, 30
functions
about, 67, 71–73
absolute value, 77–78
acceleration, 327
answers to quiz questions, 92–94
area, 441–445, 454–455
characteristics of, 68–69
composite, 70–71, 225–227
continuous, 129
demand, 379
dependent variables, 69
example questions, 73, 80, 82
exponential, 78–79, 220–221
graphs of, 74–80
horizonal transformations, 83–84
illustrating limits using, 120–121
independent variables, 69
inverse, 81–82, 234–237
inverse trig, 107
linear, 191
lines, 74–77
logarithmic, 79, 221–222
notation for, 69–70
oddball, 78
parabolic, 77–78
piecewise, 122
polynomial, 129
practice questions, 73–74, 80, 82
practice questions answers and explanations, 87–90
quiz questions, 91
rational, 123, 129, 158–159
sequences and, 583
transformations, 83–86
trigonometry, 220
vertical transformations, 85
Fundamental Theorem of Calculus
about, 446–454
area functions, 454–455
integration-differentiation connection, 456–457
statistics, 458–460

G

Gabriel's horn, 567–568
geometric series, tests for, 588–589
geometric series rule, 589
geometry
about, 27, 43
example questions, 45–46
formulas for, 44
practice questions, 46–51
practice questions answers and explanations, 56–58
quiz question answers, 61–65
quiz questions, 59–60
right triangles, 45–46
graphs
cosine, 106–107
of derivatives, 285–288
of functions, 74–80
sine, 106–107
tangent, 106–107

greatest common factor (GCF), 38
guess-and-check method, 463–464
guessing, about convergence/divergence, 600

H

harmonic series, 589–590
height, maximum and minimum, 329–330
higher-order derivatives, 237–239
hole exception, 130–131
horizontal asymptotes, 123–124
horizontal transformations, 83–84
hours-per-mile rate, 190
hypotenuse, unit circles and, 99–100

I

icons, explained, 3
identities, trigonometry, 108
implicit differentiation, 231–234
improper integrals
  about, 557
  answers to quiz questions, 576–579
  example questions, 560–561, 566
  Gabriel's horn, 567–568
  L'Hôpital's rule, 558–562
  with one or two infinite limits of integration, 565–566
  practice questions, 562, 567
  practice questions answers and explanations, 570–574
  quiz questions, 575
  with vertical asymptotes, 563–565
increasing intervals, 286
indefinite integral, 440
index of summation, 412
infinite discontinuity, 132
infinite series
  about, 17, 581
  alternating series, 602–606
  alternating series test, 603–606
  answers to quiz questions, 617–621
  comparison tests for convergence/divergence, 592–596, 606
  convergence, 586–601
  convergent series, 17–19
  direct comparison test, 592–593
  divergence, 586–601
  divergent series, 17
  example questions, 587, 592, 593–596, 599, 600, 602–603, 604, 605
  finding absolute *versus* conditional convergence, 602–603
  geometric series, 588–589
  integral comparison test, 595–596
  limit comparison test, 593–595
  nth term test, 586–588
  practice questions, 588, 591, 597–598, 601, 605–606
  practice questions answers and explanations, 607–615
  p-series, 589–590
  quiz questions, 616
  ratio test, 596–600
  root test, 596–600
  stringing sequences, 582–584
  summing series, 584–586
  telescoping series, 590–591
infinity
  about, 24
  evaluating limits at, 156–164
  solving limits at, with algebra, 161–163
inflection points
  about, 260–261
  finding, 281–284
  on functions, 287
instantaneous rate/speed
  calculating with limits, 126–128
  defined, 201
integral comparison test, 595–596
integrals
  about, 521–522
  analyzing arc length, 537–539
  answers to quiz questions, 552–556
  area between two curves, 525–529
  average value for, 522–525
  containing

integrals *(continued)*
cosecants, 489–491
cotangents, 489–491
secants, 489–491
sines and cosines, 486–489
tangents, 489–491
example questions, 523–524, 531–532
finding area with definite, 417–420
improper
about, 557
answers to quiz questions, 576–579
example questions, 560–561, 566
Gabriel's horn, 567–568
L'Hôpital's rule, 558–562
with one or two infinite limits of integration, 565–566
practice questions, 562, 567
practice questions answers and explanations, 570–574
quiz questions, 575
with vertical asymptotes, 563–565
Mean Value Theorem for, 522–525
practice questions, 525, 528–529, 535–537, 539, 542–543
practice questions answers and explanations, 544–550
quiz questions, 551
surfaces of revolution, 540–543
trigonometry, 486–491
volumes of weird solids, 529–537
integration
about, 15–17, 399–401, 439, 479
answers to quiz questions, 436–437, 477–478, 519
antidifferentiation, 439–441
approximating area
about, 403–411
with Simpson's rule, 421, 423–426
with trapezoid rule, 421–422
area functions, 441–445, 454–455
example questions, 409–410, 412–413, 424–425, 444, 451–452, 467, 490
finding
antiderivatives, 460–468
area under a curve, 401–403
area with definite integral, 417–420
area with substitution problems, 469–470
Fundamental Theorem of Calculus, 446–460
guess-and-check method, 463–464
improper integrals with one or two infinite limits of, 565–566
integrals
with one infinite limit of, 565–566
with two infinite limits of, 566
integration-differentiation connection, 456–457
partial fractions, 497–502
by parts, 479–485
practice questions, 410–411, 413–414, 417, 420, 425–426, 444–445, 452–453, 462, 464, 468, 470, 485, 489, 491, 496, 502
practice questions answers and explanations, 427–434, 471–475, 503–516
quiz questions, 435, 476, 517–518
reverse rules for antiderivatives, 460–462
statistics, 458–460
substitution method, 465–468
summation notation, 411–417
trigonometric substitution, 491–496
trigonometry integrals, 486–491
vertical asymptotes
between limits of, 564–565
at one of limits of, 563–564
vocabulary, 441
writing Riemann sums with sigma notation, 414–417
integration-by-parts, 479–485
integration-differentiation connection, 456–457
Internet resources
Cheat Sheet, 3
Technical Support, 4
Wolfram Alpha, 143
intersections problems, 340–343
inverse functions
about, 81–82
differentiating, 234–237
inverse trig functions, 107

J

jump discontinuity, 132

K

Kasube, Herbert E., 482

L

left rectangle rule, 405
left sums, approximating area with, 404–406
L'Hôpital's rule
  about, 558–562
  determining limits with, 583–584
LIATE mnemonic, 482–485
like terms
  about, 35
  equating coefficients of, 500–501
limit comparison test, 593–595
limit concept, 21–22
limits
  about, 119–120, 141
  answers to quiz questions, 139–140, 176–180
  calculating instantaneous speed with, 126–128
  calculating with calculators, 143–146
  continuity and linking, 129–131
  defined, 119
  determining with L'Hôpital's rule, 583–584
  easy, 141–143
  evaluating
    about, 141–180
    at infinity, 156–164
  example questions, 124, 133, 146, 148–150, 155, 161–163
  formal definition of, 122
  horizontal asymptotes and, 123–124
  one-sided, 121–122
  plug-and-chug problems, 142–143
  practice questions, 124–126, 133–134, 147, 150–152, 155, 156, 163–164
  practice questions answers and explanations, 135–136, 165–174
  quiz questions, 137–138, 175
  of rational functions at ± infinity, 158–159
  real-deal problems, 143–157
  sandwich (squeeze) method, 153–156
  solving
    at ± infinity with calculators, 160–161
    at infinity with algebra, 161–163
    problems with algebra, 147–152
  3333 limit mnemonic, 131–133
  using functions to illustrate, 120–121
  vertical asymptotes and, 123
line
  derivative of a, 188
  slope of a, 186–187
linear approximations, 374–378
linear factors, denominators containing, 497–498
linear functions, 191
local extrema, finding, 262–272
local maximum, 260, 262, 286
local minimum, 261, 262, 286
logarithmic functions, differentiating, 221–222
logarithms, 37

M

managing marginals, 378–381
marginal cost, 378–380
marginal profit, 378–380, 381
marginal revenue, 378–380
marginals, managing, 378–381
Math Rules icon, 3
maximum area, of corrals, 323–325
maximum height, 329–330
maximum profit, 382–383
maximum speed, 332
maximum velocity, 331
maximum volume, of boxes, 322–323
Mean Value Theorem
  about, 289–291
  for integrals and average value, 522–525

measuring
angles with radians, 98–99
instantaneous speed with limits, 126–128
related rates, 343–344
meat-slicer method, 529–532
midpoint formula, 44
midpoint rule, 409
midpoint sums, approximating area with, 408–410
miles-per-hour rate, 190
minimum height, 329–330
minimum speed, 332
minimum velocity, 331
multiplication rule, for canceling, 31–33
multiplying
by constants, 590
fractions, 28–29

N

natural log, 221
negative acceleration, 333
negative areas, 426, 450
negative numbers, 34
negative slopes, 184, 260
normal line problem, 368, 370–371
nth term test, 586–588

O

one-sided limits, 121–122
optimization problems, 321–326

P

parentheses, 227
partial fractions, 497–502
partial sums, 585
parts, integration by, 479–485
patterns, looking for, 38–39
period, 106
periodic, 106
piecewise function, 122
plug-and-chug problems, 142–143
polynomial functions, continuity of, 129
polynomials
defined, 39
first-degree, 497–498
position
about, 326–328
derivative of, 327
positive acceleration, 333
positive numbers, 34
positive slopes, 260
power rules
about, 34–35, 216–217
reverse, 461–462
practice questions
algebra, 33–34, 42–43
differentiation, 193–194, 203–204, 268, 272, 277, 280–281, 283–284, 288, 291, 325–326, 334–336, 345, 372–373, 377–378, 383–384
differentiation rules, 219, 224–225, 230–231, 233–234, 237, 238–239
functions, 73–74, 80, 82
geometry, 46–51
improper integrals, 562, 567
infinite series, 588, 591, 597–598, 601, 605–606
integrals, 525, 528–529, 535–537, 539, 542–543
integration, 410–411, 413–414, 417, 420, 425–426, 444–445, 452–453, 462, 464, 468, 470, 485, 489, 491, 496, 502
limits, 124–126, 133–134, 147, 150–152, 155, 156, 163–164
transformations, 86
trigonometry, 103–105, 108
practice questions answers and explanations
algebra, 52–56
differentiation, 205–208, 292–309, 346–361, 385–392
differentiation rules, 240–252
functions, 87–90
geometry, 56–58
improper integrals, 570–574
infinite series, 607–615

integrals, 544–550
integration, 427–434, 471–475, 503–516
limits, 135–136, 165–174
transformations, 90
trigonometry, 109–112
pre-algebra, 27. *See also* algebra
precision, 24
price function, 379
product rule, 222, 229
profit
marginal, 381
maximum, 382–383
p-series, tests for, 589–590
p-series rule, 590
Pythagorean Identities, 489
Pythagorean Theorem, 44

Q

quadratic equations, solving, 39–41
quadratic factors, denominators containing irreducible, 498–500
quadratic formula, for solving quadratic equations, 40–41
questions
example
algebra, 33, 41
differentiation, 193, 202–203, 266–267, 270–271, 276–277, 279–280, 290–291, 376–377
differentiation rules, 218, 223–224, 232–233, 236
functions, 73, 80, 82
geometry, 45–46
improper integrals, 560–561, 566
infinite series, 587, 592, 593–596, 599, 600, 602–603, 604, 605
integrals, 523–524, 531–532
integration, 409–410, 412–413, 424–425, 444, 451–452, 467, 490
limits, 124, 133, 146, 148–150, 155, 161–163
transformations, 85
trigonometry, 102, 108
practice
algebra, 33–34, 42–43
differentiation, 193–194, 203–204, 268, 272, 277, 280–281, 283–284, 288, 291, 325–326, 334–336, 345, 372–373, 377–378, 383–384
differentiation rules, 219, 224–225, 230–231, 233–234, 237, 238–239
functions, 73–74, 80, 82
geometry, 46–51
improper integrals, 562, 567
infinite series, 588, 591, 597–598, 601, 605–606
integrals, 525, 528–529, 535–537, 539, 542–543
integration, 410–411, 413–414, 417, 420, 425–426, 444–445, 452–453, 462, 464, 468, 470, 485, 489, 491, 496, 502
limits, 124–126, 133–134, 147, 150–152, 155, 156, 163–164
transformations, 86
trigonometry, 103–105, 108
quiz
algebra, 59–60
differentiation, 209–210, 310, 362, 393
differentiation rules, 253
functions, 91
geometry, 59–60
improper integrals, 575
infinite series, 616
integrals, 551
integration, 435, 476, 517–518
limits, 137–138, 175
transformations, 91
trigonometry, 113
quiz question answers
algebra, 61–65
differentiation, 211–213, 311–319, 363–365, 394–395
differentiation rules, 254–258
functions, 92–94
geometry, 61–65
improper integrals, 576–579
infinite series, 617–621
integrals, 552–556

quiz question answers *(continued)*
integration, 436–437, 477–478, 519
limits, 139–140, 176–180
transformations, 92–94
trigonometry, 114–116
quiz questions
algebra, 59–60
differentiation, 209–210, 310, 362, 393
differentiation rules, 253
functions, 91
geometry, 59–60
improper integrals, 575
infinite series, 616
integrals, 551
integration, 435, 476, 517–518
limits, 137–138, 175
transformations, 91
trigonometry, 113
quotient rule, 222–223

R

radians
degrees compared with, 99
measuring angles with, 98–99
rate, derivative as a, 15, 188–191
rate-slope connection, 190–191
ratio test, 598–600
rational functions
about, 123
continuity of, 129
limits of, at ± infinity, 158–159
real-deal limit problems, 143–157
reciprocal, 28
related rates
about, 336
blowing up balloons, 336–338
calculating, 343–344
filling up troughs, 339–340
intersections problem, 340–343
relative maximum, 260
Remember icon, 3
removable discontinuity, 132
resources, Internet
Cheat Sheet, 3
Technical Support, 4
Wolfram Alpha, 143
revenue, marginal, 378–380
reverse power rule, 461–462
reverse rules, for antiderivatives, 460–462
revolution, surfaces of, 540–542
Riemann sums
about, 409
writing with sigma notation, 414–417
right rectangle rule, 407
right sums, approximating area with, 406–408
right triangles, 45–46
rise, 187
rise per run, 15
root test, 598–600
roots
about, 35–36
simplifying, 36–37

S

sandwich (squeeze) method, 153–156
secant line, 195–197
secants
about, 496
integrals containing, 489–491
second derivative test, 268–272
second squared, 334
sequences
convergence and divergence of, 582–583
functions and, 583
stringing, 582–584
series, infinite
about, 17, 581
alternating series, 602–606
alternating series test, 603–606
answers to quiz questions, 617–621
comparison tests for convergence/divergence, 592–596, 606
convergence, 586–601
convergent series, 17–19

direct comparison test, 592–593
divergence, 586–601
divergent series, 17
example questions, 587, 592, 593–596, 599, 600, 602–603, 604, 605
finding absolute *versus* conditional convergence, 602–603
geometric series, 588–589
integral comparison test, 595–596
limit comparison test, 593–595
nth term test, 586–588
practice questions, 588, 591, 597–598, 601, 605–606
practice questions answers and explanations, 607–615
p-series, 589–590
quiz questions, 616
ratio test, 596–600
root test, 596–600
stringing sequences, 582–584
summing series, 584–586
telescoping series, 590–591
sigma notation, writing Riemann sums with, 414–417
simplifying roots, 36–37
Simpson's rule, approximating area with, 421, 423–425
sines
about, 494–495
graphing, 106–107
integrals containing, 486–489
slope of a line, 186–187
slopes
defined, 195
derivative as a, 14
formula for, 44
negative, 184, 260
positive, 260
slowing down, 333
SohCahToa mnemonic, 95–96
solids, volumes of weird, 529–537
solving
limit problems with algebra, 147–152
limits at infinity with algebra, 161–163
limits at ± infinity with calculators, 160–161
quadratic equations, 39–41
speed
about, 190
average, 332
distance traveled and, 331–332
maximum, 332
minimum, 332
velocity, acceleration and, 328–329
speeding up, 333
squares
completing the, for solving quadratic equations, 41
difference of, 38
squeeze (sandwich) method, 153–156
stationary point, 260, 262
statistics, 458–460
Sterling, Mary Jane (author)
Algebra II For Dummies, 39
straight inline problem, 8–9
stringing sequences, 582–584
stuff technique, 232
substitution, trigonometric, 491–496
substitution method, 465–468
substitution problems, finding area with, 469–470
subtracting fractions, 30
sum of cubes, 38
sum rule, 218
summation notation, 411–417
summing
fractions, 29–30
series, 584–586
sums
approximating area with left, 404–406
approximating area with midpoint, 408–410
approximating area with right, 406–408
surface area, formulas for, 44

surfaces of revolution, 540–542
sweeping-out area rate, 444

T

tangent line problem, 368–369
tangents
  about, 492–494
  graphing, 106–107
  integrals containing, 489–491
Technical Support (website), 4
teetering on corners, 261
telescoping series, tests for, 590–591
telescoping series rule, 591
tests
  alternating series, 603–606
  comparison, 592–596, 606
  direct comparison, 592–593
  for endpoints, 323
  first derivative, 264–268
  for geometric series, 588–589
  integral comparison, 595–596
  limit comparison, 593–595
  nth term, 586–588
  for p-series, 589–590
  ratio, 598–600
  root, 598–600
  second derivative, 268–272
  for telescoping series, 590–591
30°-60°-90° triangles, 45, 97
3333 limit mnemonic, 131–133
three-dimensional shapes, formulas for, 44
Tip icon, 3
top-minus-bottom method, 527
total displacement, 330
total distance traveled, 331–332
transformations
  about, 83
  answers to quiz questions, 92–94
  example questions, 85
  horizontal, 83–84
  practice questions, 86
  practice questions answers and explanations, 90
  quiz questions, 91
  vertical, 85
trapezoid rule, approximating area with, 421–422
triangles
  right, 45–46
  trigonometry, 96–97
trigonometric substitution, 491–496
trigonometry
  about, 95
  answers to quiz questions, 114–116
  differentiating trig functions, 220
  example questions, 102, 108
  graphing sine, cosine, and tangent, 106–107
  identities, 108
  integrals, 486–491
  inverse functions, 107
  practice questions, 103–105, 108
  practice questions answers and explanations, 109–112
  quiz questions, 113
  SohCahToa mnemonic, 95–96
  triangles, 96–97
  unit circle, 97–102
trinomial factoring, 39
two-dimensional shapes, formulas for, 44

U

unacceptable forms, 559–561
unit circles
  about, 97–98, 100–102
  angles in, 98
  hypotenuse, 99–100
  measuring angles with radians, 98–99

V

variety, 186
velocity
  about, 326–328
  derivative of, 327
  displacement and, 330–331
  maximum, 331
  minimum, 331
  speed, acceleration and, 328–329
vertical asymptotes
  defined, 107
  improper integrals with, 563–565
  limits and, 123
vertical tangent, 132
vertical transformations, 85
vocabulary, 441
volumes
  formulas for, 44
  of weird solids, 529–537

W

Warning icon, 3
washer method, 533–535
websites
  Cheat Sheet, 3
  Technical Support, 4
  Wolfram Alpha, 143
Wolfram Alpha, 143
writing Riemann sums, with sigma notation, 414–417

Y

Your Turn icon, 3

Z

zooming, 22–24

About the Author

A graduate of Brown University and the University of Wisconsin Law School, **Mark Ryan** has been teaching math since 1989. He runs The Math Center in Winnetka, Illinois (`www.thematthcenter.com`), a one-man math teaching and tutoring business; he helps students with all junior high and high school math courses, including calculus and statistics as well as ACT, PSAT, and SAT preparation. In high school, Ryan twice scored a perfect 800 on the math portion of the SAT, and he not only knows mathematics, he also has a gift for explaining it in plain English. He practiced law for four years before deciding he should do something he enjoys and use his natural talent for mathematics.

Calculus All-in-One For Dummies is Mark Ryan's twelfth book. His first book, *Everyday Math for Everyday Life* (Grand Central Publishing), was published in 2002. For Wiley, *Calculus For Dummies*, 1st Edition, was published in 2003; *Calculus Workbook For Dummies*, 1st Edition, in 2005; *Geometry Workbook For Dummies* in 2007; *Geometry For Dummies*, 2nd Edition, in 2008; *Calculus Essentials For Dummies* in 2010; *Geometry Essentials For Dummies* in 2011; *Calculus For Dummies*, 2nd Edition, in 2014; *Calculus Workbook For Dummies*, 2nd Edition, in 2015; *Geometry For Dummies*, 3rd Edition, in 2016; and *Calculus Workbook For Dummies*, 3rd Edition, in 2018. Ryan's math books have sold over 850,000 copies.

Ryan lives in Evanston, Illinois. For fun, he hikes, skies, plays platform tennis, travels, plays on a pub trivia team, and roots for the Chicago Blackhawks.

Dedication

To my current and former math students. Through teaching them, they taught me.

Author's Acknowledgments

I'm very grateful to my agent, Sheree Bykofsky, and her staff for all their efforts, which ultimately resulted in my writing the first edition of *Calculus For Dummies* for Wiley Publishing in 2003. Many thanks, Sheree.

This book is a testament to the high standards of everyone at Wiley. I've worked for many years with executive editor Lindsay Berg. She's intelligent, professional, down-to-earth, and has a great sense of humor. She always has an empathetic and quick understanding of my concerns as an author. And she has a special way of dealing with my oh-so-minor personality foibles with patience, skill, and finesse. It's always a pleasure to work with her. Chrissy Guthrie, editorial project manager and development editor extraordinaire, managed all phases of this complicated project, from the minor details to the big picture, with skill, efficiency, and intelligence. Marylouise Wiack, copy editor, has been invaluable in proofreading and editing the extremely complicated mathematics in the book. She has a great eye for detail and precision. Kristie Pyles, senior managing editor, and Kelsey Baird, managing editor, managed the production schedule and page proofs process. Toy Simmons, technical editor, did an excellent job checking the thousands of equations in the book. She's a calculus expert with a great eye for spotting errors. I've done four books with development editor Tim Gallan. He's a very talented and experienced editor who has a deft touch with the many aspects of taking a book from start to finish. These talented editors understand the forest, the trees, when to edit, and when not to edit. The layout and graphics team at Straive did a fantastic job with the book's complex equations and mathematical figures. Derrick Fimon and others at Wiris were a big help in bringing me up to speed in the use of MathType. (Wiris has great technical support.) This is an enormous undertaking; I'm grateful for this great team of talented professionals.

Many thanks to my bright, professional, and computer-savvy assistants, Benjamin Mumford, Randy Claussen, and Caroline DeVane. Josh Dillon did a meticulous review of the mathematics in the book. He knows calculus and how to communicate it clearly. And thanks to another assistant of mine, the multi-talented Amanda Wasielewski.

A special thanks to my brother-in-law, Steve Mardiks, and my friends Abby Lombardi, Ted Lowitz, and Barry Sullivan for their valuable advice, editing, and support. My friend, Beverly Wright, psychoacoustician and writer extraordinaire, was a big help with my contract negotiations. She gave me astute and much-needed advice and was generous with her time. Lastly, I'm grateful to my business advisor, Josh Lowitz. His insights into my writing career and my teaching and tutoring business have made him invaluable. His accessibility might make you think I was his only client instead of one of a couple dozen.

Publisher's Acknowledgments

Executive Editor: Lindsay Berg
Copy Editor: Marylouise Wiack
Technical Editor: Toy Simmons
Proofreader: Debbye Butler
Production Editor: Tamilmani Varadharaj
Cover Image: © aaaaimages/Getty Images